Lecture Notes in Statistics 106

Edited by P. Bickel, P. Diggle, S. Fienberg, K. Krickeberg,
I. Olkin, N. Wermuth, S. Zeger

Springer

New York
Berlin
Heidelberg
Barcelona
Budapest
Hong Kong
London
Milan
Paris
Santa Clara
Singapore
Tokyo

Harald Niederreiter
Peter Jau-Shyong Shiue
(Editors)

Monte Carlo and Quasi-Monte Carlo Methods in Scientific Computing

Proceedings of a conference at the
University of Nevada, Las Vegas, Nevada,
USA, June 23-25, 1994

 Springer

Harald Niederreiter
Institüt für Informationsverabeitung
Osterreichische Akademie der
Wissenschaften
Sonnenfelsgasse 19/2
A-1010 Wien
Austria

Peter Jau-Shyong Shiue
Department of Mathematical Sciences
University of Nevada, Las Vegas
Las Vegas, Nevada 89154
USA

Library of Congress Cataloging-in-Publication Data Available
Printed on acid-free paper.

Camera ready copy provided by the author.
Printed and bound by Braun-Brumfield, Ann Arbor, MI.
Printed in the United States of America.

9 8 7 6 5 4 3 2 1

ISBN 0-387-94577-6 Springer-Verlag New York Berlin Heidelberg

Preface

This volume contains the refereed proceedings of the conference "Monte Carlo and Quasi-Monte Carlo Methods in Scientific Computing" which was held at the University of Nevada, Las Vegas, in the period June 23–25, 1994. This conference was organized under the auspices of the NSF–EPSCoR program by Gary L. Mullen of Pennsylvania State University, Harald Niederreiter (Chair) of the Austrian Academy of Sciences, and Peter Jau-Shyong Shiue of the University of Nevada, Las Vegas.

Many computational tasks arising in modern science and engineering are so complex that they cannot be solved exactly, but rather have to be treated in an approximate manner through an appropriate stochastic model. Standard examples are the calculation of high-dimensional integrals, the solution of large systems of linear equations, the optimization of complicated multivariate functions, and several classical problems of statistical mechanics and computational physics. Monte Carlo methods provide a tool for analyzing such stochastic models and for setting up numerical schemes for actual computations. Quasi-Monte Carlo methods can be described as deterministic versions of Monte Carlo methods. Instead of random samples as in a Monte Carlo method, one employs judiciously chosen deterministic samples, i.e., quasirandom points. Quasi-Monte Carlo methods are known to be particularly powerful in the areas of numerical integration and global optimization, but the scope of these methods was widened in recent years.

Monte Carlo and quasi-Monte Carlo methods are entering a fascinating stage: on the one hand, the availability of more computer power makes many classical problems computationally feasible, but on the other hand it also leads to even more challenging questions. For instance, supercomputers and modern computer architectures such as parallel computers call for new types of algorithms that use the hardware more efficiently. Furthermore, the large-scale Monte Carlo and quasi-Monte Carlo calculations that can be run on present-day computers require larger and better ensembles of random samples and quasirandom points.

The conference provided an assessment of recent developments in Monte Carlo and quasi-Monte Carlo methods and was also a forum for the presentation of important new applications of these methods. One aim of the conference was to bring together proponents of both types of methods for a fruitful exchange of ideas. These proceedings contain all invited papers and a selection of the contributed papers presented at the conference. All papers in this volume have been refereed. These proceedings contain also a list of all talks and a list of all conference participants.

On behalf of all participants, we would like to thank NSF-Nevada EP-SCoR, College of Science and Mathematics, and the Department of Mathematical Sciences for their hospitality and support. Special thanks are due to Donna Fraser and Margie Wells for their tireless efforts in seeing to every detail, large or small. We are grateful to the referees of the papers for providing their expert advice so generously.

We would like to express our gratitude to Springer-Verlag for publishing this volume in their Lecture Notes in Statistics, and in particular to Martin Gilchrist and John Kimmel for the enthusiastic support they have given to this project.

Harald Niederreiter

Peter Jau-Shyong Shiue

Contents

Contributors

The following contributors presented talks. In the case of more than one author, an asterisk (*) indicates the presenter.

*NARENDRABABU BALAKRISHNAN and SHRIDHAR BENDI, "Compound Semiconductor Growth: Monte Carlo Study/Analysis"

*V. BANUNARAYANAN and Y. BAGHZOUZ, "Instantaneous Peak of Distorted Voltage Waveforms with Random Fourier Components"

RUSSEL CAFLISCH, "Smoothing and Dimension Reduction for Quasi-Random Monte Carlo Methods"

HISHAM AL-MHARMAH and *JAMES CALVIN, "Average-Case Optimal Non-Adaptive Algorithm for Global Optimization of Brownian Motion"

CHING-SHYANG CHEN, "Weighted Monte Carlo Integration for the Poisson Equation"

*JIAN CHEN and PAULA WHITLOCK, "An Implementation of a System of Distributed Pseudorandom Number Generators"

*WUN-SENG CHOU and HARALD NIEDERREITER, "On the Lattice Test for Inversive Congruential Pseudorandom Numbers"

HENRI FAURE, "Discrepancy Lower Bound in Two Dimensions"

BEN FOX, "Simulated Annealing: Folklore, Facts, and Directions"

MASANORI FUSHIMI, "Monte Carlo vs. Quasi-Monte Carlo --- Engineers' Viewpoints"

PETER GLYNN, "A Survey of the Initial Transient Problem"

GEORGE HEINE, "Smart Simulated Annealing"

FRED HICKERNELL, "A Comparison of Random and Quasirandom Sampling Methods for Multidimensional Quadrature"

CHIH-HSIANG HO, "A Simulation Study of a Change- Point Poisson Process Based on Two Well-known Test Statistics"

F.Y. HUNT, "Studying the Dynamics of Randomly Perturbed Multi-Dimensional Dissipative Maps"

MATTHEW IKLÉ, "Monte Carlo and Quasi-Monte Carlo Methods for Partial Differential Equations"

ALEXANDER KELLER, "Quasi-Monte Carlo Methods in Computer Graphics"

ABDOL R. KHOIE, "A Comparison Between a Fast and Efficient Hydrodynamic Method and Computationally Intensive Monte Carlo Method for Calculation of Intersubband and Intrasubband Scattering Rates of Electrons in Quantum Wells"

GERHARD LARCHER, "Numerical Integration of Multivariate Walsh Series"

*MICHAEL MASCAGNI, STEVEN A. CUCCARO, DANIEL V. PRYOR, and M.L. ROBINSON, "A Fast, High Quality, and Reproducible Parallel Lagged-Fibonacci Pseudorandom Number Generator"

BRADLEY MOSKOWITZ, "Improved Stochastic Simulations Using Quasi-Monte Carlo"

GARY MULLEN, "A Survey of (T,M,S)-Net Constructions"

*KENJI NAGASAKA and ICHIRO YASUNAGA, "Monte-Carlo Simulations in Genetic Algorithm"

HARALD NIEDERREITER, "The State of the Art in Uniform Pseudorandom Number Generation"

ART OWEN, "Equidistributed Latin Hypercube Samples"

ASHOK K. SINGH and *ALOK PANDEY, "Estimation of Percentage Points of a Multiple Slippage Test by Monte Carlo Simulation"

TAO PANG, "Electronic Structure of Hydrogen Ion Clusters: A Quantum Monte Carlo Study"

ANTONINI PUPPIN MACEDO, "Application of a Monte Carlo Method to the Numerical Simulation of an Integral-Differential Forest Fire Propagation Model"

*LEELA RAKESH, J. KOSUGI, and MARK MANSFIELD, "Monte Carlo Simulation of Dendrimer Molecule Modeling Using MATHEMATICA"

D.H.E. GROSS, *O. SCHAPIRO, and A. ECKER, "Microcanonical Monte-Carlo"

W.CH. SCHMID, "Optimal Polynomials for (t,m,s)-Nets and Numerical Integration of Multivariate Walsh Series"

ANITA SINGH, "Robust Estimation of Mean and Variance Using Environmental Data Sets with Below Detection Limit Observations"

JERRY SPANIER, "Quasi-Monte Carlo Methods for Particle Transport Problems"

*SALLY STEWART and ASHOK K. SINGH, "A Monte Carlo Investigation for Robustness of Some Common Statistical Procedures in Environmental Data Analysis"

TONY WARNOCK, "Sharper Error Bounds for Integration"

YI-JUN XIAO, "Estimates for the Volume of Points of (0,s)-Sequences in Base $b \geq s \geq 2$ "

Conference Participants

Dr. Malwane Ananda
Department of Mathematical Sciences
University of Nevada, Las Vegas
Las Vegas, NV 89154-4020

Dr. Jeff Angel
Department of Mathematical Sciences
University of Nevada, Las Vegas
Las Vegas, NV 89154-4020

Dr. Gennady Bachman
Department of Mathematical Sciences
University of Nevada, Las Vegas
Las Vegas, NV 89154-4020

Dr. Yahia Baghzouz
Department of Electrical Engineering
University of Nevada, Las Vegas
Las Vegas, NV 89154

Dr. Narendrababu Balakrishnan
Department of Electrical & Computer Engineering
University of Nevada, Las Vegas
Las Vegas, NV 89154

Dr. Shridhar Bendi
Department of Electrical & Computer Engineering
University of Nevada, Las Vegas
Las Vegas, NV 89154

Dr. Harold Bowman
Department of Mathematical Sciences
University of Nevada, Las Vegas
Las Vegas, NV 89154-4020

Dr. Russel Caflisch
Department of Mathematics
University of California, Los Angeles
Los Angeles, CA 90024-1555

Dr. James M. Calvin
School of Industrial & Systems Engineering
Georgia Institute of Technology
Atlanta, GA 30332-0205

Dr. C.S. Chen
Department of Mathematical Sciences
University of Nevada, Las Vegas
Las Vegas, NV 89154-4020

Dr. Jian Chen
Computer & Information Sciences Department
Brooklyn College
Brooklyn, NY 11210

Dr. Wun-Seng Chou
Institute of Mathematics
Academia Sinica
Nankang
Taipei 11529, Taiwan
Republic of China

Dr. Rohan Dalpatadu
Department of Mathematical Sciences
University of Nevada, Las Vegas
Las Vegas, NV 89154-4020

Dr. Derrick DuBose
Department of Mathematical Sciences
University of Nevada, Las Vegas
Las Vegas, NV 89154-4020

Dr. Henri Faure
U.E.R. de Mathématiques
Université de Provence, Case W
3, Place Victor Hugo
F-13331 Marseille, Cedex 3
France

Dr. Larry Foulke
Westinghouse
P.O. Box 79
West Mifflin, PA 15122

Dr. Bennett L. Fox
Dept. of Mathematics, Campus Box 170
University of Colorado
P.O. Box 173364
Denver, CO 80217-3364

Dr. Masanori Fushimi
Dept. of Math. Eng. & Inf. Physics
Faculty of Engineering
University of Tokyo
Bunkyo-Ku, Tokyo 113,
Japan

Dr. Peter W. Glynn
Department of Operations Research
Stanford University
Stanford, CA 94305-4022

Ms. Michelle Haberkorn
400 S. Hauser No. 4-C
Los Angeles, CA 90036-5526

Dr. John Halton
Computer Science Department
University of North Carolina
Sitterson Hall, CB# 3175
Chapel Hill, NC 27599-3175

Dr. George W. Heine
Department of Mathematics, Campus Box 170
University of Colorado
P.O. Box 173364
Denver, CO 80217-3364

Dr. Kurt Helmes
Department of Mathematics
College of Arts and Sciences
University of Kentucky
746 Patterson Office Tower
Lexington, KY 40506-0027

Dr. Fred Hickernell
Department of Mathematics
Hong Kong Baptist College
224 Waterloo Road
Kowloon, Hong Kong

Dr. Chih-Hsiang Ho
Department of Mathematical Sciences
University of Nevada, Las Vegas
Las Vegas, NV 89154-4020

Dr. Fern Y. Hunt
Computing & Applied Mathematics Laboratory
National Institute of Standards & Technology 101A-238
Gaithersburg, MD 20899

Dr. Matthew Iklé
Department of Mathematical Sciences
University of Nevada, Las Vegas
Las Vegas, NV 89154-4020

Dr. Alexander Keller
Fachbereich Informatik
AG Numerische Algorithmen
Universität Kaiserslautern
Postfach 3049, D-67653 Kaiserslautern
Germany

Dr. Rahim Khoie
Department of Electrical & Computer Engineering
University of Nevada, Las Vegas
Las Vegas, NV 89154

Dr. J.P. Lambert
Department of Mathematics
University of Alaska at Fairbanks
Fairbanks, Alaska 99775-1110

Dr. Gerhard Larcher
Universität Salzburg
Hellbrunnerstrasse 34
A-5020 Salzburg
Austria

Dr. Jeffery Leader
U.S. Military Academy
Department of Mathematical Sciences
West Point, NY 10996

Mr. Liming Li
Department of Mathematics
Claremont Graduate School
143 E. Tenth Street
Claremont, CA 91711-3988

Dr. Xin Li
Department of Mathematical Sciences
University of Nevada, Las Vegas
Las Vegas, NV 89154-4020

Dr. Cantian Lin
Department of Mathematical Sciences
University of Nevada, Las Vegas
Las Vegas, NV 89154-4020

Dr. Michael Mascagni
Supercomputing Research Center
Institute for Defense Analysis
17100 Science Drive
Bowie, MD 20715-4300

Dr. William Morokoff
Department of Mathematics
University of California, Los Angeles
Los Angeles, CA 90024

Dr. Bradley Moskowitz
Department of Mathematics
University of California, Los Angeles
Los Angeles, CA 90024

Dr. Gary Mullen
Mathematics Department
The Pennsylvania State University
University Park, PA 16802

Dr. Kenji Nagasaka
College of Engineering
Hosei University
Kajinocho, Koganei, Tokyo 184
Japan

Dr. Harald Niederreiter
Institute for Information Processing
Austrian Academy of Sciences
Sonnenfelsgasse 19
A-1010 Vienna
Austria

Dr. John Nocerino
U.S. EPA Financial Mgt. Center
P.O. Box 98515
Las Vegas, NV 89193-8515

Mr. Giray Okten
Department of Mathematics
143 E. Tenth Street
Claremont Graduate School
Claremont, CA 91711-3988

Dr. Art B. Owen
Department of Statistics
Stanford University
Stanford, CA 94305

Dr. Alok Pandey
Harry Reid Center for Environmental Studies
University of Nevada, Las Vegas
Las Vegas, NV 89154

Dr. Tao Pang
Department of Physics
University of Nevada, Las Vegas
Las Vegas, NV 89154

Dr. Ernest Peck, Dean
College of Science & Mathematics
University of Nevada, Las Vegas
Las Vegas, NV 89154

Mr. Dan Pick
Department of Mathematics
Claremont Graduate School
143 E. Tenth Street
Claremont, CA 91711

Ms. Maria Consuelo Pickle
Department of Mathematical Sciences
University of Nevada, Las Vegas
Las Vegas, NV 89154-4020

Dr. Antonini Puppin Macedo
Departamento de Engenharia Mecanica
Faculdade de Tecnologia
Campus Univesitario, Asa Norte
700910-900 Brasilia DF
Brazil

Dr. Leela Rakesh
Department of Mathematics
Central Michigan University
Mt. Pleasant, MI 48859

Dr. Olga Schapiro
Hahn-Meitner Institut
Bereich Theoretische Physik
Glienickerstr. 100
D-14109 Berlin
Germany

Mr. Wolfgang Schmid
Universität Salzburg
Hellbrunnerstrasse 34
A-5020 Salzburg
Austria

Mr. John Sepikas
Department of Mathematics
143 E. Tenth Street
Claremont Graduate School
Claremont, CA 91711-3988

Dr. Peter Jau-Shyong Shiue
Department of Mathematical Sciences
University of Nevada, Las Vegas
Las Vegas, NV 89154-4020

Dr. W.-K. Shiue
Department of Mathematics
Southern Illinois University, Edwardsville
Edwardsville, IL 62026-1653

Dr. Anita Singh
Lockheed Environmental Systems & Technologies Company
980 Kelly Johnson Drive
Las Vegas, NV 89119

Dr. Ashok Singh
Department of Mathematical Sciences
University of Nevada, Las Vegas
Las Vegas, NV 89154-4020

Dr. Jerry Spanier
Department of Mathematics
143 E. Tenth Street
Claremont Graduate School
Claremont, CA 91711-3988

Ms. Angela Starr
Department of Mathematical Sciences
University of Nevada, Las Vegas
Las Vegas, NV 89154-4020

Ms. Sally Stewart
Department of Mathematical Sciences
University of Nevada, Las Vegas
Las Vegas, NV 89154

Mr. Andy Tsang
Department of Mathematical Sciences
University of Nevada, Las Vegas,
Las Vegas, NV 89154

Dr. Tony Warnock
Mail Stop B284
Los Alamos National Lab
Los Alamos, NM 87545

Dr. Paula Whitlock
Computer & Information Sciences Department
Brooklyn College
Brooklyn, NY 11210

Dr. Yi-Jun Xiao
Université d'Evry-Val d'Essonne
Bd des Coquibus
F-91000 Evry
France

Dr. C.W. Xu
Department of Mathematics & Computer Sci.
Georgia Southern University
Landrum Box 8093
Statesboro, GA 30460

Modified Monte Carlo Methods Using Quasi-Random Sequences

Russel E. Caflisch *
Bradley Moskowitz †
Mathematics Department
University of California
Los Angeles, CA 90095-1555.

Abstract

Computational experiments have shown that Monte Carlo methods using quasi-random sequences lose some of their effectiveness for integration problems in which the dimension is large or the integrand is not smooth. In this paper, two modified Monte Carlo methods are developed, which regain an enhanced convergence rate. The standard rejection method involves discontinuities, corresponding to the decision to accept or reject. In place of this, a smoothed rejection method is formulated and found to be very effective when used with quasi-random sequences. Monte Carlo evaluation of Feynman-Kac path integrals involves high dimension, one dimension for each discrete time interval. Through an alternative discretization, the effective dimension of the integration domain is drastically reduced, so that quasi-random sequences are effective.

1 Introduction

Quasi-random sequences are a deterministic alternative to random or pseudo-random sequences for use in Monte Carlo methods. Whereas Monte Carlo integration using random or pseudo-random sequences has error of size $O(N^{-1/2})$, the theoretical error bounds using quasi-random points are of size $O(N^{-1} \log^d N)$ in dimension d [8].

---

*caflisch@math.ucla.edu. Research of both authors supported in part by the Army Research Office under grant number DAAH04-95-1-0155.

†moskowit@bettis.gov. Current address: Westinghouse Bettis Atomic Power Lab, Reactor Methods and Analysis, P.O. Box 79, West Mifflin, PA 15122-0079

In practice, however, this improved performance for quasi-random sequences is often not observed. The loss of accuracy for integration problems was found by Morokoff and Caflisch [2, 3, 4] to be due to two causes: high dimension of the integration domain and lack of smoothness of the integrand. These results are reviewed in Section 2.

The purpose of this paper is to show how standard Monte Carlo methods can be modified to increase the effectiveness of quasi-random sequences, through reduction of effective dimension and through smoothing. Two examples will be presented. First, in Section 3 the effective dimension for the evaluation of a Feynman-Kac integral will be reduced by modification of the simulation procedure using the Brownian bridge. The second example, presented in Section 4, is a smoothed version of the acceptance-rejection method. Additional details on some of these problems are contained in [7]

Some conclusions and prospects for further work in this direction are discussed in the final section.

2 Quasi-Random Sequences, Dimension and Smoothness

The principal error bound for Monte Carlo integration using quasi-random sequences is the Koksma-Hlawka inequality, which states that for Monte Carlo integration of a function f on the unit cube in R^d using N points, the integration error $E_N(f)$ is bounded by

$$E_N(f) \leq D_N V(f) \tag{1}$$

in which D_N is the *discrepancy* of the sequence and $V(f)$ is the variation of f in the Hardy-Krause sense [8]. Moreover, for quasi-random sequences the discrepancy D_N is of size $O(N^{-1} \log^d N)$. Roughly speaking $V(f)$ is finite if the integrand is d times differentiable in d dimensions.

The dependence of quasi-Monte Carlo performance on dimension, is most directly seen in calculations of the discrepancy for quasi-random sequences [3]. At any dimension, the discrepancy behaves like $O(N^{-1} \log^d N)$ for sufficiently large N. On the other hand, for large values of d and moderate values of N, the discrepancy of a quasi-random sequence is seen to be almost exactly that of a random sequence.

The effect of smoothness on integration error is more difficult to assess. First, note that the variation, which appears in the Koksma-Hlawka inequality is infinite unless the integrand is very smooth. On the other hand, we expect that this bound is a conservative over estimate and that less smoothness will suffice. For example, Wozniakowski [12] (cf. [3] for a simplified proof) showed that for the Brownian sheet, which is only Hölder with exponent approximately $1/2$, the average integration error is equal to the L^2 discrepancy. Note that this differs from the L^∞ discrepancy D_N and that the factor $V(f)$ in (1) is known to optimal even for some C^∞ functions [8]. Nevertheless, since $V(f)$ is infinite for almost all Brownian sheet functions, the result of [12] suggests that there could be improvements in the Koksma-Hlawka inequality.

On the other hand, computational experiments [4] show that for integrands that are discontinuous, the enhanced convergence rate of quasi-Monte Carlo integration is reduced.

3 Reduction of Effective Dimension for Feynman-Kac Integrals

In this section we show how the difficulties with high dimension can be overcome in a particular example. For a Feynman-Kac integral, the dimension is the number of time steps included in simulating the underlying Brownian motion. Although this may be quite large, a simple modification of the discretization of the Brownian path allows one to concentrate most of the variance into the first few dimensions, which significantly reduces the effective dimension and improves the accuracy of quasi-Monte Carlo integration.

The Feynman-Kac Formula provides a connection between linear parabolic differential equations and stochastic path integrals, which is similar to the method of characteristics for solving hyperbolic differential equations. For the initial value problem

$$\frac{\partial u}{\partial t} = \frac{1}{2}\frac{\partial^2 u}{\partial x^2}(x,t) + v(x,t)\,u(x,t) \ , \ \ t \geq 0 \tag{2}$$

with initial condition $u(x,0) = f(x)$, the solution is given by the Feynman-Kac formula as

$$u(x,t) = E_{x,0}\left[f(\xi(t))\,e^{\int_0^t v(\xi(r),t-r)\,dr}\right] \tag{3}$$

where ξ represents a Wiener path starting at $\xi(0) = x$.

This solution can be approximated by discretizing time, generating Wiener process sample paths, and averaging over the set of sample paths to obtain Monte Carlo estimates of the expectation above. The integral in the exponent is itself approximated for each sample path using the Trapezoidal rule.

3.1 Standard Discretization

The **Standard Discretization** of the Wiener process is defined as follows:

1. Choose (x,t) at which the solution is to be approximated.

2. Choose the number of equal time steps m and set $\Delta t = \frac{t}{m}$ and $t_i = i\,\Delta t$ for $i = 0,\ldots,m$.

3. Generate Gaussian independent variables, $\Delta W^{(i)}$ from the distribution $\mathcal{N}(0,\Delta t)$ (the normal distribution with mean 0 and variance Δt), for $i = 1,\ldots,m$.

4. The Wiener path is sampled exactly at each t_i as the following sum:

$$\xi^{(i)} = x + \sum_{j=1}^{i} \Delta W^{(j)} \tag{4}$$

The sample paths are exact at the discrete times t_i, but an approximation is involved when the path integral of v is estimated. For example, using the Trapezoidal rule the Wiener path integral is approximated as

$$\int_0^t v(\xi(r), t - r)\, dr \approx \frac{\Delta t}{2}\left[v(x, t) + 2\sum_{i=1}^{m-1} v(\xi^{(i)}, t - t_i) + v(\xi^{(m)}, 0)\right] = \tilde{I} \qquad (5)$$

The nondifferentiability of the paths leads to an error (bias) of order $O(\sqrt{\Delta t})$ for this quadrature rule. In the computations presented below, however, Δt is sufficiently small so that discretization errors are negligible in comparison to statistical errors.

Using the sample paths above, the Monte Carlo estimate of the Feynman-Kac path integral is

$$\hat{u}_N(x, t) = \frac{1}{N}\sum_{k=1}^N f(\xi_k^{(m)})\, e^{\tilde{I}_k} \qquad (6)$$

in which $\tilde{I}_k$ and $\xi_k^{(m)}$ for $i \leq k \leq N$ are N realizations of $\tilde{I}$ and $\xi^{(m)}$. If the $\xi_k^{(m)}$ are normally distributed, as described above, then

$$E(\hat{u}_N(x, t)) = u(x, t) + \text{bias error}$$

in which the bias error comes from the quadrature rules and is insignificant.

For application of quasi-random Monte Carlo, this average is expressed as a multidimensional integral. First, each normal variable $\Delta W^{(j)}$ can be represented as a transformation G of a uniformly distributed variable x_j by $\Delta W^{(j)} = G(x_j)$. Then define

$$\begin{aligned}\tilde{f}(x_1, x_2, \ldots, x_m) &= f(\xi^{(m)}) \\ \tilde{I}(x_1, x_2, \ldots, x_m) &= \tilde{I}\end{aligned}$$

in which $\xi^{(i)} = x + \sum_{j=1}^i G(x_j)$, and x_j are uniformly distributed, e.g. pseudorandom or quasi-random points. The expectation of the Monte Carlo estimate in (6) can then be written as the following integral:

$$E(\hat{u}_N(x, t)) = \int_0^1 \cdots \int_0^1 \tilde{f}(x_1, \ldots, x_m) e^{\tilde{I}(x_1, \ldots, x_m)}\, dx_1 \ldots dx_m$$

The Monte Carlo estimate of this integral is

$$\tilde{u}_N(x, t)) = \frac{1}{N}\sum_{k=1}^N \tilde{f}(x_1^k, \ldots, x_m^k) e^{\tilde{I}(x_1^k, \ldots, x_m^k)} \qquad (7)$$

The transformation G used here is that of Marsaglia [1], which is continuous and monotonic. By contrast, Box-Muller, which is discontinuous, gives poorer results for quasi-random sequences [2].

3.2 Alternative Discretization Method

When the number of dimensions, m, is moderately large, an alternative discretization method leads to significant improvements in the quasi-random integration estimates by concentrating most of the variance into the lowest dimensions.

Assume that m is a power of two, and define the alternative discretization as follows:
Alternative Discretization:

1. Choose (x, t) at which the solution is to be approximated.

2. Choose the number of equal time steps $m = 2^p$. Then let $\Delta t = \frac{t}{m}$ and $t_i = i \Delta t$ for $i = 0, \ldots, m$.

3. Generate Gaussian variables, $\Delta W^{(0)}$ distributed according to $\mathcal{N}(0, t)$ and $\Delta W^{(i)}$ distributed according to $\mathcal{N}(0, t/\alpha)$, for $i = 1, \ldots, m$, in which $\alpha = 4 \cdot 2^{int(\log_2 i)}$.

4. The Wiener path $\xi^{(i)}$ is exactly sampled at each t_i as follows:

$$
\begin{aligned}
\xi^{(0)} &= x \\
\xi^{(m)} &= \xi^{(0)} + \Delta W^{(0)} \\
\xi^{(\frac{m}{2})} &= \frac{\xi^{(0)} + \xi^{(m)}}{2} + \Delta W^{(1)} \\
\xi^{(\frac{m}{4})} &= \frac{\xi^{(0)} + \xi^{(\frac{m}{2})}}{2} + \Delta W^{(2)} \\
\xi^{(\frac{3m}{4})} &= \frac{\xi^{(\frac{m}{2})} + \xi^{(m)}}{2} + \Delta W^{(3)} \\
&\vdots
\end{aligned}
\tag{8}
$$

For this discretization method, the first step is directly from 0 to t. Then the intermediate steps are filled in by taking successive subdivisions of the time intervals into halves. Each new intermediate path position is determined using the following rule:

$$
\xi^{(j)} = \frac{\xi^{(j_1)} + \xi^{(j_2)}}{2} + \Delta W , \quad \Delta W \in \mathcal{N}(0, \frac{t_{j_2} - t_{j_1}}{4})
\tag{9}
$$

where j_1, j_2 are the indices of the nearest prior and later time steps, respectively, for which the positions have already been determined. This representation of the Wiener path in terms of past and future positions is referred to as the Brownian bridge (also called a tied-down Wiener process).

The representation (8) can be conveniently rewritten as follows:

$$
\begin{aligned}
\xi^{(0)} &= x \\
\xi^{(m)} &= x + \Delta W^{(0)} \\
\xi^{(\frac{m}{2})} &= x + \frac{\Delta W^{(0)}}{2} + \Delta W^{(1)} \\
\xi^{(\frac{m}{4})} &= x + \frac{\Delta W^{(0)}}{4} + \frac{\Delta W^{(1)}}{2} + \Delta W^{(2)} \\
\xi^{(\frac{3m}{4})} &= x + \frac{3 \Delta W^{(0)}}{4} + \frac{\Delta W^{(1)}}{2} + \Delta W^{(3)} \\
&\vdots
\end{aligned}
\tag{10}
$$

	Variance	
	Standard	Alternative
x_1	$1.0000t$	$3.1875t$
x_2	$0.8750t$	$1.6875t$
x_3	$0.7500t$	$.1875t$
x_4	$0.6250t$	$.1875t$
x_5	$0.5000t$	$.0625t$
x_6	$0.3750t$	$.0625t$
x_7	$0.2500t$	$.0625t$
x_8	$0.1250t$	$.0625t$

Table 1: Variances for Standard and Alternative Discretization Schemes, $m = 8$

The variance contributed by each dimension to the representation of the Wiener sample paths is compared for the Standard and the Alternative Discretizations are presented in Table 1 for the case $m = 8$. Note that variance due to x_k slowly decreases for the Standard Discretization, since early values of the Wiener process affect more parts of the path than later values, but that this decays is much more rapid for the Alternative Discretization.

3.3   Computational Example for the Feynman-Kac Formula

This example demonstrates the accuracy of Monte Carlo estimates of the solution of a simple linear parabolic differential equation using (6) at selected values of x and a fixed (small) time t. A comparison will be made of pseudorandom and quasi-random sequences with either the Standard discretization (4), or the Alternate discretization (9).

Example 1 *Consider the following linear parabolic differential equation:*

$$\frac{\partial f}{\partial t}(x,t) = \frac{1}{2}\frac{\partial^2 f}{\partial x^2}(x,t) + \left(\frac{1}{t+1} + \frac{1}{x^2+1} - \frac{4x^2}{(x^2+1)^2}\right)\frac{\partial f}{\partial x}(x,t)$$

with initial condition $f(x,0) = \frac{1}{x^2+1}$. The exact solution is solution: $f(x,t) = \frac{t+1}{x^2+1}$.

Estimates are computed at a fixed (small) time, T, and at eight equally spaced positions between $x = -3$ and $x = 3$ using a discretization of time into m equal steps and a sample size of N. The estimates are then compared with the exact solution using the following L^2 measure of error:

$$L2e = \sqrt{\frac{1}{8}\sum_{k=0}^{7}\left[f(x_k,T) - \hat{f}_N(x_k,T)\right]^2}$$

where $x_k = -3 + 6\,k/7$.

T is selected to be very small so that discretization errors are insignificant relative to the Monte Carlo errors.

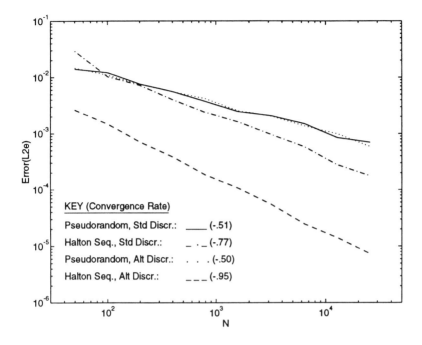

Figure 1: Log-Log Plot for Example 1, 75 runs, $T = 0.08$, $m = 32$.

A measure of the rate of decrease of the values of $L2e$ is obtained by a least squares fit of the function $cN^{-\alpha}$ to the calculated values of $L2e$. The resulting values of α for pseudorandom and quasi-random (Halton was used here) points used in both the standard and alternative discretizations are presented along with log-log plots of the data in Figures 1 for $T = .08$ ($m = 32$). The results show that the convergence rate for pseudorandom is independent of the discretization scheme. On the other hand, the performance of quasi-random Monte Carlo degrades in the standard discretization scheme, but is maintained at α nearly equal to 1 in the alternative discretization. It is important to notice that the quasi-random improvement from the alternate scheme grows as the dimension increases. This is expected, as is the fact that at 32 dimensions standard quasi-random is just barely better than pseudorandom for the values of N used here.

For standard Monte Carlo using pseudorandom sequences there is no difference between the two discretizations, since the total variance of the two methods is the same. On the other hand, the Alternative discretization method provides significantly more accurate results than the Standard discretization, when using quasi-random sequences in the Feynman-Kac formula. This improvement is due to reduction of the effective dimension of the problem.

4   Smoothing of the Rejection Method

The rejection method is one of the most commonly used sampling procedures in Monte Carlo methods. In its standard form it involves a decision to either accept or reject trial sample points. Because of the resulting discontinuity in the integrand, it is not well adapted for quasi-random sequences. In this section a smoothed rejection method, which overcomes this difficulty, will be formulated.

4.1   Importance Sampling Using the Rejection Method

One of the most common uses of the rejection method is for variance reduction through importance sampling, and this will be used here to measure the effectiveness of the rejection method. For the integral $A = \int_D f(x)dx$, introduce an importance function h which mimics the behavior of f over D but is either integrable analytically or easily numerically integrated. The sampling procedure is then altered to generate points distributed according to h instead of points which are uniformly distributed. Then, instead of evaluating $f(x)$ at each sample point, one evaluates $\frac{f(x)}{h(x)}$, which provides an unbiased estimate of the integral, since

$$E_h\left(\frac{f}{h}\right) = \int_D \frac{f(x)}{h(x)} \cdot h(x)\, dx = \int_D f(x)\, dx = A$$

The Importance Sampled Monte Carlo estimate can be written as follows:

$$\tilde{A}_N = \frac{1}{N}\sum_{i=1}^{N} \frac{f(x_i)}{h(x_i)} \quad , \quad x_i \sim h(x) \tag{11}$$

where $x_i \sim h(x)$ indicates that the sample points, $x_1, \ldots, x_N$ have a probability density given by h (note that h should be normalized so that $\int_D h\, dx = 1$). The variance of this estimate is substantially reduced if f/h has a smaller variance than f.

For a limited class of density functions h, the sample points may be generated by a transformation of the uniform distribution. For more general h, some other generation procedure must be used, such as the rejection method.

4.2   Standard Rejection Method

Here is the basic algorithm for the **Rejection Method** when the original sample is uniformly distributed in the $d-$dimensional unit cube. Note that one additional dimension is added for the extra variable y which acts as a "decision-maker":

1. Select $\gamma \geq \sup_{x \in D} h(x)$.

2. Repeat until N points have been accepted:

   (a) Sample (x_t, y_t) from $U([0, 1]^{d+1})$.

   (b) If $y_t < \gamma^{-1} h(x_t)$, accept trial point x_t.
   Otherwise, reject the trial point.

This algorithm produces a sequence of accepted points in d dimensions which are distributed according to h, as required for Importance Sampling.

The sum (11), in which the points x_i are chosen using the rejection method, can be reinterpreted as a Monte Carlo evaluation of the following integral

$$A = \gamma \int_D \int_0^1 \frac{f(x)}{h(x)} \chi(y < \gamma^{-1} h(x)) dy dx. \tag{12}$$

The direct Monte Carlo estimation of this integral is

$$\tilde{A}_N = \frac{\gamma}{N} \sum_{i=1}^N \chi \left\{ y_i < \frac{h(x_i)}{\gamma} \right\} \cdot \frac{f(x_i)}{h(x_i)} \tag{13}$$

The variance of this integral is reduced by replacing the total number of points N by the number of accepted points to get the following Monte Carlo estimator

$$\tilde{A}_N = \frac{1}{\sum_{i=1}^N \chi \left\{ y_i < \frac{h(x_i)}{\gamma} \right\}} \sum_{i=1}^N \chi \left\{ y_i < \frac{h(x_i)}{\gamma} \right\} \cdot \frac{f(x_i)}{h(x_i)} \tag{14}$$

One difference between the sum (14) and a straightforward Monte Carlo evaluation of (12), is that in (14) the number N is taken to be the number of accepted points rather than the total number of trial points. This accounts for the factor γ in (13).

Quasi-Monte Carlo cannot be effectively applied to the integral (12) because the integrand contains a characteristic function, corresponding to the decision to accept or reject. This will be demonstrated in computational examples below.

4.3 Smoothed Rejection

The Rejection Method is discontinuous as a result of the inherently discontinuous nature of the binary decision: accept or reject. This can be eliminated by allowing sample points to have associated "acceptance" weights as described below. A **Smoothed Rejection** method that retains the flexibility and advantages of the Rejection method but does not involve a discontinuous integrand will now be formulated. First, replace the integral in (12) by an equivalent smooth integral, as in

$$A = \gamma \int_D \int_0^1 \frac{f(x)}{h(x)} \chi_\delta(y, \gamma^{-1} h(x)) dy dx. \tag{15}$$

in which the smooth function χ_δ satisfies

$$\int_0^1 \chi_\delta(y, \gamma^{-1} h(x)) dy = \gamma^{-1} h(x). \tag{16}$$

The function χ_δ, which will also be referred to as the weight w, will be chosen to be piecewise linear below.

The corresponding Monte Carlo approximation corresponds to the following Smoothed Rejection procedure:

1. Select $\gamma \geq \sup_{x \in D} h(x)$ and $0 < \delta \ll 1$.

2. Repeat until weight of accepted points is within one unit of N:

   (a) Sample (x_t, y_t) from $U([0, 1]^{d+1})$.

   (b) If $y_t < \frac{h(x_t)}{\gamma} - \frac{1}{2}\delta$, then acceptance weight is $w = 1$.
   Else if $y_t > \frac{h(x_t)}{\gamma} + \frac{1}{2}\delta$, then $w = 0$.
   Else $w = \frac{1}{\delta}\left(\frac{h(x_t)}{\gamma} + \frac{1}{2}\delta - y_t\right)$.

The density of accepted points x is $f_{accept}(x)$, given by (with $w = \chi_\delta$)

$$
\begin{aligned}
f_{accept}(x) &= \frac{1 \cdot \int_0^1 w(x, y)dy}{\int_{I^d}[1 \cdot \int_0^1 w(\xi, y)dy]d\xi} \\
&= \frac{h(x)/\gamma}{1/\gamma} \\
f_{accept}(x) &= h(x)
\end{aligned}
$$

which shows that the density function h is correctly sampled.

There is some extra work required by Smoothed Rejection as compared to ordinary Rejection. First, there is the work associated with assigning each sample point a weight, and storing and using these weights. This is typically small enough to be considered insignificant. Second, there is additional work which comes from the acceptance of points with weights of less than 1; i.e., in order to reach a total acceptance weight of size N, more than N evaluations of f are required. This extra work can be minimized by setting the constant δ sufficiently small. On the other hand, if δ is made too small, the advantages of continuity will be effectively lost.

4.4 Weighted Uniform Sampling

Another alternative to the Rejection method can be formulated by eliminating the accept/reject decision entirely, and instead assigning each sample point a weight equal to its acceptance probability given by $\gamma^{-1}h(x_i)$.

A new Monte Carlo estimate is obtained as follows:

$$
\begin{aligned}
\breve{A}_N &= \frac{\sum_{i=1}^N \frac{f(x_i)}{h(x_i)} \cdot \frac{h(x_i)}{\gamma}}{\sum_{i=1}^N \frac{h(x_i)}{\gamma}} \\
&= \frac{\sum_{i=1}^N f(x_i)}{\sum_{i=1}^N h(x_i)}
\end{aligned}
$$

This sum is essentially the ratio of two Crude Monte Carlo integration estimates – an estimate of the original function, f, in the numerator, and an estimate of the importance function, h, in the denominator. The resulting estimate is biased. Nevertheless, Powell and Swann [10] and Spanier and Maize [11] have shown that this bias is negligible in comparison with the rms-error as $N \to \infty$. Moreover the positive correlation between f and h, when h has been well chosen to closely mimic the behavior of f, provides substantial variance reduction.

The bias and rms-error are given as follows:

$$bias(\check{A}_N) = \frac{A\,var(h)}{N} - \frac{cov(f,g)}{N} + O(N^{-3/2}) \tag{17}$$

$$rmse(\check{A}_N) = \frac{\sqrt{var(f) + A^2\,var(h) - 2\,A\,cov(f,h)}}{\sqrt{N}} + O(N^{-3/4}) \tag{18}$$

One advantage of Weighted Uniform Sampling, as emphasized in [10], is that sample points need not be generated for the density h. A second advantage is that the weighted uniform sampling estimate is continuous (assuming that f and h are continuous) so that quasi-random sequences may be effectively used in the sum. This will be demonstrated in computational examples below.

On the other hand, a disadvantage of Weighted Uniform Sampling is that for problems with large regions of low importance many more function evaluations will be performed within such regions than if Importance Sampling were used instead. This is reflected in a greater amount of variance reduction in the latter case, and for such problems Smoothed Rejection may be preferable to Weighted Uniform Sampling.

4.5 Computational Example

The Rejection, Smoothed Rejection and Weighted Uniform Sampling methods will now be compared an example. Consider the following integral:

$$A = \int_{I^d} f(x)\,dx$$

where I^d is the unit cube in d dimensions, and evaluate the following four estimates of this integral:

$$\text{Crude:} \quad \hat{A}_N^{(1)} = \frac{1}{N}\sum_{i=1}^{N} f(x_i) \;,\quad x_i \sim U(I^d)$$

$$\text{W.U.S.:} \quad \hat{A}_N^{(2)} = \frac{\sum_{i=1}^{N} f(x_i)}{\sum_{i=1}^{N} h(x_i)} \;,\quad x_i \sim U(I^d)$$

$$\text{Rej Meth:} \quad \hat{A}_N^{(3)} = \frac{1}{N}\sum_{i=1}^{N} \frac{f(x_i)}{h(x_i)} \;,\quad x_i \sim h(x), \text{ accepted point.}$$

$$\text{Smooth Rej:} \quad \hat{A}_N^{(4)} = \frac{1}{N}\sum_{i=1}^{N^*} w_i \frac{f(x_i)}{h(x_i)} \;,\quad (x_i, w_i) \sim h(x)$$

For the last estimate, (x_i, w_i) represents a weighted sample point with acceptance weight w_i, and N^* is chosen such that the sum of the acceptance weights is within one unit of N.

For a given value of N, take M samples of each of these estimates, denoted by $\hat{A}_{N,k}^{(j)}$ for $1 \leq k \leq M$. The average integration error will be measured through the empirical root mean square error, defined as

$$rmse(\hat{A}_N^{(j)}) \approx \sqrt{\frac{1}{M} \sum_{k=1}^{M} (\hat{A}_{N,k}^{(j)} - A)^2} \ , \ j = 1, \ldots, 4 \tag{19}$$

Note that the formula for $rmse$ error uses the exact integral A, which is known for the example below.

Since successive estimates $\hat{A}_{N,k}^{(j)}$, $k = 1, \ldots, M$ are not independent for quasi-Monte Carlo, there is no theoretical basis for these error expressions. This is in contrast to standard Monte Carlo for which the Central Limit Theorem implies that the $\hat{A}_{N,k}^{(j)}$ come approximately from a Gaussian distribution with variance given by the $rmse$.

Finally, the bias of Weighted Uniform Sampling proved to be insignificant, as expected, when using either pseudorandom or quasi-random sequences of points.

Example 2 *Consider Monte Carlo integration over $I^7 = [0, 1]^7$ of the function*

$$f_3(x_1, \ldots, x_7) = e^{1 - (\sin^2(\frac{\pi}{2} x_1) + \sin^2(\frac{\pi}{2} x_2) + \sin^2(\frac{\pi}{2} x_3))} \ \arcsin(\sin(1) + \frac{x_1 + \cdots + x_7}{200})$$

using the positive definite importance function:

$$h_3(x_1, \ldots, x_7) = \frac{1}{\eta} e^{1 - (\sin^2(\frac{\pi}{2} x_1) + \sin^2(\frac{\pi}{2} x_2) + \sin^2(\frac{\pi}{2} x_3))}$$

where η is

$$\eta = \int_{I^7} h_3(x_1, \ldots, x_7) \, dx_1 \ldots dx_7 = e \cdot \left(\int_0^1 e^{-\sin^2(\frac{\pi}{2} x)} \, dx \right)^3$$

which is easily approximated to high accuracy as a one-dimensional integral.

The resulting $rmse$ errors for Example 2 using pseudorandom and quasi-random points are presented in Figures 2 and 3, respectively.

These computational examples show that quasi-Monte Carlo works well for the Smoothed Rejection Method, but even better for Weighted Uniform Sampling.

The results confirm that Smoothed Rejection is an improvement over Ordinary Rejection when using quasi-random sequences. Furthermore, Weighted Uniform Sampling produces results that are far superior to those obtained using Importance Sampling with the Rejection Method, even with smoothing. We believe the reason for this to be that although the Smoothed Rejection method involves a continuous integrand, there is still a relatively sharp slope in acceptance weights in the transition from acceptance to rejection regions. This leads to higher variation and poorer quasi-random performance, in general, than for Weighted Uniform Sampling, where the sharp transition is completely eliminated. Another, admittedly minor, advantage to Weighted Uniform Sampling is that the extra 'decision' variable y is no longer needed, reducing the effective dimension by one.

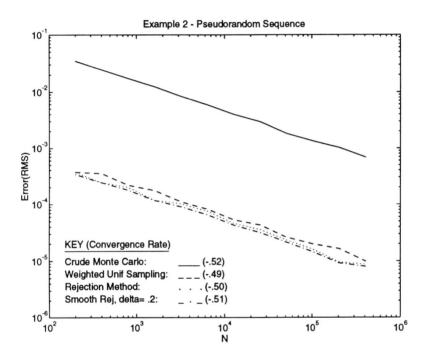

Figure 2: Pseudorandom Results, 75 runs, Example 2.

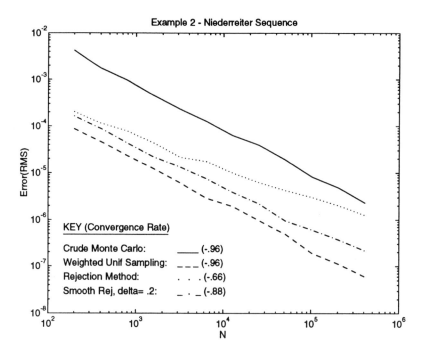

Figure 3: Quasi-Random Results, 75 runs, Example 2.

Nevertheless, Smoothed Rejection may be superior to Weighted Uniform Sampling for problems in which there are large regions of low importance.

5 Conclusions

We have presented two examples in which Monte Carlo methods could be modified to increase their effectiveness when applied with quasi-random sequences. Without these modifications, the performance of the methods with quasi-random sequences is no better than with pseudo-random sequences; after modification much of the theoretical improvement of quasi-random sequences is recovered.

There is plenty of room for further work in this direction: For example, we have not succeeded so far in developing a quasi-Monte Carlo version of the Metropolis algorithm for generation of the equilibrium of a stochastic process. We have also not tried to combine quasi-random sequences with other variance reduction techniques, such as stratification or antithetic variables.

There are several related works that deserve mention: Moskowitz [5, 6] has developed a transformation method for use of quasi-random sequences in diffusion Monte Carlo and in simulation of stochastic processes. He has applied this method to computation of the ground state energy of the hydrogen and helium atoms, as well as to other problems.

Paskov [9] has recently applied quasi-random variables to computation of integrals arising in financial derivatives. Although the integration domain has dimension 360 in his examples, the quasi-random result is significantly more accurate than the pseudo-random result.

References

[1] G. Marsaglia. Normal (Gaussian) random variables for supercomputers. *The Journal of Supercomputing*, 5:49–55, 1991.

[2] W. Morokoff and R.E. Caflisch. A Quasi-Monte Carlo approach to particle simulation of the heat equation. *SIAM Journal on Numerical Analysis*, 30:1558–1573, 1993.

[3] W. Morokoff and R.E. Caflisch. Quasi-random sequences and their discrepancies. *SIAM J. Sci. Stat. Computing*, 15:1251–1279, 1994.

[4] W. Morokoff and R.E. Caflisch. Quasi-Monte Carlo integration. *J. Comp. Phys.*, to appear, 1995.

[5] B. Moskowitz. Improved stochastic simulation using quasi-Monte Carlo: a computational study. 1994.

[6] B. Moskowitz. Quasirandom diffusion Monte Carlo. 1994.

[7] B. Moskowitz and R.E. Caflisch. Smoothness and dimension reduction in quasi-Monte Carlo methods. *J. Math. Comp. Modeling*, submitted, 1994.

[8] H. Niederreiter. *Random Number Generation and Quasi-Monte Carlo Methods.* SIAM, Philadelphia, 1992.

[9] S.H. Paskov. Computing high dimensional integrals with applications to finance. *preprint*, 1994.

[10] M.J.D. Powell and J. Swann. Weighted uniform sampling – a Monte Carlo technique for reducing variance. *J.Inst.Maths Applics*, 2:228–236, 1966.

[11] J. Spanier and E.H. Maize. Quasi-random methods for estimating integrals using relatively small samples. *SIAM Review*, 36:18–44, 1994.

[12] H. Wozniakowski. Average case complexity of multivariate integration. *Bulletin of the AMS*, 24:185–194, 1991.

SIMULATED ANNEALING: FOLKLORE, FACTS, AND DIRECTIONS

Bennett L. Fox
Mathematics Department
Campus Box 170
University of Colorado
P.O. Box 173364
Denver, CO 80217-3364
E-Mail: bfox@castle.cudenver.edu

Abstract

Simulated annealing is a (sometimes-maligned) probabilistic search method used in global optimization. Folklore has it that simulated annealing must use static (generally-large) neighborhoods, that it must blindly propose moves independently of objective-function values, that it must waste a lot of (computer) time rejecting moves, that tabu search and genetic algorithms as well as random restarting and deterministic methods are disjoint from simulated annealing, and that the canonical cooling schedule is worthwhile. All this is false. After exploding these myths, the role of search diversification in eliminating pathologies and in setting the stage for guaranteed linear speedup of parallel search is pointed out. Next we sketch a way to use low-discrepancy point sets in a preprocessor. Finally, we outline methods to handle noisy objective functions and argue that convergence results for simulated annealing have strong practical relevance in that setting.

1 Introduction

Simulated annealing generally is implemented naively. When it is, many of the criticisms of simulated annealing are valid. Here we outline a sophisticated implementation of simulated annealing and argue that the negative folklore about the naive version becomes irrelevant. We also point out directions for research.

1.1 The problem

We seek to minimize $\{c(x): x \in S\}$. The objective function c is real-valued, but the set S can be quite general. Most often, S is finite, a compact subset of finite-dimensional Euclidean space, or a hybrid of those two.

1.2   What this paper does

Both state spaces mentioned above are treated here. Section 5.4.4 discusses a hybrid state space. Much of what we say applies to general state spaces, including our discussion of sophisticated simulated annealing in Section 2. Speedup on parallel processors in discussed in Section 3.2. We assume that c is known everywhere on S, except in Section 5. There we assume that $c(x)$ can be estimated dynamically by observing certain random variables, say by Monte Carlo simulation. For S finite, a straightforward simulated-annealing algorithm discussed in Sections 5.1 and 5.2.2 converges when the range of c is restricted to a finite set, a condition which we think is weaker than others which have been proposed. We adapt our results to get asymptotically-optimal work-normalized estimators in Section 5.4.3. Our approach extends to other settings to get a confidence interval for the minimal $c(x)$. For S infinite, we propose new simulated-annealing algorithms in Sections 4 and 5.3. These handle noiseless and noisy observations, respectively. They use the framework of Section 2, with c evaluated on a low-discrepancy point set as a preprocessor. No prior knowledge of Quasi-Monte Carlo is needed to understand our description of this preprocessor. Thus, simulated annealing globally optimizes over strategies whose respective performances are simulated by Monte Carlo as well as in problems where all data are known. Our methods for the former are direct counterparts of our methods for the latter. This paper is accessible to those unfamiliar with simulated annealing. For those unfamiliar with Quasi-Monte Carlo, we try to convey some of its flavor in Section 2.3.4 and cite references for more detail.

1.3   Naive simulated annealing

In the naive version:

- there is a one-to-one correspondence between elements of S and states

- the neighborhoods are *static* and, generally, *large*

- moves are proposed *blindly*, independently of objective-function values

- many proposed moves are *explicitly* rejected

- there is no preprocessing.

Some mistakenly believe that these restrictions are necessary. None are, as the next Section details. The standard terms *neighborhood* and *proposed* are defined in the algorithm **SLOW** below. No preprocessing means that the initial state is chosen arbitrarily.

Let $t(k)$ be the *temperature* at time k. The sequence $\{t(i): i = 1, 2, ...\}$ is called the *cooling schedule*. The statement "y is *downhill* from x" means that $c(y) \le c(x)$ and y is in the neighborhood $N(x)$ of x; likewise, for *uphill*. When at state x at time k, the naive move mechanism is

SLOW(*x,k*)

> *Until exit, repeat*
> > *Set* $k \leftarrow k + 1$ {update simulated time}
> > *Generate* a proposed move to a state y in the neighborhood $N(x)$ of x, according to
> > > some distribution (generally, the uniform distribution over $N(x)$).
> > > > {tentative move}
> > *If* y is downhill from x, *then*
> > > Set $x \leftarrow y$ {accept the move}
> > > *Exit*
> > *Else*
> > > *Generate* a uniform random number U.
> > > *If* $U < exp\{[c(x)-c(y)]/t(k-1)\}$, *then*
> > > Set $x \leftarrow y$ {accept the move}
> > > *Exit*

End.

The explicit input parameters (x,k) are respectively the current state and (simulated) time. Implicit inputs are $c(\cdot)$, $t(\cdot)$, $N(\cdot)$, and the tentative-move probabilities. The neighborhood $N(x)$ of x is often specified implicitly; for example, it could consist of those y accessible from x by a fixed number of iterations of a given interchange heuristic. At exit, the generic output state is y with time updated as indicated. The acceptance probability is the right side of the inequality above, when y is uphill from x. If the time at exit is $L+1$, we say that acceptance occurs at move L with corresponding temperature $t(L)$. Any iteration of the *until-repeat* loop in which rejection occurs (i.e., no exit) is part of an explicit self-loop sequence of the form $x \rightarrow x \rightarrow \cdots \rightarrow x$. When some state has all its neighbors strictly uphill and the temperatures converge to zero, the naive method is not practical because of increasingly-long self loops — though many papers study exactly that situation; likewise, when (even if S is not discrete) near local minimizers the probability that the proposed move is strictly uphill is near one. An ostensible advantage of the naive scheme is that the objective function c need be evaluated only at states to which moves are proposed; however, we shall argue that this is illusory. In the next Section, we greatly improve this scheme.

2   Sophisticated simulated annealing

In sophisticated simulated annealing:

- the neighborhoods are *dynamic* and of *modest size*, while guaranteeing in a problem-independent way irreducibility.

- the original chain $X(k)$ is replaced by a *pruned* chain $Y(k) = X(J(k))$, where the $J(k)$'s are the successive jump epochs between pairwise distinct states [i.e., $X(J(k)-1) \neq X(J(k)) = X(J(k)+1) = \cdots = X(J(k+1)-1) \neq X(J(k+1))$].

- now a (macro)state is a collection of feasible solutions, together with respective recent histories (temporal memory) in the pruned chain.

- for each component of each (macro)state, we add a nonnegative penalty function to c to account for that component's recent history [except when that component was generated randomly — without regard to history], for example, to inhibit revisiting a recently-visited state containing a feasible solution also in the current state when S is finite or containing a feasible solution in the same region as one in the current state when S is infinite, producing an intermediate pseudo-objective function c'.

- the final pseudo-objective function c^* for a state is now the minimum of the c'-values corresponding to the feasible solutions (with their respective histories) it contains.

- when at state x at time k (i.e., simulated time in the original chain), the proposed move depends on $\{c^*(y): y \in F(x,k)\}$, where $F(x,k)$ is the current neighborhood of x and remains unchanged during (implicit) self loops (of the form $x \rightarrow x \rightarrow \cdots \rightarrow x$ in simulated time).

Among these features, there is significant synergy. This will become apparent below. The penalty function above can also reflect ranking of the importance of the variables and the respective directions for altering their values. Often, most candidates for $F(x,k)$ are chosen from a nominal neighborhood much smaller than (and generally contained in) $M(x)$, defined as the Cartesian product of the naive neighborhoods $N(s)$ for the feasible solutions s in x. The (few) additional elements of $F(x,k)$ contain feasible solutions generated randomly from all of S. Fox (1993a) and Section 4 here, respectively, streamline the construction of $F(x,k)$ when S is finite and when S is the unit cube. While $F(x,k)$ generally depends on k as well as on x, the *rule* to construct $F(x,k)$ does *not* depend on k; without this restriction, theoretical analysis would be hard. Proposed moves can depend on aspects of history not reflected in c', as Section 4.2 illustrates. Making proposed moves depend on objective-function values and on recent history is taken for granted by the tabu-search community. That explains, in part, why tabu search empirically beats *naive* simulated annealing (even when the temperatures are bounded away from zero). How it fares against sophisticated simulated annealing remains to be seen.

2.1   Macrostates and look-ahead

There is a mapping from our generic (macro)state v to S that lets us recover the corresponding best feasible solution, with respect to c', "imbedded" in v. Among the components of v, it picks out the one with minimum c'-value, breaking ties arbitrarily. Under weak conditions, the minimum of all possible c'-values equals $\min\{c(x): x \in S\}$. See Fox (1993a) for this and the link to tabu search and genetic algorithms, via the perspective of probabilistic search schemes as Markov chains on state spaces rich enough to accommodate ideas previously seen as disparate. As Fox (1993a) points out, the search is much more than a concatenation of individual searches corresponding to the respective solutions in v. For example, *crossover* (in the genetic-algorithm sense) can be used to generate candidates for $F(x,k)$:

Crossover at position k of $x = (x_1, ..., x_k, x_{k+1}, ..., x_n)$ and $y = (y_1, ..., y_k, y_{k+1}, ..., y_n)$
produces $a = (x_1, ..., x_k, y_{k+1}, ..., y_n)$ and $b = (y_1, ..., y_k, x_{k+1}, ..., x_n)$.

It is especially attractive if the problem nearly separates at position k. Another reason to have more than one feasible solution imbedded in v is to profit more from a random restarting preprocessor as Section 2.3.1 explains. To use our preprocessor of Sections 2.3.4 and 4.2.1, this feature is essential.

Making the size of $F(x,k)$ small makes intelligent tentative moves practical. One feature of such smart moves is that the tentative-move probabilities depend on all c^*-values in $F(x,k)$. A second feature is (limited) *look-ahead* as follows. The distribution used to generate (some) elements of $F(x,k)$ can be implicit. For example, after generating an element u uniformly from S, a descent path from u can be followed to pick a u' for a component of a state in $F(x,k)$ or, if c is differentiable, a (local) search can be started from u for an approximate stationary point u' of c. Section 4 refines the latter idea. For the former, the number of descent steps (each chosen from a small set of local candidates) allowed can be decreasing function of $c(u)$ — on the principle that the amount of look-ahead should be higher for better u's. If u is significantly above x, perhaps there should be no look-ahead for that u unless all (tentative) elements of $F(x,k)$ are strictly uphill from x. It probably makes sense to impose an upper bound on the overall number of descent steps from all the u's generated. There is no requirement that u be picked uniformly. Instead, its distribution may make it likely that it is picked from what might have been the neighborhood of x with naive simulated annealing. No matter how much effort goes into look-ahead, simulated time advances by exactly one unit with each proposed move.

Look-ahead meshes well with having multiple solutions in a state, because look-ahead from the current state can supply several solutions for the next state. Thus, because of the spatial memory thereby induced in macrostates, no solution found by look-ahead needs to be forgotten. Likewise, memory of selected x in S in valleys traversed (at least the deeper ones) lingers. While for discrete S look-ahead is a heuristic, it is easy to see that for continuous S the convergence rate of the best state seen so far to the set S^* of global optima (whether or not the algorithm is based on simulated annealing) is generally glacial without it. Despite this, algorithms for continuous S without look-ahead are still being proposed; see, for example, Romeijn and Smith (1994).

Another ingredient of smart simulated annealing is search *diversification*, discussed further in Section 3. One aspect comes from the multiple solutions in each state. Enforcing diversification requires that no two of these are in the same "region" of S. Fox (1993a) gives a specific method to achieve such spacing when S is finite. Section 4 here does this when S is the unit cube. Ideally, each solution in a given state should be in a distinct basin of attraction — ignoring minor, local roughness. A second aspect enriches $F(x,k)$ with elements of S that can be far from x. When S is finite, each element of S has a positive probability to enrich $F(x,k)$. When S is the unit cube, the density for putting elements in $F(x,k)$ is bounded away from zero. Diversification prevents the search from becoming myopic. That principle is incorporated in some versions of tabu search, but we apply it differently. Our ideas could also be used in tabu search. What would then distinguish *probabilistic* tabu search (without a tabu list) from our sophisticated simulated annealing is the latter's acceptance test. In our setup, there is a dynamic interplay between search *depth* (look-ahead) and search *breadth* (diversification) — much stronger than possible with cooling alone.

There are memory-work tradeoffs in deciding how much recent history to record. Generally, one explicitly records only the (dynamically-defined) oldest solution in this history with more recent history implicitly defined by recording only the necessary updates. This saves memory and work directly and makes it easier to check whether a proposed move would be to a recently-visited state, at least when that move would be to a "nearby" state. Thus, though $F(x,k)$ can potentially contain elements far from x, one should arrange that the number of such elements is likely to be small when there is significantly more work or memory associated with processing those elements. No memory problem arises in Section 4.2, where S is the unit cube. Instead of recording recently-visited states, one merely records recently-visited boxes in a certain partition of S; that is easy.

One of the reasons that $N(x)$ is generally rather large in naive simulated annealing is to assure irreducibility. In this paragraph, denote by E the elaborate state space incorporating temporal and spatial memory and denote by $M(x)$ the counterpart to $N(x)$ on E. The random elements added to (the generally much smaller) $M(x)$ in sophisticated simulated annealing to form $F(x,k)$ do more than assure irreducibility on E. They also assure that weak reversibility holds on E. {Weak reversibility means that: there is a path from x to y with maximal objective-function height $h \Leftrightarrow$ there is a path from y to x with maximal height at most h.} Without precautions, generally E does not inherit weak reversibility from S. For this, the random elements we put in $F(x,k)$ are enough. Even if neighborhood symmetry $\{x \in N(y) \Leftrightarrow y \in N(x)\}$ holds on S, generally not even weak reversibility holds on E unless the random elements are added to $M(x)$. See Fox (1993a) for a precise description of these random elements and how dynamic neighborhoods are generated using them and other candidates.

2.2   Move mechanism

2.2.1   The algorithm and its main subroutine

Consider a generic search algorithm of the form:

SEARCH(*x,k*)

Generate a *finite* set $F(x,k)$ of candidates for the next state in the pruned chain.

Optional: if $F(x,k)$ contains a state with a feasible solution at least η better {with respect to c} than any yet seen, move to it — breaking ties arbitrarily. [If S is finite, replace "at least η" by "strictly"; otherwise, set η to a positive tolerance.]

Execute **QUICKER**(*x,k,F(x,k),j*) to generate L, the transition number in the original chain when a move to a state different from x is next made [see below].

Execute **NEXT**(*x,k,F(x,k),L*) to generate that state.

Reset simulated time to $L+1$.

End.

Let $a(x,i) = P\{X(i+1) \neq x \mid X(i) = x\}$. Given $a(x,L)$ and the c^*-values of all elements in $F(x,L)$, implementing **NEXT** is easy: just generate from a discrete distribution conditioned on acceptance at L. The work to compute $a(x,i)$ is essentially free, given that we want smart tentative moves — depending on all c^*-values in $F(x,i)$. In turn, such smart moves are practical because the random elements in $F(x,i)$ make it possible to have $|F(x,i)|$ small while assuring irreducibility. Thus **QUICKER** and our dynamic neighborhoods dovetail. Fox and Heine (1995b) show, among other things, that the optional step in **SEARCH** preserves the convergence properties in Section 6. Some merely set L equal $k + 1$, thus avoiding **QUICKER**. The work thus saved is negligible. Such (random) slowing of the cooling is arguably bad in terms of first hitting time to an optimal state. If some state has all its neighbors strictly uphill, then it is not known when the convergence properties of Section 6 hold if cooling is random.

The non-trivial subroutine follows:

QUICKER(*x,k,F(x,k),j*)
    Set i ← k
    Set h ← 0   ◄
    Until exit, repeat
        Set h ← h + 1   ◄
        Generate a geometric r.v. G with parameter $a(x,i)$
                {Thus, G is the trial number of the first success
                in iid Bernoulli trials with success probability
                $a(x,i)$.}
        Set L ← i + G - 1
        Generate a uniform r.v. V

If

$$V < a(x,L)/a(x,i)$$
$$or$$
$$h > j$$

Then

Exit with L

Else

Set $i \leftarrow L + 1$

End.

Taking j equal ∞ corresponds to deleting all statements marked with a triangle. In this case, L is stochastically the same as it would have been with the naive algorithm as Fox [(1993a), (1995)] proves. Under mild conditions on the cooling schedule, Fox (1995) shows that the number of iterations of the *until-repeat* loop converges in quadratic mean to one as k goes to infinity, for all fixed x. A geometric variate can be generated in $O(1)$ time, regardless of its parameter; e.g., see Bratley, Fox, and Schrage (1987), Section 5.4.5. Thus each execution of the *until-repeat* loop requires essentially constant time, so that the computer time to generate r accepted moves is essentially proportional to r.

Regardless of the cardinality of S, we make $F(x,k)$ finite (and small). If nothing is excluded from $F(x,k)$, as above, then irreducibility and weak reversibility follow immediately. When S is finite, $F(x,k)$ is typically a superset of a fixed, nominal neighborhood $N(x)$. With our setup, we can choose $N(x)$ selectively without worrying about irreducibility. The latter we get from the random elements added to $N(x)$ to form $F(x,k)$ at step k, which depend on k and possibly on x. Tailoring the method to construct $F(x,k)$, along with specifying the tentative-move probabilities and (tabu) penalty function, to particular problems is the art of simulated annealing. The rest of **SEARCH** and **QUICKER** is problem-independent, provided that problems are encoded in a uniform way (partly to avoid problem-dependent data structures). We see that **SEARCH** is a top-level procedure, with the construction of $F(x,k)$ at the bottom. The look-ahead and diversification built in that construction strengthen the analogy with multigrid. The tabu-search community rightly puts much effort into specifying the penalty function. Many of their ingenious ideas can be adapted to simulated annealing, to the extent that a static rule to construct $F(x,k)$ and finite memory permit. Such adaptation is required to get a fair comparison of simulated annealing and tabu search. As far as we know, we are the first to propose tabu penalties for general state spaces.

2.2.2 Comments on QUICKER

While **QUICKER** is the part of **SEARCH** most amenable to mathematical analysis (Fox (1995), Heine [(1994a), (1994b)], and the remarks below), we think that its combination with the ideas set forth at the outset of Section 2 and in Section 2.1 is what makes our overall approach so attractive. As presented, **QUICKER** assumes that the temperatures are nonincreasing. While that condition could be relaxed, the corresponding modification of **QUICKER** slows it — generally significantly. Even if the speed of **QUICKER** were unaffected by non-monotone, possibly-adaptive cooling, with our diversification we see no advantage in such cooling — as discussed briefly in Section 3.

Heine (1994) shows how to choose j finite, depending on the cooling schedule and neighborhood structure, while guaranteeing certain convergence properties of the original and pruned chains, respectively. Section 6 briefly discusses these properties. If the elaborate state space is finite and every element of it has a positive probability of being in $F(x,k)$ and if a "subcanonical" cooling schedule (defined below) is used, then we can choose j equal one. If S is a compact subset of finite-dimensional Euclidean space and the density for choosing elements of $F(x,k)$ is bounded away from zero, then we can choose j equal one; results of Bélisle (1992) extend if $F(x,k)$ is forced to contain an element not uphill from x. If $|F(x,k)|$ reaches a certain threshold with no such element, this can be done by putting x in $F(x,k)$ with corresponding small tentative-move probability. In that case, the speedup from **QUICKER** is significant. Under mild conditions on the cooling schedule, Heine shows that the same convergence properties hold. If nothing is excluded from $F(x,k)$, as above, then irreducibility and weak reversibility follow immediately.

Heine (1995) sketches Richard Holley's proof (communicated personally) that it is never necessary to take j greater than 2, regardless of the contents of $F(x,k)$ relative to x. Heine also gives an example where taking j equal 2 is needed to assure convergence in probability to the set of optimal states. He conjectures that j equal 2 is always necessary if every state is visited infinitely often and there is a state with all neighbors strictly uphill.

Without **QUICKER**, simulated annealing would always be beaten by random restarting in the sense that, for all k large enough, the probability of visiting an optimal state by (simulated) time k is at least as large using random restarting. This is not true when using **QUICKER** and counting only accepted moves, as Fox (1994) shows by example. It is true that enumeration is unbeatable according to this criterion, but generally enumeration is absurd whereas random restarting and simulated annealing each have a reasonable chance to succeed.

We do not know whether the final L output from **QUICKER** can be replaced by its expectation, rounded up, and still have all the convergence properties in Section 6 hold. If L can be replaced by a random variable with the same expectation, then the final L can be replaced by its conditional expectation, rounded up, given the respective parameters of the geometric variates generated during the execution of **QUICKER**. This inhibits (anomalous) rapid cooling in the pruned chain. Using expectations this way is called (extended) conditional Monte Carlo in the simulation literature; e.g., see Bratley, Fox, and Schrage (1987), Section 2.6.

2.3   Preprocessors

We view preprocessors as a key part of our approach, not a frill. They generate the initial state $X(0)$, *intelligently*. This initializes **SEARCH** with a pair $(X(0), 0)$, which we write simply as $(X,0)$. With all preprocessors, except the one described in Section 2.3.1, their effects persist throughout the entire simulated annealing process. In a well-designed implementation, there is a problem-dependent (pre)pre/postprocessor that converts a given problem formulation to some problem-independent format and feeds it to a preprocessor below. When annealing proper terminates, its output is converted back to a format easily understood by the user.

2.3.1 Random restarting

No matter what S is, we can use random restarting in tandem with descent to generate an initial state for (smart) simulated annealing proper. Run this pair thousands of times, stratifying S as in Fox (1993a). Choose the best solution from each stratum to go into an initial state, subject to the following constraint on its size: no two solutions in it are close in a suitable metric. Since this state contains multiple feasible solutions, the work for random restarting is thereby amortized. The heuristic reasoning for this step is that, because simulated annealing has significant overhead and can make uphill moves, the initial state should have low height above the minimal value of c. Under mild conditions, the (ordinal) rank of the c'-value of the starting state puts it close to that of the best states. This holds even without the descent phase of each random restart. The restricted local neighborhoods used here can be the same as those used for look-ahead.

The first state in each random restart can be replaced by the output of the (greedy, randomized) construction phase of **GRASP**, and then the descent phase above corresponds to the local-search phase of **GRASP**. For a recent review of **GRASP**, see Feo and Resende (1995). The brief discussion there of simulated annealing deals only with what we call naive simulated annealing. As presented there, **GRASP** uses no postprocessor to local search.

2.3.2 Branch and cut or Lagrangian relaxation

Branch-and-cut algorithms or Lagrangian relaxation (e.g., see Ahuja, Magnanti, and Orlin (1993), Chapter 16, or Nemhauser and Wolsey (1988), Section II.3.6) can generate lower bounds on c (hence, a stopping criterion), suggest a starting solution (perhaps via rounding), and indicate qualitatively good proposed moves. In some settings, it may pay to use simulated annealing to generate good cuts via maximizing a dual objective function. It is also possible that simulated annealing may be a good method to use on subproblems generated by Lagrangian relaxation (say, when the running time of deterministic algorithms for these subproblems is not polynomially bounded). This reduces the number of *ad hoc* subroutines needed for probabilistic search. Instead, one writes subroutines for frequently-occurring subproblems and lets Lagrangian relaxation deal with troublesome side constraints. One could turn this procedure around, by using probabilistic search to find feasible solutions for various branch-and-bound subtrees. Still, we think, probabilistic search (as a stand-alone method or part of a hybrid) is viable only on structured subproblems where it is feasible to construct intelligent neighborhoods.

2.3.3 Gradient-based points

When S is infinite and algorithms are at hand (for example, from Luenberger (1984), part II or III) that converge to a stationary (though possibly not optimal) point relative to c, they can be stopped short of convergence and the corresponding point(s) can be added to the initial state. Such points also can be generated within the simulated annealing algorithm proper as Section 4 details.

2.3.4 Low-dispersion point sets

Suppose that S is a compact subset of finite-dimensional Euclidean space. Most people unfamiliar with Quasi-Monte Carlo do not realize that initially sampling c over a grid imbedded in S is a bad choice. If c depends strongly on only a few variables, then clearly most grid points are wasted. Even if c depends strongly on all its variables, grids leave unnecessarily large "holes" in S. See Bratley, Fox, and Schrage (1987), problem 6.2.2, for example. Likewise, a point set selected using points that try to mimic genuinely independent and uniform random numbers leaves unnecessarily large holes. In contrast, low-dispersion sets D have small holes, i.e., low values of sup {min {dist(x,y): $y \in D$}: $x \in S$}. In this paper, we use such sets only when S is the unit cube. Thus, the sizes of the holes do not depend on the units of measurement. However, if the cube is a transformation T of some other domain, then applying T^{-1} induces dependence of hole size on coordinates.

At least when S has modest dimension, a low-dispersion point set D is a good choice. It can form the core of a preprocessor as Section 4 details. If c is evaluated on D and an upper bound on its modulus of continuity or on a Lipschitz constant is known, lower bounds on c over various regions of S follow. This induces a termination criterion. Without such lower bounds, the problem is arguably ill-posed. One has to rule out pathologies such as a flat c except for a downward spike. Likewise, if S is unbounded, the problem is ill-posed unless every local minimizer is a global minimizer (and the already-mentioned provisos hold).

2.3.5 Markov-chain Monte Carlo

When S is infinite and there is no simple mapping M from the unit cube to S, the only practical procedure seems to be based on Markov-chain Monte Carlo. The latter constructs a Markov chain whose stationary distribution is the desired sampling distribution over S. For example, see Bratley, Fox, and Schrage (1987), problem 1.9.16, for the case where S is a polytope defined by linear inequalities (easily generalized to compact, convex S) and the target distribution is uniform. In such cases, especially if the dimension of S is more than a dozen say, only a crude partitioning (stratification) of S seems practical. Without a convenient M, using low-discrepancy point sets seems out of the question. Markov-chain Monte Carlo (MCMC) can be combined with random restarting as well as used in simulated annealing proper. To recast the latter combination in standard simulated-annealing format, we use (macro)states of the form (x,y) where x corresponds to c' and y is the state of the MCMC subroutine.

If MCMC is used in simulated annealing proper, then look-ahead (via gradients say) generally is essential. Otherwise, the convergence rate typically will be very slow. This bad behavior would be aggravated if **SLOW** were used, because **SLOW** gets slower as the probability of a tentative uphill move increases.

3 Diversification

We show by examples pathologies that can occur when "long" jumps across S are forbidden. Next, we indicate that, when such jumps are allowed, linear speedup occurs when independent copies of the pruned chain are run on respective parallel processors. More precisely, we forbid states from which all paths to the set S^* of optimal states require a strictly uphill move. We do this by adding x to the tentative $F(x,k)$ if all elements in the tentative $F(x,k)$ are strictly uphill from x. Heuristically, in that case it is attractive to make the probability of proposing a move to x small; so **QUICKER** still gives a significant speedup. The random elements enriching the neighborhoods then allow a path from any state to S^* which has only downhill and horizontal segments. Having multiple solutions in each state prevents the search from degenerating merely to random restarting when seemingly stalled at a local minimum, with respect to the solution (say s) in x with lowest c'-value, at very low temperatures. It also makes it more unlikely that all states in $F(x,k)$ are *strictly* uphill from x, because when x is a local minimum, typically $F(x,k)$ contains states with s as a component from the construction in Fox (1993a). That construction builds explicitly only the state visited next. A balance with local search is maintained.

Define *basins* relative to the topology induced by *local* neighborhoods of feasible solutions to the original problem. The local neighborhoods of x is an analog to the $M(x)$ of Section 2.1 when E equals S. Typically, it is a subset of the naive neighborhood $N(x)$. Roughly, a basin is a collection $\mathcal{B}$ of states such that there is a path with no strictly uphill moves from *every* state of $\mathcal{B}$ to the bottom of $\mathcal{B}$ and no strict superset of $\mathcal{B}$ has this property. When a jump occurs, it is because the element s' with smallest c'-value in $F(x,k)$ is in a basin different from s, the element of x with smallest c'-value. That can occur because s' is a random enriching element, because local search from elements of x other than s finds s', or because crossover or some other combination of solutions in x produces s'. While the short-run sample-path effects of truncated **QUICKER** and, to a lesser extent, tabu penalties are mainly local, the impact of diversification is felt globally. The computer-time effect of **QUICKER** is global. Another contrast with random restarting, especially at low temperatures, is that we are unlikely to waste time jumping up to and then descending from states with high objective-function values. Instead, we are likely to jump downhill.

3.1 Pathologies

Neighborhoods which are purely "local" lead to myopic search. To see what can go wrong when $F(x,k)$ does not (potentially) contain elements from diverse parts of S, consider the following four-state example using the naive state space with static, local neighborhoods:

EXAMPLE 1 (Fox (1995)):

state	c	neighborhood
w	2	x
x	1	w, y

y	3	x, z
z	0	y

For a fixed starting state, let N be the time to first hit z, relative to the pruned chain. With any cooling schedule where the temperatures converge to zero, it is easy to check that because of likely long sequences of $x \leftrightarrow w$ oscillations:

$$-k + E[N \mid N > k] \to \infty$$

as $k \to \infty$. Thus, roughly, the longer the search has been unsuccessful, the longer the expected remaining time to hit the sole globally-optimal state z.

Similar pathologies can be constructed for the more elaborate state spaces, incorporating temporal and spatial memory, indicated above. They scale up to huge state spaces, where the search is likely to be unsuccessful for a long time and thus likely to be unsuccessful for an even much longer time. Such examples are by no means contrived. However, they cannot occur with our construction of $F(x,k)$ above.

One might argue that the pathology is an artifact of letting the temperatures converge to zero. However, there are (at least) two arguments in favor of letting the temperatures converge to zero. First, perhaps less important when the observations are noiseless, this is necessary for the convergence results of Section 6. For those, in general, cooling must not be too fast. Second, heuristically, the longer the search has gone on, the more we want to drive it downwards — even in the short run, because our neighborhood construction prevents local "stalling". Reheating would merely stimulate a (generally slow) climb out of the local valley, whereas lowering the temperature stimulates long jumps when all local elements of $F(x,k)$ are uphill. The latter is especially desirable, intuitively, when those elements reflect look-ahead. Even if the temperatures do not converge to zero but rather to a small positive number, the pathologies above give insight into the drawbacks of myopic search and suggest that diversification is worthwhile in that setting as well.

An even more startling pathology occurs in

EXAMPLE 2 (folklore):

> Delete state w from Example 1 and use the corresponding canonical cooling schedule, where the temperature at move k in simulated time is $2/log(k+1)$.

Starting at state x and letting T be the first *simulated* time to hit z relative to the original chain, it is routine to show that $ET = \infty$ — even though $X(k) \to z$ almost surely by Borel-Cantelli. In sharp contrast, it can be shown that the expected number of moves in the pruned chain Y and so the expected *computer* time using **QUICKER** are small numbers.

Subcanonical schedules replace the numerator 2 above (in general, the maximal "depth" d^* of local, nonglobal minima) by some larger number. The speed of **QUICKER** decreases discontinuously at 2. If the numerator is near 2, **QUICKER** will have underflow problems. If in general the numerator is replaced by a number (slightly) larger than

$$\max\{c^*(x): x\in S\} - \min\{c^*(x): x\in S\},$$

then almost surely all states are visited infinitely often. A good upper bound on that difference generally is easier to find than a good upper bound on d^*. The canonical cooling schedule gets extensive analysis in theoretical studies of simulated annealing. In contrast, as far as we know, all serious practical implementations have used cooling schedules bounded away from zero. This is a practical necessity when using the naive algorithm. Teaming diversification with **SEARCH**, it is *not* a practical necessity and — as we have argued — undesirable.

3.2 Speedup with parallel processors

Now let $N(m)$ be the first time in the pruned chain when some optimal state is hit, looking globally across m independent copies of the chain. No assumption is made about cooling, other than that the temperatures approach a limit; even accepting every proposed move is not excluded. It follows easily from results of Fox (1995) that, when

- S is finite
- our neighborhood construction is used
- certain (mild) technical conditions hold,

$$\{E[N(1) \mid N(1) > k]\}/\{E[N(m) \mid N(m) > k]\} \rightarrow m,$$

i.e., the speedup is (in a well-defined sense) *linear*. From Section 3.1, it is clear that (vacuously) no such speedup is possible without some sort of neighborhood diversification. It seems likely that the condition that S be finite is not essential, but a counterpart for general S would require sophisticated technical tools (though probably nothing really new conceptually). The key to the proof in Fox (1995) is recasting, for theoretical purposes only, the dynamic neighborhoods as static neighborhoods on an expanded state space and using the spectral representation of the move matrices, conditioned on accepting all proposed moves, on this expanded state space. One of the technical conditions, alluded to above, says roughly that asymptotically the current time k is negligible relative to the expected remaining time to first hit the set S^* of optimal states. A more precise statement relates k to the limiting spectral radius of the move matrices, conditioned as above, with all rows and columns corresponding to S^* deleted. The speedup result above is obtained by letting the limiting spectral radius approach one and applying L'Hôpital's rule. In example one, the limiting spectral radius is one because the matrix reduces to two closed classes: $\{w, x\}$ and $\{y\}$.

An analogous result holds if we replace S^* by any nonempty set $Q(b)$ of the form $\{x: c^*(x) < b\}$. One would ordinarily choose b to make $Q(b)$ be a set of nearly-optimal and optimal states. When S is infinite, consideration of $Q(b)$ generally becomes not only practical but essential because S^* then typically has zero probability of being hit in any move. Heuristically, if

- *c* is smooth
- the transition kernel $K(x,A)$ is smooth at almost all x
- S is partitioned into a finite number of pieces each with a small diameter,

one can define a process on the corresponding finite-state "reduction" that is "virtually" indistinguishable for the original process — with a corresponding implication for linear speedup.

Our work on parallel speedup was stimulated by a preprint of Shonkwiler and Van Vleck (1994). There are major differences between our work and theirs. Among these, we measure speedup relative to computer time (by considering only accepted moves, because of **QUICKER's** speed) but they measure it relative to simulated time — far from the same thing.

In SIMD computers, each processor executes same instruction on any given clock tick — but on processor-dependent data. Synchronization of the processors to be compatible with this architecture is impossible when using **SLOW** but is feasible when using **QUICKER**, especially with j equal one (legitimate with our construction of $F(x,k)$) or at worst j equal two.

For random restarting, a parallel-speedup result is given in Fox (1993a), based on the Palm-Khintchine theorem. That result assumes that the number of processors gets large, but does not assume that the search has been unsuccessful for a long time; in both respects, this contrasts with what we assume for our result about parallel speedup with simulated annealing.

Laursen (1994) studies branch and bound without communication among (say) m processors. He first decomposes the original problem into m disjoint subproblems and then runs these independently on the respective processors. His measure of effectiveness is processor utilization: (apparently) the fraction of time up to solution of the original problem that all processors are busy. Processor utilization is positively correlated with speedup, but is not a well-defined function of it. Laursen tries to make each subproblem roughly equal in difficulty (as measured by computer time to solve it) via a heuristic, but there is no guarantee of success. He has encouraging empirical results for m equal 17. He has no empirical results for massively-parallel simulation (that is, a large number of processors). For that case, he proposes a method to create the m subproblems which he claims the m processors working together in a future computer architecture can do in order $\log(m)$ time. All this contrasts with our guaranteed linear speedup. In general, no theoretical analysis seems possible for any algorithm in which the processors are allowed to communicate.

4   Using low-dispersion point sets

When S is the unit cube, we mate our algorithm with the structure of (t,m,s)-nets. Ours is the first global-optimization method to exploit that structure.

4.1   Background

Section 4.2.1 details how low-dispersion point sets can be used in a preprocessor to simulated annealing proper (or to other algorithms). We freely adopt the definitions and results in Niederreiter (1992). Low-dispersion point sets on the unit cube can, in principle, be mapped into arbitrary compact (connected) subsets of Euclidean space. When S is merely a rescaling of the unit cube, the mapping is trivial. However, the general case is a hard, unsolved problem. In any case, we assume that a suitable mapping is at hand so that Theorem 6.4 in Niederreiter (1992) can be invoked. Therefore, we consider just the case where S is the unit cube. From Theorem 6.6 in Niederreiter (1992), it follows that low-discrepancy point sets have low dispersion; so we consider only the former, because good constructions for low-discrepancy point sets are known. We can use any construction which gives a (t,m,s)-net D, per Definition 4.1 in Niederreiter (1992). Section 4.2.1 below summarizes the properties of such nets. Their appeal is clear without reference to dispersion or discrepancy. Bratley, Fox, and Niederreiter (1992) point out that techniques for quickly generating base-2 Niederreiter (low-discrepancy) sequences also speed up the generation of the (low-discrepancy) base-2 (t,m,s)-nets in Hansen, Mullen, and Niederreiter (1993). See also Larcher, Lauß, Niederreiter, and Schmid (1995) and Mullen, Mahalanabis, and Niederreiter (1995).

4.2 The algorithm

Define, dynamically, the *attractiveness* of an *elementary interval* (informally: a box with faces parallel to the sides of the unit cube) as any given decreasing function of the minimum of the c-values so far computed on that interval. The current attractiveness measure summarizes relevant history. Together with the current point in S, it forms the current state — which can be augmented and redefined as in Section 4.3. Roughly, the points in a box where c is computed form a *cluster*, less *ad hoc* than clusters used in other approaches to global optimization.

4.2.1 Preprocessor

To initialize the attractiveness measure, compute c for all points in D and then, for each elementary interval J in a class C defined below, take the minimum of those c-values for the points in the intersection of D and J. Assuming that the (t,m,s)-net D uses base b {and hence that D contains exactly b^m points}, we choose C so that its members have the following properties:

- each has volume b^{t-m}
- each (therefore) contains exactly b^t points of D {by definition of the net}
- the edges of the boxes parallel to (say) the i-th axis of the unit cube $[0,1)^s$ all have length $b^{d(i)}$ and partition that axis (each corresponding interval closed on the left and open on the right)
- $d(i)$ is a nonpositive integer
- $d(1) + \ldots + d(s) = t - m$

These properties define a superset of C, holding for *all* solutions to the preceding equation. This is where the net gets its power, at least when t is small and m is large. That equation generally does not uniquely specify the $d(j)$'s. To define the partition C of $[0,1)^s$, we fix $\{d(1), \ldots, d(s)\}$ — thus making the boxes disjoint. Perhaps it is good to take the $d(j)$'s as nearly equal as possible, getting a partition into congruent cubes if s divides $t - m$. In any case, the i-th axis of the unit cube is partitioned into $b^{-d(i)}$ equal-sized pieces. Thus, it is tempting to choose $|d(i)|$ as an increasing function of the *a priori* importance of variable i — getting a partition into congruent non-cubic boxes. The properties above make it easy to enumerate the b^{m-t} boxes in the partition and to generate a variate uniformly from any given box by rescaling depending only on the s fixed values $d(1), \ldots, d(s)$. Likewise, given any point in the unit cube, it is easy to find the box in the partition which contains it. These last three operations can be (much) harder to execute if the unit cube is partitioned into boxes from the net which are not congruent. If parallel iid runs are to be made and the processors communicate (during preprocessing only), then we can partition C across the processors and broadcast the results of evaluating c on C to all processors. If the processors do not communicate, then perhaps the randomly permuted nets of Owen (1995), proposed with a different setting in mind, could be used; intuitively, a deterministic assignment of (equal-size) nets to processors such that the overall set of points is well distributed seems preferable, but we do not know a *good* way. One way extracts disjoint (t,m,s)-nets with m fixed (and $m > t$), from (t,s) sequences.

4.2.2 Provisional move mechanism

Normalize (implicitly) the attractiveness measure across all elementary intervals (boxes) in C to get a probability mass function. Next, generate a variate with that pmf (see Section 4.3) and then, given the corresponding box, generate a point z uniformly from it. Optionally, look ahead: initiate a search for stationary point of c {projecting the gradient back to the unit cube if currently at its boundary} starting from z and stopping after a finite number of iterations. Repeat this whole process a given number of times to generate $F(x,k)$, invoking **SEARCH**. Right after accepting a move, update the attractiveness measure (recursively). And so on. This optional step assumes that the objective function is piecewise smooth, a reasonable restriction. In line with our remarks in Section 2.1 about look-ahead, the effort put into local search from a given point z ought to depend on the height of z relative to that of the current state and relative to that of the best point scanned in the box containing z.

4.3   Comments

The convergence of the algorithm (in the sense of properties (i) and (i$'$) of Section 6) follows as a special case from a theorem of Bélisle (1992), even though he does not mention remembering history. Thus, unlike the method of localization of search (e.g., see Niederreiter (1992), pp. 151-152), the scheme above has the global convergence properties listed in Section 6. The price is that, with low probability, substantial time is wasted exploring uninteresting regions of S. Such wasted time is likely to be orders of magnitude larger with crude search (e.g., see Niederreiter (1992), p. 150, where it is noted that crude search generally is impractical). Our algorithm is *adaptive*: (after the preprocessing phase) the next point at which the objective function is evaluated depends on the history of the process (as all sensible algorithms do).

There are many variations to this procedure. Some can be used to bring the provisional algorithm above into final form. For example, it is probably worth inhibiting visits to recently-visited boxes via a penalty function on a state space that remembers the recent history of the sample path projected to the naive state space S; likewise, for boxes where the gradient-based subroutine always seems to be converging to the same point regardless of where it is initiated. Another modification lets the number of iterations of the gradient-based subroutine be an increasing function of the move number in the routine calling it. With a certain probability, strictly between zero and one, if that subroutine was previously used in the current box, it can be initiated from where it previously left off. Generally, this makes the probability that the proposed move is strictly uphill significantly less than one. On the unit cube, crossover always produces feasible solutions.

To generate efficiently the variate that selects the box in the partition to examine next, the algorithms of Fox (1990), Fox and Young (1991), and Matias, Vitter, and Ni (1993) can be used. They exploit the fact that the respective (unnormalized) attractiveness measures at successive moves differ on at most a few boxes, though — unless good lower bounds on the objective function are available for every box — there is no convenient majorizing vector for a pure rejection method. (The algorithm of Matias, Vitter, and Ni has the lowest computational complexity, but it is not clear how to program a practical implementation.) The algorithms of Fox (1990) and Fox and Young (1991) are efficient if and only if on most boxes the (unnormalized) current attractiveness measure equals the (unnormalized) initial attractiveness measure. (This would hold, for example, in queueing networks in heavy traffic when the attractiveness measure of a server is its service rate when busy and zero when idle; likewise, in reliability estimation when the attractiveness measure of a component is its failure rate when working and its repair rate when down, if failure rate $\ll$ repair rate.) In our setting, no matter what box-selection method is used, the attractiveness measure of any given box should be updated only when the lowest objective-function value observed in that box changes significantly — by more than a specified amount. In any case, it should never go below a user-selected positive threshold. This keeps the tentative-move density bounded away from zero, required to apply Bélisle's convergence theorem. It also guarantees that, with the following variant of our algorithm (as well as the original version), the best solution seen so far is asymptotically in S^* almost surely: wherever our algorithm calls for a uniform random number in a given box, use the next (rescaled) point from a (t,s)-sequence dedicated to that box instead. Programs to implement Niederreiter's (t,s)-sequences are available from the *Collected Algorithms of the ACM* (see Bratley, Fox, and Niederreiter (1994)).

Even if the number of stationary points is not large, a nontrivial global-optimization algorithm such as the one above is needed when the global minima are in small basins of attraction to gradient search that looks for local stationary points. Our procedure can be applied hierarchically by recursively restarting it (a limited number of times) on boxes that appear to have multiple deep basins.

5   Noisy objective functions

There is a large literature on sampling plans when the observations are noisy. Finding a scheme that converges in some reasonable sense is not hard. The trick is to show that a *heuristically-attractive* scheme converges under reasonable conditions. One approach, the bandit setup, ignores possible structure of the set S and the objective function c (and thus may be suitable if the elements of S were unrelated). Roughly, that setup has $N(x) = S$ (including x itself) for all x in S and each move can depend on the entire history of the process. So far, bandit analysis (Yakowitz and Lowe (1991) or Lai and Yakowitz (1995) for example) looks more refined than annealing analysis. Perhaps a synthesis is possible. As with noiseless observations, simulated annealing makes sense when the state space has structure exploitable by clever selection of the dynamic neighborhoods $F(x,k)$.

Our approach has nothing in common with the papers that study global optimization over a subset (sometimes unbounded) of Euclidean space by adding successively-more damped Gaussian noise to a stochastic gradient algorithm. That too has been called (continuous) simulated annealing. Very loosely, it does have that flavor even though it accepts all moves. However, we see no mathematical link.

Up to Section 5.4.4, regardless of the cardinality of S, we assume that

- $c(x)$ is a known function of $d(x)$, say, and that $d(x)$ is the expectation of a random variable $Z(x)$

- c is unknown

- when at x, iid copies of $Z(x)$ can be generated (say, by simulation)

- a consistent estimator of $d(x)$ is available, at least when x is visited infinitely often.

The state space is expanded to keep track of enough information to update d. To have the corresponding process with hypothetically-known c live on the same state space, record the same information but ignore it. Sometimes, to avoid being overly fussy, we do not distinguish carefully $c(x)$ and $d(x)$.

5.1   Straightforward approach

No matter what the cardinality of S is, a natural scheme bases its current move on the current estimates of $\{c(x): x \in S\}$ — pretending that these estimates are the respective true values. Assuming that the estimator of $c(x)$ is consistent for all x in a finite set T say, for any positive τ all estimates of $c(x)$ for x in T will be within a band of width τ around the respective true values after some finite time N almost surely. However, N is not a stopping time and, without restricting the range of c to a finite set, the process never couples (exactly) to one in which c is known.

5.2   When S is finite

5.2.1   Previous work

Gelfand and Mitter (1989) and Yan and Mukai (1992) propose convergent search schemes for noisy measurements of c when S is finite, without restricting c to a finite set.

THE GELFAND-MITTER PAPER

Roughly, Gelfand and Mitter assume that

- the tentative-move probabilities are independent of objective-function values

- $c(x) - c(y)$ is measured at time k only up to an additive Gaussian noise $W(k)$ independent of x and y

- the variance of $W(k)$ has lower order than the temperature $t(k)$ at simulated time k, for large k.

While their first condition may be reasonable in some special situations, generally it goes counter to our conception of smart simulated annealing. It is unclear how to check their second two conditions. Normally, one expects that in practice the error in measuring $c(x) - c(y)$ depends on x and y. But because some states may be visited only rarely, it is then hard to get a bound on the error variance that is valid uniformly over S, unless perhaps when all unknown parameters of c can be estimated at each state. The third Gelfand-Mitter condition guarantees that, asymptotically, the respective acceptance probabilities with and without noise differ negligibly.

The algorithm of Gelfand and Mitter *is* (implicitly) the straightforward approach. To see this, observe that $c(x) - c(y)$ + noise at move k is the estimated difference of $c(x)$ and $c(y)$ based on all observations of $Z(x)$ and $Z(y)$ up to move k. In turn, that estimated difference is the numerator in the exponent of the acceptance test in **SLOW** at move k. The rest of **SLOW** (and hence **QUICKER**) is unaffected. To analyze the resulting scheme, Gelfand and Mitter assume that the (Gaussian) noise at move k has mean zero and known variance depending on k and then compute the acceptance probability by integrating with respect to the noise distribution. The assumptions they use probably indicate that analyzing the straightforward scheme without them is hard — at least without assuming something else.

THE YAN-MUKAI PAPER

Yan and Mukai note that their scheme has no clear link to simulated annealing. It does not estimate $c(x)$ via averaging copies of $Z(x)$. Every time it needs an estimate of $c(x)$ it uses a fresh copy of $Z(x)$, ignoring previous copies. Even if the proposed move is to a downhill state, accepting it becomes increasingly hard. The number of copies of $Z(x)$ generated at visit j to x grows without bound as j increases. No counterpart of **QUICKER** is possible. They assume that the tentative-move probabilities are symmetric, running counter to what we think a good search scheme should do.

5.2.2 Our proposal

This critique of the Gelfand-Mitter and Yan-Mukai papers motivates our use of the straightforward approach, with none of the Gelfand-Mitter hypotheses. The price is restriction of the range of c (but not d) to a known finite set H say. See Section 5.4.2 for a discussion of that restriction.

Without loss of generality (by rescaling, if needed), we can assume that the set H above is a set of consecutive integers and that $c(x)$ is the nearest integer to $d(x)$ — rounding up in case of ties. A naive scheme rounds the average of the $Z(x)$'s to the nearest integer, rounding up in case of ties. This works when $d(x)$ is not a half-integer, although it may not work well when $d(x)$ is near a half-integer. When $d(x)$ is a half-integer, the rounded average of the $Z(x)$'s will not converge, instead fluctuating between the integers neighboring $d(x)$. Fox and Heine (1995a) propose two estimators which, almost surely, converge to $c(x)$, assuming that $d(x)$ is estimated consistently. One is based on the law of the iterated logarithm.

Assume that we are using an estimator that a.s. converges to $c(x)$, not necessarily one of the Fox-Heine estimators. From some time N onwards, finite a.s., the estimates of $c(x)$ will be exactly correct for all x in S. Assuming that the convergence properties in Section 6 hold starting from all initial states and times with the correct $c(x)$'s, Fox and Heine (1995b) show that they still hold with the above scheme — even though N is *not* a stopping time. In fact, this follows from their much more general result about overriding strategies for Monte Carlo simulation and coupling after some finite time to a process in which no overrides occur.

The finiteness of N follows from the finiteness of S, the only place where the latter hypothesis is used. With noisy observations, it is reasonable to think of $|S|$ in the hundreds at most.

A variation of the straightforward scheme above modifies the objective function at x while the sample variance at x exceeds a given positive threshold, making it more attractive to move to states x that have a high sample variance in their estimator of $c(x)$. Since these estimators are consistent, their sample variances converge to zero almost surely. Thus, again, we couple to a process with known c from some finite time onwards. We can likewise stimulate visits to states that have been visited less than some positive threshold.

5.3   When S is infinite

5.3.1   The algorithm

In this case, we mimic the procedure of Section 4. We find this attractive, at least heuristically. Now gradients are estimated, say, by the likelihood-ratio method or by infinitesimal perturbation analysis (e.g., see Glasserman (1991) and L'Ecuyer [(1990), (1991), (1992)]) and stationary points of d are estimated by stochastic approximation — generally, by some variant of Robbins-Munro. As before, we stop stochastic approximation short of convergence — bounding the number of iterations. If it terminates at a point v in a box V, say, belonging to the class C defined in Section 4, we may choose to let the estimator of $d(v)$ depend not only on (copies of) $Z(v)$ but also, via likelihood ratios (when mutual absolute continuity and other regularity conditions hold), on all copies of $Z(w)$ for all w in the intersection of V and D. Perhaps $Z(w)$, multiplied by the appropriate likelihood ratio, should be weighted by some decreasing function of the distance between w and v, because one suspects that the corresponding variance is an increasing function of that distance.

5.3.2 Its convergence

We try to pattern a convergence analysis after the one we use for finite S. To start, assume that all minimizers of d occur at stationary points and (hardly without loss of generality) that the number of its stationary points is finite. The first condition can be assured by imposing a penalty function which is large in a thin layer abutting the faces of the unit cube. Since the number of boxes used in our partition of the unit cube is also finite, the number of points x where we need a consistent estimator of $d(x)$ is finite. With our algorithm, it seems likely that consistency at these points can be shown; however, we have not worked out the details. Introducing the set H as above, all induced estimates of

$$\{c(x): x \text{ is a stationary point of } d\}$$

are then exactly correct from some finite time N onwards almost surely. Assume that the updates to the attractiveness measure are affected only by updates of estimates of the respective objective-function values at stationary points. Now convergence of the scheme in Section 4 from any starting state and time implies convergence of the algorithm in Section 5.3.1 by the main result in Fox and Heine (1995b). The reduction above to considering consistency at only a finite number of points is crucial, because the results in Fox and Heine [(1995a), (1995b)] would not apply otherwise.

5.3.3 Convergence rates

Recently, Yakowitz (1993) gave a globally-convergent extension of Kiefer-Wolfowitz stochastic approximation — preserving the $n^{1/3}$ convergence rate of the Cesàro average. Thus, it is an alternative to our proposal in Section 5.3.1. Because of the forced choices it makes on a deterministic subsequence of moves, that extension cannot converge in probability except in trivial cases. As Sid Yakowitz (personal communication) points out, a cure randomly chooses one element from the integers between each successive pair in that subsequence for the forced choices. The Robbins-Munro stochastic-approximation subroutine of our proposal has the (better) convergence rate $n^{1/2}$ but finding the overall convergence rate for our scheme (relative to the pruned chain) is an open problem. The convergence rate for Robbins-Munro stated above assumes that gradient estimators are unbiased. They are with the likelihood-ratio method or infinitesimal-perturbation analysis but not with finite differences. Likelihood ratios can be used to transform a gradient estimate at one point to a gradient estimate at another. If the data come from a real system rather than a simulation, then only finite differences are available.

Sometimes finite differences can be attractive even if $Z(x)$ and $Z(y)$ are generated by simulation, perhaps using common random numbers. However, one must then ask whether in seeking a (local) optimum — especially if high precision is not required (except perhaps at the estimated *global* optimum) — stochastic approximation works better if we put y at a significant step from x rather than taking y close to x in an attempt to estimate the gradient; in that situation, Kiefer-Wolfowitz may beat Robbins-Munro. Generally, the closer y is to x, the more copies of $Z(x)$ and $Z(y)$ are needed in the finite-difference estimator to attenuate noise; see L'Ecuyer and Perron (1994) and the references cited there for more precise statements. It may be better to spend that effort generating some observations at other points. Making that notion precise is, as far as we know, an open problem.

5.4   Comments

5.4.1   Consistent estimators

For S finite, suppose that each state x is visited infinitely often. See Fox and Heine (1994a) for ways to assure this when c is dynamically estimated. When S is infinite, suppose that for every positive ε a ball of radius ε centered at x is visited infinitely often. The algorithm of Section 5.3.1 assures this, when x is a stationary point of d. Information about the sample path off x or about copies of $Z(y)$ observed for y not equal to x *prior* to time t generally biases any $Z(x)$ observed *before* t as an estimator of $d(x)$. One way to get a consistent estimator of $d(x)$ ignores such information and averages $Z(x)$ over all visits to x. Otherwise, even if consistency holds, proving it probably requires a martingale argument that considers a sum with j-th summand corresponding to the j-th visit to x. That summand is the average of the copies of $Z(x)$ for that visit minus the conditional expectation of that average given the σ-field $\mathcal{F}_j$ generated by everything up to the j-th visit. Under weak moment conditions, consistency of the natural estimator of $d(x)$ follows from a standard martingale convergence theorem (e.g., see Chung (1974), Theorem 9.4.4). In turn, this implies that the induced estimators of $c(x)$ of Fox and Heine (1995a) are exactly correct from some finite point onwards, almost surely. There are settings where we must or would like to take account of the additional information in the filtration $\{\mathcal{F}_j\}$ above. This occurs when the function c depends on a set of parameters much smaller than $|S|$. It also occurs when the number of copies of $Z(x)$ generated at the current visit to x depends on everything observed up to that point. Some restriction on that number is generally needed; a counterpart of the algorithm **GENERATE** in Section 5.4.3 may be appropriate, especially if a confidence interval for the minimal $c(x)$ is desired. There is no requirement that $Z(x)$ and $Z(y)$ be independent when $x \neq y$; common random numbers can (and often should) be used. Remarkably, all our results require is that the estimator of $d(x)$ be consistent for a suitable finite set T. When S is finite, we take T equal S. In the setting of Section 5.3.2, we take T equal to the set of stationary points of d. A "visit" to a stationary point x of d must (somehow) be defined dynamically in terms of a visit of to a ball of suitably-decreasing radius centered at x.

5.4.2   The finite-range assumption

The reason for introducing the function d (which we think of as being measured to infinite precision) near the beginning of Section 5 is to avoid the assumption of finite-precision arithmetic in stating the problem setup. Such an assumption would conflict, for example, with use of uniform random variables. The referee rightly states that, from a practical standpoint, generally any algorithm pathology exhibited on the continuum will show up in a finite-precision analog of that algorithm. We restrict c to a finite set only because the Fox-Heine proofs require it. What the "minimal" restrictions are we do not know. The assumption that the range of c is a finite set H is key to coupling to a process with c known after some finite time. This set can consist, for example, of a subset of all computer-representable numbers, in which case it merely says that we are content to measure c to finite precision. It is actually more realistic than assuming $c(x)$ is simply in the reals, discussed further in the remarks in Section 5.4.3, and that $Z(x)$ is measured to infinite precision. From the viewpoint of traditional mathematics, however, it is somewhat unsatisfying. Fox and Heine (1995b) also consider an alternative to the finite-range condition, relaxing it and instead imposing a condition roughly analogous to the third Gelfand-Mitter condition and, like the latter, generally hard to check.

5.4.3   Experimental design

Our setup is broad. It includes cases where x indexes simulation strategies in a set S and $c(x)$ is the variance of the limiting distribution of a work-normalized estimator, in the sense of Fox and Glynn (1990) for example, of a parameter α of interest. Suppose that when using simulation strategy x the work-normalized estimator converges weakly to a normal distribution with variance $c(x)$ and the convergence properties of Section 6 hold, extended to noisy observations. For this subsection only, we generate the $Z(x)$'s as follows:

> Let $S(n)$ be the set of estimated optimal states at move n in the pruned chain Y, i.e., those for which the estimated $c(x)$ is minimal. Pick ε as a small positive number and k as a positive integer, possibly one. Let $Z^*(Y(j))$ be the set of all (iid) copies of $Z(Y(j))$ generated at move j of the chain Y. Let $A(n)$ be the union of $Z^*(Y(j))$ for $j = 1, ..., n$. Let $B(n)$ [resp., $C(n)$] be the sum of those $Z(x)$'s in the subset of $A(n)$ corresponding to $x \notin S(n)$ [resp., $x \in S(n)$]. The algorithm below makes $B(n)$ asymptotically negligible relative to $C(n)$.

GENERATE$(n,\varepsilon,k,Y(n),S(n),B(n-1),|A(n-1)|)$

> *If $Y(n) \in S(n)$ or if $B(n-1) < |A(n-1)|^{1/2-\varepsilon}$, then*
> *generate k iid copies of $Z(Y(n))$.*

End.

Here is the intuition for **GENERATE**. If $Y(n) \in S(n)$, then **GENERATE** makes $B(n)$ relatively smaller — good in principle. If $Y(n) \notin S(n)$ but the other condition holds, then $B(n)$ is "acceptably" small and so k more observations of $Z(Y(n))$ can be taken without making $B(n)$ "too" big.

Now let $N^*(t)$ be the number of moves in the chain Y generated up to *computer* time t. Set, when $N^*(t) > 0$,

$$N(t) = |A(N^*(t))|.$$

Thus, $N(t)$ is the number of observations up to *computer* time t, *not* counting (the at most k) observations generated after $N*(t)$.

We shall argue that

$$t^{-1/2}B(N(t)) \to 0$$

in probability as $t \to \infty$. This will follow from

(i) $S(n)$ almost surely contains exactly the optimal states for all large n (assuming the conditions of Fox and Heine [(1995a), (1995b)]).

(ii) $P\{Y(n) \in S(n)\} \to 1$ (from Fox and Heine (1995b), assuming the counterpart holds for noiseless observations).

(iii) $|A(n)| \to \infty$ almost surely (from (ii)).

(iv) $|A(n)|^{-1/2}B(n) \to 0$ almost surely (from **GENERATE**, (i), and (iii)).

(v) $|A(N(t))|$ has order $N(t)$ (from the conditions in the next paragraph).

(vi) $N(t)^{-1/2}B(N(t)) \to 0$ almost surely (from (iv) and (v)).

(vii) $N(t)$ has order t almost surely (from property (ii′) of Section 6 and the conditions in the next paragraph).

(viii) $t^{-1/2}B(N(t)) \to 0$ almost surely (from (vi) and (vii)).

The factor $t^{-1/2}$ corresponds to work normalization. It now follows from the converging-together lemma (Billingsley (1968), p. 25, for example) that the work-normalized estimator

$$t^{1/2}\{[B(N(t)) + C(N(t))]/N(t) \ - \ \alpha\} \ \Rightarrow \ \mathcal{N}(0, \min\{c(x)\colon x \in S\}),$$

where $c(x) = ET(x)\cdot\text{Var } Z(x)$ and $T(x)$ is the (random) *computer* time to generate a copy of $Z(x)$. Thus, at least when S is finite, what might have been a hard convergence argument gets easily piggybacked onto our results for simulated annealing.

The additional conditions, referred to above, assumed in this subsection are that uniformly in x and the sample points ω

(i) $EZ(x) = \alpha$

(ii) $0 < k_1 \le Z(x,\omega) \le k_2 < \infty$

(iii) $0 < k_3 \leq T(x,\omega) \leq k_4 < \infty.$

In other settings, we may define $c(x)$ as $EZ(x)$. In that case, the confidence interval following from the limiting distribution above generally is tighter (asymptotically) than one produced by multiple comparisons and the Bonferroni inequality — discussed in Bratley, Fox, and Schrage (1987), Section 3.2.2, for example. The Bonferroni-based interval explicitly accounts for randomness in the estimated

$$\arg\min \{c(x): x \in S\}$$

and so is conservative.

Relative to the bandit setup, at each move the arm x we pull (sample) is random and suboptimal arms are pulled on a random thin subsequence. The prescription of that subsequence is more complex with our adaptation of simulated annealing (via **GENERATE**) but, we think, heuristically more attractive when S has structure — in particular when its elements are (logically) related. With such structure, sensible neighborhoods can be specified. There is no analog of neighborhoods in the bandit setup.

Our approach can be used in a two-level hierarchy. At its top, one tries to minimize $c_1(x)$ for $x \in S_1$, say, where $c_1(x)$ is estimated using simulation strategies in $S_2(x)$, say. At its bottom, one tries to maximize for each $x \in S_1$ the simulation efficiency $c_2(y(x))$ for $y(x) \in S_2(x)$.

REMARKS

The upper bound in condition (ii) follows from the fact that computers have only a finite number of representable numbers (partly motivating our finite-range assumption in Section 5.4.2) and, more fundamentally, from the fact that virtually any model which literally allows unbounded output random variables is unrealistic. In addition, when the model must be handled numerically (as here) using unbounded output random variables makes statistical anomalies more likely. This is part of the attraction of *discrete-time conversion* (Fox (1993b), comment 7, and Fox and Glynn [(1986), (1990)]), in addition to variance reduction. By itself, converting to discrete time does not in general assure that the upper bound in condition (ii) holds. If $Z(x) \geq k_5 > -\infty$, the data can be rescaled to satisfy the lower bound in (ii). Thus, condition (ii) is reasonable. It implies that $d(x)$ and hence $c(x)$ have bounded ranges; likewise, for their respective estimators. The upper bound in condition (iii) is needed to make the computer program well defined, while the lower bound is a practical necessity. Generally, the upper bound in (iii) is equivalent to requiring that the number of transitions to generate $Z(x,\omega)$ be uniformly bounded in x and ω. Together with discrete-time conversion, this typically assures the upper bound in (ii). From a mathematical viewpoint, these conditions probably can be weakened at the expense of a much more sophisticated argument.

We can tailor a search scheme if there is a convenient formula for

$$\arg \min \ \{c(x)\colon x \in S\}$$

when all auxiliary parameters are known. The problem is nontrivial if not all of those parameters can be observed at every x or when the variance of the estimators of those parameters depends on x. Fox and Heine (1995a) illustrate this situation in a case where $\arg \min \ \{c(x)\colon x \in S\}$ is known to be the nearest positive integer to a given real-valued expression involving the auxiliary parameters, rounding up in case of ties.

5.4.4   Hybrid state spaces

Another indication of the scope of our setup follows. Suppose that

$$c(x) = \min \ \{f(x,y)\colon y \in B\}.$$

A trite example is mixed-integer programs where B and S correspond to the continuous and discrete variables, respectively. A more interesting example occurs when finding $c(x)$ is a nontrivial global-optimization problem in its own right. In such cases, we can consider a two-level hierarchy of simulated annealing. At its top is our original problem:

$$\min \ \{c(x)\colon x \in S\}.$$

At its bottom is a family of subproblems: $\min \ \{f(x,y)\colon y \in B\}$ for $x \in S$. Temporarily, we assume that $f(x,y)$ is evaluated without noise. At the k-th (top-level) visit to x, restart (bottom-level) simulated annealing to estimate $c(x)$, running it for $g(x,k)$ accepted moves where g is an increasing function of k. We estimate $c(x)$ by the smallest of the $f(x,y)$-values seen so far. Suppose that the range of $f(x,y)$ is a finite set. Under weak conditions, the set on which $f(x,y)$ is minimized has positive measure and then our estimator of $c(x)$ is exactly correct from some finite time onwards almost surely. Thus, Sections 5.2.2 and 5.3 apply. When $f(x,y)$ is estimated with noise, we have a subproblem of the form considered at the outset of Section 5.

6   Convergence without noise

Let S^* be the set on which c^* is minimized. The set S' below is closely related to S^* and, with our neighborhood construction which prevents all local minimizers from having only uphill neighbors, equals S^*. Four convergence properties follow:

(i) $P\{X(k) \in S^* \} \to 1$

(ii) $[I\{X(1) \in S^* \} + \cdots + I\{X(n) \in S^*\}]/n \to 1$ a.s.

(i') $P\{Y(k) \in S'\} \to 1$

(ii') $[I\{Y(1) \in S'\} + \cdots + I\{Y(n) \in S'\}]/n \to 1$ a.s.

The primes signal the pruned chain. Heine [(1994), (1995)] gives weak conditions on the cooling schedule with corresponding selection of the parameter j in **QUICKER** which imply that these four properties hold. From properties (i) and (i') respectively, expected-value counterparts of (ii) and (ii') follow. The expected-value counterpart of property (ii) may seem to say more than it does, in view of example 2 in Section 3.1. Properties (i') and (ii') show that convergence is not merely an artifact of increasingly long self-loop sequences of rejected proposed moves. The point of **QUICKER** is to generate these sequences implicitly, not explicitly.

    With our neighborhood construction, property (i) follows from Bélisle (1992) even if **NEXT** is used without **QUICKER**, just setting L equal $k + 1$. The resulting adaptive cooling generally is much slower than the nominal schedule. Because in this case the chains X and Y coincide, property (i') follows.

7   Conclusions

    We contrast folklore and facts, discuss the relevance of convergence and of the infinite-state case, and recap directions for research.

7.1   The power of sophisticated simulated annealing

    We have shown that simulated annealing is a much more flexible algorithm than naive implementations have led many to believe. Some of our proposals are clearly worthwhile, while others are heuristic and speculative. Contrary to what partisans of other techniques often assert or tacitly imply, virtually every optimization method can be combined profitably with simulated annealing. Thus, tabu penalties (in part, to inhibit short-run oscillation) and the most important feature of genetic algorithms, having multiple solutions in each state, can be imbedded in simulated annealing proper by expanding the state space. This Markov-chain setting, not previously done by others (even for tabu search or genetic algorithms separately), is vital for unification and a complete analysis that is not *ad hoc*. In contrast, the tabu-search literature has no such setting and — partly in consequence — there are few theorems; for the same reason, in the genetic-algorithm literature, all the theorems and their proofs are *ad hoc*. Other techniques can serve as a front end to simulated annealing proper. Such hybrids are illustrated by our use of low-discrepancy point sets in a preprocessor. This appears to be the first contact between simulated annealing and Quasi-Monte Carlo. Even with noisy observations, all the preprocessors indicated in Section 2.3 apply — except branch and cut.

7.2   The relevance of convergence

    Though the relevance of the convergence results in Section 6 is arguable when c is known, they can be extended to noisy observations via the results in Fox and Heine [(1995a), (1995b)] as indicated in Sections 5.2.2 and 5.3.2. When $c(x)$ must be estimated by (Monte Carlo) simulation at x and greater accuracy is desired at optimal and near-optimal states, the practical relevance of this Fox-Heine extension is clearcut. On the other hand, if the only parameters on which c depends can be estimated at every state, relevance is again arguable. The four properties in Section 6 seem to say that the search is "attracted" to S^*, but it is hard to pin down the link (if any) with the first hitting time of S^*. All we can say for now is that cooling more slowly than the canonical schedule and allowing every state to be potentially in $F(x,k)$, as we do above, eliminates the pathologies of Section 3.1, implies the speedup of Section 3.2, and is (more than) enough for the four properties in Section 6. For the huge state spaces often encountered in discrete problems with noiseless observations, the chain is unlikely to get anywhere near equilibrium in reasonable computing time. This does not affect the appeal of sophisticated simulated annealing. Chiang and Chow (1994) and Mazza (1994) study first hitting times, relative to *simulated* time, in a continuous-time version of simulated annealing. The relation to our pruned discrete-time chain is unclear. Fox (1995), remark 2, indicates how to implement the continuous-time approach and points out that *computer* time is (again) the right yardstick. A continuous-time analog of example 2 holds.

Properties (ii) and (ii′) reflect what is actually observed on a single (long) run of simulated annealing. With massively-parallel runs, properties (i) and (i′) become directly observable. These latter two properties are essential for the experiment-design application in Section 5.4.3.

7.3   The relevance of the infinite-state case

It might be argued that the case S infinite is a mathematical fiction, analogously to our rationale in Section 5.4.2 for the assumption in Sections 5.2.2 and 5.3.2 that the range of c is a finite set. However, unlike the latter situation we believe that considering S infinite leads to important practical insights. It makes gradient-based subroutines feasible. Our algorithm in Sections 4.2 and 5.1 applies to a general class of problems, with no problem-specific subroutines other than those to compute or estimate c and its gradient. This is an incentive to write programs and, with few exceptions, contrasts with the *ad hoc* way discrete problems are generally attacked by probabilistic search.

7.4   Directions for research

We have identified three major directions for research, namely to find reasonable conditions (if possible) which imply

- an expected-value version of **QUICKER** preserves the convergence properties in Section 6.

- a linear speedup of simulated annealing with parallel processing for general state spaces.

- convergence of simulated annealing for general state spaces with noisy observations and, perhaps less important, without restricting the range of c to a finite set.

For the first direction above, Heine [unpublished] has solved the problem for properties (ii) and (ii') which relate to almost-sure Cesàro convergence. And for property (i), we can round up the expectation of the geometric variate generated in **QUICKER** with j equal 1 — when our neighborhood construction is used — by Bélisle (1992). The second direction points to linear speedup for general state spaces. We know from Section 3.2 that, for finite S, a satisfactory treatment exists. Promise of linear speedup adds to the attractiveness of Monte Carlo methods, especially relative to deterministic alternatives. The latter have hardly any chance for guaranteed linear speedup, because the computations generally cannot be decoupled across processors — other than by a heuristic partitioning of the original problem into disjoint subproblems. The third direction points to global optimization in simulation, which much of the literature skirts. When S is finite, we know from Section 5.2.2 that simulated annealing provides a sturdy framework. Extension to general state spaces is sketched in Section 5.3.2, but fleshing out our proof skeleton by proving consistency of the estimators at the stationary points of d is an outstanding problem.

Finding convergence rates corresponding to (i') and (ii') and getting reasonable, easily-computed bounds on expected first hitting time to S^* *relative to the chain Y* are open problems. The discovery of **QUICKER** makes them important. It also makes analysis of the corresponding X-chain properties generally irrelevant.

Acknowledgement

We thank the referee, George Heine, and Pierre L'Ecuyer for helpful comments. The invitation to the conference on Monte Carlo and Quasi-Monte Carlo Methods in Scientific Computing, held in Las Vegas in June 1994, stimulated us to think of a connection, examined here, between Quasi-Monte Carlo and simulated annealing. We recall with pleasure that meeting and thank its organizers for their hospitality.

References

R. K. AHUJA, T. L. MAGNANTI, and J. B. ORLIN (1993). *Network Flows: Theory, Algorithms, and Applications*, Prentice-Hall, Englewood, Cliffs, New Jersey.

C. J. P. BÉLISLE (1992) Convergence Theorems for a Class of Simulated Annealing Algorithms on $\mathfrak{R}^d$. *Journal of Applied Probability* **29**, 885-895.

P. BILLINGSLEY (1968). *Convergence of Probability Measures*, Wiley, New York.

P. BRATLEY, B. L. FOX and H. NIEDERREITER (1992). Implementation and Tests of Low-Discrepancy Sequences, *ACM Transactions on Modeling and Computer Simulation* **2**, 195-213.

P. BRATLEY, B. L. FOX, and H. NIEDERREITER (1994). Algorithm 738: Programs to Generate Niederreiter's Low-discrepancy Sequences, *ACM Transactions on Mathematical Software* **20**, 494-495.

P. BRATLEY, B. L. FOX, and L. E. SCHRAGE (1987). *A Guide to Simulation*, Springer-Verlag, New York, second edition.

T.-S. CHIANG and Y. CHOW (1994). The Asymptotic Behavior of Simulated Annealing Processes with Absorption, *SIAM Journal on Control and Optimization* **32**, 1247-1265.

K. L. CHUNG (1974). *A Course in Probability Theory*, Academic Press, New York, second edition.

T. A. FEO and M. G. C. RESENDE (1995). Greedy Randomized Adaptive Search Procedures, *Journal of Global Optimization* **6**, 109-133.

B. L. FOX (1990). Generating Markov-chain Transitions Quickly: I, *ORSA Journal on Computing* **2**, 126-135.

B. L. FOX (1993a). Integrating and Accelerating Tabu Search, Simulated Annealing, and Genetic Algorithms, *Annals of Operations Research* **41**, 47-67.

B. L. FOX (1993b). Shortening Future-Event Lists, *ORSA Journal on Computing* **5**, 147-150.

B. L. FOX (1994). Random Restarting Versus Simulated Annealing, *Computers and Mathematics with Applications* **27**, No. 6, 33-35.

B. L. FOX (1995). Faster Simulated Annealing, *SIAM Journal on Optimization*, to appear.

B. L. FOX and P. W. GLYNN (1986). Discrete-time Conversion for Simulating Semi-Markov Processes, *Operations Research Letters* **5**, 191-196.

B. L. FOX and P. W. GLYNN (1990). Discrete-time Conversion for Simulating Finite-horizon Markov Processes, *SIAM Journal on Applied Mathematics* **50**, 1457-1473.

B. L. FOX and G. W. HEINE (1995a). Stable Estimators for Self-Adjusting Simulations, this volume.

B. L. FOX and G. W. HEINE (1995b). Probabilistic Search with Overrides, technical report, University of Colorado, Denver.

B. L. FOX and A. R. YOUNG (1991). Generating Markov-chain Transitions Quickly: II, *ORSA Journal on Computing* **2**, 3-11.

S. B. GELFAND and S. K. MITTER (1989). Simulated Annealing with Noisy or Imprecise Measurements, *Journal of Optimization Theory and Applications* **62**, 49-62.

P. GLASSERMAN (1989). *Gradient Estimation Via Perturbation Analysis*, Kluwer Academic Publishers, Norwalk, Massachusetts.

T. HANSEN, G. L. MULLEN, and H. NIEDERREITER (1993). Good Parameters for a Class of Node Sets in Quasi-Monte Carlo Integration, *Mathematics of Computation* **61**, 225-234.

G. W. HEINE (1994). Smart Simulated Annealing, PhD Dissertation, University of Colorado, Denver.

G. W. HEINE (1995). Convergence Properties of a Loop-Skipping Algorithm for Simulated Annealing, in preparation.

T. L. LAI and S. YAKOWITZ (1995). Machine Learning and Nonparametric Bandit Theory, *IEEE Transactions on Automatic Control*, to appear.

G. LARCHER, A. LAUß, H. NIEDERREITER, and W. Ch. SCHMID (1995). Optimal Polynomials for (t,m,s)-Nets and Numerical Integration of Multivariate Walsh Series, *SIAM Journal on Numerical Analysis*, to appear.

P. S. LAURSEN (1994). Can Parallel Branch and Bound Without Communication Be Effective?, *SIAM Journal on Optimization* **4**, 233-296.

P. L'ECUYER (1990). A Unified View of the IPA, SF, and LR Gradient Estimation Techniques, *Management Science* **36**, 1364-1383.

P. L'ECUYER (1991). An Overview of Derivative Estimation, *Proceedings of the 1991 Winter Simulation Conference*, 207-217.

P. L'ECUYER (1992). Convergence Rates for Steady-State Derivative Estimators, *Annals of Operations Research* **39**, 121-136.

P. L'ECUYER and G. PERRON (1994). On the Convergence Rates of IPA and FDC Derivative Estimators, *Operations Research* **42**, 643-656.

D. G. LUENBERGER (1984). *Linear and Nonlinear Programming*, Addison-Wesley, Reading, Massachusetts, second edition.

Y. MATIAS, J. S. VITTER, and W.-C. NI (1993). Dynamic Generation of Discrete Random Variates. *Proceedings of the Fourth ACM-SIAM Symposium on Discrete Algorithms*, 361-370.

C. MAZZA (1994). Asymptotic First Hitting-Time Distribution of Annealing Processes, *SIAM Journal on Control and Optimization* **32**, 1266-1288.

G. L. MULLEN, A. MAHALANABIS, and H. NIEDERREITER (1995). Tables of (t,m,s)-Net and (t,s)-Sequence Parameters, this volume.

G. L. NEMHAUSER and L. A. WOLSEY (1988). *Integer and Combinatorial Optimization*, Wiley, New York.

H. NIEDERREITER (1992). *Random Number Generation and Quasi-Monte Carlo Methods*, CBMS-NSF Regional Conference Series in Applied Mathematics #63, Society for Industrial and Applied Mathematics, Philadelphia.

A. B. OWEN (1995). Randomly Permuted (t,m,s)-Nets and (t,s)-Sequences, this volume.

H. E. ROMEIJN and R. L. SMITH (1994). Simulated Annealing for Constrained Optimization, *Journal of Global Optimization* **5**, 101-126.

R. SHONKWILER and E. VAN VLECK (1994). Parallel Speed-Up of Monte Carlo Methods for Global Optimization, *Journal of Complexity* **10**, 64-95.

S. YAKOWITZ (1993). A Globally Convergent Stochastic Approximation, *SIAM Journal on Control and Optimization* **31**, 30-40.

S. YAKOWITZ and W. LOWE (1991). Nonparametric Bandit Methods, *Annals of Operations Research* **28**, 297-312.

D. I. YAN and H. MUKAI (1990). Stochastic Discrete Optimization, *SIAM Journal on Control and Optimization* **30**, 594-612.

Two Approaches to the Initial Transient Problem

Peter W. Glynn[*]
Dept. of Operations Research
Stanford University
Stanford, CA 94309-4022
e-mail: glynn@leland.stanford.edu

Abstract

This paper describes two different approaches to dealing with the initial transient problem. In the first approach, the length of the "warm-up period" is determined by obtaining analytical estimates on the rate of convergence to stationarity. Specifically, we obtain an upper bound on the "second eigenvalue" of the transition matrix of a Markov chain, thereby providing one with a theoretical device that potentially can give estimates of the desired form. The second approach is data-driven, and involves using observed data from the simulation to determine an estimate of the "warm-up period". For the method we study, we are able to use a coupling argument to establish a number of important theoretical properties of the algorithm.

1 Introduction

In many applications settings, it is of interest to compute steady-state performance measures. To be specific, suppose that the system under consideration is described by a Markov process $X = (X(t) : t \geq 0)$ living on state space S. For a given $f : S \to \mathbb{R}$, the steady-state simulation problem is concerned with the estimation of the time-average limit α defined via the law of large numbers

$$\frac{1}{t} \int_0^t f(X(s)) ds \to \alpha \quad \mathrm{P}_x \text{ a.s.}$$

as $t \to \infty$ for all $x \in S$ (assuming such a limit exists), where

$$\mathrm{P}_x(\cdot) \triangleq P(\cdot | X(0) = x).$$

---

[*]This research was supported by the Army Research Office under Contract No. DAAL03-91-G-0319

Such laws of large numbers hold, in great generality, for Markov processes exhibiting some type of positive recurrence condition. In addition, for such processes, there typically exists a unique probability distribution π, known as the stationary distribution, such that

$$\alpha = \int_S f(x)\pi(dx).$$

Furthermore, π has the property that if $X(0)$ has distribution π, then X is a (strictly) stationary process. It follows that if X is initiated with distribution π, then the sample mean $\alpha(t)$ given by

$$\alpha(t) \triangleq \frac{1}{t}\int_0^t f(X(s))ds$$

is unbiased as an estimator of α.

Unfortunately, the distribution π is typically unknown and, consequently, it is generally impractical to generate $X(0)$ from π. As a result, bias is induced in $\alpha(t)$ and the initial segment of the simulation may be unrepresentative of steady-state behavior. This introduces certain complications into the estimation problem that do not arise in Monte Carlo environments in which unbiased estimators may easily be constructed. The "initial transient problem" focuses both on the effect of this initial bias, and on developing effective algorithms for mitigating the impact of this bias.

One means of attacking this problem is to note that the positive recurrent Markov processes (satisfying some sort of aperiodicity condition) typically exhibit "total variation convergence" to stationarity, by which we mean that

$$\|P(X_t \in \cdot) - P_\pi(X \in \cdot)\| \to 0$$

as $t \to \infty$, where $X_t \triangleq (X(t+s) : s \geq 0)$ is the "post-t" process, $P_\pi(\cdot)$ is the distribution under which X has initial distribution π, and $\|\cdot\|$ is the total variation norm defined by

$$\|\eta\| = \sup_A |\eta(A)|$$

for any signed measure η. Thus, if one can compute a time s for which

$$\|P(X_s \in \cdot) - P_\pi(X \in \cdot)\| < \epsilon,$$

we have an ϵ-guarantee that the post-s process is close to stationarity, and hence any data collected subsequent to s should have relatively low bias.

The remainder of this paper describes two different approaches to accomplishing this task. In section 2, we establish a new analytical bound for the total variation distance from stationarity that takes explicit advantage of the known transition structure of the system. By contrast, section 3 is concerned with developing a new method for identifying s that is purely data-driven, and takes no explicit advantage of the transition structure of the system being simulated.

2 Upper Bounds on Rates of Convergence to Stationarity

As indicated in section 1, we are concerned here with developing upper bounds on the rate of convergence to stationarity, as described via the total variation norm. For the remainder of this section, we shall assume that X is an aperiodic, irreducible, discrete-time Markov chain with finite state space (although one would expect appropriate analogs in both continuous time and general state space).

We start by noting that for $t \in \mathbb{Z}^+$,

$$
\begin{aligned}
\|\mathrm{P}(X_t \in \cdot) - \mathrm{P}_\pi(X \in \cdot)\| &= \|\mathrm{P}(X(t) \in \cdot) - \mathrm{P}_\pi(X \in \cdot)\| \\
&\leq \frac{1}{2} \max_x \sum_y |P_{xy}^t - \pi_y|,
\end{aligned}
$$

where $P = (P_{xy} : x, y \in S)$ is the transition matrix of X. Let π be the unique stationary distribution of X, and let Π be the matrix having all rows identical to π. Since $P\Pi = \Pi P = \Pi^2$, it is evident (via an inductive proof) that for $n \geq 1$,

$$
P^n - \Pi = (P - \Pi)^n.
$$

Clearly, the rate of convergence of P^n to Π is therefore related to the structure of the eigenvalues of $P - \Pi$. In particular, let $\lambda_1, \lambda_2, \ldots, \lambda_d$ be the distinct (complex-valued) eigenvalues of $P - \Pi$, with corresponding (complex-valued) eigenvectors $u_1, u_2, \ldots, u_d$. Note that

$$
\lim_{n \to \infty} \frac{1}{n} \log(\|\mathrm{P}(X_t \in \cdot) - \mathrm{P}_\pi(X \in \cdot)\|) \leq \log(\gamma) \tag{1}
$$

where $\gamma = \max(|\lambda_i| : 1 \leq i \leq d)$. Thus, a bound on γ yields a bound on the rate of convergence to stationarity.

In view of this, let (λ, u) be an eigenvalue-eigenvector pair corresponding to $P - \Pi$ so that

$$
P^n u - \Pi u = \lambda^n u,
$$

for $n \geq 1$. In other words,

$$
\mathrm{E}_x u(X_n) - \pi u = \lambda^n u(x) \tag{2}
$$

for $x \in S$. Assume $\lambda \neq 0$, and set

$$
M_n = \lambda^{-n} \left[u(X_n) - \pi u (1 - \lambda^n)(1 - \lambda)^{-1} \right].
$$

(Clearly, $|\lambda| < 1$ since $(P - \Pi)^n \to 0$, and thus $(1 - \lambda)^{-1}$ is finite.) We claim that $(M_n : n \geq 0)$ is a martingale with respect to $\mathcal{F}_n = \sigma(X_0, \ldots, X_n)$.

To verify this, note that since S is finite, M_n is integrable and adapted to $(\mathcal{F}_n : n \geq 0)$. Furthermore, (2) implies that

$$
\begin{aligned}
E_x(M_{n+1}|\mathcal{F}_n) &= \lambda^{-n-1}(\lambda u(X_n) + \pi u - \pi u(1 - \lambda^{n+1})(1 - \lambda)^{-1}) \\
&= \lambda^{-n}(u(X_n) - \pi u(1 - \lambda^n)(1 - \lambda)^{-1}) = M_n,
\end{aligned}
$$

so $(M_n : n \geq 0)$ is indeed a (complex-valued) martingale.

Let $T(x) = \inf\{n \geq 1 : X_n = x\}$ be the (first) hitting time of x. Set $D_j = M_j - M_{j-1}$ and observe that for $n \geq 1$,

$$
\begin{aligned}
E_x M_{T(x) \wedge n} &= E_x M_0 + \sum_{j=1}^{n} E_x D_j I(T(x) \geq j) \qquad (3) \\
&= E_x M_0 + \sum_{j=1}^{n} E_x I(T(x) \geq j) E_x(D_j | \mathcal{F}_{j-1}) \\
&= E_x M_0.
\end{aligned}
$$

(Thus, the optional sampling identity continues to hold despite the fact that $(M_n : n \geq 0)$ is not real valued). Note that $M_{T(x) \wedge n} \to M_{T(x)}$ a.s. (since $T(x)$ is finite valued). Furthermore, if $E_x |\lambda|^{-T(x)} < \infty$, then the Dominated Convergence Theorem, applied to (3), yields

$$
E_x M_{T(x)} = E_x M_0,
$$

so that

$$
E_x(\lambda^{-T(x)}(u(X_{T(x)}) - \pi u(1 - \lambda^{T(x)})(1 - \lambda)^{-1})) = u(x).
$$

But $u(X_{T(x)}) = u(x)$ and $E_x(1 - \lambda^{T(x)}) \neq 0$ (since $|\lambda|^{-T(x)} \geq |\lambda|^{-1} > 1$), from which it follows that if $E_x |\lambda|^{-T(x)} < \infty$,

$$
u(x) = \pi u(1 - \lambda)^{-1}.
$$

Thus, if $E_y |\lambda|^{-T(y)} < \infty$ for all $y \in S$, evidently we would obtain $u(y) = \pi u(1 - \lambda)^{-1}$ for all $y \in S$. This is a contradiction (as is easily seen by taking π of both sides). Hence, there exists $y \in S$ such that $E_y |\lambda|^{-T(y)} = +\infty$.

Let $\beta(y) = \sup\{|\lambda| : E_y |\lambda|^{-T(y)} < \infty\}$ be the radius of convergence of the probability generating function of $T(y)$. We have just shown above that there exists $y \in S$ such that $\beta(y) \leq |\lambda|^{-1}$. So $|\lambda|^{-1} \geq \min(\beta(y) : y \in S)$ or, equivalently,

$$
|\lambda| \leq \max(\beta(y)^{-1} : y \in S),
$$

yielding the bound

$$
\gamma \leq \max(\beta(y)^{-1} : y \in S).
$$

We can summarise the above discussion with the following theorem.

Theorem 1 *Let X be a finite state aperiodic irreducible discrete-time Markov chain. Then,*

$$\lim_{n\to\infty} \frac{1}{n} \log(\||P(X_t \in \cdot) - P_\pi(X \in \cdot)\||) \leq \log(\max_{y\in S} \beta(y)^{-1}),$$

where $\beta(y) = \sup\{ z : E_y z^{T(y)} < \infty \}$.

This result bounds the rate of convergence to stationarity, in terms of the rate at which the chain X returns to the various states of S. In certain settings, probability arguments can then be used to a priori dominate the $\beta(y)$'s.

The theorem above complements the many other results that have been developed in recent years to bound the rate of convergence to stationarity; see, for example, Diaconis and Stroock [3], Fill [4], and Meyn and Tweedie [8]. In certain highly structured models, these analytic tools turn out to be quite powerful, and the bounds obtained are relatively tight. However, in general, it is probably fair to say that for unstructured systems, the bounds are often quite loose and consequently of less practical value. In addition, a glance at (1) makes clear that a bound on γ does not necessarily provide a bound on the total variation distance between $P(X_t \in \cdot)$ and $P_\pi(X \in \cdot)$ (although some analytical tools give a bound also on the total variation distance).

Another criticism of the above approach is that for unbounded functions f (as often arise in engineering applications), a bound on the total variation distance does not translate into a bound on $|Ef(X(t)) - E_\pi f(X(0))|$, and hence the information obtained about the bias of $\alpha(t)$ is somewhat limited.

3   A Data-Driven Stationarity Detection Rule

In section 2, our concern was with describing upper bounds on the total variation rate of convergence that takes explicit account of the specific transition structure of the process. However, historically, it is fair to say that the most widely used methods for determining the time to stationarity have made such assessments based purely on the observed data obtained by simulating the system itself; see, for example, Conway [2] and Gafarian, Ancker, and Morisaku [5]. A principal difficulty with this approach is that very few such data-driven methods have come equipped with any theoretical guarantees; see, however, Asmussen, Glynn, and Thorisson [1] for some noteworthy exceptions.

Here, we analyze a data-driven rule first proposed in Glynn and Iglehart [6], and show that it enjoys some theoretically important properties. Our goal is to define a non-negative family of random variables ($T(t) : t \geq 0$) such that:

a $T(t) \leq t$ a.s.,

$$(4)$$

b $P(T(t) \in \cdot|X) = P(T(t) \in \cdot|X(s) : 0 \leq s \leq t)$ for $t \geq 0$,

c $||P(X_{T(t)} \in \cdot) - P_\pi(X \in \cdot)|| \to 0$ as $t \to \infty$,

d $(T(t) : t \geq 0)$ is a tight family of r.v.'s (under P).

For a simulation time horizon t, we then view $T(t)$ as the epoch at which the process X is in approximate stationarity. Condition b above states that $T(t)$ can be generated once X has been simulated to time t, whereas a forces $T(t)$ to be in the interval $[0, t]$. Condition c asserts that X is in approximate stationarity at time $T(t)$, whereas d rules out detection rules that throw out more and more data as $t \to \infty$ (e.g. $T(t) = t^{1/2}$).

The rule proposed in Glynn and Iglehart [6] is described by the following algorithm:

1. Simulate X to time t.

2. Generate a uniform r.v. U, independent of X.

3. Set $T(t) = \inf\{ s \geq 0 : X(s) = X(Ut) \}$.

Clearly $T(t) \leq t$ always holds. Set $Z(t) = X(Ut)$ and note that

$$P(Z(t) \in \cdot | X) = \pi_t(\cdot)$$

where $\pi_t(\cdot)$ is the empirical distribution of X defined by

$$\pi_t(\cdot) = \frac{1}{t} \int_0^t I(X(s) \in \cdot) ds.$$

Hence, (4)b also holds. To establish c, we need to restrict the class of processes X under consideration. Specifically, suppose that X is an irreducible, positive recurrent, continuous-time Markov chain living on a finite or countably infinite state space. Then, the law of large numbers for such processes guarantees that for each $x \in S$,

$$\pi_t(x) \triangleq \frac{1}{t} \int_0^t I(X(s) = x) ds \to \pi(x) \text{ a.s.}$$

as $t \to \infty$, where π is the (unique) stationary distribution of X. Since S is discrete, it then easily follows that

$$||\pi_t - \pi|| \to 0 \text{ a.s.} \qquad (5)$$

as $t \to \infty$.

We now use a coupling argument to complete the proof of c. For each $t \geq 0$, let $\tilde{Z}(t)$ be an S-valued r.v. having conditional distribution given by

$$P(\tilde{Z}(t) = x | X, U) = \frac{[\pi(x) - \pi_t(x)]^+}{\sum_y [\pi(y) - \pi_t(y)]^+},$$

where $[y]^+ \triangleq y \vee 0$ for $y \in \mathbb{R}$. Let U' be a uniform r.v. independent of x, U, and $\tilde{Z}(t)$ and set

$$Z^*(t) = Z(t)I(U' \le \frac{\pi(Z(t))}{\pi_t(Z(t))}) + \tilde{Z}(t)I(U' > \frac{\pi(Z(t))}{\pi_t(Z(t))}).$$

Observe that

$$
\begin{aligned}
P(Z^*(t) = x|X) &= \pi_t(x)(\frac{\pi(x)}{\pi_t(x)} \wedge 1) \\
&+ \sum_{y \in S} E(P(\tilde{Z}(t) = x|X, U)[1 - (\frac{\pi(y)}{\pi_t(y)} \wedge 1)]I(Z(t) = y)|X) \\
&= (\pi(x) \wedge \pi_t(x)) + \sum_{y \in S} P(\tilde{Z}(t) = x|X, U)[\pi_t(y) - \pi(y)]^+ \\
&= \pi(x).
\end{aligned}
$$

(We have used the easily established fact that

$$\sum_y [\pi(y) - \pi_t(y)]^+ = \sum_y [\pi_t(y) - \pi(y)]^+.)$$

Hence, $Z^*(t)$ is a r.v. having the stationary distribution that is independent of X. Set $T^*(t) = \inf\{s \ge 0 : X(s) = Z^*(t)\}$. Clearly, the aforementioned properties of $Z^*(t)$ imply that

$$P(X_{T^*(t)} \in \cdot) = P_\pi(X \in \cdot)$$

for $t \ge 0$. Since $T^*(t) = T(t)$ on $\{Z(t) = Z^*(t)\}$, it follows that

$$
\begin{aligned}
&|P(X_{T(t)} \in B) - P_\pi(X \in B)| \\
=\ &|P(X_{T(t)} \in B) - P(X_{T^*(t)} \in B)| \\
=\ &|P(X_{T(t)} \in B, Z(t) = Z^*(t)) + P(X_{T(t)} \in B, Z(t) \ne Z^*(t)) \\
&-\ P(X_{T^*(t)} \in B, Z(t) = Z^*(t)) - P(X_{T^*(t)} \in B, Z(t) \ne Z^*(t))| \\
\le\ &|P(X_{T(t)} \in B, Z(t) \ne Z^*(t)) - P(X_{T^*(t)} \in B, Z(t) \ne Z^*(t))| \\
\le\ &P(Z(t) \ne Z^*(t)).
\end{aligned}
$$

We have therefore established the following coupling inequality:

$$\|P(X_{T(t)} \in \cdot) - P_\pi(X \in \cdot)\| \le P(Z(t) \ne Z^*(t)).$$

But

$$
\begin{aligned}
P(Z(t) \ne Z^*(t)|X) &\le P(U' > \frac{\pi(Z(t))}{\pi_t(Z(t))}|X) \tag{6} \\
&= \sum_y [\pi_t(y) - \pi(y)]^+.
\end{aligned}
$$

On the other hand, the latter sum is just $||\pi_t - \pi||$, which (5) asserts goes to zero a.s. The Bounded Convergence Theorem, applied to (6), then yields the conclusion

$$P(Z(t) \neq Z^*(t)) \to 0$$

as $t \to \infty$, verifying (4)c. As for condition (4)d, observe that for $x \geq 0$,

$$
\begin{aligned}
P(T(t) \geq x) &\leq P(T(t) \geq x, Z(t) \neq Z^*(t)) + P(Z(t) \neq Z^*(t)) \\
&\leq P(T^*(t) \geq x) + P(Z(t) \neq Z^*(t)).
\end{aligned}
$$

But $T^*(t)$ has a distribution independent of t, and is finite-valued. Furthermore, $P(Z(t) \neq Z^*(t)) \to 0$ as $t \to \infty$, establishing, for each $\epsilon > 0$, existence of $x = x(\epsilon)$ and $t(\epsilon)$ such that $P(T(t) \geq x) < \epsilon$ for $t \geq t(\epsilon)$. On the other hand, over $[0, t(\epsilon)]$, the non-explosiveness of X guarantees that there exists a finite deterministic set $S_{t(\epsilon)} \subseteq S$ such that

$$P(\pi_{t(\epsilon)}(S_{t(\epsilon)}) = 1) \geq 1 - \epsilon.$$

On the event $\{ \pi_{t(\epsilon)}(S_{t(\epsilon)}) = 1 \}$ (namely, those outcomes for which X spends the entire interval $[0, t(\epsilon)]$ in $S_{t(\epsilon)}$), $T(u)$ $(0 \leq u \leq t)$ is bounded by $\max\{\min\{ s \geq 0 : X(s) = y \}, y \in S_{t(\epsilon)} \}$, which is a finite r.v. independent of u. Consequently, we can find $x'(\epsilon)$ for which $P(T(u) > x'(\epsilon)) < 2\epsilon$ for $0 \leq u \leq t(\epsilon)$. This proves the required tightness, and completes the proof of the second major theorem in this paper.

Theorem 2 *If X is an irreducible positive recurrent continuous-time Markov chain taking values in a finite or countably infinite state space, then the algorithm (1)–(3) produces a family of r.v.'s $(T(t) : t \geq 0)$ having properties (4)a – d.*

Note that for a given model, one has no guarantee that a specifically chosen time horizon t will be sufficiently large so that the asymptotics associated with (4)c are in force. While this is clearly a drawback, the same drawback is shared by (for example) most applications of the central limit theorem in a statistical environment (in which one is never certain as to whether the sample size is sufficiently large so as to guarantee a good normal approximation). Note, however, that, as in section 2, bounds on total variation distance do not directly translate into bounds on bias.

Nevertheless, we believe that the algorithm (1)–(3) has sufficient practical merit so as to be worthy of further investigation. Additional properties of this algorithm will be described in a forthcoming paper; see Glynn [7].

References

[1] Asmussen, S., Glynn, P.W., and Thorisson, H. (1992). Stationarity detection in the initial transient problem. *ACM Transactions on Modeling and Computer Simulation* 2, 130–157.

[2] Conway, R.W. (1963). Some tactical problems in digital simulation. *Manage. Sci. 10*, 47–61.

[3] Diaconis, P. and Stroock, D. (1991). Geometric bounds for eigenvalues of Markov chains. *Annals of Applied Probability 1*, 36–61.

[4] Fill, J.A. (1991). Eigenvalue bounds on convergence to stationarity for nonreversible Markov chains, with an application to the exclusion process. *Applied Probability 1*, 62–87.

[5] Gafarian, A.V., Ancker, C.J., Jr., and Morisaku, T. (1978). Evaluation of commonly used rules for detecting "steady state" in computer simulation. *Nav. Res. Logistics Quart. 25*, 511–529.

[6] Glynn, P.W. and Iglehart, D.L. (1987). A new initial bias deletion rule. *Proc. of the 1987 Winter Simulation Conference*, 318–319.

[7] Glynn, P.W. (1995). Asymptotically valid initial transient detection rules. In preparation.

[8] Meyn, S.P. and Tweedie, R.L. (1994). Computable bounds for geometric convergence rates of Markov chains. *Annals of Applied Probability 4*, 981–1011.

TABLES OF (T, M, S)-NET AND (T, S)-SEQUENCE PARAMETERS

Gary L. Mullen*
Mathematics Department
The Pennsylvania State University
University Park, PA 16802, U. S. A.
E-mail: MULLEN@MATH.PSU.EDU

Arijit Mahalanabis
Department of Computer Science & Engineering
The Pennsylvania State University
University Park, PA 16802, U. S. A.
E-mail: ARIJIT@MATH.PSU.EDU

Harald Niederreiter
Institute for Information Processing
Austrian Academy of Sciences
Sonnenfelsgasse 19
A-1010 Vienna, Austria
E-mail: NIED@QIINFO.OEAW.AC.AT

Abstract. We present a survey of the known constructions of (t,m,s)-nets and (t,s)-sequences and tabulate the best parameters arising from these constructions for various bases.

1. Introduction

Quasi-Monte Carlo methods rely on low-discrepancy point sets and sequences, i.e., on point sets and sequences that show a high degree of uniformity in their distribution. Currently, the most effective constructions of low-discrepancy point sets and sequences are obtained from the theory of (t, m, s)-nets and (t, s)-sequences in a base b. Such (t, m, s)-nets and (t, s)-sequences are s-dimensional point sets, respectively sequences, with a very regular distribution behavior; see Section 2 for the precise definitions. The aim of the constructions is to make the uniformity parameter t as small as possible. In this way one gets high-quality point sets and sequences for typical quasi-Monte Carlo applications such as multidimensional numerical integration.

1991 Mathematics Subject Classification. Primary 11K45; Secondary 65C05, 11K38.

*This author would like to thank the National Security Agency for partial support under grant agreement #MDA904-92-H-3044.

In this paper we present a brief overview of the known constructions of (t, m, s)-nets and (t, s)-sequences in Section 3, and then in Section 4 we discuss related results. The main part of the paper is Section 5 in which we provide tables of the best (t, m, s)-net and (t, s)-sequence parameters that can be obtained from the known constructions for various bases.

2. Notation and Terminology

We begin by recalling from Niederreiter [27] the fundamental concepts of (t, m, s)-nets and (t, s)-sequences. Let $s \geq 1$ be a fixed dimension. For an integer $b \geq 2$, an *elementary interval in base b* is an interval of the form

$$E = \prod_{i=1}^{s} [a^{(i)} b^{-d_i}, (a^{(i)} + 1) b^{-d_i})$$

with integers $d_i \geq 0$ and integers $0 \leq a^{(i)} < b^{d_i}$ for $1 \leq i \leq s$. For integers $0 \leq t \leq m$, a (t, m, s)-net in base b is a point set of b^m points in $[0, 1)^s$ such that every elementary interval E in base b of volume b^{t-m} contains exactly b^t points of the point set. For an integer $t \geq 0$, a sequence $x_1, x_2, \ldots$ of points in $[0, 1)^s$ is called a (t, s)-sequence in base b if for all integers $k \geq 0$ and $m > t$, the point set $\{x_n : kb^m < n \leq (k+1)b^m\}$ is a (t, m, s)-net in base b. For many properties of (t, m, s)-nets and (t, s)-sequences along with their discrepancies and dispersions, we refer to Niederreiter [27], [28]. Niederreiter [33] provides a survey of numerous quasi-Monte Carlo methods and their connections to pseudorandom number generation.

3. Constructions

We now provide a brief summary of the known constructions which are used to build nets with various parameters. In this summary we will, for the sake of space, be very brief, referring the reader to the original sources for details.

We begin by briefly discussing three propagation rules which allow one to obtain a new net from a given net. These propagation rules come from Niederreiter [27, Lemmas 2.6-2.8]. The first of these indicates how to increase the parameter t, the second indicates how to decrease the parameter s, and the third provides a method of decreasing m. These propagation rules can of course be repeatedly applied, but their effectiveness decreases upon iteration since in many cases one can obtain better parameters via the use of other constructions.

Construction 1. Every (t, m, s)-net in base b is a (u, m, s)-net in base b for $t \leq u \leq m$.

Construction 2. For $1 \leq r \leq s$, every (t, m, s)-net in base b can be transformed into a (t, m, r)-net in base b.

Construction 3. Every (t, m, s)-net in base b can be transformed into a (t, u, s)-net in base b, where $t \leq u \leq m$.

Niederreiter [27, Lemma 5.15] provides an effective method of obtaining a series of nets in base b from a given sequence in base b. In particular, if there exists a (t, s)-sequence in base b, then for any $m \geq t$ there exists a $(t, m, s + 1)$-net in base b. While this could be viewed as a fourth construction, since we must have a sequence before obtaining a net, we will simply tag the resulting net with the tag that corresponds to the construction of the original sequence. We now mention two trivial constructions.

Construction 4. For any $t \geq 0$ and any $s \geq 1$, one can construct a (t, t, s)-net in base b by taking a multiset consisting of a single point from $[0, 1)^s$ with multiplicity b^t.

Construction 5. For any $t \geq 0$ and any $s \geq 1$, one can construct a $(t, t + 1, s)$-net in base b by taking the multiset consisting of the points $(n/b, \ldots, n/b) \in [0, 1)^s$, $n = 0, 1, \ldots, b - 1$, each with multiplicity b^t.

Thus, in the following constructions we may assume $m \geq t + 2$.

Construction 6. Sobol' [38] first studied nets in base 2 through the use of (t, s)-sequences. For $s \leq 7$ his method still yields (t, s)-sequences in base 2 with the smallest value of t, and so Sobol's method provides our first non-trivial net construction.

Construction 7. In [27, Cor. 5.6] Niederreiter proves the equivalence of the existence of $(0, 2, s)$-nets in base b and the existence of $s - 2$ mutually orthogonal latin squares of order b. We refer to Dénes and Keedwell [5] for a discussion of latin squares.

Construction 8. Our next construction comes from Niederreiter [27, Thm. 6.2] where it is shown that if $b = q_1 \cdots q_u$ is the canonical factorization of b into prime powers $q_1 < \cdots < q_u$, then for any $m \geq 0$ and $s \leq q_1 + 1$, there exists a $(0, m, s)$-net in base b. In the special case where b is prime, this construction is due to Faure [6].

Construction 9. In [28] Niederreiter provides a number of constructions of (t, s)-sequences in prime-power bases. With optimized parameters this yields, for every dimension $s \geq 1$ and every prime power b, a $(T_b(s), s)$-sequence in base b, where $T_b(s)$ is defined as follows. List all monic irreducible polynomials over the finite field F_b of order b in a sequence $p_1, p_2, \ldots$ in such a way that $\deg(p_i) \leq \deg(p_h)$ whenever $i \leq h$. Let $p_1, \ldots, p_s$ be the first s terms in this sequence of polynomials. For $i = 1, \ldots, s$ let e_i denote the degree of p_i. Then $T_b(s)$ is defined by $T_b(s) = \sum_{i=1}^{s}(e_i - 1)$. We refer to Table A for the values of $T_b(s)$ for $b = 2, 3, 4, 5$ with $s \leq 30$. These (t, s)-sequences in base b are then used to construct the corresponding $(t, m, s + 1)$-nets in base b.

Construction 10. This method of construction arises from Niederreiter [28, Thm. 4]. Let $b = q_1 \cdots q_u$ be a product of prime powers. Then for every dimension $s \geq 1$ there exists a (t, s)-sequence in base b with

$$t = \max_{1 \leq j \leq u} T_{q_j}(s), \tag{1}$$

where $T_{q_j}(s)$ is defined as in Construction 9. These (t, s)-sequences in base b are then used to construct the corresponding $(t, m, s + 1)$-nets in base b.

It may be worth raising a conjecture here to the effect that in (1), for any $s \geq 1$, the maximum $T_{q_j}(s)$ occurs when q_j is the smallest prime power in the factorization of b. See Table A for evidence.

Construction 11. The next net parameters arise from tabulations of recommended parameters for pseudorandom number generation in André, Mullen, and Niederreiter [1, Tables 1-3]. These tables are restricted to $b = 2, 3 \leq s \leq 5$, and $m \leq 32$.

Construction 12. This construction is of a combinatorial nature, using $(t + 2)$-dimensional hypercubes where $t \geq 0$ is an integer. In Mullen and Whittle [23] an equivalence is shown between the existence of $(t, t + 2, s)$-nets in base b and sets of s mutually orthogonal hypercubes of dimension $t + 2$ and order b. The case where $t = 0$ was already given in Niederreiter [27, Section 5].

Construction 13. The theoretical basis for Constructions 13 and 15 is due to Niederreiter [32] and uses a digital construction of nets arising from Laurent series expansions of rational functions over the finite field F_b of order b. The general principle of digital constructions was already introduced earlier in Niederreiter [27, Section 6]. Suitable rational functions for this construction were first obtained by computer search in Hansen, Mullen, and Niederreiter [9, Table 2] for the base $b = 2$, with $m \leq 20$ for $s \leq 4$ and $m \leq 10$ for $5 \leq s \leq 12$.

Construction 14. In [20] Laywine, Mullen, and Whittle provide an algorithm for converting small sets of orthogonal latin squares into large sets of higher-dimensional orthogonal hypercubes. While these constructions are effective for hypercubes of any order, for our purposes they are most effective for bases b that are not prime powers, since optimal prime-power base hypercube constructions related to nets are given in Mullen and Whittle [23] and have already been described in Construction 12. For our tables, results of [20] will thus appear only for base $b = 10$ with $t > 0$, the case $t = 0$ being covered in Construction 7.

Construction 15. The computer search in base 2 from Construction 13 was continued by Larcher, Lauss, Niederreiter, and Schmid [12] by using a more restricted class of rational functions, with parameters having the range $m \leq 23$ and $s \leq 15$.

The following propagation rule, which generalizes the simple idea in Constructions 4 and 5, was apparently not noticed before.

Construction 16. Every (t, m, s)-net in base b can be transformed into a $(t + u, m + u, s)$-net in base b for any $u \geq 0$. This is achieved by taking the (t, m, s)-net in base b and replicating it b^u times.

4. Related Results

There are numerous other papers, and even two books, related to the construction of (t, m, s)-nets and (t, s)-sequences which do not improve any values in our tables, but which are nevertheless worth mentioning. For the sake of completeness we include a brief description of these papers and books. Sobol' [39] and Niederreiter [24, Section 3] provide an exposition of the results of Sobol' [38], while Srinivasan [40] constructs a (0,2)-sequence in base 2 in a way that is different from the earlier construction of Sobol' [38]. A computer implementation of the Sobol' sequences is described in Antonov and Saleev [2]; see also Bratley and Fox [3]. In [7] Fox presents details of a computer implementation of the Faure sequences from [6]. The paper Niederreiter [25] can be viewed as a precursor to the systematic development of nets in Niederreiter [27], although the results in [25] are not phrased in terms of nets.

A different approach from [27] is used by Niederreiter [29] to construct $(0, s)$-sequences in prime-power bases, but no new parameters are obtained. Similarly, Tezuka [41] constructs (t, s)-sequences in prime-power bases with the same parameters as in Niederreiter [28] by an alternative-approach. Bratley, Fox, and Niederreiter [4] discuss computer implementations and tests of various low-discrepancy sequences including the sequences from Niederreiter [28].

Niederreiter [34] provides considerable detail concerning digital constructions of nets in prime-power bases using rational functions over finite fields. General existence theorems for good parameters in these constructions are established in Larcher [10], Larcher, Lauss, Niederreiter, and Schmid [12], and Niederreiter [32]. The discussion in Niederreiter [26] of rational functions with partial quotients of small degree in their continued fraction expansion is related to these constructions (see [33, Thms. 4.42 and 4.46]). A method of obtaining (t, s)-sequences in prime-power bases from irrational Laurent series over finite fields is introduced in Niederreiter [33, Chapter 4], and this method is further studied in Larcher and Niederreiter [13], [14] and Niederreiter [36]. An investigation of general digital constructions of nets and sequences is carried out in Larcher, Niederreiter, and Schmid [15].

Certain orthogonal arrays are known from combinatorics to be equivalent to sets of mutually orthogonal latin squares and hypercubes (see [5, p. 191]), and in [31] Niederreiter shows an analogous equivalence between the existence of certain nets and orthogonal arrays. In [21] Mullen provides a survey of combinatorial constructions of nets using sets of mutually orthogonal hypercubes. In [30] Niederreiter discusses connections between nets and a combinatorial problem for vector spaces over finite fields, which is itself related to algebraic coding theory.

A survey of various issues involving pseudorandom numbers and low-discrepancy point sets and sequences as they relate to finite fields is given in Niederreiter [35]. André, Mullen, and Niederreiter [1] and Mullen and Niederreiter [22] calculate figures of merit for digital multistep pseudorandom numbers, and Tezuka and Fushimi [42] calculate figures of merit for generalized feedback shift-register generators for pseudorandom numbers. Such figures of merit are known to be connected with net parameters by Niederreiter [27, Section 9]. Larcher and Niederreiter [14] develop a theory of generalized (t, s)-sequences and strict (t, m, s)-nets, i.e., (t, m, s)-nets which are not $(t - 1, m, s)$-nets.

Larcher and Traunfellner [19] show that nets are very effective as node sets for the numerical integration of multivariate Walsh series by quasi-Monte Carlo methods. Related work is carried out in Larcher [11], Larcher, Niederreiter, and Schmid [15], Larcher and Schmid [16], and Larcher, Schmid, and Wolf [17], [18]. Connections between nets in base 2, Haar functions, and numerical integration were discussed earlier by Sobol' [39]. An application of nets to simulated annealing is described by Fox [8]. Randomized (t, m, s)-nets and (t, s)-sequences obtained by random permutations of digits are studied by Owen [37].

5. Tables

For each base $b = 2, 3, 5, 10$, we provide a table t_b. Because of applications in numerical integration where the value of s is usually prescribed and m is suitably chosen to guarantee a small error, it is of interest to tabulate, for given b, m, and s, a small value of t for which a (t, m, s)-net in base b can be constructed. It is known from Niederreiter [31, p. 364] that if $s \geq 2$ and $b \geq 2$ are integers, then the least value of t for which there can exist a (t, m, s)-net in base b with $m \geq t + 2$ satisfies

$$t \geq \lceil \log_b(bs - s + 1) \rceil - 2, \tag{2}$$

where $\log_b$ denotes the logarithm base b and $\lceil x \rceil$ denotes the ceiling function of x, i.e., the smallest integer $\geq x$.

In Table t_b we have provided three values at the intersection of row m and column s. The top number gives the value of the lower bound (2) for the existence of a (t, m, s)-net in base b. The middle value indicates the minimal value of t for which any of the above constructions yield a (t, m, s)-net in base b. The bottom number provides a tag which indicates the construction rule used to obtain a (t, m, s)-net in base b. It should be kept in mind that in general there may be several constructions yielding a (t, m, s)-net in base b, but as described below, the tag refers to the first construction chronologically. It may be helpful to explain why in column $s = 50$ of the t_b tables, many of the values listed come from propagation rule 2 which decreases the value of s even though $s = 50$ is the largest value of s in the tables. This happens because in each such case, the computer temporarily has in its memory a larger value of s.

For some applications it is necessary to have values of s that exceed those given in the Tables t_b. Mullen and Whittle [23, Cor. 5] provides an upper bound on s for the existence of a (t, m, s)-net base b. In particular, for $t \geq 0$ and $m \geq t + 2$, a (t, m, s)-net in base b can exist only if $s \leq (b^{t+2} - 1)/(b - 1)$. It may be worth pointing out that the lower bound (2) for t follows from this upper bound for s. Because base 2 is the most important base for applications, we have included a Table s_2, where we have, as in the above Tables t_b, provided three values at the intersection of row t and column m. The top value indicates the largest value of s for which any of the above constructions yield a (t, m, s)-net in base 2. The middle value indicates the above upper bound on s for $b = 2$, while the bottom value is once again a tag which refers to the construction method for the construction of a (t, m, s)-net in base 2. Due to space limitations we have restricted the range of values to at most three digits. As a result we have used 999 to represent the fact that the value of s is arbitrarily large; for example as in the cases of (t, t, s) and $(t, t + 1, s)$-nets in base b. If $s \geq 998$, but is bounded, we denote this by 998. This occurs for example in the case of an $(8, 10, s)$-net in base 2 where $s = 2^{10} - 1$, which of course exceeds 998.

Entries in the tables have been entered chronologically by date of publication. For papers appearing in the same year, results have been entered alphabetically by the first author. Software was written so that a set of parameters t', m, s' in base 2 can be entered for the construction of a (t', m, s')-net in base 2. Suppose in Table t_2 we have the value t and in Table s_2 we have the value s, i.e., we assume that it is known how to construct a (t, m, s)-net in base 2. If $t' \geq t$ and $s' \leq s$, so that the new parameters do not improve the current values of t or s, then nothing is changed in Table t_2 or Table s_2. On the other hand, if the new value t' improves the current value of t, i.e., if $t' < t$, then the old value of t is replaced by t', and a new tag is entered corresponding to the new construction. We proceed similarly if the new value $s' > s$. In addition, in either case, Constructions 1, 2, and 3 are immediately applied with the values t' and s', and if any of the propagation rules result in further improvements anywhere in Table t_2 or in Table s_2, these improvements and the corresponding tags are entered. Moreover, each propagation rule from Constructions 1, 2, and 3 is iterated until no further improvements are obtained in either Table t_2 or s_2. The propagation rule from Construction 16 is applied in an analogous way.

For bases $b = 3, 5, 10$ we proceed in a similar fashion for the Tables t_b, but for a lack of space, for these bases we do not include the corresponding tables for the parameter s. For each of these bases, however, one can obtain for given t and m, the largest $s \leq 50$ for which it is known how to construct a (t, m, s)-net in base b. To do this one simply looks in row m of Table t_b, and finds the rightmost occurrence of a number not exceeding t. The value of s at the top of this column provides the largest value of $s \leq 50$ for which it is known how to construct a (t, m, s)-net in base b.

T TABLE BASE 2

m/s	1	2	3	4	5	6	7	8	9	10	11	12	13	14	15	16	17	18	19	20	21	22	23	24	25
1	0	0	0	0	0	0	0	0	0	0	0	0	0	0	0	0	0	0	0	0	0	0	0	0	0
	0	0	0	0	0	0	0	0	0	0	0	0	0	0	0	0	0	0	0	0	0	0	0	0	0
	5	5	5	5	5	5	5	5	5	5	5	5	5	5	5	5	5	5	5	5	5	5	5	5	5
2	0	0	0	1	1	1	1	1	1	1	1	1	1	1	1	1	1	1	1	1	1	1	1	1	1
	0	0	0	1	1	1	1	1	1	1	1	1	1	1	1	1	1	1	1	1	1	1	1	1	1
	2	6	6	5	5	5	5	5	5	5	5	5	5	5	5	5	5	5	5	5	5	5	5	5	5
3	0	0	0	1	1	1	1	2	2	2	2	2	2	2	2	2	2	2	2	2	2	2	2	2	2
	0	0	0	1	1	1	1	2	2	2	2	2	2	2	2	2	2	2	2	2	2	2	2	2	2
	2	6	6	6	11	2	12	5	5	5	5	5	5	5	5	5	5	5	5	5	5	5	5	5	5
4	0	0	0	1	1	1	1	2	2	2	2	2	2	2	2	3	3	3	3	3	3	3	3	3	3
	0	0	0	1	1	1	2	2	2	2	2	2	2	2	2	3	3	3	3	3	3	3	3	3	3
	2	6	6	6	13	13	16	2	2	2	2	2	2	2	12	5	5	5	5	5	5	5	5	5	5
5	0	0	0	1	1	1	1	2	2	2	2	2	2	2	2	3	3	3	3	3	3	3	3	3	3
	0	0	0	1	2	2	2	2	2	2	3	3	3	3	3	3	3	3	3	3	3	3	3	3	3
	2	6	6	6	11	16	13	13	13	16	16	16	16	2	2	2	2	2	2	2	2	2	2	2	2
6	0	0	0	1	1	1	1	2	2	2	2	2	2	2	2	3	3	3	3	3	3	3	3	3	3
	0	0	0	1	2	2	3	3	3	4	4	4	4	4	4	4	4	4	4	4	4	4	4	4	4
	2	6	6	6	13	13	16	16	16	16	16	16	16	16	16	16	16	16	16	16	16	16	16	16	16
7	0	0	0	1	1	1	1	2	2	2	2	2	2	2	2	3	3	3	3	3	3	3	3	3	3
	0	0	0	1	2	2	3	3	4	4	4	4	5	5	5	5	5	5	5	5	5	5	5	5	5
	2	6	6	6	13	13	13	16	16	16	13	13	16	16	16	16	16	16	16	16	16	16	16	16	16
8	0	0	0	1	1	1	1	2	2	2	2	2	2	2	2	3	3	3	3	3	3	3	3	3	3
	0	0	0	1	2	3	4	4	4	4	5	6	6	6	6	6	6	6	6	6	6	6	6	6	6
	2	6	6	6	13	16	16	16	13	13	13	16	16	16	16	16	16	16	16	16	16	16	16	16	16
9	0	0	0	1	1	1	1	2	2	2	2	2	2	2	2	3	3	3	3	3	3	3	3	3	3
	0	0	0	1	3	4	4	5	5	5	5	6	6	6	6	7	7	7	7	7	7	7	7	7	7
	2	6	6	6	6	13	16	16	16	16	15	15	15	16	16	16	16	16	16	16	16	16	16	16	16
10	0	0	0	1	1	1	1	2	2	2	2	2	2	2	2	3	3	3	3	3	3	3	3	3	3
	0	0	0	1	3	4	4	5	6	6	6	6	7	7	7	8	8	8	8	8	8	8	8	8	8
	2	6	6	6	13	13	13	16	16	16	13	16	16	16	16	16	16	16	16	16	16	16	16	16	16
11	0	0	0	1	1	1	1	2	2	2	2	2	2	2	2	3	3	3	3	3	3	3	3	3	3
	0	0	0	1	3	4	5	6	6	6	7	7	8	8	8	9	9	9	9	9	9	9	9	9	9
	2	6	6	6	15	16	16	15	15	16	16	16	16	16	16	16	16	16	16	16	16	16	16	16	16
12	0	0	0	1	1	1	1	2	2	2	2	2	2	2	2	3	3	3	3	3	3	3	3	3	3
	0	0	0	1	3	5	5	6	7	7	8	8	8	8	10	10	10	10	10	10	10	10	10	10	10
	2	6	6	6	6	6	3	16	16	16	16	3	3	16	16	16	16	16	16	16	16	16	16	16	16
13	0	0	0	1	1	1	1	2	2	2	2	2	2	2	2	3	3	3	3	3	3	3	3	3	3
	0	0	0	1	3	5	5	6	7	7	8	8	8	8	8	11	11	11	11	11	11	11	11	11	11
	2	6	6	6	6	6	15	15	15	15	15	15	15	16	16	16	16	16	16	16	16	16	16	16	16
14	0	0	0	1	1	1	1	2	2	2	2	2	2	2	2	3	3	3	3	3	3	3	3	3	3
	0	0	0	1	3	5	5	6	8	8	9	9	9	9	9	12	12	12	12	12	12	12	12	12	12
	2	6	6	6	6	15	3	16	16	16	16	16	16	16	16	16	16	16	16	16	16	16	16	16	16
15	0	0	0	1	1	1	1	2	2	2	2	2	2	2	2	3	3	3	3	3	3	3	3	3	3
	0	0	0	1	3	5	6	6	8	8	9	9	9	9	9	13	13	13	13	13	13	13	13	13	13
	2	6	6	6	6	16	15	3	15	15	15	15	15	16	16	16	16	16	16	16	16	16	16	16	16
16	0	0	0	1	1	1	1	2	2	2	2	2	2	2	2	3	3	3	3	3	3	3	3	3	3
	0	0	0	1	3	5	6	6	8	8	9	9	10	10	10	14	14	14	14	14	14	14	14	14	14
	2	6	6	6	6	6	15	15	3	3	16	16	16	16	16	16	16	16	16	16	16	16	16	16	16
17	0	0	0	1	1	1	1	2	2	2	2	2	2	2	2	3	3	3	3	3	3	3	3	3	3
	0	0	0	1	3	5	7	7	8	8	9	9	11	11	11	15	15	15	15	15	15	15	15	15	15
	2	6	6	6	6	6	16	16	15	15	15	15	16	16	16	16	16	16	16	16	16	16	16	16	16
18	0	0	0	1	1	1	1	2	2	2	2	2	2	2	2	3	3	3	3	3	3	3	3	3	3
	0	0	0	1	3	5	7	7	9	9	9	9	12	12	12	16	16	16	16	16	16	16	16	16	16
	2	6	6	6	6	15	16	16	15	15	16	16	16	16	16	16	16	16	16	16	16	16	16	16	16
19	0	0	0	1	1	1	1	2	2	2	2	2	2	2	2	3	3	3	3	3	3	3	3	3	3
	0	0	0	1	3	5	7	7	10	10	10	12	12	13	17	17	17	17	17	17	17	17	17	17	17
	2	6	6	6	6	3	16	16	16	3	3	16	16	16	16	16	16	16	16	16	16	16	16	16	16
20	0	0	0	1	1	1	1	2	2	2	2	2	2	2	2	3	3	3	3	3	3	3	3	3	3
	0	0	0	1	3	5	7	7	10	11	11	11	12	12	13	18	18	18	18	18	18	18	18	18	18
	2	6	6	6	6	3	16	16	16	3	3	16	16	16	16	16	16	16	16	16	16	16	16	16	16
21	0	0	0	1	1	1	1	2	2	2	2	2	2	2	2	3	3	3	3	3	3	3	3	3	3
	0	0	0	1	3	5	7	7	10	11	11	12	12	13	13	19	19	19	19	19	19	19	19	19	19
	2	6	6	6	6	6	16	16	3	15	15	15	16	16	16	16	16	16	16	16	16	16	16	16	16
22	0	0	0	1	1	1	1	2	2	2	2	2	2	2	2	3	3	3	3	3	3	3	3	3	3
	0	0	0	1	3	5	7	7	10	11	11	13	13	13	13	20	20	20	20	20	20	20	20	20	20
	2	6	6	6	6	15	15	15	15	16	16	16	16	16	16	16	16	16	16	16	16	16	16	16	16
23	0	0	0	1	1	1	1	2	2	2	2	2	2	2	2	3	3	3	3	3	3	3	3	3	3
	0	0	0	1	3	5	8	8	11	12	12	14	14	14	14	21	21	21	21	21	21	21	21	21	21
	2	6	6	6	6	6	16	16	16	16	16	16	16	16	16	16	16	16	16	16	16	16	16	16	16
24	0	0	0	1	1	1	1	2	2	2	2	2	2	2	2	3	3	3	3	3	3	3	3	3	3
	0	0	0	1	3	5	8	9	12	13	13	15	15	15	15	22	22	22	22	22	22	22	22	22	22
	2	6	6	6	6	6	16	16	16	16	16	16	16	16	16	16	16	16	16	16	16	16	16	16	16
25	0	0	0	1	1	1	1	2	2	2	2	2	2	2	2	3	3	3	3	3	3	3	3	3	3
	0	0	0	1	3	5	8	10	13	14	14	16	16	16	16	23	23	23	23	23	23	23	23	23	23
	2	6	6	6	6	6	6	16	16	16	16	16	16	16	16	16	16	16	16	16	16	16	16	16	16

T TABLE BASE 2

m/s	26	27	28	29	30	31	32	33	34	35	36	37	38	39	40	41	42	43	44	45	46	47	48	49	50
1	0	0	0	0	0	0	0	0	0	0	0	0	0	0	0	0	0	0	0	0	0	0	0	0	0
	0	0	0	0	0	0	0	0	0	0	0	0	0	0	0	0	0	0	0	0	0	0	0	0	0
	5	5	5	5	5	5	5	5	5	5	5	5	5	5	5	5	5	5	5	5	5	5	5	5	5
2	1	1	1	1	1	1	1	1	1	1	1	1	1	1	1	1	1	1	1	1	1	1	1	1	1
	1	1	1	1	1	1	1	1	1	1	1	1	1	1	1	1	1	1	1	1	1	1	1	1	1
	5	5	5	5	5	5	5	5	5	5	5	5	5	5	5	5	5	5	5	5	5	5	5	5	5
3	2	2	2	2	2	2	2	2	2	2	2	2	2	2	2	2	2	2	2	2	2	2	2	2	2
	2	2	2	2	2	2	2	2	2	2	2	2	2	2	2	2	2	2	2	2	2	2	2	2	2
	5	5	5	5	5	5	5	5	5	5	5	5	5	5	5	5	5	5	5	5	5	5	5	5	5
4	3	3	3	3	3	3	3	3	3	3	3	3	3	3	3	3	3	3	3	3	3	3	3	3	3
	3	3	3	3	3	3	3	3	3	3	3	3	3	3	3	3	3	3	3	3	3	3	3	3	3
	5	5	5	5	5	5	5	5	5	5	5	5	5	5	5	5	5	5	5	5	5	5	5	5	5
5	3	3	3	3	3	3	4	4	4	4	4	4	4	4	4	4	4	4	4	4	4	4	4	4	4
	3	3	3	3	3	3	4	4	4	4	4	4	4	4	4	4	4	4	4	4	4	4	4	4	4
	2	2	2	2	2	12	5	5	5	5	5	5	5	5	5	5	5	5	5	5	5	5	5	5	5
6	3	3	3	3	3	3	4	4	4	4	4	4	4	4	4	4	4	4	4	4	4	4	4	4	4
	4	4	4	4	4	4	4	4	4	4	4	4	4	4	4	4	4	4	4	4	4	4	4	4	4
	16	16	16	16	16	16	2	2	2	2	2	2	2	2	2	2	2	2	2	2	2	2	2	2	2
7	3	3	3	3	3	3	4	4	4	4	4	4	4	4	4	4	4	4	4	4	4	4	4	4	4
	5	5	5	5	5	5	5	5	5	5	5	5	5	5	5	5	5	5	5	5	5	5	5	5	5
	16	16	16	16	16	16	16	16	16	16	16	16	16	16	16	16	16	16	16	16	16	16	16	16	16
8	3	3	3	3	3	3	4	4	4	4	4	4	4	4	4	4	4	4	4	4	4	4	4	4	4
	6	6	6	6	6	6	6	6	6	6	6	6	6	6	6	6	6	6	6	6	6	6	6	6	6
	16	16	16	16	16	16	16	16	16	16	16	16	16	16	16	16	16	16	16	16	16	16	16	16	16
9	3	3	3	3	3	3	4	4	4	4	4	4	4	4	4	4	4	4	4	4	4	4	4	4	4
	7	7	7	7	7	7	7	7	7	7	7	7	7	7	7	7	7	7	7	7	7	7	7	7	7
	16	16	16	16	16	16	16	16	16	16	16	16	16	16	16	16	16	16	16	16	16	16	16	16	16
10	3	3	3	3	3	3	4	4	4	4	4	4	4	4	4	4	4	4	4	4	4	4	4	4	4
	8	8	8	8	8	8	8	8	8	8	8	8	8	8	8	8	8	8	8	8	8	8	8	8	8
	16	16	16	16	16	16	16	16	16	16	16	16	16	16	16	16	16	16	16	16	16	16	16	16	16
11	3	3	3	3	3	3	4	4	4	4	4	4	4	4	4	4	4	4	4	4	4	4	4	4	4
	9	9	9	9	9	9	9	9	9	9	9	9	9	9	9	9	9	9	9	9	9	9	9	9	9
	16	16	16	16	16	16	16	16	16	16	16	16	16	16	16	16	16	16	16	16	16	16	16	16	16
12	3	3	3	3	3	3	4	4	4	4	4	4	4	4	4	4	4	4	4	4	4	4	4	4	4
	10	10	10	10	10	10	10	10	10	10	10	10	10	10	10	10	10	10	10	10	10	10	10	10	10
	16	16	16	16	16	16	16	16	16	16	16	16	16	16	16	16	16	16	16	16	16	16	16	16	16
13	3	3	3	3	3	3	4	4	4	4	4	4	4	4	4	4	4	4	4	4	4	4	4	4	4
	11	11	11	11	11	11	11	11	11	11	11	11	11	11	11	11	11	11	11	11	11	11	11	11	11
	16	16	16	16	16	16	16	16	16	16	16	16	16	16	16	16	16	16	16	16	16	16	16	16	16
14	3	3	3	3	3	3	4	4	4	4	4	4	4	4	4	4	4	4	4	4	4	4	4	4	4
	12	12	12	12	12	12	12	12	12	12	12	12	12	12	12	12	12	12	12	12	12	12	12	12	12
	16	16	16	16	16	16	16	16	16	16	16	16	16	16	16	16	16	16	16	16	16	16	16	16	16
15	3	3	3	3	3	3	4	4	4	4	4	4	4	4	4	4	4	4	4	4	4	4	4	4	4
	13	13	13	13	13	13	13	13	13	13	13	13	13	13	13	13	13	13	13	13	13	13	13	13	13
	16	16	16	16	16	16	16	16	16	16	16	16	16	16	16	16	16	16	16	16	16	16	16	16	16
16	3	3	3	3	3	3	4	4	4	4	4	4	4	4	4	4	4	4	4	4	4	4	4	4	4
	14	14	14	14	14	14	14	14	14	14	14	14	14	14	14	14	14	14	14	14	14	14	14	14	14
	16	16	16	16	16	16	16	16	16	16	16	16	16	16	16	16	16	16	16	16	16	16	16	16	16
17	3	3	3	3	3	3	4	4	4	4	4	4	4	4	4	4	4	4	4	4	4	4	4	4	4
	15	15	15	15	15	15	15	15	15	15	15	15	15	15	15	15	15	15	15	15	15	15	15	15	15
	16	16	16	16	16	16	16	16	16	16	16	16	16	16	16	16	16	16	16	16	16	16	16	16	16
18	3	3	3	3	3	3	4	4	4	4	4	4	4	4	4	4	4	4	4	4	4	4	4	4	4
	16	16	16	16	16	16	16	16	16	16	16	16	16	16	16	16	16	16	16	16	16	16	16	16	16
	16	16	16	16	16	16	16	16	16	16	16	16	16	16	16	16	16	16	16	16	16	16	16	16	16
19	3	3	3	3	3	3	4	4	4	4	4	4	4	4	4	4	4	4	4	4	4	4	4	4	4
	17	17	17	17	17	17	17	17	17	17	17	17	17	17	17	17	17	17	17	17	17	17	17	17	17
	16	16	16	16	16	16	16	16	16	16	16	16	16	16	16	16	16	16	16	16	16	16	16	16	16
20	3	3	3	3	3	3	4	4	4	4	4	4	4	4	4	4	4	4	4	4	4	4	4	4	4
	18	18	18	18	18	18	18	18	18	18	18	18	18	18	18	18	18	18	18	18	18	18	18	18	18
	16	16	16	16	16	16	16	16	16	16	16	16	16	16	16	16	16	16	16	16	16	16	16	16	16
21	3	3	3	3	3	3	4	4	4	4	4	4	4	4	4	4	4	4	4	4	4	4	4	4	4
	19	19	19	19	19	19	19	19	19	19	19	19	19	19	19	19	19	19	19	19	19	19	19	19	19
	16	16	16	16	16	16	16	16	16	16	16	16	16	16	16	16	16	16	16	16	16	16	16	16	16
22	3	3	3	3	3	3	4	4	4	4	4	4	4	4	4	4	4	4	4	4	4	4	4	4	4
	20	20	20	20	20	20	20	20	20	20	20	20	20	20	20	20	20	20	20	20	20	20	20	20	20
	16	16	16	16	16	16	16	16	16	16	16	16	16	16	16	16	16	16	16	16	16	16	16	16	16
23	3	3	3	3	3	3	4	4	4	4	4	4	4	4	4	4	4	4	4	4	4	4	4	4	4
	21	21	21	21	21	21	21	21	21	21	21	21	21	21	21	21	21	21	21	21	21	21	21	21	21
	16	16	16	16	16	16	16	16	16	16	16	16	16	16	16	16	16	16	16	16	16	16	16	16	16
24	3	3	3	3	3	3	4	4	4	4	4	4	4	4	4	4	4	4	4	4	4	4	4	4	4
	22	22	22	22	22	22	22	22	22	22	22	22	22	22	22	22	22	22	22	22	22	22	22	22	22
	16	16	16	16	16	16	16	16	16	16	16	16	16	16	16	16	16	16	16	16	16	16	16	16	16
25	3	3	3	3	3	3	4	4	4	4	4	4	4	4	4	4	4	4	4	4	4	4	4	4	4
	23	23	23	23	23	23	23	23	23	23	23	23	23	23	23	23	23	23	23	23	23	23	23	23	23
	16	16	16	16	16	16	16	16	16	16	16	16	16	16	16	16	16	16	16	16	16	16	16	16	16

T TABLE BASE 2

m/s	1	2	3	4	5	6	7	8	9	10	11	12	13	14	15	16	17	18	19	20	21	22	23	24	25
26	0	0	0	1	1	1	1	2	2	2	2	2	2	2	2	3	3	3	3	3	3	3	3	3	3
	0	0	0	1	3	5	8	11	14	15	15	17	17	17	17	24	24	24	24	24	24	24	24	24	24
	2	6	6	6	6	6	6	6	9	16	16	16	16	16	16	16	16	16	16	16	16	16	16	16	16
27	0	0	0	1	1	1	1	2	2	2	2	2	2	2	2	3	3	3	3	3	3	3	3	3	3
	0	0	0	1	3	5	8	11	14	16	16	18	18	18	18	25	25	25	25	25	25	25	25	25	25
	2	6	6	6	6	6	6	6	9	16	16	16	16	16	16	16	16	16	16	16	16	16	16	16	16
28	0	0	0	1	1	1	1	2	2	2	2	2	2	2	2	3	3	3	3	3	3	3	3	3	3
	0	0	0	1	3	5	8	11	14	17	17	19	19	19	19	26	26	26	26	26	26	26	26	26	26
	2	6	6	6	6	6	6	6	9	16	16	16	16	16	16	16	16	16	16	16	16	16	16	16	16
29	0	0	0	1	1	1	1	2	2	2	2	2	2	2	2	3	3	3	3	3	3	3	3	3	3
	0	0	0	1	3	5	8	11	14	18	18	20	20	20	20	27	27	27	27	27	27	27	27	27	27
	2	6	6	6	6	6	6	6	9	9	16	16	16	16	16	16	16	16	16	16	16	16	16	16	16
30	0	0	0	1	1	1	1	2	2	2	2	2	2	2	2	3	3	3	3	3	3	3	3	3	3
	0	0	0	1	3	5	8	11	14	18	19	21	21	21	21	28	28	28	28	28	28	28	28	28	28
	2	6	6	6	6	6	6	6	9	9	16	16	16	16	16	16	16	16	16	16	16	16	16	16	16
31	0	0	0	1	1	1	1	2	2	2	2	2	2	2	2	3	3	3	3	3	3	3	3	3	3
	0	0	0	1	3	5	8	11	14	18	20	22	22	22	22	29	29	29	29	29	29	29	29	29	29
	2	6	6	6	6	6	6	6	9	9	16	16	16	16	16	16	16	16	16	16	16	16	16	16	16
32	0	0	0	1	1	1	1	2	2	2	2	2	2	2	2	3	3	3	3	3	3	3	3	3	3
	0	0	0	1	3	5	8	11	14	18	21	23	23	23	23	30	30	30	30	30	30	30	30	30	30
	2	6	6	6	6	6	6	6	9	9	16	16	16	16	16	16	16	16	16	16	16	16	16	16	16
33	0	0	0	1	1	1	1	2	2	2	2	2	2	2	2	3	3	3	3	3	3	3	3	3	3
	0	0	0	1	3	5	8	11	14	18	22	24	24	24	24	31	31	31	31	31	31	31	31	31	31
	2	6	6	6	6	6	6	6	9	9	9	16	16	16	16	16	16	16	16	16	16	16	16	16	16
34	0	0	0	1	1	1	1	2	2	2	2	2	2	2	2	3	3	3	3	3	3	3	3	3	3
	0	0	0	1	3	5	8	11	14	18	22	25	25	25	25	32	32	32	32	32	32	32	32	32	32
	2	6	6	6	6	6	6	6	9	9	16	16	16	16	16	16	16	16	16	16	16	16	16	16	16
35	0	0	0	1	1	1	1	2	2	2	2	2	2	2	2	3	3	3	3	3	3	3	3	3	3
	0	0	0	1	3	5	8	11	14	18	22	26	26	26	26	33	33	33	33	33	33	33	33	33	33
	2	6	6	6	6	6	6	6	9	9	9	16	16	16	16	16	16	16	16	16	16	16	16	16	16
36	0	0	0	1	1	1	1	2	2	2	2	2	2	2	2	3	3	3	3	3	3	3	3	3	3
	0	0	0	1	3	5	8	11	14	18	22	26	27	27	27	34	34	34	34	34	34	34	34	34	34
	2	6	6	6	6	6	6	6	9	9	9	16	16	16	16	16	16	16	16	16	16	16	16	16	16
37	0	0	0	1	1	1	1	2	2	2	2	2	2	2	2	3	3	3	3	3	3	3	3	3	3
	0	0	0	1	3	5	8	11	14	18	22	26	28	28	28	35	35	35	35	35	35	35	35	35	35
	2	6	6	6	6	6	6	6	9	9	9	16	16	16	16	16	16	16	16	16	16	16	16	16	16
38	0	0	0	1	1	1	1	2	2	2	2	2	2	2	2	3	3	3	3	3	3	3	3	3	3
	0	0	0	1	3	5	8	11	14	18	22	26	29	29	29	36	36	36	36	36	36	36	36	36	36
	2	6	6	6	6	6	6	6	9	9	9	16	16	16	16	16	16	16	16	16	16	16	16	16	16
39	0	0	0	1	1	1	1	2	2	2	2	2	2	2	2	3	3	3	3	3	3	3	3	3	3
	0	0	0	1	3	5	8	11	14	18	22	26	30	30	30	37	37	37	37	37	37	37	37	37	37
	2	6	6	6	6	6	6	6	9	9	9	16	16	16	16	16	16	16	16	16	16	16	16	16	16
40	0	0	0	1	1	1	1	2	2	2	2	2	2	2	2	3	3	3	3	3	3	3	3	3	3
	0	0	0	1	3	5	8	11	14	18	22	26	30	31	31	38	38	38	38	38	38	38	38	38	38
	2	6	6	6	6	6	6	6	9	9	9	16	16	16	16	16	16	16	16	16	16	16	16	16	16
41	0	0	0	1	1	1	1	2	2	2	2	2	2	2	2	3	3	3	3	3	3	3	3	3	3
	0	0	0	1	3	5	8	11	14	18	22	26	30	32	32	39	39	39	39	39	39	39	39	39	39
	2	6	6	6	6	6	6	6	9	9	9	16	16	16	16	16	16	16	16	16	16	16	16	16	16
42	0	0	0	1	1	1	1	2	2	2	2	2	2	2	2	3	3	3	3	3	3	3	3	3	3
	0	0	0	1	3	5	8	11	14	18	22	26	30	33	33	40	40	40	40	40	40	40	40	40	40
	2	6	6	6	6	6	6	6	9	9	9	9	16	16	16	16	16	16	16	16	16	16	16	16	16
43	0	0	0	1	1	1	1	2	2	2	2	2	2	2	2	3	3	3	3	3	3	3	3	3	3
	0	0	0	1	3	5	8	11	14	18	22	26	30	34	34	41	41	41	41	41	41	41	41	41	41
	2	6	6	6	6	6	6	6	9	9	9	9	16	16	16	16	16	16	16	16	16	16	16	16	16
44	0	0	0	1	1	1	1	2	2	2	2	2	2	2	2	3	3	3	3	3	3	3	3	3	3
	0	0	0	1	3	5	8	11	14	18	22	26	30	34	35	42	42	42	42	42	42	42	42	42	42
	2	6	6	6	6	6	6	6	9	9	9	9	16	16	16	16	16	16	16	16	16	16	16	16	16
45	0	0	0	1	1	1	1	2	2	2	2	2	2	2	2	3	3	3	3	3	3	3	3	3	3
	0	0	0	1	3	5	8	11	14	18	22	26	30	34	36	43	43	43	43	43	43	43	43	43	43
	2	6	6	6	6	6	6	6	9	9	9	9	9	16	16	16	16	16	16	16	16	16	16	16	16
46	0	0	0	1	1	1	1	2	2	2	2	2	2	2	2	3	3	3	3	3	3	3	3	3	3
	0	0	0	1	3	5	8	11	14	18	22	26	30	34	37	43	44	44	44	44	44	44	44	44	44
	2	6	6	6	6	6	6	6	9	9	9	9	9	16	16	16	16	16	16	16	16	16	16	16	16
47	0	0	0	1	1	1	1	2	2	2	2	2	2	2	2	3	3	3	3	3	3	3	3	3	3
	0	0	0	1	3	5	8	11	14	18	22	26	30	34	38	43	45	45	45	45	45	45	45	45	45
	2	6	6	6	6	6	6	6	9	9	9	9	9	16	16	16	16	16	16	16	16	16	16	16	16
48	0	0	0	1	1	1	1	2	2	2	2	2	2	2	2	3	3	3	3	3	3	3	3	3	3
	0	0	0	1	3	5	8	11	14	18	22	26	30	34	38	43	46	46	46	46	46	46	46	46	46
	2	6	6	6	6	6	6	6	9	9	9	9	9	16	16	16	16	16	16	16	16	16	16	16	16
49	0	0	0	1	1	1	1	2	2	2	2	2	2	2	2	3	3	3	3	3	3	3	3	3	3
	0	0	0	1	3	5	8	11	14	18	22	26	30	34	38	43	47	47	47	47	47	47	47	47	47
	2	6	6	6	6	6	6	6	9	9	9	9	9	16	16	16	16	16	16	16	16	16	16	16	16
50	0	0	0	1	1	1	1	2	2	2	2	2	2	2	2	3	3	3	3	3	3	3	3	3	3
	0	0	0	1	3	5	8	11	14	18	22	26	30	34	38	43	48	48	48	48	48	48	48	48	48
	3	6	6	6	6	6	6	6	9	9	9	9	9	9	9	9	9	16	16	16	16	16	16	16	16

T TABLE BASE 2

m/s	26	27	28	29	30	31	32	33	34	35	36	37	38	39	40	41	42	43	44	45	46	47	48	49	50
26	3	3	3	3	3	3	4	4	4	4	4	4	4	4	4	4	4	4	4	4	4	4	4	4	4
	24	24	24	24	24	24	24	24	24	24	24	24	24	24	24	24	24	24	24	24	24	24	24	24	24
	16	16	16	16	16	16	16	16	16	16	16	16	16	16	16	16	16	16	16	16	16	16	16	16	16
27	3	3	3	3	3	3	4	4	4	4	4	4	4	4	4	4	4	4	4	4	4	4	4	4	4
	25	25	25	25	25	25	25	25	25	25	25	25	25	25	25	25	25	25	25	25	25	25	25	25	25
	16	16	16	16	16	16	16	16	16	16	16	16	16	16	16	16	16	16	16	16	16	16	16	16	16
28	3	3	3	3	3	3	4	4	4	4	4	4	4	4	4	4	4	4	4	4	4	4	4	4	4
	26	26	26	26	26	26	26	26	26	26	26	26	26	26	26	26	26	26	26	26	26	26	26	26	26
	16	16	16	16	16	16	16	16	16	16	16	16	16	16	16	16	16	16	16	16	16	16	16	16	16
29	3	3	3	3	3	3	4	4	4	4	4	4	4	4	4	4	4	4	4	4	4	4	4	4	4
	27	27	27	27	27	27	27	27	27	27	27	27	27	27	27	27	27	27	27	27	27	27	27	27	27
	16	16	16	16	16	16	16	16	16	16	16	16	16	16	16	16	16	16	16	16	16	16	16	16	16
30	3	3	3	3	3	3	4	4	4	4	4	4	4	4	4	4	4	4	4	4	4	4	4	4	4
	28	28	28	28	28	28	28	28	28	28	28	28	28	28	28	28	28	28	28	28	28	28	28	28	28
	16	16	16	16	16	16	16	16	16	16	16	16	16	16	16	16	16	16	16	16	16	16	16	16	16
31	3	3	3	3	3	3	4	4	4	4	4	4	4	4	4	4	4	4	4	4	4	4	4	4	4
	29	29	29	29	29	29	29	29	29	29	29	29	29	29	29	29	29	29	29	29	29	29	29	29	29
	16	16	16	16	16	16	16	16	16	16	16	16	16	16	16	16	16	16	16	16	16	16	16	16	16
32	3	3	3	3	3	3	4	4	4	4	4	4	4	4	4	4	4	4	4	4	4	4	4	4	4
	30	30	30	30	30	30	30	30	30	30	30	30	30	30	30	30	30	30	30	30	30	30	30	30	30
	16	16	16	16	16	16	16	16	16	16	16	16	16	16	16	16	16	16	16	16	16	16	16	16	16
33	3	3	3	3	3	3	4	4	4	4	4	4	4	4	4	4	4	4	4	4	4	4	4	4	4
	31	31	31	31	31	31	31	31	31	31	31	31	31	31	31	31	31	31	31	31	31	31	31	31	31
	16	16	16	16	16	16	16	16	16	16	16	16	16	16	16	16	16	16	16	16	16	16	16	16	16
34	3	3	3	3	3	3	4	4	4	4	4	4	4	4	4	4	4	4	4	4	4	4	4	4	4
	32	32	32	32	32	32	32	32	32	32	32	32	32	32	32	32	32	32	32	32	32	32	32	32	32
	16	16	16	16	16	16	16	16	16	16	16	16	16	16	16	16	16	16	16	16	16	16	16	16	16
35	3	3	3	3	3	3	4	4	4	4	4	4	4	4	4	4	4	4	4	4	4	4	4	4	4
	33	33	33	33	33	33	33	33	33	33	33	33	33	33	33	33	33	33	33	33	33	33	33	33	33
	16	16	16	16	16	16	16	16	16	16	16	16	16	16	16	16	16	16	16	16	16	16	16	16	16
36	3	3	3	3	3	3	4	4	4	4	4	4	4	4	4	4	4	4	4	4	4	4	4	4	4
	34	34	34	34	34	34	34	34	34	34	34	34	34	34	34	34	34	34	34	34	34	34	34	34	34
	16	16	16	16	16	16	16	16	16	16	16	16	16	16	16	16	16	16	16	16	16	16	16	16	16
37	3	3	3	3	3	3	4	4	4	4	4	4	4	4	4	4	4	4	4	4	4	4	4	4	4
	35	35	35	35	35	35	35	35	35	35	35	35	35	35	35	35	35	35	35	35	35	35	35	35	35
	16	16	16	16	16	16	16	16	16	16	16	16	16	16	16	16	16	16	16	16	16	16	16	16	16
38	3	3	3	3	3	3	4	4	4	4	4	4	4	4	4	4	4	4	4	4	4	4	4	4	4
	36	36	36	36	36	36	36	36	36	36	36	36	36	36	36	36	36	36	36	36	36	36	36	36	36
	16	16	16	16	16	16	16	16	16	16	16	16	16	16	16	16	16	16	16	16	16	16	16	16	16
39	3	3	3	3	3	3	4	4	4	4	4	4	4	4	4	4	4	4	4	4	4	4	4	4	4
	37	37	37	37	37	37	37	37	37	37	37	37	37	37	37	37	37	37	37	37	37	37	37	37	37
	16	16	16	16	16	16	16	16	16	16	16	16	16	16	16	16	16	16	16	16	16	16	16	16	16
40	3	3	3	3	3	3	4	4	4	4	4	4	4	4	4	4	4	4	4	4	4	4	4	4	4
	38	38	38	38	38	38	38	38	38	38	38	38	38	38	38	38	38	38	38	38	38	38	38	38	38
	16	16	16	16	16	16	16	16	16	16	16	16	16	16	16	16	16	16	16	16	16	16	16	16	16
41	3	3	3	3	3	3	4	4	4	4	4	4	4	4	4	4	4	4	4	4	4	4	4	4	4
	39	39	39	39	39	39	39	39	39	39	39	39	39	39	39	39	39	39	39	39	39	39	39	39	39
	16	16	16	16	16	16	16	16	16	16	16	16	16	16	16	16	16	16	16	16	16	16	16	16	16
42	3	3	3	3	3	3	4	4	4	4	4	4	4	4	4	4	4	4	4	4	4	4	4	4	4
	40	40	40	40	40	40	40	40	40	40	40	40	40	40	40	40	40	40	40	40	40	40	40	40	40
	16	16	16	16	16	16	16	16	16	16	16	16	16	16	16	16	16	16	16	16	16	16	16	16	16
43	3	3	3	3	3	3	4	4	4	4	4	4	4	4	4	4	4	4	4	4	4	4	4	4	4
	41	41	41	41	41	41	41	41	41	41	41	41	41	41	41	41	41	41	41	41	41	41	41	41	41
	16	16	16	16	16	16	16	16	16	16	16	16	16	16	16	16	16	16	16	16	16	16	16	16	16
44	3	3	3	3	3	3	4	4	4	4	4	4	4	4	4	4	4	4	4	4	4	4	4	4	4
	42	42	42	42	42	42	42	42	42	42	42	42	42	42	42	42	42	42	42	42	42	42	42	42	42
	16	16	16	16	16	16	16	16	16	16	16	16	16	16	16	16	16	16	16	16	16	16	16	16	16
45	3	3	3	3	3	3	4	4	4	4	4	4	4	4	4	4	4	4	4	4	4	4	4	4	4
	43	43	43	43	43	43	43	43	43	43	43	43	43	43	43	43	43	43	43	43	43	43	43	43	43
	16	16	16	16	16	16	16	16	16	16	16	16	16	16	16	16	16	16	16	16	16	16	16	16	16
46	3	3	3	3	3	3	4	4	4	4	4	4	4	4	4	4	4	4	4	4	4	4	4	4	4
	44	44	44	44	44	44	44	44	44	44	44	44	44	44	44	44	44	44	44	44	44	44	44	44	44
	16	16	16	16	16	16	16	16	16	16	16	16	16	16	16	16	16	16	16	16	16	16	16	16	16
47	3	3	3	3	3	3	4	4	4	4	4	4	4	4	4	4	4	4	4	4	4	4	4	4	4
	45	45	45	45	45	45	45	45	45	45	45	45	45	45	45	45	45	45	45	45	45	45	45	45	45
	16	16	16	16	16	16	16	16	16	16	16	16	16	16	16	16	16	16	16	16	16	16	16	16	16
48	3	3	3	3	3	3	4	4	4	4	4	4	4	4	4	4	4	4	4	4	4	4	4	4	4
	46	46	46	46	46	46	46	46	46	46	46	46	46	46	46	46	46	46	46	46	46	46	46	46	46
	16	16	16	16	16	16	16	16	16	16	16	16	16	16	16	16	16	16	16	16	16	16	16	16	16
49	3	3	3	3	3	3	4	4	4	4	4	4	4	4	4	4	4	4	4	4	4	4	4	4	4
	47	47	47	47	47	47	47	47	47	47	47	47	47	47	47	47	47	47	47	47	47	47	47	47	47
	16	16	16	16	16	16	16	16	16	16	16	16	16	16	16	16	16	16	16	16	16	16	16	16	16
50	3	3	3	3	3	3	4	4	4	4	4	4	4	4	4	4	4	4	4	4	4	4	4	4	4
	48	48	48	48	48	48	48	48	48	48	48	48	48	48	48	48	48	48	48	48	48	48	48	48	48
	16	16	16	16	16	16	16	16	16	16	16	16	16	16	16	16	16	16	16	16	16	16	16	16	16

S TABLE BASE 2

t/m	1	2	3	4	5	6	7	8	9	10	11	12	13	14	15	16	17	18	19	20	21	22	23	24	25
0	999	3	3	3	3	3	3	3	3	3	3	3	3	3	3	3	3	3	3	3	3	3	3	3	3
	999	3	3	3	3	3	3	3	3	3	3	3	3	3	3	3	3	3	3	3	3	3	3	3	3
	5	6	6	6	6	6	6	6	6	6	6	6	6	6	6	6	6	6	6	6	6	6	6	6	6
1	999	999	7	6	6	4	4	4	4	4	4	4	4	4	4	4	4	4	4	4	4	4	4	4	4
	999	999	7	7	7	7	7	7	7	7	7	7	7	7	7	7	7	7	7	7	7	7	7	7	7
	4	5	12	13	6	6	0	0	6	6	6	6	6	6	6	6	6	6	6	6	6	6	6	6	6
2		999	999	15	10	6	6	5	4	4	4	4	4	4	4	4	4	4	4	4	4	4	4	4	4
		999	999	15	15	15	15	15	15	15	15	15	15	15	15	15	15	15	15	15	15	15	15	15
		4	5	12	13	13	13	13	16	16	16	16	16	16	16	16	16	16	16	16	16	16	16	16
3			999	999	31	10	8	6	5	5	5	5	5	5	5	5	5	5	5	5	5	5	5	5	5
			999	999	31	31	31	31	31	31	31	31	31	31	31	31	31	31	31	31	31	31	31	31
			4	5	12	16	13	16	6	6	6	6	6	6	6	6	6	6	6	6	6	6	6	6
4				999	999	63	12	11	7	7	6	5	5	5	5	5	5	5	5	5	5	5	5	5	5
				999	999	63	63	63	63	63	63	63	63	63	63	63	63	63	63	63	63	63	63	63
				4	5	12	13	13	15	16	16	16	16	16	16	16	16	16	16	16	16	16	16	16
5					999	999	127	12	11	8	7	7	7	6	6	6	6	6	6	6	6	6	6	6	6
					999	999	127	127	127	127	127	127	127	127	127	127	127	127	127	127	127	127	127	127
					4	5	12	16	16	13	16	6	6	6	6	6	6	6	6	6	6	6	6	6
6						999	999	255	15	12	10	8	8	8	8	8	6	6	6	6	6	6	6	6	6
						999	999	255	255	255	255	255	255	255	255	255	255	255	255	255	255	255	255	255
						4	5	12	15	13	15	3	15	3	15	6	6	6	6	6	6	6	6	6
7							999	999	511	15	12	10	10	8	8	8	8	8	8	8	8	8	6	6	6
							999	999	511	511	511	511	511	511	511	511	511	511	511	511	511	511	511	511
							4	5	12	16	16	15	16	1	16	3	3	3	3	15	16	16	16	16
8								999	999	998	15	15	15	10	10	10	10	8	8	8	8	8	7	7	7
								999	999	998	998	998	998	998	998	998	998	998	998	998	998	998	998	998
								4	5	12	16	16	15	16	16	3	15	16	15	15	3	1	16	6
9									999	999	998	15	15	15	12	12	12	8	8	8	8	8	8	8	7
									999	999	998	998	998	998	998	998	998	998	998	998	998	998	998	998
									4	5	12	16	1	16	16	3	15	16	16	16	16	16	16	16
10										999	999	998	15	15	15	15	12	12	12	9	9	9	8	8	8
										999	999	998	998	998	998	998	998	998	998	998	998	998	998	998
										4	5	12	16	16	16	16	1	16	16	3	15	16	16	16
11											999	999	998	15	15	15	15	12	12	12	11	11	11	9	8
											999	999	998	998	998	998	998	998	998	998	998	998	998	998
											4	5	12	16	16	16	16	1	16	16	3	15	16	16
12												999	999	998	15	15	15	15	14	14	14	11	11	11	6
												999	999	998	998	998	998	998	998	998	998	998	998	998
												4	5	12	16	16	16	16	3	15	1	16	16	16
13													999	999	998	15	15	15	15	15	15	11	11	11	9
													999	999	998	998	998	998	998	998	998	998	998	998
													4	5	12	15	16	16	16	16	3	15	16	16
14														999	999	998	15	15	15	15	15	15	15	11	11
														999	999	998	998	998	998	998	998	998	998	998
														4	5	12	16	16	16	16	16	16	1	16
15															999	999	998	15	15	15	15	15	15	15	11
															999	999	998	998	998	998	998	998	998	998
															4	5	12	16	16	16	16	16	16	16
16																999	999	998	15	15	15	15	15	15	15
																999	999	998	998	998	998	998	998	998
																4	5	12	16	16	16	16	16	16
17																	999	999	998	15	15	15	15	15	15
																	999	999	998	998	998	998	998	998
																	4	5	12	16	16	16	16	16
18																		999	999	998	15	15	15	15	15
																		999	999	998	998	998	998	998
																		4	5	12	16	16	16	16
19																			999	999	998	15	15	15	15
																			999	999	998	998	998	998
																			4	5	12	16	16	16
20																				999	999	998	15	15	15
																				999	999	998	998	998
																				4	5	12	16	16
21																					999	999	998	15	15
																					999	999	998	998
																					4	5	12	16
22																						999	999	998	15
																						999	999	998
																						4	5	12
23																							999	999	998
																							999	999	998
																							4	5	12
24																								999	999
																								999	999
																								4	5
25																									999
																									999
																									4

S TABLE BASE 2

t/m	26	27	28	29	30	31	32	33	34	35	36	37	38	39	40	41	42	43	44	45	46	47	48	49	50
0	3	3	3	3	3	3	3	3	3	3	3	3	3	3	3	3	3	3	3	3	3	3	3	3	3
	3	3	3	3	3	3	3	3	3	3	3	3	3	3	3	3	3	3	3	3	3	3	3	3	3
	6	6	6	6	6	6	6	6	6	6	6	6	6	6	6	6	6	6	6	6	6	6	6	6	6
1	4	4	4	4	4	4	4	4	4	4	4	4	4	4	4	4	4	4	4	4	4	4	4	4	4
	7	7	7	7	7	7	7	7	7	7	7	7	7	7	7	7	7	7	7	7	7	7	7	7	7
	6	6	6	6	6	6	6	6	6	6	6	6	6	6	6	6	6	6	6	6	6	6	6	6	6
2	4	4	4	4	4	4	4	4	4	4	4	4	4	4	4	4	4	4	4	4	4	4	4	4	4
	15	15	15	15	15	15	15	15	15	15	15	15	15	15	15	15	15	15	15	15	15	15	15	15	15
	16	16	16	16	16	16	16	16	16	16	16	16	16	16	16	16	16	16	16	16	16	16	16	16	16
3	5	5	5	5	5	5	5	5	5	5	5	5	5	5	5	5	5	5	5	5	5	5	5	5	5
	31	31	31	31	31	31	31	31	31	31	31	31	31	31	31	31	31	31	31	31	31	31	31	31	31
	6	6	6	6	6	6	6	6	6	6	6	6	6	6	6	6	6	6	6	6	6	6	6	6	6
4	5	5	5	5	5	5	5	5	5	5	5	5	5	5	5	5	5	5	5	5	5	5	5	5	5
	63	63	63	63	63	63	63	63	63	63	63	63	63	63	63	63	63	63	63	63	63	63	63	63	63
	16	16	16	16	16	16	16	16	16	16	16	16	16	16	16	16	16	16	16	16	16	16	16	16	16
5	6	6	6	6	6	6	6	6	6	6	6	6	6	6	6	6	6	6	6	6	6	6	6	6	6
	127	127	127	127	127	127	127	127	127	127	127	127	127	127	127	127	127	127	127	127	127	127	127	127	127
	6	6	6	6	6	6	6	6	6	6	6	6	6	6	6	6	6	6	6	6	6	6	6	6	6
6	6	6	6	6	6	6	6	6	6	6	6	6	6	6	6	6	6	6	6	6	6	6	6	6	6
	255	255	255	255	255	255	255	255	255	255	255	255	255	255	255	255	255	255	255	255	255	255	255	255	255
	16	16	16	16	16	16	16	16	16	16	16	16	16	16	16	16	16	16	16	16	16	16	16	16	16
7	6	6	6	6	6	6	6	6	6	6	6	6	6	6	6	6	6	6	6	6	6	6	6	6	6
	511	511	511	511	511	511	511	511	511	511	511	511	511	511	511	511	511	511	511	511	511	511	511	511	511
	16	16	16	16	16	16	16	16	16	16	16	16	16	16	16	16	16	16	16	16	16	16	16	16	16
8	7	7	7	7	7	7	7	7	7	7	7	7	7	7	7	7	7	7	7	7	7	7	7	7	7
	998	998	998	998	998	998	998	998	998	998	998	998	998	998	998	998	998	998	998	998	998	998	998	998	998
	6	6	6	6	6	6	6	6	6	6	6	6	6	6	6	6	6	6	6	6	6	6	6	6	6
9	7	7	7	7	7	7	7	7	7	7	7	7	7	7	7	7	7	7	7	7	7	7	7	7	7
	998	998	998	998	998	998	998	998	998	998	998	998	998	998	998	998	998	998	998	998	998	998	998	998	998
	16	16	16	16	16	16	16	16	16	16	16	16	16	16	16	16	16	16	16	16	16	16	16	16	16
10	7	7	7	7	7	7	7	7	7	7	7	7	7	7	7	7	7	7	7	7	7	7	7	7	7
	998	998	998	998	998	998	998	998	998	998	998	998	998	998	998	998	998	998	998	998	998	998	998	998	998
	16	16	16	16	16	16	16	16	16	16	16	16	16	16	16	16	16	16	16	16	16	16	16	16	16
11	8	8	8	8	8	8	8	8	8	8	8	8	8	8	8	8	8	8	8	8	8	8	8	8	8
	998	998	998	998	998	998	998	998	998	998	998	998	998	998	998	998	998	998	998	998	998	998	998	998	998
	6	6	6	6	6	6	6	6	6	6	6	6	6	6	6	6	6	6	6	6	6	6	6	6	6
12	8	8	8	8	8	8	8	8	8	8	8	8	8	8	8	8	8	8	8	8	8	8	8	8	8
	998	998	998	998	998	998	998	998	998	998	998	998	998	998	998	998	998	998	998	998	998	998	998	998	998
	16	16	16	16	16	16	16	16	16	16	16	16	16	16	16	16	16	16	16	16	16	16	16	16	16
13	8	8	8	8	8	8	8	8	8	8	8	8	8	8	8	8	8	8	8	8	8	8	8	8	8
	998	998	998	998	998	998	998	998	998	998	998	998	998	998	998	998	998	998	998	998	998	998	998	998	998
	16	16	16	16	16	16	16	16	16	16	16	16	16	16	16	16	16	16	16	16	16	16	16	16	16
14	9	9	9	9	9	9	9	9	9	9	9	9	9	9	9	9	9	9	9	9	9	9	9	9	9
	998	998	998	998	998	998	998	998	998	998	998	998	998	998	998	998	998	998	998	998	998	998	998	998	998
	9	9	9	9	9	9	9	9	9	9	9	9	9	9	9	9	9	9	9	9	9	9	9	9	9
15	11	9	9	9	9	9	9	9	9	9	9	9	9	9	9	9	9	9	9	9	9	9	9	9	9
	998	998	998	998	998	998	998	998	998	998	998	998	998	998	998	998	998	998	998	998	998	998	998	998	998
	16	16	16	16	16	16	16	16	16	16	16	16	16	16	16	16	16	16	16	16	16	16	16	16	16
16	11	11	9	9	9	9	9	9	9	9	9	9	9	9	9	9	9	9	9	9	9	9	9	9	9
	998	998	998	998	998	998	998	998	998	998	998	998	998	998	998	998	998	998	998	998	998	998	998	998	998
	16	16	16	16	16	16	16	16	16	16	16	16	16	16	16	16	16	16	16	16	16	16	16	16	16
17	15	11	11	9	9	9	9	9	9	9	9	9	9	9	9	9	9	9	9	9	9	9	9	9	9
	998	998	998	998	998	998	998	998	998	998	998	998	998	998	998	998	998	998	998	998	998	998	998	998	998
	16	16	16	16	16	16	16	16	16	16	16	16	16	16	16	16	16	16	16	16	16	16	16	16	16
18	15	15	11	10	10	10	10	10	10	10	10	10	10	10	10	10	10	10	10	10	10	10	10	10	10
	998	998	998	998	998	998	998	998	998	998	998	998	998	998	998	998	998	998	998	998	998	998	998	998	998
	16	16	16	9	9	9	9	9	9	9	9	9	9	9	9	9	9	9	9	9	9	9	9	9	9
19	15	15	15	11	11	10	10	10	10	10	10	10	10	10	10	10	10	10	10	10	10	10	10	10	10
	998	998	998	998	998	998	998	998	998	998	998	998	998	998	998	998	998	998	998	998	998	998	998	998	998
	16	16	16	16	16	16	16	16	16	16	16	16	16	16	16	16	16	16	16	16	16	16	16	16	16
20	15	15	15	15	11	10	10	10	10	10	10	10	10	10	10	10	10	10	10	10	10	10	10	10	10
	998	998	998	998	998	998	998	998	998	998	998	998	998	998	998	998	998	998	998	998	998	998	998	998	998
	16	16	16	16	16	16	16	16	16	16	16	16	16	16	16	16	16	16	16	16	16	16	16	16	16
21	15	15	15	15	15	11	11	10	10	10	10	10	10	10	10	10	10	10	10	10	10	10	10	10	10
	998	998	998	998	998	998	998	998	998	998	998	998	998	998	998	998	998	998	998	998	998	998	998	998	998
	16	16	16	16	16	16	16	16	16	16	16	16	16	16	16	16	16	16	16	16	16	16	16	16	16
22	15	15	15	15	15	15	11	11	11	11	11	11	11	11	11	11	11	11	11	11	11	11	11	11	11
	998	998	998	998	998	998	998	998	998	998	998	998	998	998	998	998	998	998	998	998	998	998	998	998	998
	16	16	16	16	16	16	9	9	9	9	9	9	9	9	9	9	9	9	9	9	9	9	9	9	9
23	15	15	15	15	15	15	15	11	11	11	11	11	11	11	11	11	11	11	11	11	11	11	11	11	11
	998	998	998	998	998	998	998	998	998	998	998	998	998	998	998	998	998	998	998	998	998	998	998	998	998
	16	16	16	16	16	16	16	16	16	16	16	16	16	16	16	16	16	16	16	16	16	16	16	16	16
24	998	998	15	15	15	15	15	15	15	11	11	11	11	11	11	11	11	11	11	11	11	11	11	11	11
	998	998	998	998	998	998	998	998	998	998	998	998	998	998	998	998	998	998	998	998	998	998	998	998	998
	12	16	16	16	16	16	16	16	16	16	16	16	16	16	16	16	16	16	16	16	16	16	16	16	16
25	999	998	15	15	15	15	15	15	15	11	11	11	11	11	11	11	11	11	11	11	11	11	11	11	11
	999	998	998	998	998	998	998	998	998	998	998	998	998	998	998	998	998	998	998	998	998	998	998	998	998
	5	12	16	16	16	16	16	16	16	16	16	16	16	16	16	16	16	16	16	16	16	16	16	16	16

S TABLE BASE 2

t/m	26	27	28	29	30	31	32	33	34	35	36	37	38	39	40	41	42	43	44	45	46	47	48	49	50
26	999	999	998	15	15	15	15	15	15	15	12	12	12	12	12	12	12	12	12	12	12	12	12	12	12
	999	999	998	998	998	998	998	998	998	998	998	998	998	998	998	998	998	998	998	998	998	998	998	998	998
	4	5	12	16	16	16	16	16	16	16	9	9	9	9	9	9	9	9	9	9	9	9	9	9	9
27		999	999	998	15	15	15	15	15	15	15	12	12	12	12	12	12	12	12	12	12	12	12	12	12
		999	999	998	998	998	998	998	998	998	998	998	998	998	998	998	998	998	998	998	998	998	998	998	998
		4	5	12	16	16	16	16	16	16	16	16	16	16	16	16	16	16	16	16	16	10	10	10	10
28			999	999	998	15	15	15	15	15	15	15	12	12	12	12	12	12	12	12	12	12	12	12	12
			999	999	998	998	998	998	998	998	998	998	998	998	998	998	998	998	998	998	998	998	998	998	998
			4	5	12	16	16	16	16	16	16	16	16	16	16	16	16	16	16	16	16	16	16	16	16
29				999	999	998	15	15	15	15	15	15	15	12	12	12	12	12	12	12	12	12	12	12	12
				999	999	998	998	998	998	998	998	998	998	998	998	998	998	998	998	998	998	998	998	998	998
				4	5	12	16	16	16	16	16	16	16	16	16	16	16	16	16	16	16	16	16	16	16
30					999	999	998	15	15	15	15	15	15	15	13	13	13	13	13	13	13	13	13	13	13
					999	999	998	998	998	998	998	998	998	998	998	998	998	998	998	998	998	998	998	998	998
					4	5	12	16	16	16	16	16	16	16	9	9	9	9	9	9	9	9	9	9	9
31						999	999	998	15	15	15	15	15	15	15	13	13	13	13	13	13	13	13	13	13
						999	999	998	998	998	998	998	998	998	998	998	998	998	998	998	998	998	998	998	998
						4	5	12	16	16	16	16	16	16	16	16	16	16	16	16	16	16	16	16	16
32							999	999	998	15	15	15	15	15	15	15	13	13	13	13	13	13	13	13	13
							999	999	998	998	998	998	998	998	998	998	998	998	998	998	998	998	998	998	998
							4	5	12	16	16	16	16	16	16	16	16	16	16	16	16	16	16	16
33								999	999	998	15	15	15	15	15	15	15	13	13	13	13	13	13	13	13
								999	999	998	998	998	998	998	998	998	998	998	998	998	998	998	998	998	998
								4	5	12	16	16	16	16	16	16	16	16	16	16	16	16	16	16
34									999	999	998	15	15	15	15	15	15	15	14	14	14	14	14	14	14
									999	999	998	998	998	998	998	998	998	998	998	998	998	998	998	998	998
									4	5	12	16	16	16	16	16	16	16	9	9	9	9	9	9	9
35										999	999	998	15	15	15	15	15	15	15	14	14	14	14	14	14
										999	999	998	998	998	998	998	998	998	998	998	998	998	998	998	998
										4	5	12	16	16	16	16	16	16	16	16	16	16	16	16	16
36											999	999	998	15	15	15	15	15	15	15	14	14	14	14	14
											999	999	998	998	998	998	998	998	998	998	998	998	998	998	998
											4	5	12	16	16	16	16	16	16	16	16	16	16	16	16
37												999	999	998	15	15	15	15	15	15	15	14	14	14	14
												999	999	998	998	998	998	998	998	998	998	998	998	998	998
												4	5	12	16	16	16	16	16	16	16	16	16	16	16
38													999	999	998	15	15	15	15	15	15	15	15	15	15
													999	999	998	998	998	998	998	998	998	998	998	998	998
													4	5	12	16	16	16	16	16	16	16	9	9	9
39														999	999	998	15	15	15	15	15	15	15	15	15
														999	999	998	998	998	998	998	998	998	998	998	998
														4	5	12	16	16	16	16	16	16	16	16	16
40															999	999	998	15	15	15	15	15	15	15	15
															999	999	998	998	998	998	998	998	998	998	998
															4	5	12	16	16	16	16	16	16	16	16
41																999	999	998	15	15	15	15	15	15	15
																999	999	998	998	998	998	998	998	998	998
																4	5	12	16	16	16	16	16	16	16
42																	999	999	998	15	15	15	15	15	15
																	999	999	998	998	998	998	998	998	998
																	4	5	12	16	16	16	16	16	16
43																		999	999	998	16	16	16	16	16
																		999	999	998	998	998	998	998	998
																		4	5	12	9	9	9	9	9
44																			999	999	998	16	16	16	16
																			999	999	998	998	998	998	998
																			4	5	12	16	16	16	16
45																				999	999	998	16	16	16
																				999	999	998	998	998	998
																				4	5	12	16	16	16
46																					999	999	998	16	16
																					999	999	998	998	998
																					4	5	12	16	16
47																						999	999	998	16
																						999	999	998	998
																						4	5	12	16
48																							999	999	998
																							999	999	998
																							4	5	12
49																								999	999
																								999	999
																								4	5
50																									999
																									999
																									1

T TABLE BASE 3

m/s	1	2	3	4	5	6	7	8	9	10	11	12	13	14	15	16	17	18	19	20	21	22	23	24	25
1	0	0	0	0	0	0	0	0	0	0	0	0	0	0	0	0	0	0	0	0	0	0	0	0	0
	0	0	0	0	0	0	0	0	0	0	0	0	0	0	0	0	0	0	0	0	0	0	0	0	0
	5	5	5	5	5	5	5	5	5	5	5	5	5	5	5	5	5	5	5	5	5	5	5	5	5
2	0	0	0	0	1	1	1	1	1	1	1	1	1	1	1	1	1	1	1	1	1	1	1	1	1
	0	0	0	0	1	1	1	1	1	1	1	1	1	1	1	1	1	1	1	1	1	1	1	1	1
	2	2	2	7	5	5	5	5	5	5	5	5	5	5	5	5	5	5	5	5	5	5	5	5	5
3	0	0	0	0	1	1	1	1	1	1	1	1	1	2	2	2	2	2	2	2	2	2	2	2	2
	0	0	0	0	1	1	1	1	1	1	1	1	1	2	2	2	2	2	2	2	2	2	2	2	2
	2	2	2	8	9	2	2	2	2	2	2	12	5	5	5	5	5	5	5	5	5	5	5	5	5
4	0	0	0	0	1	1	1	1	1	1	1	1	1	2	2	2	2	2	2	2	2	2	2	2	2
	0	0	0	0	1	1	1	1	1	1	1	1	1	2	2	2	2	2	2	2	2	2	2	2	2
	2	2	2	8	9	9	16	16	16	16	16	16	2	2	2	2	2	2	2	2	2	2	2	2	2
5	0	0	0	0	1	1	1	1	1	1	1	1	1	2	2	2	2	2	2	2	2	2	2	2	2
	0	0	0	0	1	2	3	3	3	3	3	3	3	2	2	2	2	2	2	2	2	2	2	2	2
	2	2	2	8	9	9	9	16	16	16	16	16	16	3	3	3	3	3	3	3	3	3	3	3	3
6	0	0	0	0	1	1	1	1	1	1	1	1	1	2	2	2	2	2	2	2	2	2	2	2	2
	0	0	0	0	1	2	3	4	4	4	4	4	4	4	4	4	4	4	4	4	4	4	4	4	4
	2	2	2	8	9	9	9	16	16	16	16	16	16	16	16	16	16	16	16	16	16	16	16	16	16
7	0	0	0	0	1	1	1	1	1	1	1	1	1	2	2	2	2	2	2	2	2	2	2	2	2
	0	0	0	0	1	2	3	5	5	5	5	5	5	5	5	5	5	5	5	5	5	5	5	5	5
	2	2	2	8	9	9	9	9	16	16	16	16	16	16	16	16	16	16	16	16	16	16	16	16	16
8	0	0	0	0	1	1	1	1	1	1	1	1	1	2	2	2	2	2	2	2	2	2	2	2	2
	0	0	0	0	1	2	3	5	6	6	6	6	6	6	6	6	6	6	6	6	6	6	6	6	6
	2	2	2	8	9	9	9	16	16	16	16	16	16	16	16	16	16	16	16	16	16	16	16	16	16
9	0	0	0	0	1	1	1	1	1	1	1	1	1	2	2	2	2	2	2	2	2	2	2	2	2
	0	0	0	0	1	2	3	5	7	7	7	7	7	7	7	7	7	7	7	7	7	7	7	7	7
	2	2	2	8	9	9	9	9	16	16	16	16	16	16	16	16	16	16	16	16	16	16	16	16	16
10	0	0	0	0	1	1	1	1	1	1	1	1	1	2	2	2	2	2	2	2	2	2	2	2	2
	0	0	0	0	1	2	3	5	7	8	8	8	8	8	8	8	8	8	8	8	8	8	8	8	8
	2	2	2	8	9	9	9	9	16	16	16	16	16	16	16	16	16	16	16	16	16	16	16	16	16
11	0	0	0	0	1	1	1	1	1	1	1	1	1	2	2	2	2	2	2	2	2	2	2	2	2
	0	0	0	0	1	2	3	5	7	9	9	9	9	9	9	9	9	9	9	9	9	9	9	9	9
	2	2	2	8	9	9	9	9	16	16	16	16	16	16	16	16	16	16	16	16	16	16	16	16	16
12	0	0	0	0	1	1	1	1	1	1	1	1	1	2	2	2	2	2	2	2	2	2	2	2	2
	0	0	0	0	1	2	3	5	7	9	10	10	10	10	10	10	10	10	10	10	10	10	10	10	10
	2	2	2	8	9	9	9	9	9	16	16	16	16	16	16	16	16	16	16	16	16	16	16	16	16
13	0	0	0	0	1	1	1	1	1	1	1	1	1	2	2	2	2	2	2	2	2	2	2	2	2
	0	0	0	0	1	2	3	5	7	9	11	11	11	11	11	11	11	11	11	11	11	11	11	11	11
	2	2	2	8	9	9	9	9	9	16	16	16	16	16	16	16	16	16	16	16	16	16	16	16	16
14	0	0	0	0	1	1	1	1	1	1	1	1	1	2	2	2	2	2	2	2	2	2	2	2	2
	0	0	0	0	1	2	3	5	7	9	11	12	12	12	12	12	12	12	12	12	12	12	12	12	12
	2	2	2	8	9	9	9	9	9	16	16	16	16	16	16	16	16	16	16	16	16	16	16	16	16
15	0	0	0	0	1	1	1	1	1	1	1	1	1	2	2	2	2	2	2	2	2	2	2	2	2
	0	0	0	0	1	2	3	5	7	9	11	13	13	13	13	13	13	13	13	13	13	13	13	13	13
	2	2	2	8	9	9	9	9	9	16	16	16	16	16	16	16	16	16	16	16	16	16	16	16	16
16	0	0	0	0	1	1	1	1	1	1	1	1	1	2	2	2	2	2	2	2	2	2	2	2	2
	0	0	0	0	1	2	3	5	7	9	11	13	14	14	14	14	14	14	14	14	14	14	14	14	14
	2	2	2	8	9	9	9	9	9	16	16	16	16	16	16	16	16	16	16	16	16	16	16	16	16
17	0	0	0	0	1	1	1	1	1	1	1	1	1	2	2	2	2	2	2	2	2	2	2	2	2
	0	0	0	0	1	2	3	5	7	9	11	13	15	15	15	15	15	15	15	15	15	15	15	15	15
	2	2	2	8	9	9	9	9	9	9	16	16	16	16	16	16	16	16	16	16	16	16	16	16	16
18	0	0	0	0	1	1	1	1	1	1	1	1	1	2	2	2	2	2	2	2	2	2	2	2	2
	0	0	0	0	1	2	3	5	7	9	11	13	15	16	16	16	16	16	16	16	16	16	16	16	16
	2	2	2	8	9	9	9	9	9	9	16	16	16	16	16	16	16	16	16	16	16	16	16	16	16
19	0	0	0	0	1	1	1	1	1	1	1	1	1	2	2	2	2	2	2	2	2	2	2	2	2
	0	0	0	0	1	2	3	5	7	9	11	13	15	17	17	17	17	17	17	17	17	17	17	17	17
	2	2	2	8	9	9	9	9	9	9	16	16	16	16	16	16	16	16	16	16	16	16	16	16	16
20	0	0	0	0	1	1	1	1	1	1	1	1	1	2	2	2	2	2	2	2	2	2	2	2	2
	0	0	0	0	1	2	3	5	7	9	11	13	15	17	18	18	18	18	18	18	18	18	18	18	18
	2	2	2	8	9	9	9	9	9	9	9	16	16	16	16	16	16	16	16	16	16	16	16	16	16
21	0	0	0	0	1	1	1	1	1	1	1	1	1	2	2	2	2	2	2	2	2	2	2	2	2
	0	0	0	0	1	2	3	5	7	9	11	13	15	17	19	19	19	19	19	19	19	19	19	19	19
	2	2	2	8	9	9	9	9	9	9	9	16	16	16	16	16	16	16	16	16	16	16	16	16	16
22	0	0	0	0	1	1	1	1	1	1	1	1	1	2	2	2	2	2	2	2	2	2	2	2	2
	0	0	0	0	1	2	3	5	7	9	11	13	15	17	19	20	20	20	20	20	20	20	20	20	20
	2	2	2	8	9	9	9	9	9	9	9	16	16	16	16	16	16	16	16	16	16	16	16	16	16
23	0	0	0	0	1	1	1	1	1	1	1	1	1	2	2	2	2	2	2	2	2	2	2	2	2
	0	0	0	0	1	2	3	5	7	9	11	13	15	17	19	21	21	21	21	21	21	21	21	21	21
	2	2	2	8	9	9	9	9	9	9	9	16	16	16	16	16	16	16	16	16	16	16	16	16	16
24	0	0	0	0	1	1	1	1	1	1	1	1	1	2	2	2	2	2	2	2	2	2	2	2	2
	0	0	0	0	1	2	3	5	7	9	11	13	15	17	19	22	22	22	22	22	22	22	22	22	22
	2	2	2	8	9	9	9	9	9	9	9	16	16	16	16	16	16	16	16	16	16	16	16	16	16
25	0	0	0	0	1	1	1	1	1	1	1	1	1	2	2	2	2	2	2	2	2	2	2	2	2
	0	0	0	0	1	2	3	5	7	9	11	13	15	17	19	22	23	23	23	23	23	23	23	23	23
	2	2	2	8	9	9	9	9	9	9	9	16	16	16	16	16	16	16	16	16	16	16	16	16	16

T TABLE BASE 3

m/s	26	27	28	29	30	31	32	33	34	35	36	37	38	39	40	41	42	43	44	45	46	47	48	49	50
1	0	0	0	0	0	0	0	0	0	0	0	0	0	0	0	0	0	0	0	0	0	0	0	0	0
	0	0	0	0	0	0	0	0	0	0	0	0	0	0	0	0	0	0	0	0	0	0	0	0	0
	5	5	5	5	5	5	5	5	5	5	5	5	5	5	5	5	5	5	5	5	5	5	5	5	5
2	1	1	1	1	1	1	1	1	1	1	1	1	1	1	1	1	1	1	1	1	1	1	1	1	1
	1	1	1	1	1	1	1	1	1	1	1	1	1	1	1	1	1	1	1	1	1	1	1	1	1
	5	5	5	5	5	5	5	5	5	5	5	5	5	5	5	5	5	5	5	5	5	5	5	5	5
3	2	2	2	2	2	2	2	2	2	2	2	2	2	2	2	2	2	2	2	2	2	2	2	2	2
	2	2	2	2	2	2	2	2	2	2	2	2	2	2	2	2	2	2	2	2	2	2	2	2	2
	5	5	5	5	5	5	5	5	5	5	5	5	5	5	5	5	5	5	5	5	5	5	5	5	5
4	2	2	2	2	2	2	2	2	2	2	2	2	2	2	3	3	3	3	3	3	3	3	3	3	3
	2	2	2	2	2	2	2	2	2	2	2	2	2	2	3	3	3	3	3	3	3	3	3	3	3
	2	2	2	2	2	2	2	2	2	2	2	2	2	2	12	5	5	5	5	5	5	5	5	5	5
5	2	2	2	2	2	2	2	2	2	2	2	2	2	2	3	3	3	3	3	3	3	3	3	3	3
	3	3	3	3	3	3	3	3	3	3	3	3	3	3	3	3	3	3	3	3	3	3	3	3	3
	16	16	16	16	16	16	16	16	16	16	16	16	16	16	2	2	2	2	2	2	2	2	2	2	2
6	2	2	2	2	2	2	2	2	2	2	2	2	2	2	3	3	3	3	3	3	3	3	3	3	3
	4	4	4	4	4	4	4	4	4	4	4	4	4	4	4	4	4	4	4	4	4	4	4	4	4
	16	16	16	16	16	16	16	16	16	16	16	16	16	16	16	16	16	16	16	16	16	16	16	16	16
7	2	2	2	2	2	2	2	2	2	2	2	2	2	2	3	3	3	3	3	3	3	3	3	3	3
	5	5	5	5	5	5	5	5	5	5	5	5	5	5	5	5	5	5	5	5	5	5	5	5	5
	16	16	16	16	16	16	16	16	16	16	16	16	16	16	16	16	16	16	16	16	16	16	16	16	16
8	2	2	2	2	2	2	2	2	2	2	2	2	2	2	3	3	3	3	3	3	3	3	3	3	3
	6	6	6	6	6	6	6	6	6	6	6	6	6	6	6	6	6	6	6	6	6	6	6	6	6
	16	16	16	16	16	16	16	16	16	16	16	16	16	16	16	16	16	16	16	16	16	16	16	16	16
9	2	2	2	2	2	2	2	2	2	2	2	2	2	2	3	3	3	3	3	3	3	3	3	3	3
	7	7	7	7	7	7	7	7	7	7	7	7	7	7	7	7	7	7	7	7	7	7	7	7	7
	16	16	16	16	16	16	16	16	16	16	16	16	16	16	16	16	16	16	16	16	16	16	16	16	16
10	2	2	2	2	2	2	2	2	2	2	2	2	2	2	3	3	3	3	3	3	3	3	3	3	3
	8	8	8	8	8	8	8	8	8	8	8	8	8	8	8	8	8	8	8	8	8	8	8	8	8
	16	16	16	16	16	16	16	16	16	16	16	16	16	16	16	16	16	16	16	16	16	16	16	16	16
11	2	2	2	2	2	2	2	2	2	2	2	2	2	2	3	3	3	3	3	3	3	3	3	3	3
	9	9	9	9	9	9	9	9	9	9	9	9	9	9	9	9	9	9	9	9	9	9	9	9	9
	16	16	16	16	16	16	16	16	16	16	16	16	16	16	16	16	16	16	16	16	16	16	16	16	16
12	2	2	2	2	2	2	2	2	2	2	2	2	2	2	3	3	3	3	3	3	3	3	3	3	3
	10	10	10	10	10	10	10	10	10	10	10	10	10	10	10	10	10	10	10	10	10	10	10	10	10
	16	16	16	16	16	16	16	16	16	16	16	16	16	16	16	16	16	16	16	16	16	16	16	16	16
13	2	2	2	2	2	2	2	2	2	2	2	2	2	2	3	3	3	3	3	3	3	3	3	3	3
	11	11	11	11	11	11	11	11	11	11	11	11	11	11	11	11	11	11	11	11	11	11	11	11	11
	16	16	16	16	16	16	16	16	16	16	16	16	16	16	16	16	16	16	16	16	16	16	16	16	16
14	2	2	2	2	2	2	2	2	2	2	2	2	2	2	3	3	3	3	3	3	3	3	3	3	3
	12	12	12	12	12	12	12	12	12	12	12	12	12	12	12	12	12	12	12	12	12	12	12	12	12
	16	16	16	16	16	16	16	16	16	16	16	16	16	16	16	16	16	16	16	16	16	16	16	16	16
15	2	2	2	2	2	2	2	2	2	2	2	2	2	2	3	3	3	3	3	3	3	3	3	3	3
	13	13	13	13	13	13	13	13	13	13	13	13	13	13	13	13	13	13	13	13	13	13	13	13	13
	16	16	16	16	16	16	16	16	16	16	16	16	16	16	16	16	16	16	16	16	16	16	16	16	16
16	2	2	2	2	2	2	2	2	2	2	2	2	2	2	3	3	3	3	3	3	3	3	3	3	3
	14	14	14	14	14	14	14	14	14	14	14	14	14	14	14	14	14	14	14	14	14	14	14	14	14
	16	16	16	16	16	16	16	16	16	16	16	16	16	16	16	16	16	16	16	16	16	16	16	16	16
17	2	2	2	2	2	2	2	2	2	2	2	2	2	2	3	3	3	3	3	3	3	3	3	3	3
	15	15	15	15	15	15	15	15	15	15	15	15	15	15	15	15	15	15	15	15	15	15	15	15	15
	16	16	16	16	16	16	16	16	16	16	16	16	16	16	16	16	16	16	16	16	16	16	16	16	16
18	2	2	2	2	2	2	2	2	2	2	2	2	2	2	3	3	3	3	3	3	3	3	3	3	3
	16	16	16	16	16	16	16	16	16	16	16	16	16	16	16	16	16	16	16	16	16	16	16	16	16
	16	16	16	16	16	16	16	16	16	16	16	16	16	16	16	16	16	16	16	16	16	16	16	16	16
19	2	2	2	2	2	2	2	2	2	2	2	2	2	2	3	3	3	3	3	3	3	3	3	3	3
	17	17	17	17	17	17	17	17	17	17	17	17	17	17	17	17	17	17	17	17	17	17	17	17	17
	16	16	16	16	16	16	16	16	16	16	16	16	16	16	16	16	16	16	16	16	16	16	16	16	16
20	2	2	2	2	2	2	2	2	2	2	2	2	2	2	3	3	3	3	3	3	3	3	3	3	3
	18	18	18	18	18	18	18	18	18	18	18	18	18	18	18	18	18	18	18	18	18	18	18	18	18
	16	16	16	16	16	16	16	16	16	16	16	16	16	16	16	16	16	16	16	16	16	16	16	16	16
21	2	2	2	2	2	2	2	2	2	2	2	2	2	2	3	3	3	3	3	3	3	3	3	3	3
	19	19	19	19	19	19	19	19	19	19	19	19	19	19	19	19	19	19	19	19	19	19	19	19	19
	16	16	16	16	16	16	16	16	16	16	16	16	16	16	16	16	16	16	16	16	16	16	16	16	16
22	2	2	2	2	2	2	2	2	2	2	2	2	2	2	3	3	3	3	3	3	3	3	3	3	3
	20	20	20	20	20	20	20	20	20	20	20	20	20	20	20	20	20	20	20	20	20	20	20	20	20
	16	16	16	16	16	16	16	16	16	16	16	16	16	16	16	16	16	16	16	16	16	16	16	16	16
23	2	2	2	2	2	2	2	2	2	2	2	2	2	2	3	3	3	3	3	3	3	3	3	3	3
	21	21	21	21	21	21	21	21	21	21	21	21	21	21	21	21	21	21	21	21	21	21	21	21	21
	16	16	16	16	16	16	16	16	16	16	16	16	16	16	16	16	16	16	16	16	16	16	16	16	16
24	2	2	2	2	2	2	2	2	2	2	2	2	2	2	3	3	3	3	3	3	3	3	3	3	3
	22	22	22	22	22	22	22	22	22	22	22	22	22	22	22	22	22	22	22	22	22	22	22	22	22
	16	16	16	16	16	16	16	16	16	16	16	16	16	16	16	16	16	16	16	16	16	16	16	16	16
25	2	2	2	2	2	2	2	2	2	2	2	2	2	2	3	3	3	3	3	3	3	3	3	3	3
	23	23	23	23	23	23	23	23	23	23	23	23	23	23	23	23	23	23	23	23	23	23	23	23	23
	16	16	16	16	16	16	16	16	16	16	16	16	16	16	16	16	16	16	16	16	16	16	16	16	16

T TABLE BASE 3

m/s	1	2	3	4	5	6	7	8	9	10	11	12	13	14	15	16	17	18	19	20	21	22	23	24	25
26	0	0	0	0	1	1	1	1	1	1	1	1	1	2	2	2	2	2	2	2	2	2	2	2	2
	0	0	0	0	1	2	3	5	7	9	11	13	15	17	19	22	24	24	24	24	24	24	24	24	24
	2	2	2	8	9	9	9	9	9	9	9	9	9	9	9	9	16	16	16	16	16	16	16	16	16
27	0	0	0	0	1	1	1	1	1	1	1	1	1	2	2	2	2	2	2	2	2	2	2	2	2
	0	0	0	0	1	2	3	5	7	9	11	13	15	17	19	22	25	25	25	25	25	25	25	25	25
	2	2	2	8	9	9	9	9	9	9	9	9	9	9	9	9	16	16	16	16	16	16	16	16	16
28	0	0	0	0	1	1	1	1	1	1	1	1	1	2	2	2	2	2	2	2	2	2	2	2	2
	0	0	0	0	1	2	3	5	7	9	11	13	15	17	19	22	25	26	26	26	26	26	26	26	26
	2	2	2	8	9	9	9	9	9	9	9	9	9	9	9	9	9	16	16	16	16	16	16	16	16
29	0	0	0	0	1	1	1	1	1	1	1	1	1	2	2	2	2	2	2	2	2	2	2	2	2
	0	0	0	0	1	2	3	5	7	9	11	13	15	17	19	22	25	27	27	27	27	27	27	27	27
	2	2	2	8	9	9	9	9	9	9	9	9	9	9	9	9	9	16	16	16	16	16	16	16	16
30	0	0	0	0	1	1	1	1	1	1	1	1	1	2	2	2	2	2	2	2	2	2	2	2	2
	0	0	0	0	1	2	3	5	7	9	11	13	15	17	19	22	25	28	28	28	28	28	28	28	28
	2	2	2	8	9	9	9	9	9	9	9	9	9	9	9	9	9	16	16	16	16	16	16	16	16
31	0	0	0	0	1	1	1	1	1	1	1	1	1	2	2	2	2	2	2	2	2	2	2	2	2
	0	0	0	0	1	2	3	5	7	9	11	13	15	17	19	22	25	28	29	29	29	29	29	29	29
	2	2	2	8	9	9	9	9	9	9	9	9	9	9	9	9	9	9	16	16	16	16	16	16	16
32	0	0	0	0	1	1	1	1	1	1	1	1	1	2	2	2	2	2	2	2	2	2	2	2	2
	0	0	0	0	1	2	3	5	7	9	11	13	15	17	19	22	25	28	30	30	30	30	30	30	30
	2	2	2	8	9	9	9	9	9	9	9	9	9	9	9	9	9	9	16	16	16	16	16	16	16
33	0	0	0	0	1	1	1	1	1	1	1	1	1	2	2	2	2	2	2	2	2	2	2	2	2
	0	0	0	0	1	2	3	5	7	9	11	13	15	17	19	22	25	28	31	31	31	31	31	31	31
	2	2	2	8	9	9	9	9	9	9	9	9	9	9	9	9	9	9	16	16	16	16	16	16	16
34	0	0	0	0	1	1	1	1	1	1	1	1	1	2	2	2	2	2	2	2	2	2	2	2	2
	0	0	0	0	1	2	3	5	7	9	11	13	15	17	19	22	25	28	31	32	32	32	32	32	32
	2	2	2	8	9	9	9	9	9	9	9	9	9	9	9	9	9	9	9	16	16	16	16	16	16
35	0	0	0	0	1	1	1	1	1	1	1	1	1	2	2	2	2	2	2	2	2	2	2	2	2
	0	0	0	0	1	2	3	5	7	9	11	13	15	17	19	22	25	28	31	33	33	33	33	33	33
	2	2	2	8	9	9	9	9	9	9	9	9	9	9	9	9	9	9	9	16	16	16	16	16	16
36	0	0	0	0	1	1	1	1	1	1	1	1	1	2	2	2	2	2	2	2	2	2	2	2	2
	0	0	0	0	1	2	3	5	7	9	11	13	15	17	19	22	25	28	31	34	34	34	34	34	34
	2	2	2	8	9	9	9	9	9	9	9	9	9	9	9	9	9	9	9	16	16	16	16	16	16
37	0	0	0	0	1	1	1	1	1	1	1	1	1	2	2	2	2	2	2	2	2	2	2	2	2
	0	0	0	0	1	2	3	5	7	9	11	13	15	17	19	22	25	28	31	34	35	35	35	35	35
	2	2	2	8	9	9	9	9	9	9	9	9	9	9	9	9	9	9	9	9	16	16	16	16	16
38	0	0	0	0	1	1	1	1	1	1	1	1	1	2	2	2	2	2	2	2	2	2	2	2	2
	0	0	0	0	1	2	3	5	7	9	11	13	15	17	19	22	25	28	31	34	36	36	36	36	36
	2	2	2	8	9	9	9	9	9	9	9	9	9	9	9	9	9	9	9	9	16	16	16	16	16
39	0	0	0	0	1	1	1	1	1	1	1	1	1	2	2	2	2	2	2	2	2	2	2	2	2
	0	0	0	0	1	2	3	5	7	9	11	13	15	17	19	22	25	28	31	34	37	37	37	37	37
	2	2	2	8	9	9	9	9	9	9	9	9	9	9	9	9	9	9	9	9	16	16	16	16	16
40	0	0	0	0	1	1	1	1	1	1	1	1	1	2	2	2	2	2	2	2	2	2	2	2	2
	0	0	0	0	1	2	3	5	7	9	11	13	15	17	19	22	25	28	31	34	37	38	38	38	38
	2	2	2	8	9	9	9	9	9	9	9	9	9	9	9	9	9	9	9	9	9	16	16	16	16
41	0	0	0	0	1	1	1	1	1	1	1	1	1	2	2	2	2	2	2	2	2	2	2	2	2
	0	0	0	0	1	2	3	5	7	9	11	13	15	17	19	22	25	28	31	34	37	39	39	39	39
	2	2	2	8	9	9	9	9	9	9	9	9	9	9	9	9	9	9	9	9	9	16	16	16	16
42	0	0	0	0	1	1	1	1	1	1	1	1	1	2	2	2	2	2	2	2	2	2	2	2	2
	0	0	0	0	1	2	3	5	7	9	11	13	15	17	19	22	25	28	31	34	37	40	40	40	40
	2	2	2	8	9	9	9	9	9	9	9	9	9	9	9	9	9	9	9	9	9	16	16	16	16
43	0	0	0	0	1	1	1	1	1	1	1	1	1	2	2	2	2	2	2	2	2	2	2	2	2
	0	0	0	0	1	2	3	5	7	9	11	13	15	17	19	22	25	28	31	34	37	40	41	41	41
	2	2	2	8	9	9	9	9	9	9	9	9	9	9	9	9	9	9	9	9	9	9	16	16	16
44	0	0	0	0	1	1	1	1	1	1	1	1	1	2	2	2	2	2	2	2	2	2	2	2	2
	0	0	0	0	1	2	3	5	7	9	11	13	15	17	19	22	25	28	31	34	37	40	42	42	42
	2	2	2	8	9	9	9	9	9	9	9	9	9	9	9	9	9	9	9	9	9	9	16	16	16
45	0	0	0	0	1	1	1	1	1	1	1	1	1	2	2	2	2	2	2	2	2	2	2	2	2
	0	0	0	0	1	2	3	5	7	9	11	13	15	17	19	22	25	28	31	34	37	40	43	43	43
	2	2	2	8	9	9	9	9	9	9	9	9	9	9	9	9	9	9	9	9	9	9	16	16	16
46	0	0	0	0	1	1	1	1	1	1	1	1	1	2	2	2	2	2	2	2	2	2	2	2	2
	0	0	0	0	1	2	3	5	7	9	11	13	15	17	19	22	25	28	31	34	37	40	43	44	44
	2	2	2	8	9	9	9	9	9	9	9	9	9	9	9	9	9	9	9	9	9	9	9	16	16
47	0	0	0	0	1	1	1	1	1	1	1	1	1	2	2	2	2	2	2	2	2	2	2	2	2
	0	0	0	0	1	2	3	5	7	9	11	13	15	17	19	22	25	28	31	34	37	40	43	45	45
	2	2	2	8	9	9	9	9	9	9	9	9	9	9	9	9	9	9	9	9	9	9	9	16	16
48	0	0	0	0	1	1	1	1	1	1	1	1	1	2	2	2	2	2	2	2	2	2	2	2	2
	0	0	0	0	1	2	3	5	7	9	11	13	15	17	19	22	25	28	31	34	37	40	43	46	46
	2	2	2	8	9	9	9	9	9	9	9	9	9	9	9	9	9	9	9	9	9	9	9	16	16
49	0	0	0	0	1	1	1	1	1	1	1	1	1	2	2	2	2	2	2	2	2	2	2	2	2
	0	0	0	0	1	2	3	5	7	9	11	13	15	17	19	22	25	28	31	34	37	40	43	46	47
	2	2	2	8	9	9	9	9	9	9	9	9	9	9	9	9	9	9	9	9	9	9	9	9	16
50	0	0	0	0	1	1	1	1	1	1	1	1	1	2	2	2	2	2	2	2	2	2	2	2	2
	0	0	0	0	1	2	3	5	7	9	11	13	15	17	19	22	25	28	31	34	37	40	43	46	48
	3	3	3	8	9	9	9	9	9	9	9	9	9	9	9	9	9	9	9	9	9	9	9	9	16

T TABLE BASE 3

m/s	26	27	28	29	30	31	32	33	34	35	36	37	38	39	40	41	42	43	44	45	46	47	48	49	50
26	2	2	2	2	2	2	2	2	2	2	2	2	2	2	2	3	3	3	3	3	3	3	3	3	3
	24	24	24	24	24	24	24	24	24	24	24	24	24	24	24	24	24	24	24	24	24	24	24	24	24
	16	16	16	16	16	16	16	16	16	16	16	16	16	16	16	16	16	16	16	16	16	16	16	16	16
27	2	2	2	2	2	2	2	2	2	2	2	2	2	2	2	3	3	3	3	3	3	3	3	3	3
	25	25	25	25	25	25	25	25	25	25	25	25	25	25	25	25	25	25	25	25	25	25	25	25	25
	16	16	16	16	16	16	16	16	16	16	16	16	16	16	16	16	16	16	16	16	16	16	16	16	16
28	2	2	2	2	2	2	2	2	2	2	2	2	2	2	2	3	3	3	3	3	3	3	3	3	3
	26	26	26	26	26	26	26	26	26	26	26	26	26	26	26	26	26	26	26	26	26	26	26	26	26
	16	16	16	16	16	16	16	16	16	16	16	16	16	16	16	16	16	16	16	16	16	16	16	16	16
29	2	2	2	2	2	2	2	2	2	2	2	2	2	2	2	3	3	3	3	3	3	3	3	3	3
	27	27	27	27	27	27	27	27	27	27	27	27	27	27	27	27	27	27	27	27	27	27	27	27	27
	16	16	16	16	16	16	16	16	16	16	16	16	16	16	16	16	16	16	16	16	16	16	16	16	16
30	2	2	2	2	2	2	2	2	2	2	2	2	2	2	2	3	3	3	3	3	3	3	3	3	3
	28	28	28	28	28	28	28	28	28	28	28	28	28	28	28	28	28	28	28	28	28	28	28	28	28
	16	16	16	16	16	16	16	16	16	16	16	16	16	16	16	16	16	16	16	16	16	16	16	16	16
31	2	2	2	2	2	2	2	2	2	2	2	2	2	2	2	3	3	3	3	3	3	3	3	3	3
	29	29	29	29	29	29	29	29	29	29	29	29	29	29	29	29	29	29	29	29	29	29	29	29	29
	16	16	16	16	16	16	16	16	16	16	16	16	16	16	16	16	16	16	16	16	16	16	16	16	16
32	2	2	2	2	2	2	2	2	2	2	2	2	2	2	2	3	3	3	3	3	3	3	3	3	3
	30	30	30	30	30	30	30	30	30	30	30	30	30	30	30	30	30	30	30	30	30	30	30	30	30
	16	16	16	16	16	16	16	16	16	16	16	16	16	16	16	16	16	16	16	16	16	16	16	16	16
33	2	2	2	2	2	2	2	2	2	2	2	2	2	2	2	3	3	3	3	3	3	3	3	3	3
	31	31	31	31	31	31	31	31	31	31	31	31	31	31	31	31	31	31	31	31	31	31	31	31	31
	16	16	16	16	16	16	16	16	16	16	16	16	16	16	16	16	16	16	16	16	16	16	16	16	16
34	2	2	2	2	2	2	2	2	2	2	2	2	2	2	2	3	3	3	3	3	3	3	3	3	3
	32	32	32	32	32	32	32	32	32	32	32	32	32	32	32	32	32	32	32	32	32	32	32	32	32
	16	16	16	16	16	16	16	16	16	16	16	16	16	16	16	16	16	16	16	16	16	16	16	16	16
35	2	2	2	2	2	2	2	2	2	2	2	2	2	2	2	3	3	3	3	3	3	3	3	3	3
	33	33	33	33	33	33	33	33	33	33	33	33	33	33	33	33	33	33	33	33	33	33	33	33	33
	16	16	16	16	16	16	16	16	16	16	16	16	16	16	16	16	16	16	16	16	16	16	16	16	16
36	2	2	2	2	2	2	2	2	2	2	2	2	2	2	2	3	3	3	3	3	3	3	3	3	3
	34	34	34	34	34	34	34	34	34	34	34	34	34	34	34	34	34	34	34	34	34	34	34	34	34
	16	16	16	16	16	16	16	16	16	16	16	16	16	16	16	16	16	16	16	16	16	16	16	16	16
37	2	2	2	2	2	2	2	2	2	2	2	2	2	2	2	3	3	3	3	3	3	3	3	3	3
	35	35	35	35	35	35	35	35	35	35	35	35	35	35	35	35	35	35	35	35	35	35	35	35	35
	16	16	16	16	16	16	16	16	16	16	16	16	16	16	16	16	16	16	16	16	16	16	16	16	16
38	2	2	2	2	2	2	2	2	2	2	2	2	2	2	2	3	3	3	3	3	3	3	3	3	3
	36	36	36	36	36	36	36	36	36	36	36	36	36	36	36	36	36	36	36	36	36	36	36	36	36
	16	16	16	16	16	16	16	16	16	16	16	16	16	16	16	16	16	16	16	16	16	16	16	16	16
39	2	2	2	2	2	2	2	2	2	2	2	2	2	2	2	3	3	3	3	3	3	3	3	3	3
	37	37	37	37	37	37	37	37	37	37	37	37	37	37	37	37	37	37	37	37	37	37	37	37	37
	16	16	16	16	16	16	16	16	16	16	16	16	16	16	16	16	16	16	16	16	16	16	16	16	16
40	2	2	2	2	2	2	2	2	2	2	2	2	2	2	2	3	3	3	3	3	3	3	3	3	3
	38	38	38	38	38	38	38	38	38	38	38	38	38	38	38	38	38	38	38	38	38	38	38	38	38
	16	16	16	16	16	16	16	16	16	16	16	16	16	16	16	16	16	16	16	16	16	16	16	16	16
41	2	2	2	2	2	2	2	2	2	2	2	2	2	2	2	3	3	3	3	3	3	3	3	3	3
	39	39	39	39	39	39	39	39	39	39	39	39	39	39	39	39	39	39	39	39	39	39	39	39	39
	16	16	16	16	16	16	16	16	16	16	16	16	16	16	16	16	16	16	16	16	16	16	16	16	16
42	2	2	2	2	2	2	2	2	2	2	2	2	2	2	2	3	3	3	3	3	3	3	3	3	3
	40	40	40	40	40	40	40	40	40	40	40	40	40	40	40	40	40	40	40	40	40	40	40	40	40
	16	16	16	16	16	16	16	16	16	16	16	16	16	16	16	16	16	16	16	16	16	16	16	16	16
43	2	2	2	2	2	2	2	2	2	2	2	2	2	2	2	3	3	3	3	3	3	3	3	3	3
	41	41	41	41	41	41	41	41	41	41	41	41	41	41	41	41	41	41	41	41	41	41	41	41	41
	16	16	16	16	16	16	16	16	16	16	16	16	16	16	16	16	16	16	16	16	16	16	16	16	16
44	2	2	2	2	2	2	2	2	2	2	2	2	2	2	2	3	3	3	3	3	3	3	3	3	3
	42	42	42	42	42	42	42	42	42	42	42	42	42	42	42	42	42	42	42	42	42	42	42	42	42
	16	16	16	16	16	16	16	16	16	16	16	16	16	16	16	16	16	16	16	16	16	16	16	16	16
45	2	2	2	2	2	2	2	2	2	2	2	2	2	2	2	3	3	3	3	3	3	3	3	3	3
	43	43	43	43	43	43	43	43	43	43	43	43	43	43	43	43	43	43	43	43	43	43	43	43	43
	16	16	16	16	16	16	16	16	16	16	16	16	16	16	16	16	16	16	16	16	16	16	16	16	16
46	2	2	2	2	2	2	2	2	2	2	2	2	2	2	2	3	3	3	3	3	3	3	3	3	3
	44	44	44	44	44	44	44	44	44	44	44	44	44	44	44	44	44	44	44	44	44	44	44	44	44
	16	16	16	16	16	16	16	16	16	16	16	16	16	16	16	16	16	16	16	16	16	16	16	16	16
47	2	2	2	2	2	2	2	2	2	2	2	2	2	2	2	3	3	3	3	3	3	3	3	3	3
	45	45	45	45	45	45	45	45	45	45	45	45	45	45	45	45	45	45	45	45	45	45	45	45	45
	16	16	16	16	16	16	16	16	16	16	16	16	16	16	16	16	16	16	16	16	16	16	16	16	16
48	2	2	2	2	2	2	2	2	2	2	2	2	2	2	2	3	3	3	3	3	3	3	3	3	3
	46	46	46	46	46	46	46	46	46	46	46	46	46	46	46	46	46	46	46	46	46	46	46	46	46
	16	16	16	16	16	16	16	16	16	16	16	16	16	16	16	16	16	16	16	16	16	16	16	16	16
49	2	2	2	2	2	2	2	2	2	2	2	2	2	2	2	3	3	3	3	3	3	3	3	3	3
	47	47	47	47	47	47	47	47	47	47	47	47	47	47	47	47	47	47	47	47	47	47	47	47	47
	16	16	16	16	16	16	16	16	16	16	16	16	16	16	16	16	16	16	16	16	16	16	16	16	16
50	2	2	2	2	2	2	2	2	2	2	2	2	2	2	2	3	3	3	3	3	3	3	3	3	3
	48	48	48	48	48	48	48	48	48	48	48	48	48	48	48	48	48	48	48	48	48	48	48	48	48
	16	16	16	16	16	16	16	16	16	16	16	16	16	16	16	16	16	16	16	16	16	16	16	16	16

T TABLE BASE 5

m/s	1	2	3	4	5	6	7	8	9	10	11	12	13	14	15	16	17	18	19	20	21	22	23	24	25
1	0	0	0	0	0	0	0	0	0	0	0	0	0	0	0	0	0	0	0	0	0	0	0	0	0
	0	0	0	0	0	0	0	0	0	0	0	0	0	0	0	0	0	0	0	0	0	0	0	0	0
	5	5	5	5	5	5	5	5	5	5	5	5	5	5	5	5	5	5	5	5	5	5	5	5	5
2	0	0	0	0	0	0	1	1	1	1	1	1	1	1	1	1	1	1	1	1	1	1	1	1	1
	0	0	0	0	0	0	1	1	1	1	1	1	1	1	1	1	1	1	1	1	1	1	1	1	1
	2	2	2	2	2	7	5	5	5	5	5	5	5	5	5	5	5	5	5	5	5	5	5	5	5
3	0	0	0	0	0	0	1	1	1	1	1	1	1	1	1	1	1	1	1	1	1	1	1	1	1
	0	0	0	0	0	1	1	1	1	1	1	1	1	1	1	1	1	1	1	1	1	1	1	1	1
	2	2	2	2	2	8	9	2	2	2	2	2	2	2	2	2	2	2	2	2	2	2	2	2	2
4	0	0	0	0	0	0	1	1	1	1	1	1	1	1	1	1	1	1	1	1	1	1	1	1	1
	0	0	0	0	0	1	2	2	2	2	2	2	2	2	2	2	2	2	2	2	2	2	2	2	2
	2	2	2	2	2	8	9	9	16	16	16	16	16	16	16	16	16	16	16	16	16	16	16	16	16
5	0	0	0	0	0	0	1	1	1	1	1	1	1	1	1	1	1	1	1	1	1	1	1	1	1
	0	0	0	0	0	1	2	3	3	3	3	3	3	3	3	3	3	3	3	3	3	3	3	3	3
	2	2	2	2	2	8	9	9	16	16	16	16	16	16	16	16	16	16	16	16	16	16	16	16	16
6	0	0	0	0	0	0	1	1	1	1	1	1	1	1	1	1	1	1	1	1	1	1	1	1	1
	0	0	0	0	0	1	2	3	4	4	4	4	4	4	4	4	4	4	4	4	4	4	4	4	4
	2	2	2	2	2	8	9	9	9	16	16	16	16	16	16	16	16	16	16	16	16	16	16	16	16
7	0	0	0	0	0	0	1	1	1	1	1	1	1	1	1	1	1	1	1	1	1	1	1	1	1
	0	0	0	0	0	1	2	3	4	5	5	5	5	5	5	5	5	5	5	5	5	5	5	5	5
	2	2	2	2	2	8	9	9	9	16	16	16	16	16	16	16	16	16	16	16	16	16	16	16	16
8	0	0	0	0	0	0	1	1	1	1	1	1	1	1	1	1	1	1	1	1	1	1	1	1	1
	0	0	0	0	0	1	2	3	4	5	6	6	6	6	6	6	6	6	6	6	6	6	6	6	6
	2	2	2	2	2	8	9	9	9	9	16	16	16	16	16	16	16	16	16	16	16	16	16	16	16
9	0	0	0	0	0	0	1	1	1	1	1	1	1	1	1	1	1	1	1	1	1	1	1	1	1
	0	0	0	0	0	1	2	3	4	5	6	7	7	7	7	7	7	7	7	7	7	7	7	7	7
	2	2	2	2	2	8	9	9	9	9	9	16	16	16	16	16	16	16	16	16	16	16	16	16	16
10	0	0	0	0	0	0	1	1	1	1	1	1	1	1	1	1	1	1	1	1	1	1	1	1	1
	0	0	0	0	0	1	2	3	4	5	6	7	8	8	8	8	8	8	8	8	8	8	8	8	8
	2	2	2	2	2	8	9	9	9	9	9	9	16	16	16	16	16	16	16	16	16	16	16	16	16
11	0	0	0	0	0	0	1	1	1	1	1	1	1	1	1	1	1	1	1	1	1	1	1	1	1
	0	0	0	0	0	1	2	3	4	5	6	7	8	9	9	9	9	9	9	9	9	9	9	9	9
	2	2	2	2	2	8	9	9	9	9	9	9	9	16	16	16	16	16	16	16	16	16	16	16	16
12	0	0	0	0	0	0	1	1	1	1	1	1	1	1	1	1	1	1	1	1	1	1	1	1	1
	0	0	0	0	0	1	2	3	4	5	6	7	8	9	10	10	10	10	10	10	10	10	10	10	10
	2	2	2	2	2	8	9	9	9	9	9	9	9	9	16	16	16	16	16	16	16	16	16	16	16
13	0	0	0	0	0	0	1	1	1	1	1	1	1	1	1	1	1	1	1	1	1	1	1	1	1
	0	0	0	0	0	1	2	3	4	5	6	7	8	9	10	11	11	11	11	11	11	11	11	11	11
	2	2	2	2	2	8	9	9	9	9	9	9	9	9	9	16	16	16	16	16	16	16	16	16	16
14	0	0	0	0	0	0	1	1	1	1	1	1	1	1	1	1	1	1	1	1	1	1	1	1	1
	0	0	0	0	0	1	2	3	4	5	6	7	8	9	10	12	12	12	12	12	12	12	12	12	12
	2	2	2	2	2	8	9	9	9	9	9	9	9	9	9	16	16	16	16	16	16	16	16	16	16
15	0	0	0	0	0	0	1	1	1	1	1	1	1	1	1	1	1	1	1	1	1	1	1	1	1
	0	0	0	0	0	1	2	3	4	5	6	7	8	9	10	12	13	13	13	13	13	13	13	13	13
	2	2	2	2	2	8	9	9	9	9	9	9	9	9	9	16	16	16	16	16	16	16	16	16	16
16	0	0	0	0	0	0	1	1	1	1	1	1	1	1	1	1	1	1	1	1	1	1	1	1	1
	0	0	0	0	0	1	2	3	4	5	6	7	8	9	10	12	14	14	14	14	14	14	14	14	14
	2	2	2	2	2	8	9	9	9	9	9	9	9	9	9	16	16	16	16	16	16	16	16	16	16
17	0	0	0	0	0	0	1	1	1	1	1	1	1	1	1	1	1	1	1	1	1	1	1	1	1
	0	0	0	0	0	1	2	3	4	5	6	7	8	9	10	12	14	15	15	15	15	15	15	15	15
	2	2	2	2	2	8	9	9	9	9	9	9	9	9	9	16	16	16	16	16	16	16	16	16	16
18	0	0	0	0	0	0	1	1	1	1	1	1	1	1	1	1	1	1	1	1	1	1	1	1	1
	0	0	0	0	0	1	2	3	4	5	6	7	8	9	10	12	14	16	16	16	16	16	16	16	16
	2	2	2	2	2	8	9	9	9	9	9	9	9	9	9	9	16	16	16	16	16	16	16	16	16
19	0	0	0	0	0	0	1	1	1	1	1	1	1	1	1	1	1	1	1	1	1	1	1	1	1
	0	0	0	0	0	1	2	3	4	5	6	7	8	9	10	12	14	16	17	17	17	17	17	17	17
	2	2	2	2	2	8	9	9	9	9	9	9	9	9	9	9	16	16	16	16	16	16	16	16	16
20	0	0	0	0	0	0	1	1	1	1	1	1	1	1	1	1	1	1	1	1	1	1	1	1	1
	0	0	0	0	0	1	2	3	4	5	6	7	8	9	10	12	14	16	18	18	18	18	18	18	18
	2	2	2	2	2	8	9	9	9	9	9	9	9	9	9	9	16	16	16	16	16	16	16	16	16
21	0	0	0	0	0	0	1	1	1	1	1	1	1	1	1	1	1	1	1	1	1	1	1	1	1
	0	0	0	0	0	1	2	3	4	5	6	7	8	9	10	12	14	16	18	19	19	19	19	19	19
	2	2	2	2	2	8	9	9	9	9	9	9	9	9	9	9	16	16	16	16	16	16	16	16	16
22	0	0	0	0	0	0	1	1	1	1	1	1	1	1	1	1	1	1	1	1	1	1	1	1	1
	0	0	0	0	0	1	2	3	4	5	6	7	8	9	10	12	14	16	18	20	20	20	20	20	20
	2	2	2	2	2	8	9	9	9	9	9	9	9	9	9	9	16	16	16	16	16	16	16	16	16
23	0	0	0	0	0	0	1	1	1	1	1	1	1	1	1	1	1	1	1	1	1	1	1	1	1
	0	0	0	0	0	1	2	3	4	5	6	7	8	9	10	12	14	16	18	20	21	21	21	21	21
	2	2	2	2	2	8	9	9	9	9	9	9	9	9	9	9	16	16	16	16	16	16	16	16	16
24	0	0	0	0	0	0	1	1	1	1	1	1	1	1	1	1	1	1	1	1	1	1	1	1	1
	0	0	0	0	0	1	2	3	4	5	6	7	8	9	10	12	14	16	18	20	22	22	22	22	22
	2	2	2	2	2	8	9	9	9	9	9	9	9	9	9	9	16	16	16	16	16	16	16	16	16
25	0	0	0	0	0	0	1	1	1	1	1	1	1	1	1	1	1	1	1	1	1	1	1	1	1
	0	0	0	0	0	1	2	3	4	5	6	7	8	9	10	12	14	16	18	20	22	23	23	23	23
	2	2	2	2	2	8	9	9	9	9	9	9	9	9	9	9	16	16	16	16	16	16	16	16	16

T TABLE BASE 5

m/s	26	27	28	29	30	31	32	33	34	35	36	37	38	39	40	41	42	43	44	45	46	47	48	49	50
1	0	0	0	0	0	0	0	0	0	0	0	0	0	0	0	0	0	0	0	0	0	0	0	0	0
	0	0	0	0	0	0	0	0	0	0	0	0	0	0	0	0	0	0	0	0	0	0	0	0	0
	5	5	5	5	5	5	5	5	5	5	5	5	5	5	5	5	5	5	5	5	5	5	5	5	5
2	1	1	1	1	1	1	1	1	1	1	1	1	1	1	1	1	1	1	1	1	1	1	1	1	1
	1	1	1	1	1	1	1	1	1	1	1	1	1	1	1	1	1	1	1	1	1	1	1	1	1
	5	5	5	5	5	5	5	5	5	5	5	5	5	5	5	5	5	5	5	5	5	5	5	5	5
3	1	1	1	1	1	1	2	2	2	2	2	2	2	2	2	2	2	2	2	2	2	2	2	2	2
	1	1	1	1	1	1	2	2	2	2	2	2	2	2	2	2	2	2	2	2	2	2	2	2	2
	2	2	2	2	12	5	5	5	5	5	5	5	5	5	5	5	5	5	5	5	5	5	5	5	5
4	1	1	1	1	1	1	2	2	2	2	2	2	2	2	2	2	2	2	2	2	2	2	2	2	2
	2	2	2	2	2	2	2	2	2	2	2	2	2	2	2	2	2	2	2	2	2	2	2	2	2
	16	16	16	16	16	16	2	2	2	2	2	2	2	2	2	2	2	2	2	2	2	2	2	2	2
5	1	1	1	1	1	1	2	2	2	2	2	2	2	2	2	2	2	2	2	2	2	2	2	2	2
	3	3	3	3	3	3	3	3	3	3	3	3	3	3	3	3	3	3	3	3	3	3	3	3	3
	16	16	16	16	16	16	16	16	16	16	16	16	16	16	16	16	16	16	16	16	16	16	16	16	16
6	1	1	1	1	1	1	2	2	2	2	2	2	2	2	2	2	2	2	2	2	2	2	2	2	2
	4	4	4	4	4	4	4	4	4	4	4	4	4	4	4	4	4	4	4	4	4	4	4	4	4
	16	16	16	16	16	16	16	16	16	16	16	16	16	16	16	16	16	16	16	16	16	16	16	16	16
7	1	1	1	1	1	1	2	2	2	2	2	2	2	2	2	2	2	2	2	2	2	2	2	2	2
	5	5	5	5	5	5	5	5	5	5	5	5	5	5	5	5	5	5	5	5	5	5	5	5	5
	16	16	16	16	16	16	16	16	16	16	16	16	16	16	16	16	16	16	16	16	16	16	16	16	16
8	1	1	1	1	1	1	2	2	2	2	2	2	2	2	2	2	2	2	2	2	2	2	2	2	2
	6	6	6	6	6	6	6	6	6	6	6	6	6	6	6	6	6	6	6	6	6	6	6	6	6
	16	16	16	16	16	16	16	16	16	16	16	16	16	16	16	16	16	16	16	16	16	16	16	16	16
9	1	1	1	1	1	1	2	2	2	2	2	2	2	2	2	2	2	2	2	2	2	2	2	2	2
	7	7	7	7	7	7	7	7	7	7	7	7	7	7	7	7	7	7	7	7	7	7	7	7	7
	16	16	16	16	16	16	16	16	16	16	16	16	16	16	16	16	16	16	16	16	16	16	16	16	16
10	1	1	1	1	1	1	2	2	2	2	2	2	2	2	2	2	2	2	2	2	2	2	2	2	2
	8	8	8	8	8	8	8	8	8	8	8	8	8	8	8	8	8	8	8	8	8	8	8	8	8
	16	16	16	16	16	16	16	16	16	16	16	16	16	16	16	16	16	16	16	16	16	16	16	16	16
11	1	1	1	1	1	1	2	2	2	2	2	2	2	2	2	2	2	2	2	2	2	2	2	2	2
	9	9	9	9	9	9	9	9	9	9	9	9	9	9	9	9	9	9	9	9	9	9	9	9	9
	16	16	16	16	16	16	16	16	16	16	16	16	16	16	16	16	16	16	16	16	16	16	16	16	16
12	1	1	1	1	1	1	2	2	2	2	2	2	2	2	2	2	2	2	2	2	2	2	2	2	2
	10	10	10	10	10	10	10	10	10	10	10	10	10	10	10	10	10	10	10	10	10	10	10	10	10
	16	16	16	16	16	16	16	16	16	16	16	16	16	16	16	16	16	16	16	16	16	16	16	16	16
13	1	1	1	1	1	1	2	2	2	2	2	2	2	2	2	2	2	2	2	2	2	2	2	2	2
	11	11	11	11	11	11	11	11	11	11	11	11	11	11	11	11	11	11	11	11	11	11	11	11	11
	16	16	16	16	16	16	16	16	16	16	16	16	16	16	16	16	16	16	16	16	16	16	16	16	16
14	1	1	1	1	1	1	2	2	2	2	2	2	2	2	2	2	2	2	2	2	2	2	2	2	2
	12	12	12	12	12	12	12	12	12	12	12	12	12	12	12	12	12	12	12	12	12	12	12	12	12
	16	16	16	16	16	16	16	16	16	16	16	16	16	16	16	16	16	16	16	16	16	16	16	16	16
15	1	1	1	1	1	1	2	2	2	2	2	2	2	2	2	2	2	2	2	2	2	2	2	2	2
	13	13	13	13	13	13	13	13	13	13	13	13	13	13	13	13	13	13	13	13	13	13	13	13	13
	16	16	16	16	16	16	16	16	16	16	16	16	16	16	16	16	16	16	16	16	16	16	16	16	16
16	1	1	1	1	1	1	2	2	2	2	2	2	2	2	2	2	2	2	2	2	2	2	2	2	2
	14	14	14	14	14	14	14	14	14	14	14	14	14	14	14	14	14	14	14	14	14	14	14	14	14
	16	16	16	16	16	16	16	16	16	16	16	16	16	16	16	16	16	16	16	16	16	16	16	16	16
17	1	1	1	1	1	1	2	2	2	2	2	2	2	2	2	2	2	2	2	2	2	2	2	2	2
	15	15	15	15	15	15	15	15	15	15	15	15	15	15	15	15	15	15	15	15	15	15	15	15	15
	16	16	16	16	16	16	16	16	16	16	16	16	16	16	16	16	16	16	16	16	16	16	16	16	16
18	1	1	1	1	1	1	2	2	2	2	2	2	2	2	2	2	2	2	2	2	2	2	2	2	2
	16	16	16	16	16	16	16	16	16	16	16	16	16	16	16	16	16	16	16	16	16	16	16	16	16
	16	16	16	16	16	16	16	16	16	16	16	16	16	16	16	16	16	16	16	16	16	16	16	16	16
19	1	1	1	1	1	1	2	2	2	2	2	2	2	2	2	2	2	2	2	2	2	2	2	2	2
	17	17	17	17	17	17	17	17	17	17	17	17	17	17	17	17	17	17	17	17	17	17	17	17	17
	16	16	16	16	16	16	16	16	16	16	16	16	16	16	16	16	16	16	16	16	16	16	16	16	16
20	1	1	1	1	1	1	2	2	2	2	2	2	2	2	2	2	2	2	2	2	2	2	2	2	2
	18	18	18	18	18	18	18	18	18	18	18	18	18	18	18	18	18	18	18	18	18	18	18	18	18
	16	16	16	16	16	16	16	16	16	16	16	16	16	16	16	16	16	16	16	16	16	16	16	16	16
21	1	1	1	1	1	1	2	2	2	2	2	2	2	2	2	2	2	2	2	2	2	2	2	2	2
	19	19	19	19	19	19	19	19	19	19	19	19	19	19	19	19	19	19	19	19	19	19	19	19	19
	16	16	16	16	16	16	16	16	16	16	16	16	16	16	16	16	16	16	16	16	16	16	16	16	16
22	1	1	1	1	1	1	2	2	2	2	2	2	2	2	2	2	2	2	2	2	2	2	2	2	2
	20	20	20	20	20	20	20	20	20	20	20	20	20	20	20	20	20	20	20	20	20	20	20	20	20
	16	16	16	16	16	16	16	16	16	16	16	16	16	16	16	16	16	16	16	16	16	16	16	16	16
23	1	1	1	1	1	1	2	2	2	2	2	2	2	2	2	2	2	2	2	2	2	2	2	2	2
	21	21	21	21	21	21	21	21	21	21	21	21	21	21	21	21	21	21	21	21	21	21	21	21	21
	16	16	16	16	16	16	16	16	16	16	16	16	16	16	16	16	16	16	16	16	16	16	16	16	16
24	1	1	1	1	1	1	2	2	2	2	2	2	2	2	2	2	2	2	2	2	2	2	2	2	2
	22	22	22	22	22	22	22	22	22	22	22	22	22	22	22	22	22	22	22	22	22	22	22	22	22
	16	16	16	16	16	16	16	16	16	16	16	16	16	16	16	16	16	16	16	16	16	16	16	16	16
25	1	1	1	1	1	1	2	2	2	2	2	2	2	2	2	2	2	2	2	2	2	2	2	2	2
	23	23	23	23	23	23	23	23	23	23	23	23	23	23	23	23	23	23	23	23	23	23	23	23	23
	16	16	16	16	16	16	16	16	16	16	16	16	16	16	16	16	16	16	16	16	16	16	16	16	16

T TABLE BASE 5

m/s	1	2	3	4	5	6	7	8	9	10	11	12	13	14	15	16	17	18	19	20	21	22	23	24	25
26	0	0	0	0	0	0	1	1	1	1	1	1	1	1	1	1	1	1	1	1	1	1	1	1	1
	0	0	0	0	0	0	1	2	3	4	5	6	7	8	9	10	12	14	16	18	20	22	24	24	24
	2	2	2	2	2	8	9	9	9	9	9	9	9	9	9	9	9	9	9	9	9	9	9	16	16
27	0	0	0	0	0	0	1	1	1	1	1	1	1	1	1	1	1	1	1	1	1	1	1	1	1
	0	0	0	0	0	0	1	2	3	4	5	6	7	8	9	10	12	14	16	18	20	22	24	25	25
	2	2	2	2	2	8	9	9	9	9	9	9	9	9	9	9	9	9	9	9	9	9	9	16	16
28	0	0	0	0	0	0	1	1	1	1	1	1	1	1	1	1	1	1	1	1	1	1	1	1	1
	0	0	0	0	0	0	1	2	3	4	5	6	7	8	9	10	12	14	16	18	20	22	24	26	26
	2	2	2	2	2	8	9	9	9	9	9	9	9	9	9	9	9	9	9	9	9	9	9	9	16
29	0	0	0	0	0	0	1	1	1	1	1	1	1	1	1	1	1	1	1	1	1	1	1	1	1
	0	0	0	0	0	0	1	2	3	4	5	6	7	8	9	10	12	14	16	18	20	22	24	26	27
	2	2	2	2	2	8	9	9	9	9	9	9	9	9	9	9	9	9	9	9	9	9	9	9	16
30	0	0	0	0	0	0	1	1	1	1	1	1	1	1	1	1	1	1	1	1	1	1	1	1	1
	0	0	0	0	0	0	1	2	3	4	5	6	7	8	9	10	12	14	16	18	20	22	24	26	28
	2	2	2	2	2	8	9	9	9	9	9	9	9	9	9	9	9	9	9	9	9	9	9	9	9
31	0	0	0	0	0	0	1	1	1	1	1	1	1	1	1	1	1	1	1	1	1	1	1	1	1
	0	0	0	0	0	0	1	2	3	4	5	6	7	8	9	10	12	14	16	18	20	22	24	26	28
	2	2	2	2	2	8	9	9	9	9	9	9	9	9	9	9	9	9	9	9	9	9	9	9	9
32	0	0	0	0	0	0	1	1	1	1	1	1	1	1	1	1	1	1	1	1	1	1	1	1	1
	0	0	0	0	0	0	1	2	3	4	5	6	7	8	9	10	12	14	16	18	20	22	24	26	28
	2	2	2	2	2	8	9	9	9	9	9	9	9	9	9	9	9	9	9	9	9	9	9	9	9
33	0	0	0	0	0	0	1	1	1	1	1	1	1	1	1	1	1	1	1	1	1	1	1	1	1
	0	0	0	0	0	0	1	2	3	4	5	6	7	8	9	10	12	14	16	18	20	22	24	26	28
	2	2	2	2	2	8	9	9	9	9	9	9	9	9	9	9	9	9	9	9	9	9	9	9	9
34	0	0	0	0	0	0	1	1	1	1	1	1	1	1	1	1	1	1	1	1	1	1	1	1	1
	0	0	0	0	0	0	1	2	3	4	5	6	7	8	9	10	12	14	16	18	20	22	24	26	28
	2	2	2	2	2	8	9	9	9	9	9	9	9	9	9	9	9	9	9	9	9	9	9	9	9
35	0	0	0	0	0	0	1	1	1	1	1	1	1	1	1	1	1	1	1	1	1	1	1	1	1
	0	0	0	0	0	0	1	2	3	4	5	6	7	8	9	10	12	14	16	18	20	22	24	26	28
	2	2	2	2	2	8	9	9	9	9	9	9	9	9	9	9	9	9	9	9	9	9	9	9	9
36	0	0	0	0	0	0	1	1	1	1	1	1	1	1	1	1	1	1	1	1	1	1	1	1	1
	0	0	0	0	0	0	1	2	3	4	5	6	7	8	9	10	12	14	16	18	20	22	24	26	28
	2	2	2	2	2	8	9	9	9	9	9	9	9	9	9	9	9	9	9	9	9	9	9	9	9
37	0	0	0	0	0	0	1	1	1	1	1	1	1	1	1	1	1	1	1	1	1	1	1	1	1
	0	0	0	0	0	0	1	2	3	4	5	6	7	8	9	10	12	14	16	18	20	22	24	26	28
	2	2	2	2	2	8	9	9	9	9	9	9	9	9	9	9	9	9	9	9	9	9	9	9	9
38	0	0	0	0	0	0	1	1	1	1	1	1	1	1	1	1	1	1	1	1	1	1	1	1	1
	0	0	0	0	0	0	1	2	3	4	5	6	7	8	9	10	12	14	16	18	20	22	24	26	28
	2	2	2	2	2	8	9	9	9	9	9	9	9	9	9	9	9	9	9	9	9	9	9	9	9
39	0	0	0	0	0	0	1	1	1	1	1	1	1	1	1	1	1	1	1	1	1	1	1	1	1
	0	0	0	0	0	0	1	2	3	4	5	6	7	8	9	10	12	14	16	18	20	22	24	26	28
	2	2	2	2	2	8	9	9	9	9	9	9	9	9	9	9	9	9	9	9	9	9	9	9	9
40	0	0	0	0	0	0	1	1	1	1	1	1	1	1	1	1	1	1	1	1	1	1	1	1	1
	0	0	0	0	0	0	1	2	3	4	5	6	7	8	9	10	12	14	16	18	20	22	24	26	28
	2	2	2	2	2	8	9	9	9	9	9	9	9	9	9	9	9	9	9	9	9	9	9	9	9
41	0	0	0	0	0	0	1	1	1	1	1	1	1	1	1	1	1	1	1	1	1	1	1	1	1
	0	0	0	0	0	0	1	2	3	4	5	6	7	8	9	10	12	14	16	18	20	22	24	26	28
	2	2	2	2	2	8	9	9	9	9	9	9	9	9	9	9	9	9	9	9	9	9	9	9	9
42	0	0	0	0	0	0	1	1	1	1	1	1	1	1	1	1	1	1	1	1	1	1	1	1	1
	0	0	0	0	0	0	1	2	3	4	5	6	7	8	9	10	12	14	16	18	20	22	24	26	28
	2	2	2	2	2	8	9	9	9	9	9	9	9	9	9	9	9	9	9	9	9	9	9	9	9
43	0	0	0	0	0	0	1	1	1	1	1	1	1	1	1	1	1	1	1	1	1	1	1	1	1
	0	0	0	0	0	0	1	2	3	4	5	6	7	8	9	10	12	14	16	18	20	22	24	26	28
	2	2	2	2	2	8	9	9	9	9	9	9	9	9	9	9	9	9	9	9	9	9	9	9	9
44	0	0	0	0	0	0	1	1	1	1	1	1	1	1	1	1	1	1	1	1	1	1	1	1	1
	0	0	0	0	0	0	1	2	3	4	5	6	7	8	9	10	12	14	16	18	20	22	24	26	28
	2	2	2	2	2	8	9	9	9	9	9	9	9	9	9	9	9	9	9	9	9	9	9	9	9
45	0	0	0	0	0	0	1	1	1	1	1	1	1	1	1	1	1	1	1	1	1	1	1	1	1
	0	0	0	0	0	0	1	2	3	4	5	6	7	8	9	10	12	14	16	18	20	22	24	26	28
	2	2	2	2	2	8	9	9	9	9	9	9	9	9	9	9	9	9	9	9	9	9	9	9	9
46	0	0	0	0	0	0	1	1	1	1	1	1	1	1	1	1	1	1	1	1	1	1	1	1	1
	0	0	0	0	0	0	1	2	3	4	5	6	7	8	9	10	12	14	16	18	20	22	24	26	28
	2	2	2	2	2	8	9	9	9	9	9	9	9	9	9	9	9	9	9	9	9	9	9	9	9
47	0	0	0	0	0	0	1	1	1	1	1	1	1	1	1	1	1	1	1	1	1	1	1	1	1
	0	0	0	0	0	0	1	2	3	4	5	6	7	8	9	10	12	14	16	18	20	22	24	26	28
	2	2	2	2	2	8	9	9	9	9	9	9	9	9	9	9	9	9	9	9	9	9	9	9	9
48	0	0	0	0	0	0	1	1	1	1	1	1	1	1	1	1	1	1	1	1	1	1	1	1	1
	0	0	0	0	0	0	1	2	3	4	5	6	7	8	9	10	12	14	16	18	20	22	24	26	28
	2	2	2	2	2	8	9	9	9	9	9	9	9	9	9	9	9	9	9	9	9	9	9	9	9
49	0	0	0	0	0	0	1	1	1	1	1	1	1	1	1	1	1	1	1	1	1	1	1	1	1
	0	0	0	0	0	0	1	2	3	4	5	6	7	8	9	10	12	14	16	18	20	22	24	26	28
	2	2	2	2	2	8	9	9	9	9	9	9	9	9	9	9	9	9	9	9	9	9	9	9	9
50	0	0	0	0	0	0	1	1	1	1	1	1	1	1	1	1	1	1	1	1	1	1	1	1	1
	0	0	0	0	0	0	1	2	3	4	5	6	7	8	9	10	12	14	16	18	20	22	24	26	28
	3	3	3	3	3	8	9	9	9	9	9	9	9	9	9	9	9	9	9	9	9	9	9	9	9

T TABLE BASE 5

m/s	26	27	28	29	30	31	32	33	34	35	36	37	38	39	40	41	42	43	44	45	46	47	48	49	50
26	1	1	1	1	1	1	2	2	2	2	2	2	2	2	2	2	2	2	2	2	2	2	2	2	2
	24	24	24	24	24	24	24	24	24	24	24	24	24	24	24	24	24	24	24	24	24	24	24	24	24
	16	16	16	16	16	16	16	16	16	16	16	16	16	16	16	16	16	16	16	16	16	16	16	16	16
27	1	1	1	1	1	1	2	2	2	2	2	2	2	2	2	2	2	2	2	2	2	2	2	2	2
	25	25	25	25	25	25	25	25	25	25	25	25	25	25	25	25	25	25	25	25	25	25	25	25	25
	16	16	16	16	16	16	16	16	16	16	16	16	16	16	16	16	16	16	16	16	16	16	16	16	16
28	1	1	1	1	1	1	2	2	2	2	2	2	2	2	2	2	2	2	2	2	2	2	2	2	2
	26	26	26	26	26	26	26	26	26	26	26	26	26	26	26	26	26	26	26	26	26	26	26	26	26
	16	16	16	16	16	16	16	16	16	16	16	16	16	16	16	16	16	16	16	16	16	16	16	16	16
29	1	1	1	1	1	1	2	2	2	2	2	2	2	2	2	2	2	2	2	2	2	2	2	2	2
	27	27	27	27	27	27	27	27	27	27	27	27	27	27	27	27	27	27	27	27	27	27	27	27	27
	16	16	16	16	16	16	16	16	16	16	16	16	16	16	16	16	16	16	16	16	16	16	16	16	16
30	1	1	1	1	1	1	2	2	2	2	2	2	2	2	2	2	2	2	2	2	2	2	2	2	2
	28	28	28	28	28	28	28	28	28	28	28	28	28	28	28	28	28	28	28	28	28	28	28	28	28
	16	16	16	16	16	16	16	16	16	16	16	16	16	16	16	16	16	16	16	16	16	16	16	16	16
31	1	1	1	1	1	1	2	2	2	2	2	2	2	2	2	2	2	2	2	2	2	2	2	2	2
	29	29	29	29	29	29	29	29	29	29	29	29	29	29	29	29	29	29	29	29	29	29	29	29	29
	16	16	16	16	16	16	16	16	16	16	16	16	16	16	16	16	16	16	16	16	16	16	16	16	16
32	1	1	1	1	1	1	2	2	2	2	2	2	2	2	2	2	2	2	2	2	2	2	2	2	2
	30	30	30	30	30	30	30	30	30	30	30	30	30	30	30	30	30	30	30	30	30	30	30	30	30
	9	16	16	16	16	16	16	16	16	16	16	16	16	16	16	16	16	16	16	16	16	16	16	16	16
33	1	1	1	1	1	1	2	2	2	2	2	2	2	2	2	2	2	2	2	2	2	2	2	2	2
	30	31	31	31	31	31	31	31	31	31	31	31	31	31	31	31	31	31	31	31	31	31	31	31	31
	9	16	16	16	16	16	16	16	16	16	16	16	16	16	16	16	16	16	16	16	16	16	16	16	16
34	1	1	1	1	1	1	2	2	2	2	2	2	2	2	2	2	2	2	2	2	2	2	2	2	2
	30	32	32	32	32	32	32	32	32	32	32	32	32	32	32	32	32	32	32	32	32	32	32	32	32
	9	9	16	16	16	16	16	16	16	16	16	16	16	16	16	16	16	16	16	16	16	16	16	16	16
35	1	1	1	1	1	1	2	2	2	2	2	2	2	2	2	2	2	2	2	2	2	2	2	2	2
	30	32	33	33	33	33	33	33	33	33	33	33	33	33	33	33	33	33	33	33	33	33	33	33	33
	9	9	16	16	16	16	16	16	16	16	16	16	16	16	16	16	16	16	16	16	16	16	16	16	16
36	1	1	1	1	1	1	2	2	2	2	2	2	2	2	2	2	2	2	2	2	2	2	2	2	2
	30	32	34	34	34	34	34	34	34	34	34	34	34	34	34	34	34	34	34	34	34	34	34	34	34
	9	9	9	16	16	16	16	16	16	16	16	16	16	16	16	16	16	16	16	16	16	16	16	16	16
37	1	1	1	1	1	1	2	2	2	2	2	2	2	2	2	2	2	2	2	2	2	2	2	2	2
	30	32	34	35	35	35	35	35	35	35	35	35	35	35	35	35	35	35	35	35	35	35	35	35	35
	9	9	9	16	16	16	16	16	16	16	16	16	16	16	16	16	16	16	16	16	16	16	16	16	16
38	1	1	1	1	1	1	2	2	2	2	2	2	2	2	2	2	2	2	2	2	2	2	2	2	2
	30	32	34	36	36	36	36	36	36	36	36	36	36	36	36	36	36	36	36	36	36	36	36	36	36
	9	9	9	9	16	16	16	16	16	16	16	16	16	16	16	16	16	16	16	16	16	16	16	16	16
39	1	1	1	1	1	1	2	2	2	2	2	2	2	2	2	2	2	2	2	2	2	2	2	2	2
	30	32	34	36	37	37	37	37	37	37	37	37	37	37	37	37	37	37	37	37	37	37	37	37	37
	9	9	9	9	16	16	16	16	16	16	16	16	16	16	16	16	16	16	16	16	16	16	16	16	16
40	1	1	1	1	1	1	2	2	2	2	2	2	2	2	2	2	2	2	2	2	2	2	2	2	2
	30	32	34	36	38	38	38	38	38	38	38	38	38	38	38	38	38	38	38	38	38	38	38	38	38
	9	9	9	9	9	16	16	16	16	16	16	16	16	16	16	16	16	16	16	16	16	16	16	16	16
41	1	1	1	1	1	1	2	2	2	2	2	2	2	2	2	2	2	2	2	2	2	2	2	2	2
	30	32	34	36	38	39	39	39	39	39	39	39	39	39	39	39	39	39	39	39	39	39	39	39	39
	9	9	9	9	9	16	16	16	16	16	16	16	16	16	16	16	16	16	16	16	16	16	16	16	16
42	1	1	1	1	1	1	2	2	2	2	2	2	2	2	2	2	2	2	2	2	2	2	2	2	2
	30	32	34	36	38	40	40	40	40	40	40	40	40	40	40	40	40	40	40	40	40	40	40	40	40
	9	9	9	9	9	9	16	16	16	16	16	16	16	16	16	16	16	16	16	16	16	16	16	16	16
43	1	1	1	1	1	1	2	2	2	2	2	2	2	2	2	2	2	2	2	2	2	2	2	2	2
	30	32	34	36	38	40	41	41	41	41	41	41	41	41	41	41	41	41	41	41	41	41	41	41	41
	9	9	9	9	9	9	16	16	16	16	16	16	16	16	16	16	16	16	16	16	16	16	16	16	16
44	1	1	1	1	1	1	2	2	2	2	2	2	2	2	2	2	2	2	2	2	2	2	2	2	2
	30	32	34	36	38	40	42	42	42	42	42	42	42	42	42	42	42	42	42	42	42	42	42	42	42
	9	9	9	9	9	9	16	16	16	16	16	16	16	16	16	16	16	16	16	16	16	16	16	16	16
45	1	1	1	1	1	1	2	2	2	2	2	2	2	2	2	2	2	2	2	2	2	2	2	2	2
	30	32	34	36	38	40	43	43	43	43	43	43	43	43	43	43	43	43	43	43	43	43	43	43	43
	9	9	9	9	9	9	16	16	16	16	16	16	16	16	16	16	16	16	16	16	16	16	16	16	16
46	1	1	1	1	1	1	2	2	2	2	2	2	2	2	2	2	2	2	2	2	2	2	2	2	2
	30	32	34	36	38	40	44	44	44	44	44	44	44	44	44	44	44	44	44	44	44	44	44	44	44
	9	9	9	9	9	9	16	16	16	16	16	16	16	16	16	16	16	16	16	16	16	16	16	16	16
47	1	1	1	1	1	1	2	2	2	2	2	2	2	2	2	2	2	2	2	2	2	2	2	2	2
	30	32	34	36	38	40	45	45	45	45	45	45	45	45	45	45	45	45	45	45	45	45	45	45	45
	9	9	9	9	9	9	16	16	16	16	16	16	16	16	16	16	16	16	16	16	16	16	16	16	16
48	1	1	1	1	1	1	2	2	2	2	2	2	2	2	2	2	2	2	2	2	2	2	2	2	2
	30	32	34	36	38	40	46	46	46	46	46	46	46	46	46	46	46	46	46	46	46	46	46	46	46
	9	9	9	9	9	9	16	16	16	16	16	16	16	16	16	16	16	16	16	16	16	16	16	16	16
49	1	1	1	1	1	1	2	2	2	2	2	2	2	2	2	2	2	2	2	2	2	2	2	2	2
	30	32	34	36	38	40	47	47	47	47	47	47	47	47	47	47	47	47	47	47	47	47	47	47	47
	9	9	9	9	9	9	16	16	16	16	16	16	16	16	16	16	16	16	16	16	16	16	16	16	16
50	1	1	1	1	1	1	2	2	2	2	2	2	2	2	2	2	2	2	2	2	2	2	2	2	2
	30	32	34	36	38	40	48	48	48	48	48	48	48	48	48	48	48	48	48	48	48	48	48	48	48
	9	9	9	9	9	9	16	16	16	16	16	16	16	16	16	16	16	16	16	16	16	16	16	16	16

T TABLE BASE 10

m/s	1	2	3	4	5	6	7	8	9	10	11	12	13	14	15	16	17	18	19	20	21	22	23	24	25
1	0	0	0	0	0	0	0	0	0	0	0	0	0	0	0	0	0	0	0	0	0	0	0	0	0
	0	0	0	0	0	0	0	0	0	0	0	0	0	0	0	0	0	0	0	0	0	0	0	0	0
	5	5	5	5	5	5	5	5	5	5	5	5	5	5	5	5	5	5	5	5	5	5	5	5	5
2	0	0	0	0	0	0	0	0	0	0	1̄	1̄	1̄	1̄	1̄	1̄	1̄	1̄	1̄	1̄	1̄	1̄	1̄	1̄	1̄
	0	0	0	0	1	1	1	1	1	1	1	1	1	1	1	1	1	1	1	1	1	1	1	1	1
	2	2	2	7	5	5	5	5	5	5	5	5	5	5	5	5	5	5	5	5	5	5	5	5	5
3	0	0	0	0	0	0	0	0	0	0	1̄	1̄	1̄	1̄	1̄	1̄	1̄	1̄	1̄	1̄	1̄	1̄	1̄	1̄	1̄
	0	0	0	1	1	1	1	1	1	1	1	1	1	2	2	2	2	2	2	2	2	2	2	2	2
	2	2	8	16	2	2	2	2	2	2	2	14	5	5	5	5	5	5	5	5	5	5	5	5	5
4	0	0	0	0	0	0	0	0	0	0	1̄	1̄	1̄	1̄	1̄	1̄	1̄	1̄	1̄	1̄	1̄	1̄	1̄	1̄	1̄
	0	0	0	1	2	2	2	2	2	2	2	2	2	2	2	2	2	2	2	2	2	2	2	2	2
	2	2	8	10	16	16	16	16	16	16	16	16	2	2	2	2	2	2	2	2	2	2	2	2	2
5	0	0	0	0	0	0	0	0	0	0	1̄	1̄	1̄	1̄	1̄	1̄	1̄	1̄	1̄	1̄	1̄	1̄	1̄	1̄	1̄
	0	0	0	1	3	3	3	3	3	3	3	3	3	3	3	3	3	3	3	3	3	3	3	3	3
	2	2	8	10	16	16	16	16	16	16	16	16	16	16	16	16	16	16	16	16	16	16	16	16	16
6	0	0	0	0	0	0	0	0	0	0	1̄	1̄	1̄	1̄	1̄	1̄	1̄	1̄	1̄	1̄	1̄	1̄	1̄	1̄	1̄
	0	0	0	1	4	4	4	4	4	4	4	4	4	4	4	4	4	4	4	4	4	4	4	4	4
	2	2	8	10	16	16	16	16	16	16	16	16	16	16	16	16	16	16	16	16	16	16	16	16	16
7	0	0	0	0	0	0	0	0	0	0	1̄	1̄	1̄	1̄	1̄	1̄	1̄	1̄	1̄	1̄	1̄	1̄	1̄	1̄	1̄
	0	0	0	1	5	5	5	5	5	5	5	5	5	5	5	5	5	5	5	5	5	5	5	5	5
	2	2	8	10	2	10	16	16	16	16	16	16	16	16	16	16	16	16	16	16	16	16	16	16	16
8	0	0	0	0	0	0	0	0	0	0	1̄	1̄	1̄	1̄	1̄	1̄	1̄	1̄	1̄	1̄	1̄	1̄	1̄	1̄	1̄
	0	0	0	1	5	5	6	6	6	6	6	6	6	6	6	6	6	6	6	6	6	6	6	6	6
	2	2	8	10	2	10	16	16	16	16	16	16	16	16	16	16	16	16	16	16	16	16	16	16	16
9	0	0	0	0	0	0	0	0	0	0	1̄	1̄	1̄	1̄	1̄	1̄	1̄	1̄	1̄	1̄	1̄	1̄	1̄	1̄	1̄
	0	0	0	1	5	5	7	7	7	7	7	7	7	7	7	7	7	7	7	7	7	7	7	7	7
	2	2	8	10	2	10	16	16	16	16	16	16	16	16	16	16	16	16	16	16	16	16	16	16	16
10	0	0	0	0	0	0	0	0	0	0	1̄	1̄	1̄	1̄	1̄	1̄	1̄	1̄	1̄	1̄	1̄	1̄	1̄	1̄	1̄
	0	0	0	1	5	5	8	8	8	8	8	8	8	8	8	8	8	8	8	8	8	8	8	8	8
	2	2	8	10	2	10	16	16	16	16	16	16	16	16	16	16	16	16	16	16	16	16	16	16	16
11	0	0	0	0	0	0	0	0	0	0	1̄	1̄	1̄	1̄	1̄	1̄	1̄	1̄	1̄	1̄	1̄	1̄	1̄	1̄	1̄
	0	0	0	1	5	5	8	9	9	9	9	9	9	9	9	9	9	9	9	9	9	9	9	9	9
	2	2	8	10	2	10	10	16	16	16	16	16	16	16	16	16	16	16	16	16	16	16	16	16	16
12	0	0	0	0	0	0	0	0	0	0	1̄	1̄	1̄	1̄	1̄	1̄	1̄	1̄	1̄	1̄	1̄	1̄	1̄	1̄	1̄
	0	0	0	1	5	5	8	10	10	10	10	10	10	10	10	10	10	10	10	10	10	10	10	10	10
	2	2	8	10	2	10	16	16	16	16	16	16	16	16	16	16	16	16	16	16	16	16	16	16	16
13	0	0	0	0	0	0	0	0	0	0	1̄	1̄	1̄	1̄	1̄	1̄	1̄	1̄	1̄	1̄	1̄	1̄	1̄	1̄	1̄
	0	0	0	1	5	5	8	11	11	11	11	11	11	11	11	11	11	11	11	11	11	11	11	11	11
	2	2	8	10	2	10	16	16	16	16	16	16	16	16	16	16	16	16	16	16	16	16	16	16	16
14	0	0	0	0	0	0	0	0	0	0	1̄	1̄	1̄	1̄	1̄	1̄	1̄	1̄	1̄	1̄	1̄	1̄	1̄	1̄	1̄
	0	0	0	1	5	5	8	11	12	12	12	12	12	12	12	12	12	12	12	12	12	12	12	12	12
	2	2	8	10	2	10	16	16	16	16	16	16	16	16	16	16	16	16	16	16	16	16	16	16	16
15	0	0	0	0	0	0	0	0	0	0	1̄	1̄	1̄	1̄	1̄	1̄	1̄	1̄	1̄	1̄	1̄	1̄	1̄	1̄	1̄
	0	0	0	1	5	5	8	11	13	13	13	13	13	13	13	13	13	13	13	13	13	13	13	13	13
	2	2	8	10	2	10	16	16	16	16	16	16	16	16	16	16	16	16	16	16	16	16	16	16	16
16	0	0	0	0	0	0	0	0	0	0	1̄	1̄	1̄	1̄	1̄	1̄	1̄	1̄	1̄	1̄	1̄	1̄	1̄	1̄	1̄
	0	0	0	1	5	5	8	11	14	14	14	14	14	14	14	14	14	14	14	14	14	14	14	14	14
	2	2	8	10	2	10	10	10	16	16	16	16	16	16	16	16	16	16	16	16	16	16	16	16	16
17	0	0	0	0	0	0	0	0	0	0	1̄	1̄	1̄	1̄	1̄	1̄	1̄	1̄	1̄	1̄	1̄	1̄	1̄	1̄	1̄
	0	0	0	1	5	5	8	11	14	15	15	15	15	15	15	15	15	15	15	15	15	15	15	15	15
	2	2	8	10	2	10	10	10	16	16	16	16	16	16	16	16	16	16	16	16	16	16	16	16	16
18	0	0	0	0	0	0	0	0	0	0	1̄	1̄	1̄	1̄	1̄	1̄	1̄	1̄	1̄	1̄	1̄	1̄	1̄	1̄	1̄
	0	0	0	1	5	5	8	11	14	16	16	16	16	16	16	16	16	16	16	16	16	16	16	16	16
	2	2	8	10	2	10	10	10	16	16	16	16	16	16	16	16	16	16	16	16	16	16	16	16	16
19	0	0	0	0	0	0	0	0	0	0	1̄	1̄	1̄	1̄	1̄	1̄	1̄	1̄	1̄	1̄	1̄	1̄	1̄	1̄	1̄
	0	0	0	1	5	5	8	11	14	17	17	17	17	17	17	17	17	17	17	17	17	17	17	17	17
	2	2	8	10	2	10	10	10	16	16	16	16	16	16	16	16	16	16	16	16	16	16	16	16	16
20	0	0	0	0	0	0	0	0	0	0	1̄	1̄	1̄	1̄	1̄	1̄	1̄	1̄	1̄	1̄	1̄	1̄	1̄	1̄	1̄
	0	0	0	1	5	5	8	11	14	18	18	18	18	18	18	18	18	18	18	18	18	18	18	18	18
	2	2	8	10	2	10	10	10	16	16	16	16	16	16	16	16	16	16	16	16	16	16	16	16	16
21	0	0	0	0	0	0	0	0	0	0	1̄	1̄	1̄	1̄	1̄	1̄	1̄	1̄	1̄	1̄	1̄	1̄	1̄	1̄	1̄
	0	0	0	1	5	5	8	11	14	18	19	19	19	19	19	19	19	19	19	19	19	19	19	19	19
	2	2	8	10	2	10	10	10	16	16	16	16	16	16	16	16	16	16	16	16	16	16	16	16	16
22	0	0	0	0	0	0	0	0	0	0	1̄	1̄	1̄	1̄	1̄	1̄	1̄	1̄	1̄	1̄	1̄	1̄	1̄	1̄	1̄
	0	0	0	1	5	5	8	11	14	18	20	20	20	20	20	20	20	20	20	20	20	20	20	20	20
	2	2	8	10	2	10	10	10	16	16	16	16	16	16	16	16	16	16	16	16	16	16	16	16	16
23	0	0	0	0	0	0	0	0	0	0	1̄	1̄	1̄	1̄	1̄	1̄	1̄	1̄	1̄	1̄	1̄	1̄	1̄	1̄	1̄
	0	0	0	1	5	5	8	11	14	18	21	21	21	21	21	21	21	21	21	21	21	21	21	21	21
	2	2	8	10	2	10	10	10	16	16	16	16	16	16	16	16	16	16	16	16	16	16	16	16	16
24	0	0	0	0	0	0	0	0	0	0	1̄	1̄	1̄	1̄	1̄	1̄	1̄	1̄	1̄	1̄	1̄	1̄	1̄	1̄	1̄
	0	0	0	1	5	5	8	11	14	18	22	22	22	22	22	22	22	22	22	22	22	22	22	22	22
	2	2	8	10	2	10	10	10	16	16	16	16	16	16	16	16	16	16	16	16	16	16	16	16	16
25	0	0	0	0	0	0	0	0	0	0	1̄	1̄	1̄	1̄	1̄	1̄	1̄	1̄	1̄	1̄	1̄	1̄	1̄	1̄	1̄
	0	0	0	1	5	5	8	11	14	18	22	23	23	23	23	23	23	23	23	23	23	23	23	23	23
	2	2	8	10	2	10	10	10	10	10	16	16	16	16	16	16	16	16	16	16	16	16	16	16	16

T TABLE BASE 10

m/s	26	27	28	29	30	31	32	33	34	35	36	37	38	39	40	41	42	43	44	45	46	47	48	49	50
1	0	0	0	0	0	0	0	0	0	0	0	0	0	0	0	0	0	0	0	0	0	0	0	0	0
	0	0	0	0	0	0	0	0	0	0	0	0	0	0	0	0	0	0	0	0	0	0	0	0	0
	5	5	5	5	5	5	5	5	5	5	5	5	5	5	5	5	5	5	5	5	5	5	5	5	5
2	1	1	1	1	1	1	1	1	1	1	1	1	1	1	1	1	1	1	1	1	1	1	1	1	1
	1	1	1	1	1	1	1	1	1	1	1	1	1	1	1	1	1	1	1	1	1	1	1	1	1
	5	5	5	5	5	5	5	5	5	5	5	5	5	5	5	5	5	5	5	5	5	5	5	5	5
3	1	1	1	1	1	1	1	1	1	1	1	1	1	1	1	1	1	1	1	1	1	1	1	1	1
	2	2	2	2	2	2	2	2	2	2	2	2	2	2	2	2	2	2	2	2	2	2	2	2	2
	5	5	5	5	5	5	5	5	5	5	5	5	5	5	5	5	5	5	5	5	5	5	5	5	5
4	1	1	1	1	1	1	1	1	1	1	1	1	1	1	1	1	1	1	1	1	1	1	1	1	1
	2	2	2	2	2	2	2	2	2	2	2	2	2	2	3	3	3	3	3	3	3	3	3	3	3
	2	2	2	2	2	2	2	2	2	2	2	2	2	14	5	5	5	5	5	5	5	5	5	5	5
5	1	1	1	1	1	1	1	1	1	1	1	1	1	1	1	1	1	1	1	1	1	1	1	1	1
	3	3	3	3	3	3	3	3	3	3	3	3	3	3	3	3	3	3	3	3	3	3	3	3	3
	16	16	16	16	16	16	16	16	16	16	16	16	16	16	2	2	2	2	2	2	2	2	2	2	2
6	1	1	1	1	1	1	1	1	1	1	1	1	1	1	1	1	1	1	1	1	1	1	1	1	1
	4	4	4	4	4	4	4	4	4	4	4	4	4	4	4	4	4	4	4	4	4	4	4	4	4
	16	16	16	16	16	16	16	16	16	16	16	16	16	16	16	16	16	16	16	16	16	16	16	16	16
7	1	1	1	1	1	1	1	1	1	1	1	1	1	1	1	1	1	1	1	1	1	1	1	1	1
	5	5	5	5	5	5	5	5	5	5	5	5	5	5	5	5	5	5	5	5	5	5	5	5	5
	16	16	16	16	16	16	16	16	16	16	16	16	16	16	16	16	16	16	16	16	16	16	16	16	16
8	1	1	1	1	1	1	1	1	1	1	1	1	1	1	1	1	1	1	1	1	1	1	1	1	1
	6	6	6	6	6	6	6	6	6	6	6	6	6	6	6	6	6	6	6	6	6	6	6	6	6
	16	16	16	16	16	16	16	16	16	16	16	16	16	16	16	16	16	16	16	16	16	16	16	16	16
9	1	1	1	1	1	1	1	1	1	1	1	1	1	1	1	1	1	1	1	1	1	1	1	1	1
	7	7	7	7	7	7	7	7	7	7	7	7	7	7	7	7	7	7	7	7	7	7	7	7	7
	16	16	16	16	16	16	16	16	16	16	16	16	16	16	16	16	16	16	16	16	16	16	16	16	16
10	1	1	1	1	1	1	1	1	1	1	1	1	1	1	1	1	1	1	1	1	1	1	1	1	1
	8	8	8	8	8	8	8	8	8	8	8	8	8	8	8	8	8	8	8	8	8	8	8	8	8
	16	16	16	16	16	16	16	16	16	16	16	16	16	16	16	16	16	16	16	16	16	16	16	16	16
11	1	1	1	1	1	1	1	1	1	1	1	1	1	1	1	1	1	1	1	1	1	1	1	1	1
	9	9	9	9	9	9	9	9	9	9	9	9	9	9	9	9	9	9	9	9	9	9	9	9	9
	16	16	16	16	16	16	16	16	16	16	16	16	16	16	16	16	16	16	16	16	16	16	16	16	16
12	1	1	1	1	1	1	1	1	1	1	1	1	1	1	1	1	1	1	1	1	1	1	1	1	1
	10	10	10	10	10	10	10	10	10	10	10	10	10	10	10	10	10	10	10	10	10	10	10	10	10
	16	16	16	16	16	16	16	16	16	16	16	16	16	16	16	16	16	16	16	16	16	16	16	16	16
13	1	1	1	1	1	1	1	1	1	1	1	1	1	1	1	1	1	1	1	1	1	1	1	1	1
	11	11	11	11	11	11	11	11	11	11	11	11	11	11	11	11	11	11	11	11	11	11	11	11	11
	16	16	16	16	16	16	16	16	16	16	16	16	16	16	16	16	16	16	16	16	16	16	16	16	16
14	1	1	1	1	1	1	1	1	1	1	1	1	1	1	1	1	1	1	1	1	1	1	1	1	1
	12	12	12	12	12	12	12	12	12	12	12	12	12	12	12	12	12	12	12	12	12	12	12	12	12
	16	16	16	16	16	16	16	16	16	16	16	16	16	16	16	16	16	16	16	16	16	16	16	16	16
15	1	1	1	1	1	1	1	1	1	1	1	1	1	1	1	1	1	1	1	1	1	1	1	1	1
	13	13	13	13	13	13	13	13	13	13	13	13	13	13	13	13	13	13	13	13	13	13	13	13	13
	16	16	16	16	16	16	16	16	16	16	16	16	16	16	16	16	16	16	16	16	16	16	16	16	16
16	1	1	1	1	1	1	1	1	1	1	1	1	1	1	1	1	1	1	1	1	1	1	1	1	1
	14	14	14	14	14	14	14	14	14	14	14	14	14	14	14	14	14	14	14	14	14	14	14	14	14
	16	16	16	16	16	16	16	16	16	16	16	16	16	16	16	16	16	16	16	16	16	16	16	16	16
17	1	1	1	1	1	1	1	1	1	1	1	1	1	1	1	1	1	1	1	1	1	1	1	1	1
	15	15	15	15	15	15	15	15	15	15	15	15	15	15	15	15	15	15	15	15	15	15	15	15	15
	16	16	16	16	16	16	16	16	16	16	16	16	16	16	16	16	16	16	16	16	16	16	16	16	16
18	1	1	1	1	1	1	1	1	1	1	1	1	1	1	1	1	1	1	1	1	1	1	1	1	1
	16	16	16	16	16	16	16	16	16	16	16	16	16	16	16	16	16	16	16	16	16	16	16	16	16
	16	16	16	16	16	16	16	16	16	16	16	16	16	16	16	16	16	16	16	16	16	16	16	16	16
19	1	1	1	1	1	1	1	1	1	1	1	1	1	1	1	1	1	1	1	1	1	1	1	1	1
	17	17	17	17	17	17	17	17	17	17	17	17	17	17	17	17	17	17	17	17	17	17	17	17	17
	16	16	16	16	16	16	16	16	16	16	16	16	16	16	16	16	16	16	16	16	16	16	16	16	16
20	1	1	1	1	1	1	1	1	1	1	1	1	1	1	1	1	1	1	1	1	1	1	1	1	1
	18	18	18	18	18	18	18	18	18	18	18	18	18	18	18	18	18	18	18	18	18	18	18	18	18
	16	16	16	16	16	16	16	16	16	16	16	16	16	16	16	16	16	16	16	16	16	16	16	16	16
21	1	1	1	1	1	1	1	1	1	1	1	1	1	1	1	1	1	1	1	1	1	1	1	1	1
	19	19	19	19	19	19	19	19	19	19	19	19	19	19	19	19	19	19	19	19	19	19	19	19	19
	16	16	16	16	16	16	16	16	16	16	16	16	16	16	16	16	16	16	16	16	16	16	16	16	16
22	1	1	1	1	1	1	1	1	1	1	1	1	1	1	1	1	1	1	1	1	1	1	1	1	1
	20	20	20	20	20	20	20	20	20	20	20	20	20	20	20	20	20	20	20	20	20	20	20	20	20
	16	16	16	16	16	16	16	16	16	16	16	16	16	16	16	16	16	16	16	16	16	16	16	16	16
23	1	1	1	1	1	1	1	1	1	1	1	1	1	1	1	1	1	1	1	1	1	1	1	1	1
	21	21	21	21	21	21	21	21	21	21	21	21	21	21	21	21	21	21	21	21	21	21	21	21	21
	16	16	16	16	16	16	16	16	16	16	16	16	16	16	16	16	16	16	16	16	16	16	16	16	16
24	1	1	1	1	1	1	1	1	1	1	1	1	1	1	1	1	1	1	1	1	1	1	1	1	1
	22	22	22	22	22	22	22	22	22	22	22	22	22	22	22	22	22	22	22	22	22	22	22	22	22
	16	16	16	16	16	16	16	16	16	16	16	16	16	16	16	16	16	16	16	16	16	16	16	16	16
25	1	1	1	1	1	1	1	1	1	1	1	1	1	1	1	1	1	1	1	1	1	1	1	1	1
	23	23	23	23	23	23	23	23	23	23	23	23	23	23	23	23	23	23	23	23	23	23	23	23	23
	16	16	16	16	16	16	16	16	16	16	16	16	16	16	16	16	16	16	16	16	16	16	16	16	16

T TABLE BASE 10

m/s	1	2	3	4	5	6	7	8	9	10	11	12	13	14	15	16	17	18	19	20	21	22	23	24	25
26	0	0	0	0	0	0	0	0	0	0	0	I	I	I	I	I	I	I	I	I	I	I	I	I	I
	0	0	0	1	5	5	8	11	14	18	22	24	24	24	24	24	24	24	24	24	24	24	24	24	24
	2	2	8	10	2	10	10	10	10	10	10	16	16	16	16	16	16	16	16	16	16	16	16	16	16
27	0	0	0	0	0	0	0	0	0	0	0	I	I	I	I	I	I	I	I	I	I	I	I	I	I
	0	0	0	1	5	5	8	11	14	18	22	25	25	25	25	25	25	25	25	25	25	25	25	25	25
	2	2	8	10	2	10	10	10	10	10	10	16	16	16	16	16	16	16	16	16	16	16	16	16	16
28	0	0	0	0	0	0	0	0	0	0	0	I	I	I	I	I	I	I	I	I	I	I	I	I	I
	0	0	0	1	5	5	8	11	14	18	22	26	26	26	26	26	26	26	26	26	26	26	26	26	26
	2	2	8	10	2	10	10	10	10	10	10	16	16	16	16	16	16	16	16	16	16	16	16	16	16
29	0	0	0	0	0	0	0	0	0	0	0	I	I	I	I	I	I	I	I	I	I	I	I	I	I
	0	0	0	1	5	5	8	11	14	18	22	26	27	27	27	27	27	27	27	27	27	27	27	27	27
	2	2	8	10	2	10	10	10	10	10	10	16	16	16	16	16	16	16	16	16	16	16	16	16	16
30	0	0	0	0	0	0	0	0	0	0	0	I	I	I	I	I	I	I	I	I	I	I	I	I	I
	0	0	0	1	5	5	8	11	14	18	22	26	28	28	28	28	28	28	28	28	28	28	28	28	28
	2	2	8	10	2	10	10	10	10	10	10	16	16	16	16	16	16	16	16	16	16	16	16	16	16
31	0	0	0	0	0	0	0	0	0	0	0	I	I	I	I	I	I	I	I	I	I	I	I	I	I
	0	0	0	1	5	5	8	11	14	18	22	26	29	29	29	29	29	29	29	29	29	29	29	29	29
	2	2	8	10	2	10	10	10	10	10	10	16	16	16	16	16	16	16	16	16	16	16	16	16	16
32	0	0	0	0	0	0	0	0	0	0	0	I	I	I	I	I	I	I	I	I	I	I	I	I	I
	0	0	0	1	5	5	8	11	14	18	22	26	30	30	30	30	30	30	30	30	30	30	30	30	30
	2	2	8	10	2	10	10	10	10	10	10	16	16	16	16	16	16	16	16	16	16	16	16	16	16
33	0	0	0	0	0	0	0	0	0	0	0	I	I	I	I	I	I	I	I	I	I	I	I	I	I
	0	0	0	1	5	5	8	11	14	18	22	26	30	31	31	31	31	31	31	31	31	31	31	31	31
	2	2	8	10	2	10	10	10	10	10	10	16	16	16	16	16	16	16	16	16	16	16	16	16	16
34	0	0	0	0	0	0	0	0	0	0	0	I	I	I	I	I	I	I	I	I	I	I	I	I	I
	0	0	0	1	5	5	8	11	14	18	22	26	30	32	32	32	32	32	32	32	32	32	32	32	32
	2	2	8	10	2	10	10	10	10	10	10	16	16	16	16	16	16	16	16	16	16	16	16	16	16
35	0	0	0	0	0	0	0	0	0	0	0	I	I	I	I	I	I	I	I	I	I	I	I	I	I
	0	0	0	1	5	5	8	11	14	18	22	26	30	33	33	33	33	33	33	33	33	33	33	33	33
	2	2	8	10	2	10	10	10	10	10	10	16	16	16	16	16	16	16	16	16	16	16	16	16	16
36	0	0	0	0	0	0	0	0	0	0	0	I	I	I	I	I	I	I	I	I	I	I	I	I	I
	0	0	0	1	5	5	8	11	14	18	22	26	30	34	34	34	34	34	34	34	34	34	34	34	34
	2	2	8	10	2	10	10	10	10	10	10	16	16	16	16	16	16	16	16	16	16	16	16	16	16
37	0	0	0	0	0	0	0	0	0	0	0	I	I	I	I	I	I	I	I	I	I	I	I	I	I
	0	0	0	1	5	5	8	11	14	18	22	26	30	34	35	35	35	35	35	35	35	35	35	35	35
	2	2	8	10	2	10	10	10	10	10	10	16	16	16	16	16	16	16	16	16	16	16	16	16	16
38	0	0	0	0	0	0	0	0	0	0	0	I	I	I	I	I	I	I	I	I	I	I	I	I	I
	0	0	0	1	5	5	8	11	14	18	22	26	30	34	36	36	36	36	36	36	36	36	36	36	36
	2	2	8	10	2	10	10	10	10	10	10	16	16	16	16	16	16	16	16	16	16	16	16	16	16
39	0	0	0	0	0	0	0	0	0	0	0	I	I	I	I	I	I	I	I	I	I	I	I	I	I
	0	0	0	1	5	5	8	11	14	18	22	26	30	34	37	37	37	37	37	37	37	37	37	37	37
	2	2	8	10	2	10	10	10	10	10	10	16	16	16	16	16	16	16	16	16	16	16	16	16	16
40	0	0	0	0	0	0	0	0	0	0	0	I	I	I	I	I	I	I	I	I	I	I	I	I	I
	0	0	0	1	5	5	8	11	14	18	22	26	30	34	38	38	38	38	38	38	38	38	38	38	38
	2	2	8	10	2	10	10	10	10	10	10	16	16	16	16	16	16	16	16	16	16	16	16	16	16
41	0	0	0	0	0	0	0	0	0	0	0	I	I	I	I	I	I	I	I	I	I	I	I	I	I
	0	0	0	1	5	5	8	11	14	18	22	26	30	34	38	39	39	39	39	39	39	39	39	39	39
	2	2	8	10	2	10	10	10	10	10	10	16	16	16	16	16	16	16	16	16	16	16	16	16	16
42	0	0	0	0	0	0	0	0	0	0	0	I	I	I	I	I	I	I	I	I	I	I	I	I	I
	0	0	0	1	5	5	8	11	14	18	22	26	30	34	38	40	40	40	40	40	40	40	40	40	40
	2	2	8	10	2	10	10	10	10	10	10	16	16	16	16	16	16	16	16	16	16	16	16	16	16
43	0	0	0	0	0	0	0	0	0	0	0	I	I	I	I	I	I	I	I	I	I	I	I	I	I
	0	0	0	1	5	5	8	11	14	18	22	26	30	34	38	41	41	41	41	41	41	41	41	41	41
	2	2	8	10	2	10	10	10	10	10	10	16	16	16	16	16	16	16	16	16	16	16	16	16	16
44	0	0	0	0	0	0	0	0	0	0	0	I	I	I	I	I	I	I	I	I	I	I	I	I	I
	0	0	0	1	5	5	8	11	14	18	22	26	30	34	38	42	42	42	42	42	42	42	42	42	42
	2	2	8	10	2	10	10	10	10	10	10	16	16	16	16	16	16	16	16	16	16	16	16	16	16
45	0	0	0	0	0	0	0	0	0	0	0	I	I	I	I	I	I	I	I	I	I	I	I	I	I
	0	0	0	1	5	5	8	11	14	18	22	26	30	34	38	43	43	43	43	43	43	43	43	43	43
	2	2	8	10	2	10	10	10	10	10	10	16	16	16	16	16	16	16	16	16	16	16	16	16	16
46	0	0	0	0	0	0	0	0	0	0	0	I	I	I	I	I	I	I	I	I	I	I	I	I	I
	0	0	0	1	5	5	8	11	14	18	22	26	30	34	38	43	44	44	44	44	44	44	44	44	44
	2	2	8	10	2	10	10	10	10	10	10	16	16	16	16	16	16	16	16	16	16	16	16	16	16
47	0	0	0	0	0	0	0	0	0	0	0	I	I	I	I	I	I	I	I	I	I	I	I	I	I
	0	0	0	1	5	5	8	11	14	18	22	26	30	34	38	43	45	45	45	45	45	45	45	45	45
	2	2	8	10	2	10	10	10	10	10	10	16	16	16	16	16	16	16	16	16	16	16	16	16	16
48	0	0	0	0	0	0	0	0	0	0	0	I	I	I	I	I	I	I	I	I	I	I	I	I	I
	0	0	0	1	5	5	8	11	14	18	22	26	30	34	38	43	46	46	46	46	46	46	46	46	46
	2	2	8	10	2	10	10	10	10	10	10	16	16	16	16	16	16	16	16	16	16	16	16	16	16
49	0	0	0	0	0	0	0	0	0	0	0	I	I	I	I	I	I	I	I	I	I	I	I	I	I
	0	0	0	1	5	5	8	11	14	18	22	26	30	34	38	43	47	47	47	47	47	47	47	47	47
	2	2	8	10	2	10	10	10	10	10	10	16	16	16	16	16	16	16	16	16	16	16	16	16	16
50	0	0	0	0	0	0	0	0	0	0	0	I	I	I	I	I	I	I	I	I	I	I	I	I	I
	0	0	0	1	5	5	8	11	14	18	22	26	30	34	38	43	48	48	48	48	48	48	48	48	48
	3	3	3	10	2	10	10	10	10	10	10	10	10	10	10	10	10	16	16	16	16	16	16	16	16

T TABLE BASE 10

Each cell contains three stacked values: an overlined "$\overline{1}$" (top), a middle value, and "16" (bottom).

m/s	26	27	28	29	30	31	32	33	34	35	36	37	38	39	40	41	42	43	44	45	46	47	48	49	50
26		$\overline{1}$ 24 16	$\overline{1}$ 24 16	$\overline{1}$ 24 16	$\overline{1}$ 24 16	$\overline{1}$ 24 16	$\overline{1}$ 24 16	$\overline{1}$ 24 16	$\overline{1}$ 24 16	$\overline{1}$ 24 16	$\overline{1}$ 24 16	$\overline{1}$ 24 16	$\overline{1}$ 24 16	$\overline{1}$ 24 16	$\overline{1}$ 24 16	$\overline{1}$ 24 16	$\overline{1}$ 24 16	$\overline{1}$ 24 16	$\overline{1}$ 24 16	$\overline{1}$ 24 16	$\overline{1}$ 24 16	$\overline{1}$ 24 16	$\overline{1}$ 24 16	$\overline{1}$ 24 16	$\overline{1}$ 24 16
27	$\overline{1}$ 25 16	$\overline{1}$ 25 16	$\overline{1}$ 25 16	$\overline{1}$ 25 16	$\overline{1}$ 25 16	$\overline{1}$ 25 16	$\overline{1}$ 25 16	$\overline{1}$ 25 16	$\overline{1}$ 25 16	$\overline{1}$ 25 16	$\overline{1}$ 25 16	$\overline{1}$ 25 16	$\overline{1}$ 25 16	$\overline{1}$ 25 16	$\overline{1}$ 25 16	$\overline{1}$ 25 16	$\overline{1}$ 25 16	$\overline{1}$ 25 16	$\overline{1}$ 25 16	$\overline{1}$ 25 16	$\overline{1}$ 25 16	$\overline{1}$ 25 16	$\overline{1}$ 25 16	$\overline{1}$ 25 16	$\overline{1}$ 25 16
28	$\overline{1}$ 26 16	$\overline{1}$ 26 16	$\overline{1}$ 26 16	$\overline{1}$ 26 16	$\overline{1}$ 26 16	$\overline{1}$ 26 16	$\overline{1}$ 26 16	$\overline{1}$ 26 16	$\overline{1}$ 26 16	$\overline{1}$ 26 16	$\overline{1}$ 26 16	$\overline{1}$ 26 16	$\overline{1}$ 26 16	$\overline{1}$ 26 16	$\overline{1}$ 26 16	$\overline{1}$ 26 16	$\overline{1}$ 26 16	$\overline{1}$ 26 16	$\overline{1}$ 26 16	$\overline{1}$ 26 16	$\overline{1}$ 26 16	$\overline{1}$ 26 16	$\overline{1}$ 26 16	$\overline{1}$ 26 16	$\overline{1}$ 26 16
29	$\overline{1}$ 27 16	$\overline{1}$ 27 16	$\overline{1}$ 27 16	$\overline{1}$ 27 16	$\overline{1}$ 27 16	$\overline{1}$ 27 16	$\overline{1}$ 27 16	$\overline{1}$ 27 16	$\overline{1}$ 27 16	$\overline{1}$ 27 16	$\overline{1}$ 27 16	$\overline{1}$ 27 16	$\overline{1}$ 27 16	$\overline{1}$ 27 16	$\overline{1}$ 27 16	$\overline{1}$ 27 16	$\overline{1}$ 27 16	$\overline{1}$ 27 16	$\overline{1}$ 27 16	$\overline{1}$ 27 16	$\overline{1}$ 27 16	$\overline{1}$ 27 16	$\overline{1}$ 27 16	$\overline{1}$ 27 16	$\overline{1}$ 27 16
30	$\overline{1}$ 28 16	$\overline{1}$ 28 16	$\overline{1}$ 28 16	$\overline{1}$ 28 16	$\overline{1}$ 28 16	$\overline{1}$ 28 16	$\overline{1}$ 28 16	$\overline{1}$ 28 16	$\overline{1}$ 28 16	$\overline{1}$ 28 16	$\overline{1}$ 28 16	$\overline{1}$ 28 16	$\overline{1}$ 28 16	$\overline{1}$ 28 16	$\overline{1}$ 28 16	$\overline{1}$ 28 16	$\overline{1}$ 28 16	$\overline{1}$ 28 16	$\overline{1}$ 28 16	$\overline{1}$ 28 16	$\overline{1}$ 28 16	$\overline{1}$ 28 16	$\overline{1}$ 28 16	$\overline{1}$ 28 16	$\overline{1}$ 28 16
31	$\overline{1}$ 29 16	$\overline{1}$ 29 16	$\overline{1}$ 29 16	$\overline{1}$ 29 16	$\overline{1}$ 29 16	$\overline{1}$ 29 16	$\overline{1}$ 29 16	$\overline{1}$ 29 16	$\overline{1}$ 29 16	$\overline{1}$ 29 16	$\overline{1}$ 29 16	$\overline{1}$ 29 16	$\overline{1}$ 29 16	$\overline{1}$ 29 16	$\overline{1}$ 29 16	$\overline{1}$ 29 16	$\overline{1}$ 29 16	$\overline{1}$ 29 16	$\overline{1}$ 29 16	$\overline{1}$ 29 16	$\overline{1}$ 29 16	$\overline{1}$ 29 16	$\overline{1}$ 29 16	$\overline{1}$ 29 16	$\overline{1}$ 29 16
32	$\overline{1}$ 30 16	$\overline{1}$ 30 16	$\overline{1}$ 30 16	$\overline{1}$ 30 16	$\overline{1}$ 30 16	$\overline{1}$ 30 16	$\overline{1}$ 30 16	$\overline{1}$ 30 16	$\overline{1}$ 30 16	$\overline{1}$ 30 16	$\overline{1}$ 30 16	$\overline{1}$ 30 16	$\overline{1}$ 30 16	$\overline{1}$ 30 16	$\overline{1}$ 30 16	$\overline{1}$ 30 16	$\overline{1}$ 30 16	$\overline{1}$ 30 16	$\overline{1}$ 30 16	$\overline{1}$ 30 16	$\overline{1}$ 30 16	$\overline{1}$ 30 16	$\overline{1}$ 30 16	$\overline{1}$ 30 16	$\overline{1}$ 30 16
33	$\overline{1}$ 31 16	$\overline{1}$ 31 16	$\overline{1}$ 31 16	$\overline{1}$ 31 16	$\overline{1}$ 31 16	$\overline{1}$ 31 16	$\overline{1}$ 31 16	$\overline{1}$ 31 16	$\overline{1}$ 31 16	$\overline{1}$ 31 16	$\overline{1}$ 31 16	$\overline{1}$ 31 16	$\overline{1}$ 31 16	$\overline{1}$ 31 16	$\overline{1}$ 31 16	$\overline{1}$ 31 16	$\overline{1}$ 31 16	$\overline{1}$ 31 16	$\overline{1}$ 31 16	$\overline{1}$ 31 16	$\overline{1}$ 31 16	$\overline{1}$ 31 16	$\overline{1}$ 31 16	$\overline{1}$ 31 16	$\overline{1}$ 31 16
34	$\overline{1}$ 32 16	$\overline{1}$ 32 16	$\overline{1}$ 32 16	$\overline{1}$ 32 16	$\overline{1}$ 32 16	$\overline{1}$ 32 16	$\overline{1}$ 32 16	$\overline{1}$ 32 16	$\overline{1}$ 32 16	$\overline{1}$ 32 16	$\overline{1}$ 32 16	$\overline{1}$ 32 16	$\overline{1}$ 32 16	$\overline{1}$ 32 16	$\overline{1}$ 32 16	$\overline{1}$ 32 16	$\overline{1}$ 32 16	$\overline{1}$ 32 16	$\overline{1}$ 32 16	$\overline{1}$ 32 16	$\overline{1}$ 32 16	$\overline{1}$ 32 16	$\overline{1}$ 32 16	$\overline{1}$ 32 16	$\overline{1}$ 32 16
35	$\overline{1}$ 33 16	$\overline{1}$ 33 16	$\overline{1}$ 33 16	$\overline{1}$ 33 16	$\overline{1}$ 33 16	$\overline{1}$ 33 16	$\overline{1}$ 33 16	$\overline{1}$ 33 16	$\overline{1}$ 33 16	$\overline{1}$ 33 16	$\overline{1}$ 33 16	$\overline{1}$ 33 16	$\overline{1}$ 33 16	$\overline{1}$ 33 16	$\overline{1}$ 33 16	$\overline{1}$ 33 16	$\overline{1}$ 33 16	$\overline{1}$ 33 16	$\overline{1}$ 33 16	$\overline{1}$ 33 16	$\overline{1}$ 33 16	$\overline{1}$ 33 16	$\overline{1}$ 33 16	$\overline{1}$ 33 16	$\overline{1}$ 33 16
36	$\overline{1}$ 34 16	$\overline{1}$ 34 16	$\overline{1}$ 34 16	$\overline{1}$ 34 16	$\overline{1}$ 34 16	$\overline{1}$ 34 16	$\overline{1}$ 34 16	$\overline{1}$ 34 16	$\overline{1}$ 34 16	$\overline{1}$ 34 16	$\overline{1}$ 34 16	$\overline{1}$ 34 16	$\overline{1}$ 34 16	$\overline{1}$ 34 16	$\overline{1}$ 34 16	$\overline{1}$ 34 16	$\overline{1}$ 34 16	$\overline{1}$ 34 16	$\overline{1}$ 34 16	$\overline{1}$ 34 16	$\overline{1}$ 34 16	$\overline{1}$ 34 16	$\overline{1}$ 34 16	$\overline{1}$ 34 16	$\overline{1}$ 34 16
37	$\overline{1}$ 35 16	$\overline{1}$ 35 16	$\overline{1}$ 35 16	$\overline{1}$ 35 16	$\overline{1}$ 35 16	$\overline{1}$ 35 16	$\overline{1}$ 35 16	$\overline{1}$ 35 16	$\overline{1}$ 35 16	$\overline{1}$ 35 16	$\overline{1}$ 35 16	$\overline{1}$ 35 16	$\overline{1}$ 35 16	$\overline{1}$ 35 16	$\overline{1}$ 35 16	$\overline{1}$ 35 16	$\overline{1}$ 35 16	$\overline{1}$ 35 16	$\overline{1}$ 35 16	$\overline{1}$ 35 16	$\overline{1}$ 35 16	$\overline{1}$ 35 16	$\overline{1}$ 35 16	$\overline{1}$ 35 16	$\overline{1}$ 35 16
38	$\overline{1}$ 36 16	$\overline{1}$ 36 16	$\overline{1}$ 36 16	$\overline{1}$ 36 16	$\overline{1}$ 36 16	$\overline{1}$ 36 16	$\overline{1}$ 36 16	$\overline{1}$ 36 16	$\overline{1}$ 36 16	$\overline{1}$ 36 16	$\overline{1}$ 36 16	$\overline{1}$ 36 16	$\overline{1}$ 36 16	$\overline{1}$ 36 16	$\overline{1}$ 36 16	$\overline{1}$ 36 16	$\overline{1}$ 36 16	$\overline{1}$ 36 16	$\overline{1}$ 36 16	$\overline{1}$ 36 16	$\overline{1}$ 36 16	$\overline{1}$ 36 16	$\overline{1}$ 36 16	$\overline{1}$ 36 16	$\overline{1}$ 36 16
39	$\overline{1}$ 37 16	$\overline{1}$ 37 16	$\overline{1}$ 37 16	$\overline{1}$ 37 16	$\overline{1}$ 37 16	$\overline{1}$ 37 16	$\overline{1}$ 37 16	$\overline{1}$ 37 16	$\overline{1}$ 37 16	$\overline{1}$ 37 16	$\overline{1}$ 37 16	$\overline{1}$ 37 16	$\overline{1}$ 37 16	$\overline{1}$ 37 16	$\overline{1}$ 37 16	$\overline{1}$ 37 16	$\overline{1}$ 37 16	$\overline{1}$ 37 16	$\overline{1}$ 37 16	$\overline{1}$ 37 16	$\overline{1}$ 37 16	$\overline{1}$ 37 16	$\overline{1}$ 37 16	$\overline{1}$ 37 16	$\overline{1}$ 37 16
40	$\overline{1}$ 38 16	$\overline{1}$ 38 16	$\overline{1}$ 38 16	$\overline{1}$ 38 16	$\overline{1}$ 38 16	$\overline{1}$ 38 16	$\overline{1}$ 38 16	$\overline{1}$ 38 16	$\overline{1}$ 38 16	$\overline{1}$ 38 16	$\overline{1}$ 38 16	$\overline{1}$ 38 16	$\overline{1}$ 38 16	$\overline{1}$ 38 16	$\overline{1}$ 38 16	$\overline{1}$ 38 16	$\overline{1}$ 38 16	$\overline{1}$ 38 16	$\overline{1}$ 38 16	$\overline{1}$ 38 16	$\overline{1}$ 38 16	$\overline{1}$ 38 16	$\overline{1}$ 38 16	$\overline{1}$ 38 16	$\overline{1}$ 38 16
41	$\overline{1}$ 39 16	$\overline{1}$ 39 16	$\overline{1}$ 39 16	$\overline{1}$ 39 16	$\overline{1}$ 39 16	$\overline{1}$ 39 16	$\overline{1}$ 39 16	$\overline{1}$ 39 16	$\overline{1}$ 39 16	$\overline{1}$ 39 16	$\overline{1}$ 39 16	$\overline{1}$ 39 16	$\overline{1}$ 39 16	$\overline{1}$ 39 16	$\overline{1}$ 39 16	$\overline{1}$ 39 16	$\overline{1}$ 39 16	$\overline{1}$ 39 16	$\overline{1}$ 39 16	$\overline{1}$ 39 16	$\overline{1}$ 39 16	$\overline{1}$ 39 16	$\overline{1}$ 39 16	$\overline{1}$ 39 16	$\overline{1}$ 39 16
42	$\overline{1}$ 40 16	$\overline{1}$ 40 16	$\overline{1}$ 40 16	$\overline{1}$ 40 16	$\overline{1}$ 40 16	$\overline{1}$ 40 16	$\overline{1}$ 40 16	$\overline{1}$ 40 16	$\overline{1}$ 40 16	$\overline{1}$ 40 16	$\overline{1}$ 40 16	$\overline{1}$ 40 16	$\overline{1}$ 40 16	$\overline{1}$ 40 16	$\overline{1}$ 40 16	$\overline{1}$ 40 16	$\overline{1}$ 40 16	$\overline{1}$ 40 16	$\overline{1}$ 40 16	$\overline{1}$ 40 16	$\overline{1}$ 40 16	$\overline{1}$ 40 16	$\overline{1}$ 40 16	$\overline{1}$ 40 16	$\overline{1}$ 40 16
43	$\overline{1}$ 41 16	$\overline{1}$ 41 16	$\overline{1}$ 41 16	$\overline{1}$ 41 16	$\overline{1}$ 41 16	$\overline{1}$ 41 16	$\overline{1}$ 41 16	$\overline{1}$ 41 16	$\overline{1}$ 41 16	$\overline{1}$ 41 16	$\overline{1}$ 41 16	$\overline{1}$ 41 16	$\overline{1}$ 41 16	$\overline{1}$ 41 16	$\overline{1}$ 41 16	$\overline{1}$ 41 16	$\overline{1}$ 41 16	$\overline{1}$ 41 16	$\overline{1}$ 41 16	$\overline{1}$ 41 16	$\overline{1}$ 41 16	$\overline{1}$ 41 16	$\overline{1}$ 41 16	$\overline{1}$ 41 16	$\overline{1}$ 41 16
44	$\overline{1}$ 42 16	$\overline{1}$ 42 16	$\overline{1}$ 42 16	$\overline{1}$ 42 16	$\overline{1}$ 42 16	$\overline{1}$ 42 16	$\overline{1}$ 42 16	$\overline{1}$ 42 16	$\overline{1}$ 42 16	$\overline{1}$ 42 16	$\overline{1}$ 42 16	$\overline{1}$ 42 16	$\overline{1}$ 42 16	$\overline{1}$ 42 16	$\overline{1}$ 42 16	$\overline{1}$ 42 16	$\overline{1}$ 42 16	$\overline{1}$ 42 16	$\overline{1}$ 42 16	$\overline{1}$ 42 16	$\overline{1}$ 42 16	$\overline{1}$ 42 16	$\overline{1}$ 42 16	$\overline{1}$ 42 16	$\overline{1}$ 42 16
45	$\overline{1}$ 43 16	$\overline{1}$ 43 16	$\overline{1}$ 43 16	$\overline{1}$ 43 16	$\overline{1}$ 43 16	$\overline{1}$ 43 16	$\overline{1}$ 43 16	$\overline{1}$ 43 16	$\overline{1}$ 43 16	$\overline{1}$ 43 16	$\overline{1}$ 43 16	$\overline{1}$ 43 16	$\overline{1}$ 43 16	$\overline{1}$ 43 16	$\overline{1}$ 43 16	$\overline{1}$ 43 16	$\overline{1}$ 43 16	$\overline{1}$ 43 16	$\overline{1}$ 43 16	$\overline{1}$ 43 16	$\overline{1}$ 43 16	$\overline{1}$ 43 16	$\overline{1}$ 43 16	$\overline{1}$ 43 16	$\overline{1}$ 43 16
46	$\overline{1}$ 44 16	$\overline{1}$ 44 16	$\overline{1}$ 44 16	$\overline{1}$ 44 16	$\overline{1}$ 44 16	$\overline{1}$ 44 16	$\overline{1}$ 44 16	$\overline{1}$ 44 16	$\overline{1}$ 44 16	$\overline{1}$ 44 16	$\overline{1}$ 44 16	$\overline{1}$ 44 16	$\overline{1}$ 44 16	$\overline{1}$ 44 16	$\overline{1}$ 44 16	$\overline{1}$ 44 16	$\overline{1}$ 44 16	$\overline{1}$ 44 16	$\overline{1}$ 44 16	$\overline{1}$ 44 16	$\overline{1}$ 44 16	$\overline{1}$ 44 16	$\overline{1}$ 44 16	$\overline{1}$ 44 16	$\overline{1}$ 44 16
47	$\overline{1}$ 45 16	$\overline{1}$ 45 16	$\overline{1}$ 45 16	$\overline{1}$ 45 16	$\overline{1}$ 45 16	$\overline{1}$ 45 16	$\overline{1}$ 45 16	$\overline{1}$ 45 16	$\overline{1}$ 45 16	$\overline{1}$ 45 16	$\overline{1}$ 45 16	$\overline{1}$ 45 16	$\overline{1}$ 45 16	$\overline{1}$ 45 16	$\overline{1}$ 45 16	$\overline{1}$ 45 16	$\overline{1}$ 45 16	$\overline{1}$ 45 16	$\overline{1}$ 45 16	$\overline{1}$ 45 16	$\overline{1}$ 45 16	$\overline{1}$ 45 16	$\overline{1}$ 45 16	$\overline{1}$ 45 16	$\overline{1}$ 45 16
48	$\overline{1}$ 46 16	$\overline{1}$ 46 16	$\overline{1}$ 46 16	$\overline{1}$ 46 16	$\overline{1}$ 46 16	$\overline{1}$ 46 16	$\overline{1}$ 46 16	$\overline{1}$ 46 16	$\overline{1}$ 46 16	$\overline{1}$ 46 16	$\overline{1}$ 46 16	$\overline{1}$ 46 16	$\overline{1}$ 46 16	$\overline{1}$ 46 16	$\overline{1}$ 46 16	$\overline{1}$ 46 16	$\overline{1}$ 46 16	$\overline{1}$ 46 16	$\overline{1}$ 46 16	$\overline{1}$ 46 16	$\overline{1}$ 46 16	$\overline{1}$ 46 16	$\overline{1}$ 46 16	$\overline{1}$ 46 16	$\overline{1}$ 46 16
49	$\overline{1}$ 47 16	$\overline{1}$ 47 16	$\overline{1}$ 47 16	$\overline{1}$ 47 16	$\overline{1}$ 47 16	$\overline{1}$ 47 16	$\overline{1}$ 47 16	$\overline{1}$ 47 16	$\overline{1}$ 47 16	$\overline{1}$ 47 16	$\overline{1}$ 47 16	$\overline{1}$ 47 16	$\overline{1}$ 47 16	$\overline{1}$ 47 16	$\overline{1}$ 47 16	$\overline{1}$ 47 16	$\overline{1}$ 47 16	$\overline{1}$ 47 16	$\overline{1}$ 47 16	$\overline{1}$ 47 16	$\overline{1}$ 47 16	$\overline{1}$ 47 16	$\overline{1}$ 47 16	$\overline{1}$ 47 16	$\overline{1}$ 47 16
50	$\overline{1}$ 48 16	$\overline{1}$ 48 16	$\overline{1}$ 48 16	$\overline{1}$ 48 16	$\overline{1}$ 48 16	$\overline{1}$ 48 16	$\overline{1}$ 48 16	$\overline{1}$ 48 16	$\overline{1}$ 48 16	$\overline{1}$ 48 16	$\overline{1}$ 48 16	$\overline{1}$ 48 16	$\overline{1}$ 48 16	$\overline{1}$ 48 16	$\overline{1}$ 48 16	$\overline{1}$ 48 16	$\overline{1}$ 48 16	$\overline{1}$ 48 16	$\overline{1}$ 48 16	$\overline{1}$ 48 16	$\overline{1}$ 48 16	$\overline{1}$ 48 16	$\overline{1}$ 48 16	$\overline{1}$ 48 16	$\overline{1}$ 48 16

Since many of the net constructions come from (t, s)-sequence constructions, we provide in Table A the currently best known sequence parameters which are taken from Niederreiter [28, Tables II-V].

Table A

s	$T_2(s)$	$T_3(s)$	$T_4(s)$	$T_5(s)$
1	0	0	0	0
2	0	0	0	0
3	1	0	0	0
4	3	1	0	0
5	5	2	1	0
6	8	3	2	1
7	11	5	3	2
8	14	7	4	3
9	18	9	5	4
10	22	11	6	5
11	26	13	8	6
12	30	15	10	7
13	34	17	12	8
14	38	19	14	9
15	43	22	16	10
16	48	25	18	12
17	53	28	20	14
18	58	31	22	16
19	63	34	24	18
20	68	37	26	20
21	73	40	28	22
22	78	43	30	24
23	83	46	32	26
24	89	49	34	28
25	95	52	36	30
26	101	55	38	32
27	107	58	40	34
28	113	61	42	36
29	119	64	44	38
30	125	67	46	40

In an effort to make these tables as useful and as up-to-date as possible, we will be happy to incorporate new results or new constructions as they become available. Results may be submitted either in hardcopy or electronic form and should be sent to the first author. Copies of the latest tables either in hardcopy or electronic form will be available upon request from the first author.

REFERENCES

[1] D. A. ANDRÉ, G. L. MULLEN, AND H. NIEDERREITER, Figures of merit for digital multistep pseudorandom numbers, Math. Comp., 54 (1990), pp. 737-748.

[2] I. A. ANTONOV AND V. M. SALEEV, An economic method of computing LP_τ-sequences, Zh. Vychisl. Mat. i Mat. Fiz., 19 (1979), pp. 243-245 (in Russian).

[3] P. BRATLEY AND B. L. FOX, Algorithm 659: Implementing Sobol's quasirandom sequence generator, ACM Trans. Math. Software, 14 (1988), pp. 88-100.

[4] P. BRATLEY, B. L. FOX, AND H. NIEDERREITER, Implementation and tests of low-discrepancy sequences, ACM Trans. Modeling and Computer Simulation, 2 (1992), pp. 195-213.

[5] J. DÉNES AND A. D. KEEDWELL, Latin Squares and Their Applications, Academic Press, New York, 1974.

[6] H. FAURE, Discrépance de suites associées à un système de numération (en dimension s), Acta Arith., 41 (1982), pp. 337-351.

[7] B. L. FOX, Algorithm 647: Implementation and relative efficiency of quasirandom sequence generators, ACM Trans. Math. Software, 12 (1986), pp. 362-376.

[8] B. L. FOX, Simulated annealing: folklore, facts, and directions, this volume.

[9] T. HANSEN, G. L. MULLEN, AND H. NIEDERREITER, Good parameters for a class of node sets in quasi-Monte Carlo integration, Math. Comp., 61 (1993), pp. 225-234.

[10] G. LARCHER, Nets obtained from rational functions over finite fields, Acta Arith., 63 (1993), pp. 1-13.

[11] G. LARCHER, A class of low-discrepancy point-sets and its application to numerical integration by number-theoretical methods, Grazer Math. Berichte, 318 (1993), pp. 69-80.

[12] G. LARCHER, A. LAUSS, H. NIEDERREITER, AND W. C. SCHMID, Optimal polynomials for (t, m, s)-nets and numerical integration of multivariate Walsh series, SIAM J. Numer. Analysis, to appear.

[13] G. LARCHER AND H. NIEDERREITER, Kronecker-type sequences and nonarchimedean diophantine approximations, Acta Arith., 63 (1993), pp. 379-396.

[14] G. LARCHER AND H. NIEDERREITER, Generalized (t, s)-sequences, Kronecker-type sequences, and diophantine approximations of formal Laurent series, Trans. Amer. Math. Soc., to appear.

[15] G. LARCHER, H. NIEDERREITER, AND W. C. SCHMID, Digital nets and sequences constructed over finite rings and their application to quasi-Monte Carlo integration, Monatsh. Math., to appear.

[16] G. LARCHER AND W.C. SCHMID, Multivariate Walsh series, digital nets and quasi-Monte Carlo integration, this volume.

[17] G. LARCHER, W. C. SCHMID, AND R. WOLF, Representation of functions as Walsh series to different bases and an application to the numerical integration of high-dimensional Walsh series, Math. Comp., 63 (1994), pp. 701-716.

[18] G. LARCHER, W. C. SCHMID, AND R. WOLF, Quasi-Monte Carlo methods for the numerical integration of multivariate Walsh series, Math. and Computer Modelling, to appear.

[19] G. LARCHER AND C. TRAUNFELLNER, On the numerical integration of Walsh series by number-theoretic methods, Math. Comp., 63 (1994), pp. 277-291.

[20] C. F. LAYWINE, G. L. MULLEN, AND G. WHITTLE, D-dimensional hypercubes and the Euler and MacNeish conjectures, Monatsh. Math., 119 (1995), pp. 223-238.

[21] G. L. MULLEN, Combinatorial methods in the construction of point sets with uniformity properties, Math. and Computer Modelling, to appear.

[22] G. L. MULLEN AND H. NIEDERREITER, *Optimal characteristic polynomials for digital multistep pseudorandom numbers*, Computing, 39 (1987), pp. 155-163.

[23] G. L. MULLEN AND G. WHITTLE, *Point sets with uniformity properties and orthogonal hypercubes*, Monatsh. Math., 113 (1992), pp. 265-273.

[24] H. NIEDERREITER, *Quasi-Monte Carlo methods and pseudo-random numbers*, Bull. Amer. Math. Soc., 84 (1978), pp. 957-1041.

[25] H. NIEDERREITER, *Low-discrepancy point sets*, Monatsh. Math., 102 (1986), pp. 155-167.

[26] H. NIEDERREITER, *Rational functions with partial quotients of small degree in their continued fraction expansion*, Monatsh. Math., 103 (1987), pp. 269-288.

[27] H. NIEDERREITER, *Point sets and sequences with small discrepancy*, Monatsh. Math., 104 (1987), pp. 273-337.

[28] H. NIEDERREITER, *Low-discrepancy and low-dispersion sequences*, J. Number Theory, 30 (1988), pp. 51-70.

[29] H. NIEDERREITER, *Quasi-Monte Carlo methods for multidimensional numerical integration*, in Numerical Integration III, H. Brass and G. Hämmerlin, eds., Internat. Series of Numer. Math., Vol. 85, Birkhäuser, Basel, 1988, pp. 157-171.

[30] H. NIEDERREITER, *A combinatorial problem for vector spaces over finite fields*, Discrete Math., 96 (1991), pp. 221-228.

[31] H. NIEDERREITER, *Orthogonal arrays and other combinatorial aspects in the theory of uniform point distributions in unit cubes*, Discrete Math., 106/107 (1992), pp. 361-367.

[32] H. NIEDERREITER, *Low-discrepancy point sets obtained by digital constructions over finite fields*, Czechoslovak Math. J., 42 (1992), pp. 143-166.

[33] H. NIEDERREITER, Random Number Generation and Quasi-Monte Carlo Methods, CBMS-NSF Regional Conference Series in Applied Math., Vol. 63, SIAM, Philadelphia, PA, 1992.

[34] H. NIEDERREITER, *Constructions of low-discrepancy point sets and sequences*, in Sets, Graphs and Numbers (Budapest, 1991), Colloquia Math. Soc. János Bolyai, Vol. 60, North-Holland, Amsterdam, 1992, pp. 529-559.

[35] H. NIEDERREITER, *Finite fields, pseudorandom numbers, and quasirandom points*, in Finite Fields, Coding Theory, and Advances in Communications and Computing, G. L. Mullen and P. J.-S. Shiue, eds., Marcel Dekker, New York, 1993, pp. 375-394.

[36] H. NIEDERREITER, *Low-discrepancy sequences and nonarchimedean diophantine approximations*, Studia Sci. Math. Hungarica, to appear.

[37] A. B. OWEN, *Randomly permuted (t, m, s)-nets and (t, s)-sequences*, this volume.

[38] I. M. SOBOL', *The distribution of points in a cube and the approximate evaluation of integrals*, Zh. Vychisl. Mat. i Mat. Fiz., 7 (1967), pp. 784-802 (in Russian).

[39] I. M. SOBOL', Multidimensional Quadrature Formulas and Haar Functions, Izdat. "Nauka," Moscow, 1969 (in Russian).

[40] S. SRINIVASAN, *On two-dimensional Hammersley's sequences*, J. Number Theory, 10 (1978), pp. 421-429.

[41] S. TEZUKA, *Polynomial arithmetic analogue of Halton sequences*, ACM Trans. Modeling and Computer Simulation, 3 (1993), pp. 99-107.

[42] S. TEZUKA AND M. FUSHIMI, *Calculation of Fibonacci polynomials for GFSR sequences with low discrepancies*, Math. Comp., 60 (1993), pp. 763-770.

NEW DEVELOPMENTS IN UNIFORM PSEUDORANDOM NUMBER AND VECTOR GENERATION

Harald Niederreiter

Institute for Information Processing
Austrian Academy of Sciences
Sonnenfelsgasse 19
A-1010 Vienna, Austria
E-mail: nied@qiinfo.oeaw.ac.at

Abstract. A survey of recent and new developments in the areas of uniform pseudorandom number and uniform pseudorandom vector generation is presented. The emphasis is on generators for which a detailed theoretical analysis is available.

1. Introduction

1.1. General Remarks

Pseudorandom numbers are basic ingredients of any implementation of a Monte Carlo method, and the success of a Monte Carlo calculation often stands or falls with the quality of the pseudorandom numbers that are used. Therefore, the selection of good pseudorandom number generators is a crucial issue in Monte Carlo methods.

In this survey article the attention is restricted to *uniform* pseudorandom numbers, i.e., to pseudorandom numbers that simulate the uniform distribution on the unit interval $[0, 1]$, and to their multivariate analogs, i.e., uniform pseudorandom vectors. The latter are basic tools in parallelized Monte Carlo computations and thus of growing importance in modern simulation techniques. There is already a rich expository literature on uniform pseudorandom number and vector generation; besides the classical monograph of Knuth [87] we mention *inter alia* the more recent works of Anderson [3], Dagpunar [19], Deák [20], Dieter [25], James [83], L'Ecuyer [89], [90], and Niederreiter [117], [119], [121]. These sources, taken together, cover the literature up to fairly recent times very well, and this provides a justification for the concentration of the present paper on some key topics that dominate current research activities. It should also be stressed that we deal only with pseudorandom numbers and vectors for simulation methods and that pseudorandom sequences for cryptographic purposes are not treated.

For the sake of background we present in Section 2 a brief review of classical generators for uniform pseudorandom numbers and an account of recent work on these generators, i.e., on linear congruential, multiple-recursive congruential, and shift-register generators. The currently most active research area in uniform pseudorandom number generation, as measured by the number of publications in the 1990s, is that of nonlinear generators; a detailed survey of this area is given in Section 3. New developments in linear methods for uniform pseudorandom number generation, such as the multiple-recursive matrix method and the theoretical analysis of combined generators, are covered in Section 4. Methods for the generation of uniform pseudorandom vectors are discussed in Section 5. Some conclusions are drawn in Section 6.

1.2. Tests for Randomness

We recall that a sequence of uniform pseudorandom numbers is generated by a deterministic algorithm and should simulate a sequence of independent, uniformly distributed random variables with values in $I = [0, 1]$. The standard algorithms for generating uniform pseudorandom numbers yield periodic sequences. Some of the desirable properties of sequences of uniform pseudorandom numbers are the following:

(i) large period length;
(ii) good equidistribution properties;
(iii) statistical (almost-)independence of successive pseudorandom numbers;
(iv) little intrinsic structure;
(v) a reasonably fast generation algorithm.

Uniform pseudorandom numbers have to pass a variety of statistical tests for randomness in order to be acceptable. The most detailed account of tests for randomness can be found in the book of Knuth [87, Chapter 3].

In our discussions we emphasize structural properties and statistical tests for which a theoretical analysis has been carried out. For statistical tests, this applies in particular to the *s-dimensional Kolmogorov test* which is based on the notion of discrepancy.

Definition 1. For any N points $\mathbf{w}_0, \mathbf{w}_1, \ldots, \mathbf{w}_{N-1}$ in the s-dimensional interval $I^s = [0, 1]^s, s \geq 1$, their *discrepancy* is defined by

$$D_N(\mathbf{w}_0, \mathbf{w}_1, \ldots, \mathbf{w}_{N-1}) = \sup_J |E_N(J) - V(J)|,$$

where the supremum is extended over all subintervals J of I^s with one vertex at the origin, $E_N(J)$ is N^{-1} times the number of terms among $\mathbf{w}_0, \mathbf{w}_1, \ldots, \mathbf{w}_{N-1}$ falling into J, and $V(J)$ denotes the s-dimensional volume of J.

Let $x_0, x_1, \ldots$ be an arbitrary sequence of uniform pseudorandom numbers in the interval I. For a given $s \geq 1$ we consider the sequence of s-dimensional points

(1) $$\mathbf{x}_n = (x_n, x_{n+1}, \ldots, x_{n+s-1}) \in I^s \quad \text{for } n = 0, 1, \ldots$$

and we define

(2) $$D_N^{(s)} = D_N(\mathbf{x}_0, \mathbf{x}_1, \ldots, \mathbf{x}_{N-1})$$

to be the discrepancy of the first N terms of this sequence. This is the statistical test quantity for the s-dimensional Kolmogorov test. For $s = 1$ we get a test for equidistribution in I, which is also called the *uniformity test*, and for $s \geq 2$ we get a test for the statistical independence of s successive pseudorandom numbers, which is also called the *s-dimensional serial test*. On statistical grounds, the test should actually be applied to nonoverlapping tuples, but for the generators of interest here the analysis is practically the same for the overlapping case in (1) and the nonoverlapping case.

Various probabilistic results on the distribution of the discrepancy of random points are available. The most widely known result is the *law of the iterated logarithm* for the discrepancy, according to which we have

$$\overline{\lim}_{N\to\infty} \frac{(2N)^{1/2} D_N(\mathbf{w}_0, \mathbf{w}_1, \ldots, \mathbf{w}_{N-1})}{(\log\log N)^{1/2}} = 1 \quad \text{a.e.}$$

for random sequences $\mathbf{w}_0, \mathbf{w}_1, \ldots$ of points in I^s. For $s = 1$ this was shown by Chung [14] and for $s \geq 2$ by Kiefer [86]. Furthermore, in the case $s = 1$ we have the *Kolmogorov-Smirnov law*

$$\lim_{N\to\infty} \text{Prob}(N^{1/2} D_N(w_0, w_1, \ldots, w_{N-1}) \leq u) = 1 - 2\sum_{k=1}^{\infty}(-1)^{k+1} e^{-2k^2 u^2} \quad \text{for all } u > 0$$

and for random sequences $w_0, w_1, \ldots$ of numbers in I. For arbitrary $s \geq 1$ and fixed N the following result was established by Kiefer [86]: for every $\varepsilon > 0$ there exists a constant $c > 0$ depending only on s and ε such that

$$\text{Prob}(N^{1/2} D_N(\mathbf{w}_0, \mathbf{w}_1, \ldots, \mathbf{w}_{N-1}) \leq u) \geq 1 - c e^{-(2-\varepsilon)u^2} \quad \text{for all } u \geq 0$$

and for random points $\mathbf{w}_0, \mathbf{w}_1, \ldots, \mathbf{w}_{N-1} \in I^s$. In simple terms, the upshot of these results is that the discrepancy $D_N(\mathbf{w}_0, \mathbf{w}_1, \ldots, \mathbf{w}_{N-1})$ of random points $\mathbf{w}_0, \mathbf{w}_1, \ldots, \mathbf{w}_{N-1} \in I^s$ is roughly of the order of magnitude $N^{-1/2}$.

Throughout the paper we use the following notation: for a purely periodic sequence of arbitrary elements $t_n, n = 0, 1, \ldots$, we write $\text{per}(t_n)$ for the least period length of the sequence.

2. Classical Generators

2.1. The Linear Congruential Method

This is *the* classical method for the generation of uniform pseudorandom numbers. In this method we choose a large integer M, called the *modulus*, and then generate a sequence $y_0, y_1, \ldots$ of integers in $Z_M := \{0, 1, \ldots, M-1\}$ by the recursion

$$y_{n+1} \equiv ay_n + c \bmod M \quad \text{for } n = 0, 1, \ldots$$

with the initial value y_0, where a and c are suitable constants taken from Z_M and where we assume $\gcd(a, M) = 1$ to get a purely periodic sequence. The *linear congruential pseudorandom numbers* $x_0, x_1, \ldots$ in I are obtained by the normalization

$$x_n = \frac{1}{M} y_n \quad \text{for } n = 0, 1, \ldots.$$

It is clear that we always have $\text{per}(x_n) \leq M$. On the other hand, conditions are known which guarantee that $\text{per}(x_n) = M$. For instance, this happens if M is a power of

$2, a \equiv 1 \bmod 4$, and c is odd (see [87, Section 3.2]). Another situation of interest arises if M is prime, a is a primitive root modulo $M, c = 0$, and $y_0 \neq 0$, in which case we have $\mathrm{per}(x_n) = M - 1$. For surveys of the structural and statistical properties of linear congruential pseudorandom numbers we refer to Knuth [87, Chapter 3] and Niederreiter [119, Chapter 7]. Tables of recommended parameters in the linear congruential method can be found in Fishman [73], L'Ecuyer [88], and L'Ecuyer, Blouin, and Couture [94].

The linear congruential method has been investigated extensively, and this research exposed several deficiencies of this method which are well known by now. We list a few of these:

(i) The modulus M, and therefore $\mathrm{per}(x_n)$, are bounded in terms of the word length of the machine. For instance, with a 32-bit processor we have $\mathrm{per}(x_n) \leq M \leq 2^{32}$, unless we want to use costly multiple-precision arithmetic.

(ii) The simple nature of the generation algorithm implies a lot of regularity in sequences of linear congruential pseudorandom numbers. For instance, the discrepancy $D_N^{(s)}$ in (2) with $N = \mathrm{per}(x_n)$ tends to be much smaller than the law of the iterated logarithm would dictate (compare with [119, Chapter 7]). The situation is somewhat better for parts of the period, i.e., for $N < \mathrm{per}(x_n)$; see the results of Levin [98], Niederreiter [107], and Shparlinskii [133]. Furthermore, various long-range correlations in sequences of linear congruential pseudorandom numbers have also been discovered; we refer to de Matteis, Eichenauer-Herrmann, and Grothe [21], de Matteis and Pagnutti [22], Eichenauer-Herrmann and Grothe [58], and Percus and Percus [130], [131].

(iii) The points $\mathbf{x}_n$ in (1) obtained from linear congruential pseudorandom numbers have an unfavorable lattice structure which was first pointed out independently by Marsaglia [101] and Wood [146]. For a short proof of this lattice structure see [119, Theorem 7.6], and for a detailed discussion we refer to [87, Section 3.3].

2.2. The Multiple-Recursive Congruential Method

One attempt at overcoming the deficiencies of the linear congruential method is based on the use of higher-order linear recursions for pseudorandom number generation. Let $M \geq 2$ be a modulus, let $k \geq 1$ be an integer that serves as the order of the linear recursion, and choose $a_0, \ldots, a_{k-1} \in Z_M$ with $\gcd(a_0, M) = 1$. Then we generate a sequence $y_0, y_1, \ldots$ of elements of Z_M by the recursion

$$(3) \qquad y_{n+k} \equiv \sum_{h=0}^{k-1} a_h y_{n+h} \bmod M \quad \text{for } n = 0, 1, \ldots$$

with initial values $y_0, \ldots, y_{k-1}$. We obtain *multiple-recursive congruential pseudorandom numbers* $x_0, x_1, \ldots$ in I by the normalization

$$x_n = \frac{1}{M} y_n \quad \text{for } n = 0, 1, \ldots.$$

In practice, M is again chosen to be large. The case $k = 1$ yields a linear congruential method, so it is usually assumed that $k \geq 2$, although this is not needed for the theory. The sequence $x_0, x_1, \ldots$ is purely periodic with $\mathrm{per}(x_n) \leq M^k - 1$. This upper bound can be attained in the following way. Choose $M = p$ to be a prime modulus and associate with the recursion (3) its *characteristic polynomial*

$$f(x) = x^k - \sum_{h=0}^{k-1} a_h x^h \in F_p[x],$$

where we have identified Z_p with the finite field F_p of order p. In view of $\text{per}(x_n) = \text{per}(y_n)$, the following well-known result (see [100, Theorem 6.33]) then yields a case in which $\text{per}(x_n) = p^k - 1$.

Lemma 1. *Let $M = p$ be a prime, let the characteristic polynomial $f(x)$ of the recursion (3) be a primitive polynomial over the finite field F_p, and suppose that the initial values $y_0, \ldots, y_{k-1}$ are not all 0. Then the sequence $y_0, y_1, \ldots$ generated by (3) satisfies $\text{per}(y_n) = p^k - 1$.*

Here we recall that for an arbitrary finite field F_q of order q, a monic polynomial $f(x) \in F_q[x]$ of degree $k \geq 1$ is called a *primitive polynomial* over F_q if $f(0) \neq 0$ and if $q^k - 1$ is the least positive integer m such that $f(x)$ divides $x^m - 1$ in $F_q[x]$, and that for every $k \geq 1$ there exists a primitive polynomial over F_q of degree k (compare with [100, Chapter 3]).

In the case described in Lemma 1, multiple-recursive congruential pseudorandom numbers have the attractive property that for every dimension $s \leq k$ the points $\mathbf{x}_n, 0 \leq n \leq p^k - 2$, in (1) show an almost perfect equidistribution (see [119, pp. 174-175]). Again in this case, the lattice structure of multiple-recursive congruential pseudorandom numbers was investigated by Dieter [24] and Grube [80]. A more general result on the lattice structure will be presented in Section 5.2. For larger values of k, the disadvantages incurred by the lattice structure become less pronounced. Recommended parameters in the multiple-recursive congruential method are tabulated in L'Ecuyer, Blouin, and Couture [94]. The lattice structure of points with more general lags than in (1) was recently studied by L'Ecuyer [91].

For the case where the modulus M is a power of 2, the period lengths of sequences of multiple-recursive congruential pseudorandom numbers were analyzed by Brent [9] and conditions for reaching the maximum possible period length for given M and k were provided. The work of Parker [129] on period lengths puts an emphasis on the special recursion $y_{n+k} \equiv y_{n+k-1} + y_n \bmod M$ with a general modulus M.

The following variant of the multiple-recursive congruential method, called the *add-with-carry* (or *AWC*) *generator*, was introduced a few years ago by Marsaglia and Zaman [104]. This generator is based on the recursion

$$y_{n+k} \equiv y_{n+r} + y_n + c_n \bmod M \quad \text{for } n = 0, 1, \ldots$$

with $1 \leq r < k$, where c_n is a carry bit which is obtained recursively by starting from an initial value $c_0 \in \{0, 1\}$ and putting $c_{n+1} = 1$ if $y_{n+r} + y_n + c_n \geq M$ and $c_{n+1} = 0$ otherwise. There is also a subtractive analog which is called the *subtract-with-borrow* (or *SWB*) *generator*. A survey of AWC/SWB generators can be found in Marsaglia [103]. Statistical anomalies of these generators were pointed out soon after the appearance of [104] by Ferrenberg, Landau, and Wong [72]. The theoretical explanation for the defects of these generators was first given by Tezuka and L'Ecuyer [141] and later in a more detailed form by Tezuka, L'Ecuyer, and Couture [142]. The point is that, with a large precision, AWC/SWB pseudorandom numbers can also be generated by suitable linear congruential generators, so that for practical purposes the AWC/SWB method is just an efficient way of implementing certain linear congruential generators. As a consequence, AWC/SWB pseudorandom numbers lead to practically the same lattice structure as linear congruential pseudorandom numbers. The lattice structures produced by AWC/SWB generators were analyzed by Couture and L'Ecuyer [16], and it was found that these lattice structures are extremely coarse already in moderately large dimensions. L'Ecuyer [91] has shown that if more general lags than in (1) are considered, then AWC/SWB generators can yield lattice structures that are extremely coarse already in small dimensions, such as dimension 3.

2.3. Shift-Register Methods

These methods for generating pseudorandom numbers are based on kth-order linear recursions as in (3), but with a *small* prime modulus $M = p$, usually $p = 2$. To obtain large periods, it is customary to choose the characteristic polynomial $f(x) \in F_p[x]$ of the recursion (3) to be a primitive polynomial over F_p. If we suppose also that the initial values $y_0, \ldots, y_{k-1}$ are not all 0, then we have $\text{per}(y_n) = p^k - 1$ by Lemma 1. Various techniques are now applied to transform the sequence $y_0, y_1, \ldots$ of elements of Z_p into a sequence of uniform pseudorandom numbers.

One such technique is the *digital multistep method* in which we choose an integer m with $2 \leq m \leq k$ and put

$$x_n = \sum_{j=1}^{m} y_{mn+j-1} p^{-j} \in [0,1) \quad \text{for } n = 0, 1, \ldots.$$

If in addition to the conditions above we assume that $\gcd(m, p^k - 1) = 1$, then the sequence $x_0, x_1, \ldots$ is purely periodic with $\text{per}(x_n) = p^k - 1$. The numbers x_n are called *digital multistep pseudorandom numbers*.

Another technique of deriving uniform pseudorandom numbers from the sequence $y_0, y_1, \ldots$ is the *generalized feedback shift-register* (or *GFSR*) *method*. Here we choose integers $m \geq 2$ and $h_1, \ldots, h_m \geq 0$ and put

$$(4) \qquad x_n = \sum_{j=1}^{m} y_{n+h_j} p^{-j} \in [0,1) \quad \text{for } n = 0, 1, \ldots.$$

If we assume again that $f(x)$ is a primitive polynomial over F_p and that $y_0, \ldots, y_{k-1}$ are not all 0, then the sequence $x_0, x_1, \ldots$ is purely periodic with $\text{per}(x_n) = p^k - 1$. The numbers x_n are called *GFSR pseudorandom numbers*. A convenient implementation of GFSR pseudorandom numbers is based on the vectors $Y_n = (y_{n+h_1}, \ldots, y_{n+h_m}) \in Z_p^m, n = 0, 1, \ldots$, which in view of (3) can be obtained by the vector recursion

$$(5) \qquad Y_{n+k} \equiv \sum_{h=0}^{k-1} a_h Y_{n+h} \bmod p \quad \text{for } n = 0, 1, \ldots.$$

If Y_n is interpreted as a block of p-ary digits, then we immediately get x_n from (4).

Digital multistep and GFSR pseudorandom numbers, also collectively called *shift-register pseudorandom numbers*, have been analyzed extensively. We refer to L'Ecuyer

[90] and Niederreiter [113], [117], [119, Chapter 9] for recent surveys of work on shift-register pseudorandom numbers. There is a sense in which digital multistep and GFSR pseudorandom numbers can actually be viewed as being equivalent (see Fushimi [75]), so there is a certain unity in this subject (compare also with L'Ecuyer [90, Section 3]). Shift-register pseudorandom numbers have the following attractive properties: (i) the pseudorandom numbers are generated by a fast algorithm; (ii) the least period length is not bounded in terms of the word length of the machine; (iii) the pseudorandom numbers do not, in general, lead to a coarse lattice structure. On the other hand, the points $\mathbf{x}_n$ in (1) obtained from shift-register pseudorandom numbers show a strong uniformity in their distribution, which is captured by the following notion.

Definition 2. Let $0 \leq t \leq m$ and $b \geq 2$ be integers. A (t,m,s)-net in base b is a point set of b^m points in $[0,1)^s$ such that every interval of the form

$$\prod_{i=1}^{s} [a_i b^{-d_i}, (a_i + 1) b^{-d_i})$$

with integers $d_i \geq 0$ and $0 \leq a_i < b^{d_i}$ for $1 \leq i \leq s$ and of volume b^{t-m} contains exactly b^t points of the point set.

We consider again the case where the characteristic polynomial $f(x)$ is a primitive polynomial over F_p and where $y_0, \ldots, y_{k-1}$ are not all 0, and in the digital multistep method we assume also that $\gcd(m, p^k - 1) = 1$. As we have seen, the sequence $x_0, x_1, \ldots$ of shift-register pseudorandom numbers then satisfies $\mathrm{per}(x_n) = p^k - 1 =: T$. For dimensions $s > k/m$ the p^k points $\mathbf{0}, \mathbf{x}_0, \mathbf{x}_1, \ldots, \mathbf{x}_{T-1}$ then form a (t, k, s)-net in base p, as was first shown by Niederreiter [109, Section 9] (see also [119, Chapter 9]). Here the value of t depends in a known way on s, m, and the polynomial $f(x)$ in the case of the digital multistep method, and on $s, h_1, \ldots, h_m$, and $f(x)$ in the case of the GFSR method. Tables of parameters yielding a small value of t can be found in André, Mullen, and Niederreiter [4] for the digital multistep method and in Tezuka and Fushimi [139] for the GFSR method.

Results on the s-dimensional serial test for shift-register pseudorandom numbers are summarized in [119, Chapter 9]. Some of these results rely on the general discrepancy bounds in terms of exponential sums discussed in [119, Section 3.2]. An elegant approach to these general discrepancy bounds using Walsh functions was recently described by Hellekalek [81]; see also Tezuka [136] for related work. Walsh functions also serve as the basis for the Walsh spectral test introduced by Yuen [147] and, in a somewhat different form, by Tezuka [135].

As noted above, shift-register pseudorandom numbers do not, in general, lead to a coarse (euclidean) lattice structure, i.e., the corresponding points $\mathbf{x}_n$ in (1) do not all lie in a coarse $\mathbf{Z}$-submodule of $\mathbf{R}^s$. However, in a model based on generating functions, shift-register pseudorandom numbers lead to "lattices" in spaces of formal Laurent series over the finite field F_p, where here a "lattice" is an $F_p[x]$-submodule of such a space. This was first pointed out by Tezuka [137]. This type of lattice structure was used by Tezuka and L'Ecuyer [140] to assess the quality of shift-register pseudorandom numbers and to obtain recommended parameters. Couture, L'Ecuyer, and Tezuka [18] were able to relate this type of lattice structure with equidistribution properties of the points $\mathbf{x}_n$, and this approach was pursued further by Tezuka [138].

Among further recent work on shift-register methods we mention the paper of Fushimi [76] in which special types of recursions (5) for $p = 2$ are considered. Aluru, Prabhu, and Gustafson [2] studied the parallelization of GFSR generators. Shift-register generators for which the characteristic polynomial of (3) is a product of several primitive polynomials were investigated by Tezuka and L'Ecuyer [140] and Wang and Compagner [143]. An interesting extension of GFSR generators, the *twisted GFSR generator*, was introduced by Matsumoto and Kurita [105] in the case $p = 2$. Here we use instead of (5) the vector recursion

$$Y_{n+k} \equiv Y_{n+r} + Y_n A \bmod 2 \quad \text{for } n = 0, 1, \ldots,$$

where $1 \leq r < k, A$ is an $m \times m$ matrix with entries in Z_2, and the Y_n are in Z_2^m. The pseudorandom numbers $x_n \in [0, 1)$ are obtained by viewing Y_n as a block of binary digits which gives the (finite) binary expansion of x_n. With a suitable choice of k, r, and A we can achieve $\mathrm{per}(x_n) = 2^{km} - 1$. The twisted GFSR generator is a special case of the multiple-recursive matrix method to be discussed in Section 4.1.

3. Nonlinear Generators

3.1. The General Nonlinear Congruential Method

Nonlinear methods of uniform pseudorandom number generation have been introduced in an attempt to break up the coarse lattice structure inherent in the linear congruential method (compare with Section 2.1). We use the general framework described in [119, Section 8.1]. Thus, in a *first-order congruential method* we choose a large modulus M and generate a sequence $y_0, y_1, \ldots$ of integers in Z_M by the first-order recursion

(6) $$y_{n+1} \equiv f(y_n) \bmod M \quad \text{for } n = 0, 1, \ldots$$

with the initial value y_0, where f is a fixed integer-valued function on Z_M. The pseudorandom numbers $x_0, x_1, \ldots$ in I are obtained by the normalization

$$x_n = \frac{1}{M} y_n \quad \text{for } n = 0, 1, \ldots.$$

All such methods different from the linear congruential method are called *nonlinear congruential methods*, and the corresponding pseudorandom numbers $x_0, x_1, \ldots$ are called *nonlinear congruential pseudorandom numbers*. Surveys of nonlinear congruential methods can be found in Eichenauer-Herrmann [48] and Niederreiter [118], [119, Chapter 8], [122].

General nonlinear congruential methods with a prime modulus $M = p$ were first proposed by Eichenauer, Grothe, and Lehn [27]. Here it is customary to restrict the attention to the case where the sequence $y_0, y_1, \ldots$ is purely periodic with $\mathrm{per}(y_n) = p$. If we identify Z_p with the finite field F_p of order p, then there exists a uniquely determined polynomial $g \in F_p[x]$ with $d := \deg(g) < p$ such that

$$y_n = g(n) \in F_p \quad \text{for all } n \in F_p.$$

The number d plays an important role in the theory. In the following, we can assume that $p \geq 5$ and $3 \leq d \leq p - 2$ (compare with [119, p. 178]).

The lattice test defined below can be viewed as a special case of a lattice test introduced by Marsaglia [102]. For $s \geq 1$ we consider F_p^s as an s-dimensional vector space over F_p.

Definition 3. For given $s \geq 1$ a sequence $y_0, y_1, \ldots$ of elements of F_p passes the s-dimensional lattice test if the vectors $\mathbf{y}_n - \mathbf{y}_0, n = 1, 2, \ldots$, span F_p^s, where

$$\mathbf{y}_n = (y_n, y_{n+1}, \ldots, y_{n+s-1}) \in F_p^s \quad \text{for } n = 0, 1, \ldots.$$

It is easy to see that a sequence $y_0, y_1, \ldots \in F_p$ generated by the linear congruential method with modulus p can pass the s-dimensional lattice test at most for $s = 1$. For nonlinear congruential methods we have a much better behavior if the number d is large, as the following result of Eichenauer, Grothe, and Lehn [27] (see also Niederreiter [110]) shows.

Theorem 1. A sequence $y_0, y_1, \ldots$ generated by a nonlinear congruential method with modulus p passes the s-dimensional lattice test if and only if $s \leq d$.

While serving a useful purpose, the lattice test is rather weak and should be employed only for the elimination of bad generators; see an example constructed by Eichenauer and Niederreiter [32]. For the case where M is a composite prime power, the lattice structure (in the sense of Section 2.1) of nonlinear congruential pseudorandom numbers was analyzed by Eichenauer-Herrmann [40], and it was shown that in situations of practical interest there is again a nontrivial lattice structure in evidence. Thus, from the viewpoint of the lattice structure it is preferable to work with prime moduli in nonlinear congruential methods.

The uniformity test and the s-dimensional serial test for nonlinear congruential pseudorandom numbers with a prime modulus p were first studied by Niederreiter [111]. The results were expressed as upper and lower bounds for the discrepancy $D_N^{(s)}$ in (2). Later refinements were obtained by Eichenauer-Herrmann [43], [46], [51] and Eichenauer-Herrmann and Niederreiter [66]. The latter paper considers also more general lags than in the points of (1). We can summarize the upper bounds for $D_N^{(s)}$ as follows.

Theorem 2. For any nonlinear congruential pseudorandom numbers with prime modulus p we have

$$D_p^{(s)} = O(dp^{-1/2}(\log p)^s) \text{ for } 2 \leq s \leq d,$$

$$D_N^{(s)} = O(dN^{-1}p^{1/2}(\log p)^{s+1}) \text{ for } 1 \leq N < p \text{ and } 1 \leq s \leq d - 1.$$

The lower bounds for $D_N^{(s)}$ show that the orders of magnitude of the upper bounds in Theorem 2 are in general best possible up to the logarithmic factors and that the upper bounds cannot (or need not) hold for dimensions s higher than those allowed in Theorem 2.

A general technique that is of interest for the convenient implementation of nonlinear congruential generators is the *compound method*. For the specific case of nonlinear congruential methods this technique works as follows. Choose distinct primes $p_1, \ldots, p_r$, and for $1 \leq j \leq r$ let $x_0^{(j)}, x_1^{(j)}, \ldots$ be a purely periodic sequence of nonlinear congruential pseudorandom numbers with modulus p_j and $\mathrm{per}(x_n^{(j)}) = p_j$. Now define a sequence $x_0, x_1, \ldots$ of pseudorandom numbers in I by

$$x_n \equiv \sum_{j=1}^{r} x_n^{(j)} \bmod 1 \quad \text{for } n = 0, 1, \ldots.$$

This sequence is purely periodic with $\mathrm{per}(x_n) = p_1 \cdots p_r$. The compound method extends the well-known approach of Wichmann and Hill [144] for linear congruential generators. The uniformity test and the s-dimensional serial test for the sequence $x_0, x_1, \ldots$ of pseudorandom numbers were studied by Eichenauer-Herrmann in [46] and in a more general framework in [51].

Two special nonlinear congruential methods, namely the inversive congruential method and the quadratic congruential method, were already studied before the paper of Eichenauer, Grothe, and Lehn [27] which introduced the general nonlinear congruential method. The inversive congruential method will be discussed in the remaining parts of Section 3. The *quadratic congruential method* was proposed by Knuth [87, Section 3.2.2] and proceeds by using the recursion (6) with the quadratic polynomial $f(x) = ax^2 + bx + c$, where $a, b, c \in Z_M$ are constants. Under suitable conditions involving a, b, c, and M the corresponding sequence $x_0, x_1, \ldots$ of *quadratic congruential pseudorandom numbers* is purely periodic with $\mathrm{per}(x_n) = M$ (see [30], [87, p.34]). It is customary to concentrate on the case of a prime-power modulus $M = p^\alpha$ with p prime and $\alpha \geq 2$. Then, for instance, for $M = 2^\alpha$ we have $\mathrm{per}(x_n) = M$ if and only if a is even, $b \equiv a + 1 \bmod 4$, and c is odd. An investigation of the lattice structure inherent in quadratic congruential pseudorandom numbers was carried out by Eichenauer and Lehn [30].

In the case $M = p^\alpha$ and $\mathrm{per}(x_n) = M$ the s-dimensional serial test for quadratic congruential pseudorandom numbers was studied in several papers. Eichenauer-Herrmann and Niederreiter [63] established upper and lower bounds for the discrepancy $D_M^{(2)}$ of pairs. The upper bound was later improved by Eichenauer-Herrmann [50] for $p = 2$ and by Eichenauer-Herrmann and Niederreiter [69] for $p \geq 3$. For instance, in the latter paper it is shown that

$$D_M^{(2)} = O\left(M^{-1/2}(p^{1/2} + p^{-1/2}(\log M)^2)\right).$$

For $p = 2$ the discrepancy of the nonoverlapping pairs $(x_{2n}, x_{2n+1}), n = 0, 1, \ldots, (M/2) - 1$, was analyzed by Eichenauer-Herrmann [55]. The behavior of quadratic congruential pseudorandom numbers under the 2-dimensional serial test is in good accordance with the law of the iterated logarithm for discrepancies (compare with Section 1.2). The situation changes when the dimension $s = 3$ is considered. Then there are conditions under which $D_M^{(3)}$ can be at least of an order of magnitude $M^{-1/3}$; see Eichenauer-Herrmann [49] for $p = 2$ and Eichenauer-Herrmann [38] for small primes $p \geq 5$. However, if for $p = 2$ the average is taken over all odd $c \in Z_M$, then we get on the average $D_M^{(3)} = O(M^{-1/2}(\log M)^3)$ by a result of Eichenauer-Herrmann [50]. An analogous result for $p \geq 3$ was shown by the same author in [54].

3.2. Recursive Inversive Generators

A common way of achieving nonlinearity in pseudorandom number generation employs the operation of multiplicative inversion in suitable algebraic structures. Methods based on this principle are collectively called *inversive methods*. Surveys of inversive methods can be found in Eichenauer-Herrmann [35], [42], [48], Lehn [96], [97], and Niederreiter [117], [119, Chapter 8], [121].

Chronologically the first inversive method is the *recursive inversive congruential method* introduced by Eichenauer and Lehn [29]. This method was originally proposed only for prime moduli $p \geq 5$ and is a special nonlinear congruential method. For $c \in Z_p$ we define $\overline{c} \in Z_p$ by $c\overline{c} \equiv 1 \bmod p$ if $c \neq 0$ and $\overline{c} = 0$ if $c = 0$. Now we choose parameters $a, b \in Z_p$ with $a \neq 0$ and an initial value $y_0 \in Z_p$. Then the sequence $y_0, y_1, \ldots$ of elements of Z_p is generated by the recursion

(7) $$y_{n+1} \equiv a\overline{y}_n + b \bmod p \quad \text{for } n = 0, 1, \ldots,$$

and the pseudorandom numbers $x_0, x_1, \ldots$ in I are obtained by the normalization

$$x_n = \frac{1}{p}y_n \quad \text{for } n = 0, 1, \ldots.$$

The numbers x_n are called *recursive inversive congruential pseudorandom numbers*. The sequence $x_0, x_1, \ldots$ is purely periodic with $\text{per}(x_n) \leq p$. To formulate conditions under which $\text{per}(x_n) = p$, it is convenient to again identify Z_p with the finite field F_p of order p. If we now choose $a, b \in F_p$ such that $x^2 - bx - a \in F_p[x]$ is a primitive polynomial over F_p (see Section 2.2 for this notion), then $\text{per}(x_n) = p$ according to a result of Eichenauer and Lehn [29]. Primitive polynomials of this type are tabulated in Grothe [79]. To obtain a necessary and sufficient condition for $\text{per}(x_n) = p$, the following concept is crucial.

Definition 4. Let F_q be an arbitrary finite field of order q and let $a, b \in F_q$ with $a \neq 0$. Then the polynomial $x^2 - bx - a$ is called an *IMP polynomial* over F_q if the quotient of its roots is an element of order $q + 1$ in the multiplicative group of the finite field F_{q^2}.

It was then proved by Flahive and Niederreiter [74] that we have $\text{per}(x_n) = p$ if and only if $a, b \in F_p$ are such that $x^2 - bx - a$ is an IMP polynomial over F_p. A detailed study of IMP polynomials over an arbitrary finite field F_q was carried out by Chou [12]. In particular, it was shown in this paper that the total number of IMP polynomials over F_q is $(q-1)\phi(q+1)/2$, where ϕ is Euler's totient function, and that the property of $x^2 - bx - a$ being an IMP polynomial over F_q depends only on the element $b^2 a^{-1} \in F_q$. The possible values of $\text{per}(x_n)$ for a sequence $x_0, x_1, \ldots$ of recursive inversive congruential pseudorandom numbers were analyzed in Chou [10].

Eichenauer, Grothe, and Lehn [27] studied the performance of the sequence $y_0, y_1, \ldots$ generated by (7) under the s-dimensional lattice test (see Definition 3). They showed that if $\text{per}(x_n) = p$, then the s-dimensional lattice test is passed for all $s \leq (p-1)/2$. This was improved to $s \leq (p+1)/2$ by Niederreiter [110] and to $s \leq (p+3)/2$ for $p \equiv 3 \bmod 4$ by Flahive and Niederreiter [74]. The optimal behavior, namely that the s-dimensional lattice test is passed for all $s \leq p - 2$, was analyzed by Chou and Niederreiter [13]. Several criteria for the optimal behavior were established, and it was also shown that if p is a Mersenne prime, then there exist parameters $a, b \in F_p$ such that $\text{per}(x_n) = p$ and the sequence $y_0, y_1, \ldots$ passes the s-dimensional lattice test for all $s \leq p - 2$.

The following remarkable nonlinearity property was proved by Eichenauer-Herrmann [33]: if F_p^s is the s-dimensional affine space over F_p, then for $s \geq 2$ and every sequence $y_0, y_1, \ldots$ generated by (7) with $\text{per}(y_n) = p$, any hyperplane in F_p^s contains at most s of the points $(y_n, y_{n+1}, \ldots, y_{n+s-1}) \in F_p^s$ with $0 \leq n \leq p - 1$ and $y_n y_{n+1} \cdots y_{n+s-2} \neq 0$.

This result is optimal in the sense that there do exist hyperplanes in F_p^s that contain exactly s of these points, as long as that many points are available in the full period, i.e., as long as $s \leq (p+1)/2$.

The s-dimensional serial test for recursive inversive congruential pseudorandom numbers was first investigated by Niederreiter [112], and the following upper bound for the discrepancy $D_p^{(s)}$ in (2) for the full period was established (see also [119, Theorem 8.7]).

Theorem 3. *For any sequence $x_0, x_1, \ldots$ of recursive inversive congruential pseudorandom numbers with prime modulus p and $\mathrm{per}(x_n) = p$ we have*

$$D_p^{(s)} = O(p^{-1/2}(\log p)^s) \quad \text{for } s \geq 2.$$

The upper bound in Theorem 3 is in general best possible up to the logarithmic factor. In fact, it was shown by Niederreiter [115] that if we consider $a, b \in F_p$ for which $x^2 - bx - a$ is a primitive polynomial over F_p (so that in particular $\mathrm{per}(x_n) = p$), then for a positive fraction of all such pairs (a, b) the discrepancy $D_p^{(s)}$ is at least of the order of magnitude $p^{-1/2}$ for all $s \geq 2$. Eichenauer-Herrmann [45] extended this result to the case where $x^2 - bx - a$ is an IMP polynomial over F_p. Altogether, the order of magnitude of $D_p^{(s)}$ is in good accordance with the law of the iterated logarithm in Section 1.2.

The recursive inversive congruential method can also be used with a composite modulus M. Let G_M be the set of $c \in Z_M$ with $\gcd(c, M) = 1$. For $c \in G_M$ let $\bar{c}$ be the unique element of G_M with $c\bar{c} \equiv 1 \bmod M$. With parameters $a \in G_M$ and $b \in Z_M$ and an initial value $y_0 \in G_M$ we generate a sequence $y_0, y_1, \ldots$ of elements of G_M by the recursion

$$y_{n+1} \equiv a\bar{y}_n + b \bmod M \quad \text{for } n = 0, 1, \ldots.$$

The integers a, b must be chosen in such a way that each y_n is guaranteed to be in G_M. The *recursive inversive congruential pseudorandom numbers* $x_0, x_1, \ldots$ are defined by the usual normalization $x_n = y_n/M$ for $n = 0, 1, \ldots$. The sequence $x_0, x_1, \ldots$ is purely periodic with $\mathrm{per}(x_n) \leq \phi(M)$, where ϕ is again Euler's totient function. In the case of greatest practical interest, namely when M is a power of 2, these pseudorandom numbers were introduced by Eichenauer, Lehn, and Topuzoğlu [31]. It was shown in this paper that if $M \geq 8$ is a power of 2, then $\mathrm{per}(x_n) = \phi(M) = M/2$ if and only if $a \equiv 1 \bmod 4$ and $b \equiv 2 \bmod 4$. This result was generalized to arbitrary prime-power moduli by Eichenauer-Herrmann and Topuzoğlu [70], and an explicit construction of parameters yielding the maximum value of $\mathrm{per}(x_n)$ for odd prime-power moduli was given in Eichenauer-Herrmann [36]. A detailed study of the possible values of $\mathrm{per}(x_n)$ for arbitrary composite moduli M was carried out by Chou [11].

For composite prime-power moduli M, recursive inversive congruential pseudorandom numbers yield a lattice structure. This lattice structure was analyzed by Eichenauer-Herrmann, Grothe, Niederreiter, and Topuzoğlu [61] in the case where M is a power of 2 and by Eichenauer-Herrmann [40] in the case where M is a power of an arbitrary prime.

Niederreiter [112] studied the 2-dimensional serial test for recursive inversive congruential pseudorandom numbers in the case where M is a power of 2 and $\mathrm{per}(x_n) = M/2$. It was shown that the discrepancy $D_{M/2}^{(2)}$ of the points $(x_n, x_{n+1}) \in I^2, 0 \leq n \leq (M/2)-1$, satisfies $D_{M/2}^{(2)} = O(M^{-1/2}(\log M)^2)$. Eichenauer-Herrmann and Niederreiter [64] proved that for a positive fraction of all pairs (a, b) of parameters yielding $\mathrm{per}(x_n) = M/2$, the discrepancy $D_{M/2}^{(s)}$ is at least of the order of magnitude $M^{-1/2}$ for all $s \geq 2$. Similar upper and lower bounds for the discrepancy of the nonoverlapping pairs $(x_{2n}, x_{2n+1}), 0 \leq n \leq (M/4)-1$, were obtained in Eichenauer-Herrmann and Niederreiter [65]. An upper bound of the order of magnitude $M^{-1/2}(\log M)^2$ was established by Eichenauer-Herrmann [37] for the discrepancy of the pairs $(x_n, x_{n+l}), 0 \leq n \leq (M/2) - 1$, for any odd integer $l \geq 1$.

For odd prime-power moduli M, the behavior of recursive inversive congruential pseudorandom numbers under the s-dimensional serial test was studied by Eichenauer-Herrmann [34] for $s = 2$ and by the same author in [41] for $s = 3$. For $s = 2$ the performance is satisfactory, but for $s = 3$ there are cases where the discrepancy is too large to be in line with the law of the iterated logarithm.

Eichenauer-Herrmann and Grothe [59] introduced the following variant of the recursive inversive congruential method with the modulus $M \geq 8$ being a power of 2. For fixed $a \in G_M$ and $b \in Z_M$ the integer-valued function f on Z_M is defined by

$$f(2^k c) = 2^k a \bar{c} + b$$

with integers $k \geq 0$ and $c \in G_M$, where 0 is identified with M. Then an initial value $y_0 \in Z_M$ is chosen and the sequence $y_0, y_1, \ldots$ of elements of Z_M is generated by the recursion

$$y_{n+1} \equiv f(y_n) \bmod M \quad \text{for } n = 0, 1, \ldots.$$

Finally, the corresponding pseudorandom numbers are obtained by $x_n = y_n/M \in I$ for $n = 0, 1, \ldots$, and the sequence $x_0, x_1, \ldots$ is purely periodic. The advantage of this variant is that $\mathrm{per}(x_n)$ can attain the larger value M, as opposed to the maximum value $M/2$ in the original method. In fact, it was shown in [59] that $\mathrm{per}(x_n) = M$ if and only if $a \equiv 1 \bmod 4$ and $b \equiv 1 \bmod 2$. The paper [59] contains also a study of the 2-dimensional serial test for these pseudorandom numbers. The performance is similar to that for the original method.

Huber [82] proposed another variant of the recursive inversive congruential method for an arbitrary modulus M. With parameters $a \in G_M$ and $b \in Z_M$ and an initial value $y_0 \in Z_M$ we now generate a sequence $y_0, y_1, \ldots$ of elements of Z_M by the recursion

$$y_{n+1} \equiv a y_n^{\phi(M)-1} + b \bmod M \quad \text{for } n = 0, 1, \ldots,$$

and pseudorandom numbers are obtained as usual by putting $x_n = y_n/M$ for $n = 0, 1, \ldots$. If M is an odd prime, then the recursion above agrees with the recursion (7). Huber [82] proved several results on $\mathrm{per}(x_n)$, in particular, that if $M = p_1 \cdots p_r$ is a squarefree modulus with distinct primes $p_1, \ldots, p_r$, then $\mathrm{per}(x_n) = M$ if and only if $x^2 - bx - a$ is an IMP polynomial over F_{p_j} for $1 \leq j \leq r$. A detailed analysis of the periodicity properties of the sequence $x_0, x_1, \ldots$ for an arbitrary modulus M was carried out by Chou [11]. For squarefree M, Eichenauer-Herrmann [47] showed that Huber's pseudorandom numbers allow a convenient implementation by compounding standard recursive inversive congruential pseudorandom numbers with the prime factors p_j of M as moduli (compare with Section 3.1 for the compound method). On this basis, the s-dimensional serial test for Huber's pseudorandom numbers was studied by Eichenauer-Herrmann [47] and Eichenauer-Herrmann and Emmerich [57]. The latter paper contains also results on parts of the period.

Still another variant of the recursive inversive congruential method was introduced by Kato, Wu, and Yanagihara [84]. Here the modulus M is a power of 2, and we choose parameters $a, b, c \in Z_M$ and an initial value $y_0 \in G_M$. Then a sequence $y_0, y_1, \ldots$ of elements of G_M is generated by the recursion

$$(8) \qquad y_{n+1} \equiv a\overline{y}_n + b + cy_n \bmod M \quad \text{for } n = 0, 1, \ldots .$$

The integers a, b, c must be chosen in such a way that each y_n is guaranteed to be in G_M. Pseudorandom numbers are obtained by $x_n = y_n/M$ for $n = 0, 1, \ldots$. A necessary and sufficient condition for $\mathrm{per}(x_n) = M/2$ is given in [84], and the 2-dimensional serial test for these pseudorandom numbers is studied in [85]. We remark that if the recursion (8) is used with a prime modulus $M = p \geq 5$ and with $a \not\equiv 0 \bmod p$ and $c \not\equiv 0 \bmod p$, then we cannot achieve $\mathrm{per}(y_n) = p$ (and thus $\mathrm{per}(x_n) = p$) as in the standard recursive inversive congruential method. For if we had $\mathrm{per}(y_n) = p$, then the map $\psi : F_p \to F_p$ defined by $\psi(y) = a\overline{y} + b + cy$ for $y \in F_p$ would have to be surjective. But $\psi(y) = \psi(a\overline{c}y)$ for all $y \in F_p$, and for $p \geq 5$ there exists a $y \in F_p$ with $y \neq a\overline{c}y$ in F_p, hence ψ is not surjective. Therefore, the method based on (8) does not seem to be of interest for prime moduli p in the case $ac \not\equiv 0 \bmod p$ where it differs from the linear congruential method and the standard recursive inversive congruential method.

A multiple-recursive inversive congruential method with prime moduli was proposed by Eichenauer, Grothe, Lehn and Topuzoğlu [28], but apart from a study of periodicity properties no further analysis was carried out.

Among the inversive methods discussed in this section, it appears that the generator based on the recursion (7) is the most promising. Here no nontrivial lattice structure is observed; on the contrary, we have the strong nonlinearity property established by Eichenauer-Herrmann [33]. The performance under the s-dimensional serial test is satisfactory even for relatively large values of s. Moreover, the method allows a wide choice of parameters: as soon as $a, b \in F_p$ have been chosen in such a way that $x^2 - bx - a$ is an IMP polynomial over F_p (or, in particular, a primitive polynomial over F_p), then all properties listed above are satisfied. Extensive statistical testing of pseudorandom numbers generated by inversive methods is currently undertaken by a research team at the University of Salzburg (Austria) led by Peter Hellekalek. The first – and by the way very encouraging – results are already available in the papers of Auer and Hellekalek [5], [6].

3.3. Explicit Inversive Generators

An inversive method that is quite different from the recursive methods considered in Section 3.2 is the *explicit inversive congruential method*. This method is due to Eichenauer-Herrmann [39], and the original proposal employed a prime modulus $p \geq 5$. We again identify Z_p and F_p, and for $c \in F_p$ we define as before $\overline{c} = c^{-1} \in F_p$ if $c \neq 0$ and $\overline{c} = 0 \in F_p$ if $c = 0$. Then we choose $a, b \in F_p$ with $a \neq 0$ and explicitly define

$$(9) \qquad y_n = \overline{an + b} \in F_p \quad \text{for } n = 0, 1, \ldots .$$

Now *explicit inversive congruential pseudorandom numbers* $x_0, x_1, \ldots$ in I are obtained by setting

$$x_n = \frac{1}{p} y_n \quad \text{for } n = 0, 1, \ldots .$$

It is clear that the sequence $x_0, x_1, \ldots$ is purely periodic with $\mathrm{per}(x_n) = p$. This method may also be viewed as a nonlinear congruential method with modulus $M = p$ and feedback function $f(c) = \overline{c + a}$ for $c \in F_p$ in (6). A brief survey of the explicit inversive congruential method is presented in Eichenauer-Herrmann [48].

Since the y_n in (9) are also given by $y_n = g(n) = (an + b)^{p-2} \in F_p$ for $n = 0, 1, \ldots$, we have $d = \deg(g) = p - 2$, and so by Theorem 1 the sequence $y_0, y_1, \ldots$ passes the s-dimensional lattice test (in the sense of Definition 3) exactly for all $s \leq p - 2$. We emphasize that this is the optimal behavior under the lattice test. The following strong nonlinearity property was shown by Niederreiter [120], [125]: if F_p^s is the s-dimensional affine space over F_p, then for $s \geq 2$ and every sequence $y_0, y_1, \ldots$ defined by (9), any hyperspace in F_p^s contains at most s of the points $(y_n, y_{n+1}, \ldots, y_{n+s-1}) \in F_p^s$ with $0 \leq n \leq p - 1$ and $y_n y_{n+1} \cdots y_{n+s-1} \neq 0$. As for the similar result of Eichenauer-Herrmann [33] for the recursive inversive congruential method mentioned in Section 3.2, this result for the explicit inversive congruential method is best possible in the sense that there do exist hyperplanes in F_p^s that contain exactly s of these points, as long as that many points are available in the full period, i.e., as long as $s \leq (p - 1)/2$.

Eichenauer-Herrmann [39] studied the s-dimensional serial test for explicit inversive congruential pseudorandom numbers over the full period. Points $\mathbf{x}_n \in I^s$ with more general lags than those in (1) were considered and an upper bound for the discrepancy of $\mathbf{x}_0, \mathbf{x}_1, \ldots, \mathbf{x}_{p-1}$ of the order of magnitude $p^{-1/2}(\log p)^s$ was established. Furthermore, it was shown that for a positive fraction of all $a \in F_p$ with $a \neq 0$ the corresponding discrepancies are at least of the order of magnitude $p^{-1/2}$ for all $s \geq 2$. The uniformity test and the s-dimensional serial test for parts of the period were investigated by Niederreiter [120], [125] and upper bounds for discrepancies were proved. Corresponding lower bounds for discrepancies can be found in Eichenauer-Herrmann and Niederreiter [67].

An explicit inversive congruential method for a modulus M which is a power of 2 was introduced by Eichenauer-Herrmann and Ickstadt [62]. For every odd integer c we define $\overline{c} \in G_M$ by $c\overline{c} \equiv 1 \bmod M$. Then we choose integers $a, b \in Z_M$ with $a \equiv 2 \bmod 4$ and $b \equiv 1 \bmod 2$ and explicitly define

$$y_n = \overline{an + b} \in G_M \quad \text{for } n = 0, 1, \ldots.$$

Pseudorandom numbers $x_0, x_1, \ldots$ in I are obtained by the normalization $x_n = y_n/M$ for $n = 0, 1, \ldots$. The condition $a \equiv 2 \bmod 4$ guarantees that the sequence $x_0, x_1, \ldots$ is purely periodic with $\mathrm{per}(x_n) = M/2$. In [62] it is shown that the order of magnitude of the discrepancy $D_{M/2}^{(2)}$ in (2) is between $M^{-1/2}$ and $M^{-1/2}(\log M)^2$. However, for all dimensions $s \geq 3$ the discrepancy $D_{M/2}^{(s)}$ is at least of the order of magnitude $M^{-1/3}$. Thus, the s-dimensional serial test is passed only for $s = 2$. Discrepancy bounds for the nonoverlapping pairs $(x_{2n}, x_{2n+1}), 0 \leq n \leq (M/4) - 1$, were established in Eichenauer-Herrmann [52]. For M again a power of 2, Eichenauer-Herrmann [53] proposed a modification of the explicit inversive congruential method which yields purely periodic sequences $x_0, x_1, \ldots$ of pseudorandom numbers with $\mathrm{per}(x_n) = M$. This modified method was further analyzed by the same author in [56]. Compound versions of the explicit inversive congruential method were studied in Eichenauer-Herrmann [44], [51].

In our age of parallel computing, it is an important advantage of explicit inversive congruential pseudorandom numbers that they easily lend themselves to the generation of many parallel streams of pseudorandom numbers with mutual statistical independence properties. This was first pointed out by Niederreiter [120]; see also [125]. To generate s parallel streams of pseudorandom numbers, we choose a large prime modulus p and elements $a_1, \ldots, a_s, b_1, \ldots, b_s \in F_p$ with $a_i \neq 0$ for $1 \leq i \leq s$. Then as in (9) we define

$$y_n^{(i)} = \overline{a_i n + b_i} \in F_p \quad \text{for } n = 0, 1, \ldots \text{ and } 1 \leq i \leq s,$$

and the ith stream consists of the pseudorandom numbers $x_n^{(i)} = y_n^{(i)}/p$ for $n = 0, 1, \ldots$. The strong nonlinearity property for the explicit inversive congruential method with prime moduli, which we have stated earlier, holds also in this more general setting, under the condition that the elements $b_1 \overline{a}_1, \ldots, b_s \overline{a}_s \in F_p$ are distinct. Every hyperplane in $F_p^s, s \geq 2$, contains then at most s of the points $(y_n^{(1)}, \ldots, y_n^{(s)}) \in F_p^s$ with $0 \leq n \leq p-1$ and $y_n^{(1)} \cdots y_n^{(s)} \neq 0$.

The mutual statistical independence properties of the s parallel streams of pseudorandom numbers defined above can be analyzed by discrepancy bounds. We assume again that $b_1 \overline{a}_1, \ldots, b_s \overline{a}_s$ are distinct elements of F_p. Note that this condition can be satisfied as long as $s \leq p$. We consider the points

$$\mathbf{u}_n = \left(x_n^{(1)}, \ldots, x_n^{(s)} \right) \in I^s \quad \text{for } n = 0, 1, \ldots.$$

The following upper bounds for the discrepancy of these points were shown in Niederreiter [120], [125].

Theorem 4. *If* $b_1 \overline{a}_1, \ldots, b_s \overline{a}_s \in F_p$ *are distinct, then*

$$D_p(\mathbf{u}_0, \mathbf{u}_1, \ldots, \mathbf{u}_{p-1}) = O\left(p^{-1/2} (\log p)^s \right) \text{ for } s \geq 2,$$

$$D_N(\mathbf{u}_0, \mathbf{u}_1, \ldots, \mathbf{u}_{N-1}) = O\left(N^{-1} p^{1/2} (\log p)^{s+1} \right) \text{ for } 1 \leq N < p \text{ and } s \geq 1.$$

Corresponding lower bounds for the discrepancy are proved in Eichenauer-Herrmann and Niederreiter [67] and demonstrate that the bounds in Theorem 4 are in general best possible up to the logarithmic factors.

We have seen that the explicit inversive congruential method with prime modulus leads to an optimal behavior under the lattice test and to a strong nonlinearity property. The generated pseudorandom numbers pass the s-dimensional serial test. Concrete results on statistical tests are available in Auer and Hellekalek [5], [6]. Furthermore, we can easily obtain a large number of parallel streams of pseudorandom numbers which satisfy the requirements of Eddy [26] on random number generators for parallel processors.

3.4. The Digital Inversive Method

The recursive and explicit inversive congruential methods with a prime modulus p lead to pseudorandom numbers with very attractive properties, but there is one disadvantage which we have not yet pointed out, namely that the calculation of one pseudorandom number takes $O(\log p)$ multiplications in the finite field F_p. Although pseudorandom number generation is usually not the bottleneck in Monte Carlo calculations, it would still be desirable to find a faster inversive method. Such a method is now available in the form of the *digital inversive method* introduced by Eichenauer-Herrmann and Niederreiter [68].

For simplicity, we describe only the binary case of the digital inversive method; see [68] for the general case. We select a precision $k \geq 1$ and let F_q be the finite field of order $q = 2^k$. We denote by F_q^* the multiplicative group of nonzero elements of F_q. For $\gamma \in F_q^*$ let $\overline{\gamma} = \gamma^{-1} \in F_q^*$ be the multiplicative inverse of γ in F_q^* and define $\overline{\gamma} = 0 \in F_q$ for $\gamma = 0 \in F_q$. Now we proceed in analogy with (7), i.e., we choose parameters $\alpha \in F_q^*$ and $\beta \in F_q$ and an initial value $\gamma_0 \in F_q$, and then we generate the sequence $\gamma_0, \gamma_1, \ldots$ of elements of F_q by the recursion

(10) $$\gamma_{n+1} = \alpha \overline{\gamma}_n + \beta \quad \text{for } n = 0, 1, \ldots .$$

This sequence is purely periodic, and we have $\text{per}(\gamma_n) = q$ if and only if $x^2 - \beta x - \alpha$ is an IMP polynomial over F_q (see Definition 4 for this concept). We observe that any primitive quadratic polynomial is an IMP polynomial over F_q.

From the sequence generated by (10), the actual pseudorandom numbers are obtained as follows. We view F_q as a k-dimensional vector space over $F_2 = Z_2 = \{0, 1\}$ and we choose a fixed ordered basis B of F_q over F_2. Then let

$$\mathbf{c}_n = \left(c_n^{(1)}, \ldots, c_n^{(k)} \right) \in Z_2^k \quad \text{for } n = 0, 1, \ldots$$

be the coordinate vector of $\gamma_n \in F_q$ relative to the ordered basis B. Now a sequence $x_0, x_1, \ldots$ of *digital inversive pseudorandom numbers* is defined by

$$x_n = \sum_{j=1}^{k} c_n^{(j)} 2^{-j} \in [0, 1) \quad \text{for } n = 0, 1, \ldots .$$

Obviously, this sequence is purely periodic, and we have $\text{per}(x_n) = q$ if and only if $x^2 - \beta x - \alpha$ is an IMP polynomial over F_q.

To get a fast implementation of this method, it is advisable, first of all, to represent the elements of F_q right away by their coordinate vectors relative to the chosen ordered basis B. Furthermore, it is advantageous to work with a special type of basis, namely a *normal basis* of F_q over F_2, i.e., an ordered basis of F_q over F_2 of the form $B = \{\lambda, \lambda^2, \lambda^4, \ldots, \lambda^{2^{k-1}}\}$ with some $\lambda \in F_q$. A normal basis of F_q over F_2 always exists. With such a normal basis B, there is a known fast algorithm for the calculation of $\overline{\gamma}$ for any $\gamma \in F_q$ (see [68] for the details). In fact, the calculation of $\overline{\gamma}$ in F_q (with $q = 2^k, k \geq 2$) takes at most $2\lfloor \log_2(k-1) \rfloor$ multiplications in F_q, where $\log_2$ is the logarithm to the base 2. In terms of the maximum period length q, this means that only $O(\log \log q)$ multiplications in F_q are needed. If the normal basis B is chosen cleverly, then multiplication in F_q also turns out to be a cheap operation. Altogether, it transpires that digital inversive pseudorandom numbers can be generated in a much faster way than recursive or explicit inversive congruential pseudorandom numbers. Concrete computational examples can be found in [68].

If $\text{per}(x_n) = q = 2^k$, then each rational number in $[0, 1)$ with fixed denominator q appears exactly once among the pseudorandom numbers $x_0, x_1, \ldots, x_{q-1}$. Therefore, the full period of the sequence $x_0, x_1, \ldots$ shows a perfect equidistribution in the interval $[0, 1)$. For $s \geq 2$ the following bound for the discrepancy in (2) was proved in [68].

Theorem 5. *For any sequence $x_0, x_1, \ldots$ of digital inversive pseudorandom numbers with* $\text{per}(x_n) = q = 2^k$ *we have*

$$D_q^{(s)} < (s-1)(2q^{-1/2} + q^{-1})(\frac{k}{2} + 1)^s \text{ for } s \geq 2.$$

There is also a companion result in [68] which shows that for a positive fraction of all possible parameters in Theorem 5 the discrepancy $D_q^{(s)}$ is at least of the order of magnitude $q^{-1/2}$ for all $s \geq 2$. Thus, digital inversive pseudorandom numbers display a satisfactory behavior under the s-dimensional serial test. Since the digital inversive method was introduced only very recently, no empirical statistical testing has yet been performed with the generated pseudorandom numbers.

4. Recent Linear Methods

4.1. The Multiple-Recursive Matrix Method

We now describe a general class of algorithms for pseudorandom number generation which was recently introduced by Niederreiter [123]. This general method provides an extension of earlier methods and also a unifying framework for studying many pseudorandom number generators using linear types of recursions. The method includes several of the methods discussed earlier as special cases, such as the multiplicative linear congruential method (with prime modulus), the multiple-recursive congruential method (with prime modulus), the GFSR method, and the twisted GFSR generator.

This general method, which is called the *multiple-recursive matrix method*, proceeds as follows. We choose a prime p and integers $k \geq 1$ and $m \geq 1$. Let $A_0, A_1, \ldots, A_{k-1}$ be $m \times m$ matrices over the finite field F_p, where A_0 is assumed to be nonsingular. Then we generate a sequence $\mathbf{z}_0, \mathbf{z}_1, \ldots$ of row vectors from F_p^m by selecting initial vectors $\mathbf{z}_0, \mathbf{z}_1, \ldots, \mathbf{z}_{k-1}$ that are not all $\mathbf{0}$ and using the kth-order vector recursion

$$(11) \qquad \mathbf{z}_{n+k} = \sum_{h=0}^{k-1} \mathbf{z}_{n+h} A_h \quad \text{for } n = 0, 1, \ldots .$$

For the sake of easier reference, we call such a sequence $\mathbf{z}_0, \mathbf{z}_1, \ldots$ a (kth-order) *recursive vector sequence* (in F_p^m). Specifically, let

$$\mathbf{z}_n = \left(z_n^{(1)}, \ldots, z_n^{(m)} \right) \in F_p^m \quad \text{for } n = 0, 1, \ldots .$$

Using the standard identification of F_p with Z_p, we can view the coordinates $z_n^{(j)}$ as integers from Z_p. Then a sequence $x_0, x_1, \ldots$ of pseudorandom numbers is defined by

$$x_n = \sum_{j=1}^{m} z_n^{(j)} p^{-j} \in [0, 1) \quad \text{for } n = 0, 1, \ldots .$$

There is a considerable amount of flexibility in this method: we may either choose p to be a large prime and m small, or we may take p to be a small prime (such as $p = 2$) and m sufficiently large to obtain a small discretization.

If we take $k = m = 1$, then the method above reduces to the multiplicative linear congruential method with modulus p. The case $k \geq 2$ and $m = 1$ yields the multiple-recursive congruential method with modulus p (see Section 2.2). If $k \geq 2$, $m \geq 2$, and each matrix A_h in (11) is a scalar multiple of the $m \times m$ identity matrix I_m over F_p, then we obtain the vector recursion (5) and thus the GFSR method. If $k \geq 2$, $m \geq 2$, and all matrices A_h in (11) except A_0 are scalar multiples of I_m, then we get a generalization of the twisted GFSR generator (see Section 2.3).

Any sequence $x_0, x_1, \ldots$ of pseudorandom numbers generated by the multiple-recursive matrix method is purely periodic with $\operatorname{per}(x_n) \leq p^{km} - 1$. It was shown in [123] that we have $\operatorname{per}(x_n) = p^{km} - 1$ if and only if the polynomial

$$\det\left(x^k I_m - \sum_{h=0}^{k-1} x^h A_h\right)$$

of degree km is a primitive polynomial over F_p. However, from this condition it is not clear whether we can actually achieve $\operatorname{per}(x_n) = p^{km} - 1$. Here the further analysis carried out in Niederreiter [126] sheds more light. For the following material, the reader requiring more background on finite fields is referred to the book [100]. Let F_q be the finite field of order $q = p^{km}$, and note that F_q can also be viewed as a vector space of dimension km over F_p.

Definition 5. For $\sigma \in F_q$ with $\sigma \neq 0$, an F_p-linear subspace V of F_q of dimension m is called σ-*splitting* if F_q is the direct sum of the F_p-linear subspaces $V, \sigma V, \sigma^2 V, \ldots, \sigma^{k-1} V$.

Since $\operatorname{per}(x_n) = \operatorname{per}(\mathbf{z}_n)$, the following result from [126] serves also as a characterization of the case $\operatorname{per}(x_n) = p^{km} - 1$.

Theorem 6. *Let*

$$\mathbf{z}_n = \left(z_n^{(1)}, \ldots, z_n^{(m)}\right) \in F_p^m \text{ for } n = 0, 1, \ldots$$

be an arbitrary sequence of elements of F_p^m. Then this sequence is a kth-order recursive vector sequence with $\operatorname{per}(\mathbf{z}_n) = p^{km} - 1$ if and only if

$$z_n^{(j)} = \operatorname{Tr}(\beta_j \sigma^n) \quad \text{for } 1 \leq j \leq m \text{ and } n \geq 0,$$

where σ is a primitive element of F_q, the elements $\beta_1, \ldots, \beta_m$ form a basis of a σ-splitting subspace of F_q, and Tr denotes the trace function from F_q to F_p.

On the basis of Theorem 6, one can now show that the maximum possible value $p^{km} - 1$ for $\operatorname{per}(x_n) = \operatorname{per}(\mathbf{z}_n)$ can actually be attained. Indeed, it suffices to prove that a σ-splitting subspace of F_q of dimension m exists in the case where σ is a primitive element of F_q. Several such examples are given in [126], the easiest example being the subfield F_{p^m}, viewed as an F_p-linear subspace of F_q. This leads to the following corollary of Theorem 6.

Corollary 1. *For any p, k, and m there exists a kth-order recursive vector sequence $\mathbf{z}_0, \mathbf{z}_1, \ldots$ in F_p^m with $\operatorname{per}(\mathbf{z}_n) = p^{km} - 1$.*

We now consider the distribution of the s-dimensional points $\mathbf{x}_n$ in (1) in the case where $\operatorname{per}(x_n) = T := p^{km} - 1$. In fact, it is easy to see (compare with [126, Corollary 3]) that as long as $s \leq k$, the points $\mathbf{x}_n$ in the full period display an almost perfect equidistribution in the following sense: among the points $\mathbf{x}_0, \mathbf{x}_1, \ldots, \mathbf{x}_{T-1}$ every nonzero point in $[0, 1)^s$ all of whose coordinates are rationals with fixed denominator p^m occurs with frequency $p^{m(k-s)}$, and the point $\mathbf{0} \in [0, 1)^s$ occurs with frequency $p^{m(k-s)} - 1$.

For dimensions $s > k$ the distribution behavior of the points $\mathbf{x}_0, \mathbf{x}_1, \ldots, \mathbf{x}_{T-1}$ depends on the specific parameters in the algorithm for generating the pseudorandom numbers.

In general, some uniformity properties are still valid, which can be expressed most conveniently in the language of the theory of nets (see Definition 2). The following result was shown in [126]: if $\text{per}(x_n) = T$ and $s > k$, then the p^{km} points $\mathbf{0}, \mathbf{x}_0, \mathbf{x}_1, \ldots, \mathbf{x}_{T-1}$ form a (t, km, s)-net in base p, where the value of t depends in a known way on s, m, and the elements $\sigma, \beta_1, \ldots, \beta_m$ in Theorem 6. In the case $\text{per}(x_n) = T$, a detailed theoretical study of the s-dimensional serial test for pseudorandom numbers generated by the multiple-recursive matrix method was carried out in [126] for the full period and also for parts of the period.

When compared with the GFSR method, the multiple-recursive matrix method is much more effective with regard to producing long periods, since the GFSR method uses a kth-order vector recursion in F_p^m to generate pseudorandom numbers $y_0, y_1, \ldots$ with m p-adic digits and $\text{per}(y_n) = p^k - 1$, whereas the multiple-recursive matrix method uses the same setup to achieve $\text{per}(x_n) = p^{km} - 1$. Further work needs to be done on suitable practical implementations of this method and on the statistical testing of the generated pseudorandom numbers.

4.2. Combined Generators

The combination of several streams of pseudorandom numbers into a new sequence of pseudorandom numbers can lead to an increased period length, to an improved statistical performance, and to a more favorable lattice structure (if applicable). The use of combined generators was advocated in the 1980s by Wichmann and Hill [144], [145]. An example of a combination of streams of pseudorandom numbers is provided by the compound method discussed in Section 3.1.

The proposals of Wichmann and Hill [144], [145] were based on the combination of linear congruential generators. More recently, combined linear congruential generators were investigated and tested by L'Ecuyer [88], L'Ecuyer and Tezuka [95], and Levitan and Sobol' [99]. Such generators can be of interest for practical work since they lead to efficient ways of implementing linear congruential generators with very large moduli. However, a careful analysis of the combined generator is needed to make sure that the lattice structure is not too bad. Combined multiple-recursive congruential generators were recently studied by Couture and L'Ecuyer [17] and L'Ecuyer [92].

In the last few years, beginning with the work of Tezuka and L'Ecuyer [140], considerable efforts were made to understand the structure of combined shift-register generators and to come up with well-founded recommendations for specific parameters. The important paper of Couture, L'Ecuyer, and Tezuka [18] provided deep theoretical tools for the analysis of such combined generators, based on the "lattice" structure they yield in spaces of formal Laurent series over finite fields (compare also with Section 2.3). These tools were later used by Tezuka [138] to analyze combinations of GFSR and twisted GFSR generators. A different approach to the analysis of combined shift-register generators was described very recently by L'Ecuyer [93]. Further work on combined shift-register generators was carried out by Wang and Compagner [143].

Deng, George, and Chu [23] set up a stochastic model for the study of combined generators. Further techniques of combining sequences of pseudorandom numbers are discussed in Collings [15] and Sherif and Dear [132].

5. Pseudorandom Vector Generation

5.1. The Matrix Method

The task in uniform pseudorandom vector generation is to simulate a sequence of independent random vector variables with the uniform distribution on $I^m = [0,1]^m, m \geq 2$, as their common multivariate distribution function. Thus, uniform pseudorandom vectors will be vectors (or points) in I^m. As mentioned in Section 1.1, uniform pseudorandom vectors are important for parallelized simulation methods. We refer to Anderson [3], Bhavsar and Isaac [8], and Eddy [26], and also to the excellent recent discussion in L'Ecuyer [90], for background on issues related to pseudorandom vector generation and parallelized simulation. Although the topic is of great practical interest, there is still relatively little work on uniform pseudorandom vectors. Brief surveys of uniform pseudorandom vector generation have been given earlier in Niederreiter [117], [119, Chapter 10]. Although there are ways of building uniform pseudorandom vectors from uniform pseudorandom numbers (see e.g. Section 3.3), we concentrate mainly on direct methods for uniform pseudorandom vector generation.

The *matrix method* for uniform pseudorandom vector generation is an analog of the linear congruential method. For given $m \geq 2$, we choose a large modulus M and an $m \times m$ matrix A with entries from Z_M. Then we generate a sequence $z_0, z_1, \ldots$ of row vectors in Z_M^m by starting from an initial vector $z_0 \neq 0$ and using the vector recursion

$$(12) \qquad\qquad z_{n+1} \equiv z_n A \bmod M \quad \text{for } n = 0, 1, \ldots.$$

From this sequence we derive the uniform pseudorandom vectors

$$u_n = \frac{1}{M} z_n \in I^m \quad \text{for } n = 0, 1, \ldots.$$

To guarantee that the sequences $z_0, z_1, \ldots$ and $u_0, u_1, \ldots$ are purely periodic, we assume that the matrix A is nonsingular modulo M, i.e., that $\gcd(\det(A), M) = 1$. The matrix method was introduced in equivalent forms by Grothe [77], Niederreiter [108], Niki [128], and Tahmi [134]. Short expository accounts of the matrix method can be found in Niederreiter [117], [119, Chapter 10].

Most of the attention in the matrix method has focused on the case where $M = p$ is a prime modulus. Then A is viewed as a matrix over the finite field F_p, and there is a simple criterion for A to yield the maximum possible value $p^m - 1$ of $\text{per}(u_n)$. Concretely, we have $\text{per}(u_n) = p^m - 1$ if and only if the characteristic polynomial of A is primitive over F_p; see Grothe [77], [78] and Niederreiter [119, Theorem 10.2]. The approaches of Niederreiter [108] and Niki [128] work only for prime moduli. A recent contribution to the latter approach is contained in Mullen [106].

Uniform pseudorandom vectors generated by the matrix method lead to a lattice structure, just like linear congruential pseudorandom numbers. This lattice structure was first pointed out by Afflerbach and Grothe [1] and Grothe [79] in the case of a prime modulus; see also [119, Theorem 10.3] for a simple proof. A generalization of this result will be discussed in Section 5.2 in the context of the multiple-recursive matrix method for pseudorandom vector generation.

For uniform pseudorandom vectors $u_0, u_1, \ldots$ there is an analog of the s-dimensional serial test (compare with Section 1.2 for the case of uniform pseudorandom numbers).

For a given $s \geq 2$, this test is based on considering the sequence of points

(13) $$\mathbf{v}_n = (\mathbf{u}_n, \mathbf{u}_{n+1}, \ldots, \mathbf{u}_{n+s-1}) \in I^{ms} \quad \text{for } n = 0, 1, \ldots$$

and analyzing the discrepancy of its finite initial segments. A detailed study of the s-dimensional serial test for pseudorandom vectors generated by the matrix method was carried out by Niederreiter [114]. An account of the results can also be found in [119, Chapter 10].

Let $\mathbf{u}_0, \mathbf{u}_1, \ldots$ be a sequence of pseudorandom vectors generated by the matrix method with modulus M. If M is a composite prime power, say $M = p^\alpha$ with p prime and $\alpha \geq 2$, and if the matrix A is nonsingular modulo M, then we have the bound $\text{per}(\mathbf{u}_n) \leq (p^m - 1)p^{\alpha-1}$; see e.g. [119, Lemma 10.7] for a proof. Eichenauer-Herrmann, Grothe, and Lehn [60] have shown by constructive means that the bound $(p^m - 1)p^{\alpha-1}$ can always be attained. Tahmi [134] studied $\text{per}(\mathbf{u}_n)$ for arbitrary moduli M.

Baya [7] studied a uniform pseudorandom vector generator in which the recursion (12) is replaced by the vector recursion

$$\mathbf{z}_{n+1} \equiv \mathbf{z}_n A + \mathbf{w}_n \mod p \quad \text{for } n = 0, 1, \ldots,$$

where $\mathbf{w}_0, \mathbf{w}_1, \ldots$ is a purely periodic sequence of row vectors in Z_p^m.

The matrix method, being a straightforward generalization of the linear congruential method, suffers from similar defects as the latter method (compare with Section 2.1). Methods for uniform pseudorandom vector generation that can serve to alleviate the deficiencies of the matrix method will be discussed in the following Sections 5.2 and 5.3.

5.2. The Multiple-Recursive Matrix Method

We discuss a generalization of the matrix method which is obtained by replacing the first-order vector recursion in (12) by a higher-order vector recursion. In fact, higher-order vector recursions with a prime modulus p were already considered in Section 4.1 in the context of the multiple-recursive matrix method for pseudorandom number generation, and for the following we will use a similar framework.

In the *multiple-recursive matrix method* for uniform pseudorandom vector generation we choose a large prime p and we generate a kth-order recursive vector sequence $\mathbf{z}_0, \mathbf{z}_1, \ldots$ in F_p^m by the recursion (11). Then we identify F_p as usual, with the set Z_p of integers and we derive a sequence $\mathbf{u}_0, \mathbf{u}_1, \ldots$ of m-dimensional pseudorandom vectors by putting

$$\mathbf{u}_n = \frac{1}{p}\mathbf{z}_n \in I^m \quad \text{for } n = 0, 1, \ldots.$$

This method was introduced by Niederreiter [123] and studied in detail in [127].

The periodicity properties of the sequence $\mathbf{u}_0, \mathbf{u}_1, \ldots$ are already implicit in the discussion in Section 4.1. Thus, the sequence is purely periodic with $\text{per}(\mathbf{u}_n) \leq p^{km} - 1$, and by Corollary 1 the value $p^{km} - 1$ can always be reached with a suitable choice of parameters. Criteria for $\text{per}(\mathbf{u}_n) = p^{km} - 1$ follow from those given in Section 4.1.

In order to study further properties of the sequence $\mathbf{u}_0, \mathbf{u}_1, \ldots$ of pseudorandom vectors generated by the multiple-recursive matrix method, we consider the points $\mathbf{v}_n$ in (13) for a given s. Suppose that we are in the case of the maximum possible period, i.e., that $\mathrm{per}(\mathbf{u}_n) = p^{km} - 1 =: T$. Then for $s \le k$ the points $\mathbf{v}_0, \mathbf{v}_1, \dots, \mathbf{v}_{T-1}$ in the full period show an almost perfect equidistribution in the following sense: among these points, every nonzero point in $[0,1)^{ms}$ all of whose coordinates are rationals with fixed denominator p occurs with frequency $p^{m(k-s)}$, and the point $\mathbf{0} \in [0,1)^{ms}$ occurs with frequency $p^{m(k-s)}-1$.

A nontrivial lattice structure of the points $\mathbf{v}_n$ in (13) arises for $s > k$. We define the $m \times m$ matrices $A_h^{(j)}$ over F_p for $0 \le h \le k-1$ and $j \ge 0$ by setting the initial values

$$A_h^{(0)} = A_h \quad \text{for } 0 \le h \le k-1,$$

where the A_h are as in the recursion (11), and then using the following recursions for $j \ge 0$:

$$A_0^{(j+1)} = A_0 A_{k-1}^{(j)},$$

$$A_h^{(j+1)} = A_{h-1}^{(j)} + A_h A_{k-1}^{(j)} \quad \text{for } 1 \le h \le k-1.$$

For given $s > k$ we introduce the $mk \times m(s-k)$ matrix

$$A^{(s)} = \begin{pmatrix} A_0^{(0)} & A_0^{(1)} & \cdots & A_0^{(s-k-1)} \\ A_1^{(0)} & A_1^{(1)} & \cdots & A_1^{(s-k-1)} \\ \vdots & \vdots & & \vdots \\ A_{k-1}^{(0)} & A_{k-1}^{(1)} & \cdots & A_{k-1}^{(s-k-1)} \end{pmatrix},$$

which we view as an integer matrix by identifying F_p with Z_p. Then we define the $ms \times ms$ matrix

$$G^{(s)} = \begin{pmatrix} p^{-1} E_{mk} & p^{-1} A^{(s)} \\ 0 & E_{m(s-k)} \end{pmatrix}$$

with rational entries, where E_r is the $r \times r$ identity matrix over $\mathbf{Z}$. Let $L^{(s)}$ be the lattice in $\mathbf{R}^{ms}$ with generator matrix $G^{(s)}$, i.e., $L^{(s)}$ consists of all $\mathbf{Z}$-linear combinations of the row vectors of $G^{(s)}$. Then the following result was shown in [127].

Theorem 7. *Let* $\mathbf{u}_0, \mathbf{u}_1, \dots$ *be a sequence of m-dimensional pseudorandom vectors generated by the multiple-recursive matrix method with* $\mathrm{per}(\mathbf{u}_n) = T = p^{km} - 1$. *If* $s > k$ *and the points* $\mathbf{v}_n$ *are given by* (13), *then we have*

$$\{\mathbf{0}, \mathbf{v}_0, \mathbf{v}_1, \dots, \mathbf{v}_{T-1}\} = L^{(s)} \cap [0,1)^{ms},$$

where $L^{(s)}$ *is the lattice in* $\mathbf{R}^{ms}$ *with generator matrix* $G^{(s)}$.

This theorem contains the result on the lattice structure obtained from pseudorandom vectors generated by the matrix method with prime modulus as a special case (namely the case $k = 1$), and thus also the corresponding result for the multiplicative linear congruential method with prime modulus (namely the case $k = m = 1$).

Again for $s > k$, the s-dimensional serial test for pseudorandom vectors generated by the multiple-recursive matrix method was studied by Niederreiter [127]. As in Section 5.1, this test is carried out by analyzing the discrepancy of the finite initial segments of the sequence of points $\mathbf{v}_n$ in (13). In fact, since the discretization error is the dominant term in this discrepancy, it is more natural to consider the discrete discrepancy defined in [119, p. 215], i.e., a discrepancy which takes into account the discrete nature of the points $\mathbf{v}_n$ (namely, that all coordinates of $\mathbf{v}_n$ are rationals with fixed denominator p). Upper and lower bounds for such discrete discrepancies, also for parts of the period, are given in [127].

Compared with the matrix method, the multiple-recursive matrix method for pseudo-random vector generation has the advantage that we can achieve $\mathrm{per}(\mathbf{u}_n) = p^{km} - 1$ and that, even for fixed p and m, this value can still be made arbitrarily large by choosing sufficiently large values of k, the order of the recursion. Furthermore, there is still a lattice structure present (as is the case for the matrix method), but it is less pronounced for larger values of k, in the sense that the determinant of the lattice $L^{(s)}$ in Theorem 7 is equal to p^{-km} and thus becomes quite small.

5.3. The Inversive Method

Nonlinear methods for the generation of pseudorandom numbers were discussed in Section 3. In recent years, efforts have been made to develop analogous nonlinear methods for pseudorandom vector generation. An extension of the general nonlinear congruential method in Section 3.1 to pseudorandom vector generation was first proposed in Niederreiter [118]. A brief description of this method can be found also in [119, Chapter 10].

Probably the most promising nonlinear method for uniform pseudorandom vector generation is the *inversive method* introduced in Niederreiter [116]. This method is an analog of the recursive inversive congruential method discussed in Section 3.2. A detailed analysis of the inversive method was carried out in Niederreiter [124]. We follow the description of the method given in that paper.

Suppose again that we want to generate m-dimensional pseudorandom vectors. Then we choose a large prime p and consider the finite field F_q of order $q = p^m$, together with its multiplicative group F_q^*. For $\gamma \in F_q^*$ let $\overline{\gamma} = \gamma^{-1} \in F_q^*$ be the multiplicative inverse of γ in F_q^* and define $\overline{\gamma} = 0 \in F_q$ for $\gamma = 0 \in F_q$. Now we select parameters $\alpha \in F_q^*$ and $\beta \in F_q$ and generate a sequence $\gamma_0, \gamma_1, \ldots$ of elements of F_q by choosing an initial value γ_0 and using the recursion

$$\gamma_{n+1} = \alpha \overline{\gamma}_n + \beta \quad \text{for } n = 0, 1, \ldots.$$

Next we view F_q as an m-dimensional vector space over F_p and let B be an ordered basis of F_q over F_p. Then for $n = 0, 1, \ldots$ we let $\mathbf{c}_n \in F_p^m$ be the coordinate vector of $\gamma_n \in F_q$ relative to B. We identify F_p with the set Z_p of integers, so that each $\mathbf{c}_n$ can be considered as an element of Z_p^m. Finally, a sequence $\mathbf{u}_0, \mathbf{u}_1, \ldots$ of m-dimensional pseudorandom vectors is obtained by putting

$$\mathbf{u}_n = \frac{1}{p}\mathbf{c}_n \in I^m \quad \text{for } n = 0, 1, \ldots.$$

This sequence is purely periodic, and we have $\mathrm{per}(\mathbf{u}_n) = q$ if and only if $x^2 - \beta x - \alpha$ is an IMP polynomial over F_q (see Definition 4 for this concept). In particular, we have $\mathrm{per}(\mathbf{u}_n) = q$ if $x^2 - \beta x - \alpha$ is a primitive polynomial over F_q (see [124] for these results).

The s-dimensional serial test for pseudorandom vectors generated by the inversive method was also studied in [124]. In a similar way as in Section 5.2, this test is set up by considering the discrete discrepancy of the points $\mathbf{v}_0, \mathbf{v}_1, \ldots, \mathbf{v}_{q-1}$ in (13) from the full period in the case $\mathrm{per}(\mathbf{u}_n) = q$. Upper and lower bounds for such discrete discrepancies are established.

More recently, Emmerich [71] introduced a compound version of the inversive method, analyzed the periodicity properties of the generated sequences of pseudorandom vectors, and established results on the s-dimensional serial test.

We recall from Section 3.3 that the explicit inversive congruential method can be used to generate many parallel streams of uniform pseudorandom numbers with mutual statistical independence properties. This can also be viewed as a method of generating attractive uniform pseudorandom vectors on parallel processors.

6. Conclusions

The areas of uniform pseudorandom number and vector generation have seen enormous progress in the 1990s, but nevertheless the state of absolute bliss – namely to arrive at *the* definitive generator - has not been reached and probably never will be. The user is thus well advised to exercise great care in the selection of pseudorandom number and vector generators. It seems to be a common feature that algorithms that are too simple, such as the linear congruential method or higher-order linear recursions with few terms, may look good on first glance because of their speed, but show serious deficiencies after a finer analysis. Therefore, a certain "complexity" in the algorithm could be a virtue, although it is of course no guarantee for an acceptable performance of the generated numbers or vectors. The best seal of quality for a generator is still a solid theoretical analysis (extending beyond periodicity properties!), combined with extensive empirical testing.

For the user, the multitude of available methods may be a boon, since it allows him/her to do different runs with generators stemming from completely different methods. If these runs yield very similar results, this should lead to higher confidence in the Monte Carlo calculation at hand. At any rate, the user should be aware of the desirable statistical properties of the pseudorandom numbers and vectors that are needed in the specific computational project, so as to avoid *a priori* those generators that are known to do badly with regard to these properties. This implies, in particular, that the conscientious user should have some notion of the strengths and weaknesses of standard generators, and it is of course up to the specialists in pseudorandom number and vector generation to provide such information.

Typical features of the generators discussed in this article have already been presented in the relevant earlier sections. In this last section we just give a very rough overview. If pseudorandom numbers with strong uniformity properties are desired, then linear congruential generators and shift-register generators are good candidates. However, the coarse lattice structure inherent in linear congruential generators makes them unsuitable in many applications. Shift-register generators, on the other hand, have the additional advantage that they can be used to produce very large period lengths. This aspect is important for Monte Carlo calculations on supercomputers. Other methods that yield very large period lengths are the multiple-recursive congruential method, the multiple-recursive matrix method, and combined generators. If true randomness with reference to the s-dimensional serial test and a lack of lattice structure are desired, then inversive generators seem to be the way to go. Among new methods, one of the most promising ones is certainly the digital inversive method, which offers the virtues of nonlinear methods together with a fast algorithm. We have barely touched upon the computer implementation of pseudorandom numbers here; for an excellent discussion of this issue we refer to the survey paper of L'Ecuyer [90].

Promising methods for uniform pseudorandom vector generation are the multiple-recursive matrix method and the inversive method, but more work needs to be done on these methods before well-founded recommendations can be made. For instance, more data on the structural and statistical properties have to be collected, concrete parameters have to be proposed on the basis of extensive statistical testing, and convenient computer implementations have to be developed.

References

[1] L. Afflerbach and H. Grothe, The lattice structure of pseudo-random vectors generated by matrix generators, *J. Comp. Appl. Math.* **23**, 127–131 (1988).

[2] S. Aluru, G.M. Prabhu, and J. Gustafson, A random number generator for parallel computers, *Parallel Comput.* **18**, 839–847 (1992).

[3] S.L. Anderson, Random number generators on vector supercomputers and other advanced architectures, *SIAM Rev.* **32**, 221–251 (1990).

[4] D.A. André, G.L. Mullen, and H. Niederreiter, Figures of merit for digital multistep pseudorandom numbers, *Math. Comp.* **54**, 737–748 (1990).

[5] T. Auer and P. Hellekalek, Independence of uniform pseudorandom numbers, part I: the theoretical background, *Proc. Internat. Workshop Parallel Numerics '94* (M. Vajteršic and P. Zinterhof, eds.), pp. 44–58, Slovak Academy of Sciences, Bratislava, 1994.

[6] T. Auer and P. Hellekalek, Independence of uniform pseudorandom numbers, part II: empirical results, *Proc. Internat. Workshop Parallel Numerics '94* (M. Vajteršic and P. Zinterhof, eds.), pp. 59–73, Slovak Academy of Sciences, Bratislava, 1994.

[7] A. Baya, Contribution à la génération de vecteurs aléatoires et à la cryptographie, Thèse, Université de Grenoble, 1990.

[8] V.C. Bhavsar and J.R. Isaac, Design and analysis of parallel Monte Carlo algorithms, *SIAM J. Sci. Statist. Comput.* **8**, s73–s95 (1987).

[9] R.P. Brent, On the periods of generalized Fibonacci recurrences, *Math. Comp.* **63**, 389–401 (1994).

[10] W.-S. Chou, The period lengths of inversive pseudorandom vector generations, *Finite Fields and Their Appl.* **1**, 126–132 (1995).

[11] W.-S. Chou, The period lengths of inversive congruential recursions, *Acta Arith.*, to appear.

[12] W.-S. Chou, On inversive maximal period polynomials over finite fields, *Appl. Algebra Engrg. Comm. Comput.*, to appear.

[13] W.-S. Chou and H. Niederreiter, On the lattice test for inversive congruential pseudorandom numbers, this volume.

[14] K.L. Chung, An estimate concerning the Kolmogoroff limit distribution, *Trans. Amer. Math. Soc.* **67**, 36–50 (1949).

[15] B.J. Collings, Compound random number generators, *J. Amer. Statist. Assoc.* **82**, 525–527 (1987).

[16] R. Couture and P. L'Ecuyer, On the lattice structure of certain linear congruential sequences related to AWC/SWB generators, *Math. Comp.* **62**, 799–808 (1994).

[17] R. Couture and P. L'Ecuyer, Orbits and lattices for linear random number generators with composite moduli, *Math. Comp.*, to appear.

[18] R. Couture, P. L'Ecuyer, and S. Tezuka, On the distribution of k-dimensional vectors for simple and combined Tausworthe sequences, *Math. Comp.* **60**, 749–761, S11–S16 (1993).

[19] J. Dagpunar, *Principles of Random Variate Generation*, Clarendon Press, Oxford, 1988.

[20] I. Deák, *Random Number Generators and Simulation*, Akadémiai Kiadó, Budapest, 1989.

[21] A. de Matteis, J. Eichenauer-Herrmann, and H. Grothe, Computation of critical distances within multiplicative congruential pseudorandom number sequences, *J. Comp. Appl. Math.* **39**, 49–55 (1992).

[22] A. de Matteis and S. Pagnutti, Parallelization of random number generators and long-range correlations, *Numer. Math.* **53**, 595–608 (1988).

[23] L.Y. Deng, E.O. George, and Y.C. Chu, On improving pseudo-random number generators, *Proc. 1991 Winter Simulation Conf.* (G.M. Clark et al., eds.), pp. 1035–1042, IEEE Press, Piscataway, NJ, 1991.

[24] U. Dieter, Probleme bei der Erzeugung gleichverteilter Zufallszahlen, *Zufallszahlen und Simulationen* (L. Afflerbach and J. Lehn, eds.), pp. 7–20, Teubner, Stuttgart, 1986.

[25] U. Dieter, Erzeugung von gleichverteilten Zufallszahlen, *Jahrbuch Überblicke Mathematik 1993*, pp. 25–44, Vieweg, Braunschweig, 1993.

[26] W.F. Eddy, Random number generators for parallel processors, *J. Comp. Appl. Math.* **31**, 63–71 (1990).

[27] J. Eichenauer, H. Grothe, and J. Lehn, Marsaglia's lattice test and non-linear congruential pseudo random number generators, *Metrika* **35**, 241–250 (1988).

[28] J. Eichenauer, H. Grothe, J. Lehn, and A. Topuzoğlu, A multiple recursive nonlinear congruential pseudo random number generator, *Manuscripta Math.* **59**, 331–346 (1987).

[29] J. Eichenauer and J. Lehn, A non-linear congruential pseudo random number generator, *Statist. Papers* **27**, 315–326 (1986).

[30] J. Eichenauer and J. Lehn, On the structure of quadratic congruential sequences, *Manuscripta Math.* **58**, 129–140 (1987).

[31] J. Eichenauer, J. Lehn, and A. Topuzoğlu, A nonlinear congruential pseudorandom number generator with power of two modulus, *Math. Comp.* **51**, 757–759 (1988).

[32] J. Eichenauer and H. Niederreiter, On Marsaglia's lattice test for pseudorandom numbers, *Manuscripta Math.* **62**, 245–248 (1988).

[33] J. Eichenauer-Herrmann, Inversive congruential pseudorandom numbers avoid the planes, *Math. Comp.* **56**, 297–301 (1991).

[34] J. Eichenauer-Herrmann, On the discrepancy of inversive congruential pseudorandom numbers with prime power modulus, *Manuscripta Math.* **71**, 153–161 (1991).

[35] J. Eichenauer-Herrmann, Inversive congruential pseudorandom numbers: a tutorial, *Internat. Statist. Rev.* **60**, 167–176 (1992).

[36] J. Eichenauer-Herrmann, Construction of inversive congruential pseudorandom number generators with maximal period length, *J. Comp. Appl. Math.* **40**, 345–349 (1992).

[37] J. Eichenauer-Herrmann, On the autocorrelation structure of inversive congruential pseudorandom number sequences, *Statist. Papers* **33**, 261–268 (1992).

[38] J. Eichenauer-Herrmann, A remark on the discrepancy of quadratic congruential pseudorandom numbers, *J. Comp. Appl. Math.* **43**, 383–387 (1992).

[39] J. Eichenauer-Herrmann, Statistical independence of a new class of inversive congruential pseudorandom numbers, *Math. Comp.* **60**, 375–384 (1993).

[40] J. Eichenauer-Herrmann, The lattice structure of nonlinear congruential pseudorandom numbers, *Metrika* **40**, 115–120 (1993).

[41] J. Eichenauer-Herrmann, On the discrepancy of inversive congruential pseudorandom numbers with prime power modulus, II, *Manuscripta Math.* **79**, 239–246 (1993).

[42] J. Eichenauer-Herrmann, Inversive congruential pseudorandom numbers, *Z. angew. Math. Mech.* **73**, T644–T647 (1993).

[43] J. Eichenauer-Herrmann, Equidistribution properties of nonlinear congruential pseudorandom numbers, *Metrika* **40**, 333–338 (1993).

[44] J. Eichenauer-Herrmann, Explicit inversive congruential pseudorandom numbers: the compound approach, *Computing* **51**, 175–182 (1993).

[45] J. Eichenauer-Herrmann, Improved lower bounds for the discrepancy of inversive congruential pseudorandom numbers, *Math. Comp.* **62**, 783–786 (1994).

[46] J. Eichenauer-Herrmann, Compound nonlinear congruential pseudorandom numbers, *Monatsh. Math.* **117**, 213–222 (1994).

[47] J. Eichenauer-Herrmann, On generalized inversive congruential pseudorandom numbers, *Math. Comp.* **63**, 293–299 (1994).

[48] J. Eichenauer-Herrmann, Pseudorandom number generation by nonlinear methods, *Internat. Statist. Rev.*, to appear.

[49] J. Eichenauer-Herrmann, On the discrepancy of quadratic congruential pseudorandom numbers with power of two modulus, *J. Comp. Appl. Math.* **53**, 371–376 (1994).

[50] J. Eichenauer-Herrmann, Quadratic congruential pseudorandom numbers: distribution of triples, *J. Comp. Appl. Math.*, to appear.

[51] J. Eichenauer-Herrmann, A unified approach to the analysis of compound pseudorandom numbers, *Finite Fields and Their Appl.* **1**, 102–114 (1995).

[52] J. Eichenauer-Herrmann, Nonoverlapping pairs of explicit inversive congruential pseudorandom numbers, *Monatsh. Math.* **119**, 49–61 (1995).

[53] J. Eichenauer-Herrmann, Modified explicit inversive congruential pseudorandom numbers with power of two modulus, *Statistics and Computing*, to appear.

[54] J. Eichenauer-Herrmann, Quadratic congruential pseudorandom numbers: distribution of triples, II, *J. Comp. Appl. Math.*, to appear.

[55] J. Eichenauer-Herrmann, Discrepancy bounds for nonoverlapping pairs of quadratic congruential pseudorandom numbers, preprint.

[56] J. Eichenauer-Herrmann, Modified explicit inversive congruential pseudorandom numbers with power of two modulus, II, preprint.

[57] J. Eichenauer-Herrmann and F. Emmerich, Compound inversive congruential pseudorandom numbers: an average-case analysis, *Math. Comp.*, to appear.

[58] J. Eichenauer-Herrmann and H. Grothe, A remark on long-range correlations in multiplicative congruential pseudo random number generators, *Numer. Math.* **56**, 609–611 (1989).

[59] J. Eichenauer-Herrmann and H. Grothe, A new inversive congruential pseudorandom number generator with power of two modulus, *ACM Trans. Modeling and Computer Simulation* **2**, 1-11 (1992).

[60] J. Eichenauer-Herrmann, H. Grothe, and J. Lehn, On the period length of pseudorandom vector sequences generated by matrix generators, *Math. Comp.* **52**, 145–148 (1989).

[61] J. Eichenauer-Herrmann, H. Grothe, H. Niederreiter, and A. Topuzoğlu, On the lattice structure of a nonlinear generator with modulus 2^{α}, *J. Comp. Appl. Math.* **31**, 81–85 (1990).

[62] J. Eichenauer-Herrmann and K. Ickstadt, Explicit inversive congruential pseudorandom numbers with power of two modulus, *Math. Comp.* **62**, 787–797 (1994).

[63] J. Eichenauer-Herrmann and H. Niederreiter, On the discrepancy of quadratic congruential pseudorandom numbers, *J. Comp. Appl. Math.* **34**, 243–249 (1991).

[64] J. Eichenauer-Herrmann and H. Niederreiter, Lower bounds for the discrepancy of inversive congruential pseudorandom numbers with power of two modulus, *Math. Comp.* **58**, 775–779 (1992).

[65] J. Eichenauer-Herrmann and H. Niederreiter, Kloosterman-type sums and the discrepancy of nonoverlapping pairs of inversive congruential pseudorandom numbers, *Acta Arith.* **65**, 185–194 (1993).

[66] J. Eichenauer-Herrmann and H. Niederreiter, On the statistical independence of nonlinear congruential pseudorandom numbers, *ACM Trans. Modeling and Computer Simulation* **4**, 89–95 (1994).

[67] J. Eichenauer-Herrmann and H. Niederreiter, Bounds for exponential sums and their applications to pseudorandom numbers, *Acta Arith.* **67**, 269–281 (1994).

[68] J. Eichenauer-Herrmann and H. Niederreiter, Digital inversive pseudorandom numbers, *ACM Trans. Modeling and Computer Simulation*, to appear.

[69] J. Eichenauer-Herrmann and H. Niederreiter, An improved upper bound for the discrepancy of quadratic congruential pseudorandom numbers, *Acta Arith.* **69**, 193–198 (1995).

[70] J. Eichenauer-Herrmann and A. Topuzoğlu, On the period length of congruential pseudorandom number sequences generated by inversions, *J. Comp. Appl. Math.* **31**, 87–96 (1990).

[71] F. Emmerich, Pseudorandom vector generation by the compound inversive method, *Math. Comp.*, to appear.

[72] A.M. Ferrenberg, D.P. Landau, and Y.J. Wong, Monte Carlo simulations: hidden errors from "good" random number generators, *Physical Review Letters* **69**, 3382–3384 (1992).

[73] G.S. Fishman, Multiplicative congruential random number generators with modulus 2^β: an exhaustive analysis for $\beta = 32$ and a partial analysis for $\beta = 48$, *Math. Comp.* **54**, 331–344 (1990).

[74] M. Flahive and H. Niederreiter, On inversive congruential generators for pseudorandom numbers, *Finite Fields, Coding Theory, and Advances in Communications and Computing* (G.L. Mullen and P.J.-S. Shiue, eds.), pp. 75–80, Dekker, New York, 1993.

[75] M. Fushimi, An equivalence relation between Tausworthe and GFSR sequences and applications, *Applied Math. Lett.* **2**, 135–137 (1989).

[76] M. Fushimi, Random number generation with the recursion $X_t = X_{t-3p} \oplus X_{t-3q}$, *J. Comp. Appl. Math.* **31**, 105–118 (1990).

[77] H. Grothe, Matrixgeneratoren zur Erzeugung gleichverteilter Zufallsvektoren, *Zufallszahlen und Simulationen* (L. Afflerbach and J. Lehn, eds.), pp. 29–34, Teubner, Stuttgart, 1986.

[78] H. Grothe, Matrix generators for pseudo-random vector generation, *Statist. Papers* **28**, 233–238 (1987).

[79] H. Grothe, Matrixgeneratoren zur Erzeugung gleichverteilter Pseudozufallsvektoren, Dissertation, Technische Hochschule Darmstadt, 1988.

[80] A. Grube, Mehrfach rekursiv-erzeugte Pseudo-Zufallszahlen, *Z. angew. Math. Mech.* **53**, T223–T225 (1973).

[81] P. Hellekalek, General discrepancy estimates: the Walsh function system, *Acta Arith.* **67**, 209–218 (1994).

[82] K. Huber, On the period length of generalized inversive pseudorandom number generators, *Appl. Algebra Engrg. Comm. Comput.* **5**, 255–260 (1994).

[83] F. James, A review of pseudorandom number generators, *Computer Physics Comm.* **60**, 329–344 (1990).

[84] T. Kato, L.-M. Wu, and N. Yanagihara, On a nonlinear congruential pseudorandom number generator, preprint, 1994.

[85] T. Kato, L.-M. Wu, and N. Yanagihara, The serial test for a nonlinear pseudorandom number generator, preprint, 1994.

[86] J. Kiefer, On large deviations of the empiric d.f. of vector chance variables and a law of the iterated logarithm, *Pacific J. Math.* **11**, 649–660 (1961).

[87] D.E. Knuth, *The Art of Computer Programming*, Vol. 2: *Seminumerical Algorithms*, 2nd ed., Addison-Wesley, Reading, MA, 1981.

[88] P. L'Ecuyer, Efficient and portable combined random number generators, *Comm. ACM* **31**, 742–749, 774 (1988).

[89] P. L'Ecuyer, Random numbers for simulation, *Comm. ACM* **33**, no. 10, 85–97 (1990).

[90] P. L'Ecuyer, Uniform random number generation, *Ann. Oper. Res.* **53**, 77–120 (1994).

[91] P. L'Ecuyer, Bad lattice structures for vectors of non-successive values produced by some linear recurrences, *ORSA J. Computing*, to appear.

[92] P. L'Ecuyer, Combined multiple recursive random number generators, *Oper. Res.*, to appear.

[93] P. L'Ecuyer, Maximally equidistributed combined Tausworthe generators, *Math. Comp.*, to appear.

[94] P. L'Ecuyer, F. Blouin, and R. Couture, A search for good multiple recursive random number generators, *ACM Trans. Modeling and Computer Simulation* **3**, 87–98 (1993).

[95] P. L'Ecuyer and S. Tezuka, Structural properties for two classes of combined random number generators, *Math. Comp.* **57**, 735–746 (1991).

[96] J. Lehn, Pseudorandom number generators, *Operations Research 91* (P. Gritzmann et al., eds.), pp. 9–13, Physica-Verlag, Heidelberg, 1992.

[97] J. Lehn, Special methods for pseudorandom number generation, *Bootstrapping and Related Techniques* (Trier, 1990), Lecture Notes in Economics and Math. Systems, Vol. 376, pp. 13–19, Springer, Berlin, 1992.

[98] M.B. Levin, On the choice of parameters in generators of pseudorandom numbers (Russian), *Dokl. Akad. Nauk SSSR* **307**, 529–534 (1989).

[99] Yu.L. Levitan and I.M. Sobol', On a pseudo-random number generator for personal computers (Russian), *Matem. Modelirovanie* **2**, no. 8, 119–126 (1990).

[100] R. Lidl and H. Niederreiter, *Introduction to Finite Fields and Their Applications*, revised ed., Cambridge University Press, Cambridge, 1994.

[101] G. Marsaglia, Random numbers fall mainly in the planes, *Proc. Nat. Acad. Sci. U.S.A.* **61**, 25–28 (1968).

[102] G. Marsaglia, The structure of linear congruential sequences, *Applications of Number Theory to Numerical Analysis* (S.K. Zaremba, ed.), pp. 249–285, Academic Press, New York, 1972.

[103] G. Marsaglia, The mathematics of random number generators, *The Unreasonable Effectiveness of Number Theory* (S.A. Burr, ed.), Proc. Symp. Applied Math., Vol. 46, pp. 73–90, American Math. Society, Providence, RI, 1992.

[104] G. Marsaglia and A. Zaman, A new class of random number generators, *Ann. Appl. Probab.* **1**, 462–480 (1991).

[105] M. Matsumoto and Y. Kurita, Twisted GFSR generators, *ACM Trans. Modeling and Computer Simulation* **2**, 179–194 (1992).

[106] G.L. Mullen, A note on a finite field pseudorandom vector generator of Niki, *Math. Japon.* **38**, 59–60 (1993).

[107] H. Niederreiter, The serial test for pseudo-random numbers generated by the linear congruential method, *Numer. Math.* **46**, 51–68 (1985).

[108] H. Niederreiter, A pseudorandom vector generator based on finite field arithmetic, *Math. Japon.* **31**, 759–774 (1986).

[109] H. Niederreiter, Point sets and sequences with small discrepancy, *Monatsh. Math.* **104**, 273–337 (1987).

[110] H. Niederreiter, Remarks on nonlinear congruential pseudorandom numbers, *Metrika* **35**, 321–328 (1988).

[111] H. Niederreiter, Statistical independence of nonlinear congruential pseudorandom numbers, *Monatsh. Math.* **106**, 149–159 (1988).

[112] H. Niederreiter, The serial test for congruential pseudorandom numbers generated by inversions, *Math. Comp.* **52**, 135–144 (1989).

[113] H. Niederreiter, Pseudorandom numbers generated from shift register sequences, *Number-Theoretic Analysis* (E. Hlawka and R.F. Tichy, eds.), Lecture Notes in Math., Vol. 1452, pp. 165–177, Springer, Berlin, 1990.

[114] H. Niederreiter, Statistical independence properties of pseudorandom vectors produced by matrix generators, *J. Comp. Appl. Math.* **31**, 139–151 (1990).

[115] H. Niederreiter, Lower bounds for the discrepancy of inversive congruential pseudorandom numbers, *Math. Comp.* **55**, 277-287 (1990).

[116] H. Niederreiter, Finite fields and their applications, *Contributions to General Algebra 7* (Vienna, 1990), pp. 251-264, Teubner, Stuttgart, 1991.

[117] H. Niederreiter, Recent trends in random number and random vector generation, *Ann. Oper. Res.* **31**, 323-345 (1991).

[118] H. Niederreiter, Nonlinear methods for pseudorandom number and vector generation, *Simulation and Optimization* (G. Pflug and U. Dieter, eds.), Lecture Notes in Economics and Math. Systems, Vol. 374, pp. 145-153, Springer, Berlin, 1992.

[119] H. Niederreiter, *Random Number Generation and Quasi-Monte Carlo Methods*, SIAM, Philadelphia, 1992.

[120] H. Niederreiter, New methods for pseudorandom number and pseudorandom vector generation, *Proc. 1992 Winter Simulation Conf.* (J.J. Swain et al., eds.), pp. 264-269, IEEE Press, Piscataway, NJ, 1992.

[121] H. Niederreiter, Finite fields, pseudorandom numbers, and quasirandom points, *Finite Fields, Coding Theory, and Advances in Communications and Computing* (G.L. Mullen and P.J.-S. Shiue, eds.), pp. 375-394, Dekker, New York, 1993.

[122] H. Niederreiter, Pseudorandom numbers and quasirandom points, *Z. angew. Math. Mech.* **73**, T648-T652 (1993).

[123] H. Niederreiter, Factorization of polynomials and some linear-algebra problems over finite fields, *Linear Algebra Appl.* **192**, 301-328 (1993).

[124] H. Niederreiter, Pseudorandom vector generation by the inversive method, *ACM Trans. Modeling and Computer Simulation* **4**, 191-212 (1994).

[125] H. Niederreiter, On a new class of pseudorandom numbers for simulation methods, *J. Comp. Appl. Math.*, to appear.

[126] H. Niederreiter, The multiple-recursive matrix method for pseudorandom number generation, *Finite Fields and Their Appl.* **1**, 3-30 (1995).

[127] H. Niederreiter, Pseudorandom vector generation by the multiple-recursive matrix method, *Math. Comp.* **64**, 279-294 (1995).

[128] N. Niki, Finite field arithmetics and multidimensional uniform pseudorandom numbers (Japanese), *Proc. Inst. Statist. Math.* **32**, 231-239 (1984).

[129] J.D. Parker, The period of the Fibonacci random number generator, *Discrete Appl. Math.* **20**, 145-164 (1988).

[130] O.E. Percus and J.K. Percus, Long range correlations in linear congruential generators, *J. Comp. Physics* **77**, 267-269 (1988).

[131] O.E. Percus and J.K. Percus, Intrinsic relations in the structure of linear congruential generators modulo 2^β, *Statist. Probab. Lett.* **15**, 381-383 (1992).

[132] Y.S. Sherif and R.G. Dear, Development of a new composite pseudo random number generator, *Microelectronics and Reliability* **30**, 545–553 (1990).

[133] I.E. Shparlinskii, A sequence of pseudorandom numbers (Russian), *Avtomat. i Telemekh.* **7**, 185–188 (1988).

[134] E.A.D.E. Tahmi, Contribution aux générateurs de vecteurs pseudo-aléatoires, Thèse, Univ. Sci. Techn. H. Boumedienne, Algiers, 1982.

[135] S. Tezuka, Walsh-spectral test for GFSR pseudorandom numbers, *Comm. ACM* **30**, 731–735 (1987).

[136] S. Tezuka, On the discrepancy of GFSR pseudorandom numbers, *J. Assoc. Comput. Mach.* **34**, 939–949 (1987).

[137] S. Tezuka, Lattice structure of pseudorandom sequences from shift register generators, *Proc. 1990 Winter Simulation Conf.* (R.E. Nance et al., eds.), pp. 266–269, IEEE Press, Piscataway, NJ, 1990.

[138] S. Tezuka, The k-dimensional distribution of combined GFSR sequences, *Math. Comp.* **62**, 809–817 (1994).

[139] S. Tezuka and M. Fushimi, Calculation of Fibonacci polynomials for GFSR sequences with low discrepancies, *Math. Comp.* **60**, 763–770 (1993).

[140] S. Tezuka and P. L'Ecuyer, Efficient and portable combined Tausworthe random number generators, *ACM Trans. Modeling and Computer Simulation* **1**, 99–112 (1991).

[141] S. Tezuka and P. L'Ecuyer, Analysis of add-with-carry and subtract-with-borrow generators, *Proc. 1992 Winter Simulation Conf.* (J.J. Swain et al., eds.), pp. 443–447, IEEE Press, Piscataway, NJ, 1992.

[142] S. Tezuka, P. L'Ecuyer, and R. Couture, On the lattice structure of the add-with-carry and subtract-with-borrow random number generators, *ACM Trans. Modeling and Computer Simulation* **3**, 315–331 (1993).

[143] D. Wang and A. Compagner, On the use of reducible polynomials as random number generators, *Math. Comp.* **60**, 363–374 (1993).

[144] B.A. Wichmann and I.D. Hill, An efficient and portable pseudo-random number generator, *Appl. Statist.* **31**, 188–190 (1982); Corrections, *ibid.* **33**, 123 (1984).

[145] B.A. Wichmann and I.D. Hill, Building a random-number generator, *Byte* **12**, no. 3, 127–128 (1987).

[146] W.W. Wood, Monte Carlo calculations for hard disks in the isothermal-isobaric ensemble, *J. Chemical Physics* **48**, 415–434 (1968).

[147] C.-K. Yuen, Testing random number generators by Walsh transform, *IEEE Trans. Comput.* **26**, 329–333 (1977).

Quasi-Monte Carlo Methods for Particle Transport Problems

Jerome Spanier
Mathematics Department
The Claremont Graduate School
Claremont, CA 91711

Abstract

Particle transport problems arise in such diverse application areas as the modeling of nuclear reactors and of semiconductor devices, and in the remote sensing of underground geologic features. Conventional Monte Carlo methods solve such problems by using pseudorandom numbers to make decisions at the microscopic level in order to draw conclusions about the macroscopic behavior of the system. Application of quasirandom (low discrepancy) sequences to such problems encounters certain difficulties that must be overcome if predictable gains over the use of pseudorandom Monte Carlo are to be realized. This paper outlines several ideas for achieving this and presents the results of "model" problem analyses and numerical tests of these ideas.

1 Introduction

Monte Carlo methods have provided effective means for solving a great many problems that seem intractable when approached with more traditional numerical methods. The key idea is to view each number to be approximated as the theoretical expected value of a random variable on a suitably defined probability measure space and then to approximate the expected value by sample means. When pseudorandom numbers (i.e., realizations of approximately uniformly and independently distributed random variables) are used to generate the sample values, one argues that the law of large numbers and the central limit theorem may be invoked to analyze the rate of convergence of each sample mean to the corresponding theoretical mean. In this way, one obtains convergence rates that are $O(N^{-\frac{1}{2}})$, where N = sample size, irrespective of the "dimensionality" of the problem under study. It is this dimension-independent convergence rate that endows the Monte Carlo method with its popularity and success for estimating high dimensional integrals and for solving the integral equations that characterize particle transport.

In the last two decades or so, a good deal of effort [1, 2, 3] has been directed to the use of so-called quasi-Monte Carlo methods. These are methods for solving such problems that rely on extremely regular distributions of "nodal" points and error analyses that are deterministic, not statistical. In emphasizing the evenness, rather than the "randomness" of the distribution of the underlying sample values, and in seeking rigorous upper bounds for the error, quasi-Monte Carlo methods re-introduce a (weak) dependence of the error on the "dimensionality" of the problem yet can achieve asymptotic rates of convergence more favorable than pseudorandom Monte Carlo. Thus, for the estimation of an s-dimensional definite integral, use of optimally placed nodal points results in an error bound of order at most $O(\log(N)^s/N)$ rather than $O(N^{-\frac{1}{2}})$ as N approaches infinity, even when the integrand function is not terribly smooth. In sufficiently hard problems (i.e., ones for which N must be very large to obtain reasonable error bounds), quasi-Monte Carlo methods are thus vastly to be preferred. Results such as these have stimulated interest in quasi-Monte Carlo methods in recent years. However, because most transport applications demand random walk simulations that are Markovian, quasirandom sequences, being highly correlated, must be used thoughtfully in the solution of such problems lest incorrect results obtain.

The particle transport problems that are the focus of attention in this paper are, in a very real sense, *infinite*-dimensional problems since they are characterized by integral equations whose Neumann series solutions are obtained by summing an infinite series of integrals of ever-increasing dimension. It is not *a priori* clear, then, whether quasi-Monte Carlo methods can be applied to such problems effectively. In this paper we will discuss how a careful use of quasirandom sequences can, in fact, produce more efficient solutions to such problems than is possible using pseudorandom sequences alone. Such quasi-Monte Carlo methods should then find wide use in applications involving the design and analysis of nuclear reactors for propulsion and electric power generation, in the modeling and analysis of semiconductor devices, and in the remote sensing of underground geologic features for oil well exploration, among other problems. This paper will restrict its attention to outlining the main theoretical results and the analysis of, and numerical experimentation with, certain key "model" transport problems that have proven useful in developing these ideas. Other papers will delve further into the theory and apply these new techniques to some of the (real - not "model") transport problems just mentioned.

2 The Basic Probability Model

In [4, 5] a useful correspondence is established between a probability model based on fairly general random walk processes and the analytic model that describes a class of Fredholm integral equations that arise in modeling the physical systems alluded to above. It is this correspondence that provides the fundamental tools for analyzing the requirements of a successful simulation of a transport problem, *whether by pseudorandom or by quasirandom methods*. That is, since our work with quasirandom Monte Carlo methods sets out initially to reproduce the *same* transport process as the one on which the pseudorandom Monte Carlo method is traditionally based, it is important to understand the connection between that transport process and the probability model that is designed to mimic it.

We begin, therefore, with the integral equation for the collision density

$$\Psi(P) = \int_{\Gamma} K(P, P')\Psi(P')dP' + S(P) \tag{1}$$

This equation, which finds use in the applications referred to above as well as in others, can be thought of in the following way: The quantity $\Psi(P')dP'$ is the total number of collisions caused (per unit time) by a steady-state distribution of particles in the volume dP' about the phase space point P'. The kernel $K(P, P')$ is a (not necessarily normalized) probability density for a particle, undergoing collision at P', to make its *next* collision at P. The integral over all of phase space Γ then represents the net particle density undergoing collision at P following a collision elsewhere in phase space. To this is added the density $S(P)$ of particles undergoing *initial* collision at P to represent the collision density $\Psi(P)$.

These quantities are subject to some natural restrictions as a result of their definition in terms of the physical model. For example, both the kernel $K(P, P')$ and the source function $S(P)$ must be nonnegative; also, $S(P)$ is usually normalized so that $\int_\Gamma S(P)dP = 1$. The collision density $\Psi(P)$ is related to the particle flux $\Phi(P)$ through the equation $\Psi(P) = \Sigma_t(P)\Phi(P)$, where $\Sigma_t(P)$ is the total macroscopic cross section at the phase space point P. The total cross section $\Sigma_t(P)$ may be regarded simply as a function whose inverse is the average distance traveled between collisions for particles originating at P and moving in an infinite medium of material with the same total cross section. Spatial heterogeneity can be accomodated in this model by varying the cross section as a function of position and by integrating along the flight path of the particle. The various collision processes that characterize each particular transport problem are specified through the definition of the kernel K.

Our objective is usually not to solve the integral Equation (1) for $\Psi(P)$ for *all* P but to obtain an estimate of one or more integrals

$$I = \int_\Gamma g(P)\Psi(P)dP \qquad (2)$$

The integral (2) may be thought of as the response of a physical "detector" with $g(P)$ (also known as the response function) describing the reaction of the detector to the collision density $\Psi(P)$. The response function is regarded as known and subject to the reasonable restriction that it, too, be nonnegative. Often it is a ratio $\Sigma_r(P)/\Sigma_t(P)$ of a reaction cross section to the total cross section at the point P; thus, it describes the probability that the reaction typified by the subscript r occurs upon collision at P.

The way in which the conventional Monte Carlo method estimates the integral (2) is to replace it with the expectation of a random variable over an appropriate probability space. Specifically, (2) is replaced by

$$I = \int_\Omega \xi d\mu \qquad (3)$$

where Ω is the sample space of all random walk histories in phase space Γ, μ is a probability measure on Ω, and ξ is an unbiased estimator of I with respect to μ. The measure μ is defined relative to a general random walk process in a fashion that we now sketch; further details may be found in [4].

A general random walk process $\{f_n, p_n\}$ is a sequence $f_n(P_1, ..., P_n)$ of continuous probability density functions on Γ^n together with a sequence $p_n(P_1, ..., P_n)$ of functions on Γ^n with

$$F_n(P_1, ..., P_k, \infty, ..., \infty) = F_k(P_1, ..., P_k) \qquad (k < n, n \geq 1)$$

where

$$F_n(P_1, ..., P_n) = \int_{-\infty}^{P_n} \cdots \int_{-\infty}^{P_1} f_n(P_1', ..., P_n')dP_1' \cdots dP_n'$$

is the distribution function of f_n and $0 \leq p_n(P_1, ..., P_n) \leq 1$, $n = 1, 2, ...$. Intuitively, $f_n(P_1, ..., P_n)$ represents the probability density of a random walk chain involving the ordered sequence of collision points $(P_1, ..., P_n)$ and $p_n(P_1, ..., P_n)$ is the probability of terminating the random walk at the state P_n following collisions at $P_1, P_2, ..., P_n$. Then $q_n = 1 - p_n$ is the probability of continuing the random walk beyond the state P_n.

The space Ω of all random walk histories has an obvious decomposition

$$\Omega = (\cup_{k=1}^{\infty}\Lambda_k) \cup \Lambda_{\infty}$$

where $\Lambda_k = \{\omega \in \Omega| \ \omega \text{ terminates after } k \text{ collisions}\}$ and Λ_{∞} consists of chains that do not terminate in a finite number of steps.

The probability measure μ is a real-valued function defined on certain subsets of Ω, to be interpreted as the probability of occurrence of that subset. It is derived from a rather technical theorem of Tulcea (see [4, 5]) from the random walk process $\{f_n, p_n\}$. For our purposes, however, μ will always be defined by an integral, the most important of which are

$$\mu(\Lambda_k) = \int_{\Gamma} \cdots \int_{\Gamma} f_k(P_1, ..., P_k) \prod_{i=1}^{k-1} q_i(P_1, ..., P_i)p_k(P_1, ..., P_k)dP_1 \cdots dP_k \quad (4)$$

The integral (4) is the expected value over Ω of the random variable which is the characteristic function of the set Λ_k. More generally, for any random variable ξ on Ω, the expected value of ξ taken over Λ_k is

$$\int_{\Lambda_k} \xi d\mu = \int_{\Gamma} \cdots \int_{\Gamma} \xi(P_1, ..., P_k)f_k(P_1, ..., P_k) \prod_{i=1}^{k-1} q_i(P_1, ..., P_i)p_k(P_1, ..., P_k)dP_1 \cdots dP_k$$

$$(5)$$

and the expected value of ξ over all of Ω is the sum over k of the integrals (5).

The condition $\mu(\Lambda_{\infty}) = 0$ is necessary to the construction of a practical computational model and in [5] and Chapter 3 of [4], there is a discussion of some natural conditions on the transport process that guarantee this. We shall merely assume such conditions henceforth and thus freely use the corresponding identity

$$1 = \sum_{k=1}^{\infty} \mu(\Lambda_k)$$

or its analytic equivalent that assures convergence of the Neumann series to the (unique) solution of the transport equation upon which the definition of μ is based. Thus if

$$\Psi(P) = \int_{\Gamma} K(P, P')\Psi(P')dP' + S(P) \quad (6)$$

is the integral equation for the collision density, the <u>analog random walk process</u> is defined by the choices

$$f_1(P) = S(P)$$

$$f_n(P_1, ..., P_n) = \prod_{l=2}^{n} \left[\frac{K(P_l, P_{l-1})}{\int_{\Gamma} K(P_l, P_{l-1})dP_l} \right] S(P_1), \qquad n \geq 2 \quad (7)$$

and

$$p_n(P_1, ..., P_n) = p(P_n) = \frac{\sum_a(P_n)}{\sum_t(P_n)} \tag{8}$$

where $\sum_a(P_n)$ is the macroscopic absorption cross section at P_n. In a non-multiplying medium one has

$$\int_\Gamma K(Q, P)dQ = \frac{\sum_s(P)}{\sum_t(P)} = 1 - p(P) = q(P) \leq 1 \tag{9}$$

where $\sum_s(P)$ is the macroscopic scattering cross section at P and $\sum_t(P) = \sum_a(P) + \sum_s(P)$. The kernel $K(Q, P)$ may be interpreted as the density of particles entering collision at P that enter their next collision at Q. Explicit formulas for the kernel K for conventional transport processes may be found in [4].

All of the foregoing discussion has been aimed at creating a mathematical model that can be used to study transport processes; i.e., processes described by the transport Equation (1). The model is a probability model (Ω, F, μ) defined in terms of the source S and kernel K of Equation (1). The probability measure μ on Ω provides a faithful recreation of the particle histories whose phase space distribution is described by the solution Ψ of Equation (1). It is for this reason that μ is called an analog measure on Ω. We have seen that the crucial feature by which the fidelity of the model (Ω, F, μ) is judged is the linkage to the transport Equation (1), through the discrete probability distribution $\{p_k\} = \{\mu(\Lambda_k)\}$ defined in Equation (4), and the definitions (7),(8). In other words, any attempt to produce particle histories whose phase space distribution is described by (1) must result in sets Λ_k whose analog measures $\mu(\Lambda_k) = p_k$ are defined by (4).

Next we outline how to estimate integrals such as (2) through the averaging of certain functions associated with the particle histories. The key to that is the use of the relation (5). For with (5) and the decomposition

$$\Omega = (\cup_{k=1}^\infty \Lambda_k) \cup \Lambda_\infty$$

of Ω into sets of histories Λ_k that terminate after k collisions, plus the set Λ_∞ of nonterminating histories, we can write

$$E[\xi] = \int_\Omega \xi d\mu = \sum_{k=1}^\infty \int_{\Lambda_k} \xi d\mu + \int_{\Lambda_\infty} \xi d\mu$$

We then place assumptions on the kernel K that assure $\mu(\Lambda_\infty) = 0$ (see [4] and [5]) and find

$$E[\xi] = \sum_{k=1}^\infty \int_{\Lambda_k} \xi d\mu \tag{10}$$

and the infinite series of integrals will also converge (and, in fact, produce the same result as summing the Neumann series for the integral Equation (1)) based on these same assumptions.

Of course, there are many estimators ξ whose expectation $E[\xi]$ with respect to the analog measure μ all produce the integral I of Equation (2). Each estimator ξ assigns a real number $\xi(\omega)$ to every particle history ω, whose value should be thought of as the weighted contribution from that history ω to I. When the analog measure μ is imposed, certain "natural" estimators ξ suggest themselves for each function g occurring in (2). For example, if

$$g(P) = \frac{\Sigma_a(P)}{\Sigma_t(P)} \chi_V(P) \tag{11}$$

where $\chi_V(x)$ is the characteristic function of the phase space volume V, then

$$\begin{aligned} I &= \int_\Gamma g(P)\Psi(P)dP \\ &= \int_V \frac{\Sigma_a(P)}{\Sigma_t(P)} \Psi(P)dP \end{aligned} \tag{12}$$

is the absorption rate in V. It is intuitively plausible to define an estimator ξ by

$$\xi(\omega) = \begin{cases} 1 \text{ if } \omega \text{ terminates in absorption in } V \\ 0 \text{ if } \omega \text{ terminates in any other way} \end{cases} \tag{13}$$

and, in fact, this is an appropriate way to estimate the integral (12), as is shown in [4]. The estimator defined in (13) is called the terminal estimator since it provides information only on the final collision of each history.

In similar fashion, we may define the collision estimator of the integral (2) which provides information on every collision within the phase volume V, not just on the final one.

There are many other useful estimators; the reader is referred to [4] or [7] for additional information.

3 The Asymptotic Frequency Model

The quasi-Monte Carlo methods that we wish to discuss replaces the probability model of the previous section with a model based on the notion of asymptotic frequency. The latter places central emphasis on the construction of sequences $x_1, x_2, \ldots$ that assure the convergence

$$\lim_{N \to \infty} \frac{1}{N} \sum_{i=1}^N f(x_i) = \int_{I^s} f(x)dx$$

for a reasonable class of functions $f(x)$. Here we have taken the domain space for our integration to be the s-dimensional unit hypercube I^s. Such sequences $x_1, x_2, \ldots$ are said to be uniformly distributed over I^s. Analogously, we will be concerned with the construction of sequences $\omega_1, \omega_2, \ldots$ in the space Ω of all particle random walk histories that guarantee the convergence

$$\lim_{N \to \infty} \frac{1}{N} \sum_{i=1}^N \xi(\omega_i) = \int_\Omega \xi d\mu \tag{14}$$

provided that ξ satisfies mild smoothness restrictions. Sequences $\omega_1, \omega_2, \ldots$ that achieve this will be called μ-uniformly distributed in Ω .

The basic correspondences needed to construct such an asymptotic frequency-based model were established in the doctoral dissertation of Paul Chelson [6]; see also [3]. We will review here just the basic outline of this theory.

The theory hinges on the definition of a mapping $G : \Omega \to I^s$ that is cumbersome to describe in detail, but easy to explain functionally. The construction of each random walk history $\omega \epsilon \Omega$ is accomplished by using each component of a (pseudorandomly or quasirandomly distributed) vector $(r_1, r_2, ..., r_s) \epsilon I^s$ to make a single decision about the random walk ω, beginning with the decisions necessary to construct a source location, energy and direction. Succeeding components of the vector $(r_1, r_2, ..., r_s)$ are then used to make the decisions necessary to construct the location, energy and direction of additional collisions that ω makes, until it terminates through absorption or escape from the physical domain of interest.[1] These decisions can be rigorously described in terms of conditional probability density functions constructed by normalizing the source density S and the transition kernel K of the integral Equation (1), and by including certain discrete event probabilities in the construction (such as the decision to continue or terminate a random walk at each collision point of phase space Γ). Full details may be found in [6] but are omitted here.

Once the map G is constructed, it provides the vehicle for analyzing the abstract space Ω , endowed with the probability measure μ, in terms of more familiar distribution functions on the real line R by means of the relationship

$$\int_{\Omega} \xi d\mu = \int_R t dF(t) \tag{15}$$

where

$$F(t) \equiv \mu\{\omega : \xi(\omega) \le t\} \tag{16}$$

(see [8]). Then, given a sequence $(\omega_1, \omega_2, ...)$ in Ω, we define a discrete probability measure μ_1^N on Ω by

$$\mu_1^N(\Lambda) \equiv \frac{1}{N} \sum_{i=1}^{N} \chi_\Lambda(\omega_i) \tag{17}$$

where χ_Λ is the characteristic function of the measurable subset Λ of Ω. Then with local μ-discrepancy in Ω defined by

$$L_N^\mu(\Lambda, \{\omega_i\}) \equiv \mu_1^N(\Lambda) - \mu(\Lambda), \tag{18}$$

it is not difficult to show that

$$L_N^\mu(\Lambda, \{\omega_i\}) = L_N(G(\Lambda), \{G(\omega_i)\}) \tag{19}$$

where L_N is ordinary local discrepancy [1] in the space I^s.

The analysis of the error

$$\delta_N(\xi) = \frac{1}{N} \sum_{i=1}^{N} \xi(\omega_i) - \int_{\Omega} \xi d\mu \tag{20}$$

is then accomplished by arguments that are analogous to those that have been successfully used in deriving the classical Koksma-Hlawka inequality for finite dimensional definite integrals [1]. We again merely sketch this derivation, details of which can be found in [6].

---

[1] It is sometimes of value to introduce *artificial* terminations of random walks ω through processes such as Russian roulette or other weight-dependent games (see [4, 7]). We shall not dwell on these possibilities here, although they can be accomodated within our model.

Our task, then, is to derive an upper bound for the absolute value of the error (20). The classical arguments make essential use of various integration by parts formulas. The argument developed by Chelson [6] also uses a version of Abel's summation formula [9] and integration by parts for Riemann-Stieltjes integrals and makes heavy use of the connections (15) and (19) between the abstract space Ω and the hypercube I^s. In skeletal form, the argument proceeds as follows:

$$\left|\delta_N(\xi)\right| = \left|\int_\Omega \xi dL_N^\mu\right| = \left|\int_R t dL_N^\mu\{\xi^{-1}(-\infty, t]\}\right| = \left|\int_R L_N^\mu(\Lambda_t) dt\right| = \left|\int_R L_N(G(\Lambda_t)) dt\right|$$
$$\leq \int_R |L_N(G(\Lambda_t))| \, dt \leq C J_N \leq \hat{C}(D_N)^{\frac{1}{s}}.$$

$$(21)$$

In (21), the second equality follows from (15) while the third equality results from the application of a fairly general integration by parts formula [9], and the fourth from (19). The quantity J_N is the induced isotropic discrepancy [1] resulting from the imbedding of Ω as a finitely convex subset of a sufficiently high dimensional Euclidean space, and the final inequality results from the application of the known result [1]

$$J_N \leq 4s D_N^{\frac{1}{s}}$$

relating the isotropic and the ordinary (extreme) discrepancy in s-dimensional space. The constants C and $\hat{C}$ in Equation (21) are independent of N but may depend on s. We observe that the string of relationships in (21) is exact until the point at which the absolute value is taken inside the integral, after which a good deal of imprecision is introduced as a result of the losses in sharpness that may occur from that point forward in the argument. This means that we expect the resulting Koksma-Hlawka type bound

$$|\delta_N(\xi)| \leq \hat{C}(D_N)^{\frac{1}{s}} \qquad (22)$$

to lack sharpness in many cases, a defect which is also present in the use of the classical Koksma-Hlawka inequalities for finite dimensional integrals of known functions.

If use of the isotropic discrepancy can be avoided in this derivation (which the author believes should be possible), then the inequality (22) can be sharpened with respect to the exponent $1/s$. However, because the extreme discrepancy D_N itself goes to zero like $(\log N)^s/N$, even a sharpened form of the inequality (22) will exhibit the reduced effectiveness that may be expected when using quasirandom sequences in the simulation of transport processes. On the other hand, the inequality (22), even in its present form, is nonetheless sufficient to guarantee convergence, as $N \to \infty$, of the error $\delta_N(\xi)$ to 0 provided that the sequence $(r_1, r_2, ...)$ is uniformly distributed in a sufficiently high dimensional unit cube I^s. According to the Equality (19), this is sufficient to ensure that the sequence $(\omega_1, \omega_2, ...)$ of random walk histories that is constructed using the sequence $(r_1, r_2, ...)$ and the mapping G, is μ-uniformly distributed in Ω, and, in fact, the discrepancies of the two sequences are identical. Thus we seek to use the lowest discrepancy sequence $(r_1, r_2, ...)$ possible so that the error bound (22) will be as small as possible.

4 Efficient Quasi-Random Methods

The discussion of the previous section would lead us to conclude that, in order to implement quasi-Monte Carlo methods for the solution of particle transport problems we must use quasirandom vector sequences with arbitrarily many "independent" components in order correctly to simulate μ-uniformly distributed particle random walk histories $\omega_1, \omega_2, \ldots$, where μ is the probability measure on Ω that is faithful to the physical transport model. Then the use of estimators ξ that are unbiased with respect to μ will clearly provide estimates of the integral (3) with errors that can be analyzed according to the generalized Koksma-Hlawka inequality (22). Initial verification of this theory was provided in [6]. Alternatively, one can use quasirandom vector sequences to construct ν-uniformly distributed random walks $\omega_1, \omega_2, \ldots$ where the Radon-Nikodym derivative $d\mu/d\nu$ exists, and then apply weighted estimators $\xi d\mu/d\nu$. These estimators will then be unbiased with respect to the ν measure and, once again, Chelson [6] showed that a generalized Koksma-Hlawka inequality applies to the analysis of the quasirandom error. This, then, provides what might be called *quasi-importance sampling* estimators; the constant $\hat{C}$ involves the variation in the sense of Hardy and Krause of the function $\xi d\mu/d\nu$, which is reduced by a choice of the probability measure ν that stresses random walk histories that contain more "information" than do analog histories generated by using the probability measure μ. In his Ph.D. dissertation [10], Earl Maize extended the same sort of quasi-Monte Carlo analysis to include weighted uniform sampling estimators as introduced by Powell and Swann [11] and generalized by Spanier [12]. Indeed, it seems clear that there is a complete parallel between pseudorandom and quasirandom estimation for transport problems based on the correspondence we have sketched in Section 3.

While the theory outlined above for applying quasi-Monte Carlo methods to transport problems is effective in many cases, it suffers from some undesirable features. First, as presented this theory requires a large supply of (essentially) infinite-dimensional quasirandom vectors of low discrepancy. This is, in itself, not a fatal flaw since it is relatively easy to construct the components of such vectors as needed in the computation. For example, given a sufficiently large table of primes, there are fast and efficient routines for generating, one at a time, the components of a Halton vector [1] ($\phi_2(n)$, $\phi_3(n)$, $\phi_5(n)$, ...) and using it to make all of the decisions needed to construct the n th random walk history. A more telling disadvantage of this scheme arises as a result of the unfavorable convergence rates that may be anticipated in problems for which the Neumann series converges very slowly. Because the dimension s of the integral associated with the kth term in the Neumann series is essentially $6k$ [2], and because Koksma-Hlawka type bounds for the error associated with such a term will be $O((\log(N))^s/N)$, one can easily see that the pseudorandom rate of convergence $O(N^{-\frac{1}{2}})$ should give *better* average error results than the quasirandom method when N is moderate (say, in the range $10^6 - 10^7$) and k is, say, 5 or more. In fact, we will find confirmation of this general observation in Section 6 of this paper in which model problem results are presented.

One may now ask: Is it possible to do better with quasi-Monte Carlo methods? More specifically, is it possible to utilize low-discrepancy sequences to construct appropriate random walk histories so that improvements over pseudo-Monte Carlo convergence rates can be expected for most - or even for all - transport problems? We seek methods that are

    a. application-free; that is, applicable to the most general class of particle transport

---

[2]because, for steady-state transport problems, the physical phase space is six dimensional: three dimensions are used to describe location and three more to characterize the (scalar) energy and (unit) direction for travel.

problems possible, and

b. able to take advantage of the very large existing literature dealing with variance reduction for pseudo-Monte Carlo (which, in the context of the Koksma-Hlawka inequality for low-discrepancy methods, should translate to *variation* reduction methods).

For the most part, general transport problems have not been attacked using low-discrepancy sequences, no doubt in part because of the defects alluded to above for these infinite dimensional applications. There is, however, a small, and quite recent literature, in which low-discrepancy sequences have been applied to specific diffusion or transport problems [2, 13, 14, 15, 21]. Methods employed by these authors have, however, usually taken advantage of some special features of the problem being solved, and their generalization to a wider class of problems does not seem readily apparent.

The approach taken here is to combine the best features of pseudorandom and quasirandom Monte Carlo methods, creating "hybrid" methods that are designed to be superior to either pseudo-Monte Carlo or conventional quasi-Monte Carlo used alone. Two such methods will be described:

A. A "mixed" strategy in which initial decisions (collisions) are made using quasirandom vector sequences, while subsequent ones are made using a pseudorandom sequence.

Moskowitz [21] has suggested the reverse strategy for certain diffusion applications. In the transport problems we study, the possibility of absorption at each collision progressively reduces the effective "mass" or "weight" of the particles. This perhaps suggests that use of quasirandom sequences for the determination of the early steps, as our mixed strategy employs, will be more effective than if used for the later ones. However, the problems studied by Moskowitz involved no absorption: all steps are then of equal "importance". Moskowitz's scheme of using pseudorandomly-generated steps to obtain a good approximation, which is then improved by the use of a small number of quasirandom steps at the end, has been shown to be effective in such problems.

B. A "scrambled" quasirandom strategy in which the aim is pseudorandomly to reorder a quasirandom scalar or low dimensional vector sequence and use it, in place of a pseudorandom sequence, to make the decisions needed to construct the mapping G of Section 3.

A similar idea has been considered by several authors [16, 17, 18, 19, 20]. However, the goal of these authors has been to improve the discrepancy of the original sequence, whereas our goal is to break up in an effective manner the correlations that are present in the low discrepancy sequence in its original order, while maintaining as low a discrepancy as possible. This, then, permits the scrambled quasirandom sequence to be effective in constructing the (Markovian) random walk histories needed in the simulation. In other words, the histories $\omega_1, \omega_2, \ldots$ will then be $\mu-$uniformly distributed, and the error analysis outlined in Section 3 can be applied.

It is quite evident that the mixed strategy A leads to the construction of $\mu-$ uniformly distributed random walk sequences $\omega_1, \omega_2, \ldots \in \Omega$. This is intuitively clear since use of both the "pure" quasirandom sequence and of the pseudorandom sequence reproduce the critical probabilities $\{p_k = \mu(\Lambda_k)\}$ of Equation (4). This can also be established by induction on K, the number of initial collisions treated quasirandomly in the mixed strategy.

Our initial approach to the scrambled strategy B was to attempt to produce an approximately completely equidistributed scalar sequence - that is, a sequence $s_1, s_2, s_3, \ldots$ such that consecutive $k-$ tuples formed from this sequence are (approximately) uniformly distributed over the $k-$ dimensional hypercube for all integers k - by suitably scrambling the elements of a scalar quasirandom sequence, such as the van der Corput sequence. Such a scalar sequence could then be used directly to replace a pseudorandom sequence in generating the steps of each particle history. Provided that the discrepancy of such a sequence could then be controlled sufficiently, use of such a sequence should yield convergence rates faster than the $O(N^{-\frac{1}{2}})$ associated with pseudorandom Monte Carlo implementation. Efforts to achieve this goal met with limited success and were then shifted to the goal of making all decisions with a single, *low-dimensional* quasirandom vector sequence, suitably reordered. The objective here was to achieve convergence as close as possible to the asymptotic rate $O(\log N)^d/N$, where d is the number of components of the quasirandom vector sequence. This general strategy appears to be very promising, but we are still experimenting with it and will report here only the main ideas used and beginning numerical results.

We let d be the number of independent decisions needed for each collision in the transport problem. Roughly speaking, then, d is the dimension of the physical phase space, Γ, although our definition of d will depend on the *implementation* of the sampling techniques. For example, use of rejection sampling methods tend to increase d (and should therefore be avoided). Our idea is to take advantage of the fact that the dimensional requirements for each collision are identical (i.e., a $d-$ dimensional vector is needed for each collision of every history) to reduce the overall dimensional requirements of the quasirandom sequence needed. Consider the diagram below:

$$
\begin{array}{lcccc}
\text{"pure"} & & & & \\
\text{quasirandom} & \text{history 1} & \text{history 2} & \circ \quad \circ \quad \circ & \text{history } N \\
\text{implementation} & & & & \\
\text{source} & \underline{q}_1^{d(1)} & \underline{q}_2^{d(1)} & & \underline{q}_N^{d(1)} \\
\text{collision 1} & \underline{q}_1^{d(2)} & \underline{q}_2^{d(2)} & & \underline{q}_N^{d(2)} \\
\text{collision 2} & \underline{q}_1^{d(3)} & \underline{q}_2^{d(3)} & & \underline{q}_N^{d(3)} \\
& \circ & \circ & & \circ \\
& \circ & \circ & & \circ \\
& \circ & \circ & & \circ \\
\text{collision m} & \underline{q}_1^{d(m)} & \underline{q}_2^{d(m)} & & \underline{q}_N^{d(m)} \\
\circ & \circ & \circ & & \circ \\
\circ & \circ & \circ & & \circ \\
\circ & \circ & \circ & & \circ \\
\end{array}
$$

In this schematic depiction of a correctly implemented transport solution based on the use of a "pure" quasirandom vector sequence as described in Section 3, $q_i^{d(j)}$ denotes the d- dimensional quasirandom vector which is used to determine the jth collision state of history i, $1 \leq i \leq N$, $1 \leq j \leq m$. Moving down each column of this array, one uses a "fresh" $d-$dimensional vector sequence in each row. For example, if the Halton sequence were being used to construct the mapping G of Section 3, the first d prime bases would be used in row 1, the next d primes would be used in row 2, etc. As we have seen in Section 3, such a scheme would correctly reproduce (asymptotically) the probabilities $\{p_k = \mu(\Lambda_k)\}$ of Equation (4) because successive $k-$ tuples of the $d-$ dimensional vectors used in each column *are* uniformly distributed over a kd-dimensional euclidean hypercube.

We now make use of the fact that pseudorandom sequences are (approximately) uniformly *and* independently distributed to "stretch" the uniformity of quasirandom vector sequences. Thus, if a way can be found pseudorandomly to couple successive $k-$ tuples of vectors drawn from a *single* $d-$ dimensional quasirandom sequence, then these $k-$ tuples should also be (approximately) uniformly distributed over the kd - dimensional hypercube and could be used to generate Markovian particle histories.

We simplify our notation at this stage by letting $q_1, q_2, ...$ denote successive elements in a fixed $d-$ dimensional quasirandom sequence (dropping both the superscript and the underline); e.g., successive entries in any fixed row of the above matrix. We also define M to be an upper bound for the largest number of collisions required by any of the N particle histories. The integer M depends on N, but we suppress this dependence in our choice of notation for simplicity. Of course, if we were simply to use the elements of this sequence as shown in the next schematic

incorrect *quasirandom* *implementation*	*history* 1	*history* 2	∘ ∘ ∘	*history* N
source	q_1	q_2		q_N
collision 1	q_{N+1}	q_{N+2}		q_{2N}
collision 2	q_{2N+1}	q_{2N+2}		q_{3N}
	∘	∘		∘
	∘	∘		∘
	∘	∘		∘
collision M	q_{MN+1}	q_{MN+2}		$q_{(M+1)N}$

then the *systematic* coupling of successive $k-$tuples down each column prevents such $k-$ tuples from being uniform in the higher, $kd-$ dimensional space.

We now denote by $P_1, P_2, ..., P_M$, M independent random permutations of the integers $1, 2, ..., N$. Our strategy will be to apply the permutation P_j to the elements of the jth row of the schematic above. The resulting schematic

permuted *quasirandom* *implementation*	*history* 1	*history* 2	∘ ∘ ∘	*history* N
source	q_1	q_2		q_N
collision 1	$q_{N+P_1(1)}$	$q_{N+P_1(2)}$		$q_{N+P_1(N)}$
collision 2	$q_{2N+P_2(1)}$	$q_{2N+P_2(2)}$		$q_{2N+P_2(N)}$
	∘	∘		∘
	∘	∘		∘
	∘	∘		∘
collision M	$q_{MN+P_M(1)}$	$q_{MN+P_M(2)}$		$q_{MN+P_M(N)}$

illustrates the manner in which the different $d-$ tuples could be used to make decisions in constructing histories in the permuted quasirandom implementation just described.

The permuted, or scrambled, ordering used here has obviated the need to use a "fresh" $d-$ dimensional quasirandom sequence for each collision, provided that N is sufficiently large. Clearly, if N is small, the correlations inherent in the single $d-$ dimensional sequence will defeat this scheme. For example, in the extreme case of only a single history, $N = 1$, and *no* scrambling occurs as a result of the algorithm described above. The various collision states would then be chosen using the $d-$ dimensional sequence *in its natural order:* $q_1, q_2, ..., q_M, ...$ which will not produce $\mu-$uniformity, in general. However, when N is large and scrambling is used in the fashion we have described, the correlation affecting each individual history should be negligible.

To implement this method, one should choose a sufficiently large N and scramble the sequence $q_1, q_2, ...$ in blocks of N $d-$tuples at a time, making certain that the permutations P_i used in different blocks are independent of each other. In order to avoid significant storage problems, it is important to apply the permutation P_i to the ith batch of $d-$tuples *as needed,* <u>without</u> precomputing and storing all $(M + 1)N$ $d-$ tuples in advance. This can be accomplished by making appropriate use of linear congruential algorithms to generate the needed pseudorandom numbers. Details about implementation will have to await a future publication.

The scrambling technique introduced here appears to be essentially the same as that suggested by Lécot [14] and discussed as well by Morokoff and Caflisch [2] and by Moskowitz [21]. These authors describe the idea in more physical terms. As well, the problems treated by these authors permit no absorption at each collision, with the result that all random walks make the same number of collisions. We have attempted to focus our attention on generic methods that can be used *universally* in transport problem applications. In so doing, we have found it more convenient to express the idea purely in terms of alterations in the basic quasirandom sequence used to generate the Markov chains, rather than through a redefinition of the physical chains themselves. This difference in point of view also makes it clear that the integer N and the number of histories need not be related, and also enables the treatment of problems in which the individual random walk histories suffer *differing* numbers of collisions, as is the case in typical transport problems. In this way, our formulation comes closer to the goal we had set of generating techniques that are universally applicable to transport problems.

We report next on a series of experiments that were run in which these ideas for combining pseudorandom and quasirandom sequences were tested. In Section 6 of this paper we provide preliminary numerical results concerning Strategy B; we hope to treat it more fully in a separate paper.

5   Model Problem Analysis for Transport Problems

Most authors, when comparing pseudorandom and quasirandom Monte Carlo methods, have illustrated their results with tests performed on multidimensional integrals. Our belief is that while much useful information can be gained from such tests, it is very much more informative to use *model transport problems* when the objective is to solve this class of *infinite*-dimensional problems. For one thing, the choice of multidimensional integrand function is always open to criticism. A perhaps more important reason to avoid drawing conclusions about transport problems from such tests is that the class of transport problems under study here, while broad enough to capture most of the important (linear) transport applications, is quite special *viewed as a problem in integration in an infinite-dimensional space.* This is so because our solution method is necessarily restricted to problems for which the Neumann series for Equation (1) converges, and the

Neumann series constitutes a very special series of multidimensional integrals. In fact, the success of Monte Carlo methods in solving such problems can be attributed, in part, to the special nature of this series and to the construction of estimating methods that take advantage of this fact. It is, thus, important to study sample transport problems whose solutions can be found analytically or through some other highly accurate alternative to Monte Carlo but which capture most of the essential features of more realistic transport problems. In this section we shall describe several such model problems that we have used extensively to test the hybrid methods described earlier. Each test problem has been selected to focus attention on one or more characterisitics of more complex problems so that these characteristics can be studied in relative isolation from each other.

5.1   Model Problem 1

Our first model problem is deceptively simple, yet it is very useful in testing many ideas. Briefly, imagine a particle moving through a slab of unit thickness and traveling distances uniformly distributed on $[0,1]$. Particles are only permitted to travel in one direction and the problem is to estimate the expected number of steps required to escape the slab. It is not difficult to see that the quantities $p_k = \mu(\Lambda_k)$ of Equation (4) are exactly given by

$$p_1 = 0, \qquad p_k = \frac{1}{k(k-2)!}, \qquad k \geq 2$$

which leads to the conclusion that the expected number of steps to escape is the transcendental number e . A crucial test, therefore, of any scheme for generating particle random walk histories (whether pseudorandomly or quasirandomly) for this problem is to determine whether the quantities p_k are accurately reproduced (approximately, of course) by the simulation.

In Table 1 we present a listing of the exact and estimated probabilities $p_k, \hat{p}_k$ when pseudorandom numbers were employed to generate step sizes. A total of 1,000,000 random walk histories were processed in this simulation to produce the approximate probabilities of the table. Notice the excellent agreement between the Monte Carlo simulation and the theory. If, for example, we were to attempt to substitute the van der Corput sequence directly for the pseudorandom sequence we would discover that we are not able faithfully to simulate the physical process. The difficulty can be traced to an inability, in that case, to reproduce the probabilities p_k adequately. Thus, because the numbers in the van der Corput sequence alternate between quantities smaller than $1/2$ and those larger than $1/2$, *every* random walk history escapes the unit slab of material in *at most* three steps, so $\hat{p}_k = 0$ for $k > 3$. Thus, the van der Corput sequence, used in its original order, is not capable of producing random walk histories in which the fundamental quantities p_k are correctly simulated and so, cannot be used to solve this problem without some sort of modification, such as the scrambling strategy of Section 4.

It is convenient to introduce the notion of the "effective" dimension of a transport problem. By this we shall mean the product of the average number of collisions per history and the dimension d of the phase space. Clearly, the effective dimension depends on the methods used to implement the simulation. For example, use of importance sampling may well alter the the distribution of the probabilities $\mu(\Lambda_k)$ of Equation (4), even though the *expected weight density* of collisions is left unchanged. Nevertheless, the effective dimension does provide a rough gauge of the dimensional requirements

Number of Steps, k	Exact Probability, p_k	Estimate of Probability, $\hat{p}_k$	Error
1	0.00000E+00	0.00000E+00	0.00000E+00
2	5.00000E-01	4.98583E-01	1.41701E-03
3	3.33333E-01	3.34535E-01	1.20166E-03
4	1.25000E-01	1.24834E-01	1.65999E-04
5	3.33333E-02	3.35700E-02	2.36664E-04
6	6.94444E-03	7.11600E-03	1.71556E-04
7	1.19048E-03	1.18300E-03	7.47619E-06
8	1.73611E-04	1.54000E-04	1.96111E-05

Table 1: Exact and Estimated Probabilities $p_k, \hat{p}_k$ for Model Problem 1.

of transport problems with convergent Neumann series. In Model Problem 1, when implemented in a straightforward analogue fashion, this effective dimension is e because $d = 1$ and e is the expected value of the number of steps to escape.

5.2 Model Problem 2

A second model problem, Model Problem 2, is equivalent to a matrix equation and is especially useful for studying the effect of the rate of convergence of the Neumann series on our hybrid methods. It results from examining transport in an infinite, homogeneous material in which particles can exist in only a finite number, E, of energy states. Particle collisions then can result either in absorption or in transition from the current energy state to any other. When these restrictions are applied to the transport Equation (1) it is not difficult to see ([4]) that the result is the equation

$$\underline{x} = H\underline{x} + \underline{a} \tag{23}$$

in which $\underline{x}$ and $\underline{a}$ are $E-$ dimensional vectors and H is an $E \times E$ matrix. The integral I of Equation (2) is then replaced by the inner product

$$I = \langle \underline{g}, \underline{x} \rangle \tag{24}$$

where $\underline{g}$ is a known vector that plays the role of a detector "function" in this discrete version of the transport problem. By varying the amount of absorption per collision, it is possible to create Neumann series whose convergence characteristics can be controlled. Because of this, Model Problem 2 affords an ideal opportunity to study the influence of this convergence on the effectiveness of the various strategies we would like to test. Using the terminology introduced earlier, the "effective" dimension can be varied easily in Model Problem 2 simply by adjusting the amount of absorption per collision.

The third model problem to be discussed next provides an opportunity to study variation reduction in combination with our recommended quasirandom strategies. Model Problem 3, which extends the simple Model Problem 1 by introducing absorption, back scattering and variable thickness for the material to be penetrated, poses the most exacting test yet of all of our ideas for improvement.

5.3   MODEL PROBLEM 3

A vertical slab x cm thick and of unit total macroscopic cross section is assumed to contain material that may either scatter or absorb particles impinging on it from the left . Energy dependence has been suppressed to keep the analysis as simple as possible. The scattering probability per collision is denoted by P_s and the problem is to estimate the fraction, P_T , of particles *eventually* transmitted through the right face of the slab. When the slab thickness is large (as in a typical nuclear shielding calculation), $P_T \ll 1$ and error-reducing non-analog (importance sampling) schemes are essential in order to obtain credible estimates of P_T by simulation.

The problem, in this generality, has been analyzed in [22]. In particular, using an invariant imbedding approach, it is possible to derive differential equations for the means and variances of a variety of interesting analog and non-analog estimators for Model Problem 3. The closed form solutions of these equations then provide exact results against which the sample averages of simulations can be compared. Such problems provide very severe tests for our hybrid methods when the input parameters are varied to make the probability of transmission extremely small (i.e., when the "effective" dimension is very large).

6   Model Problem Results

We first used Model Problems 1 and 2 to evaluate our new "mixed" Strategy A, as described in Section 4. Figure 1 compares errors obtained in solving Model Problem 1 using three different methods for selecting step sizes:

  (A) pseudorandom;
  (B) "pure" quasirandom (utilizing the Halton vector sequence); and
  (C) utilizing our "mixed" Strategy A.[3]

We expect the pseudorandom rate of convergence to be worst in Model Problem 1, because of the relatively rapid convergence of the Neumann series. The "pure" quasi-random strategy is much better than pseudorandom here, not surprisingly, since so few collisions are needed in Model Problem 1, but our "mixed" strategy is nearly as good. Furthermore, both of these quasirandomly-based methods produce errors (after one million random walks) of the order 10^{-4}, compared with errors on the order 1.5×10^{-3} for pseudorandom Monte Carlo. Thus, the mixed strategy produces sizable gains over pseudorandom Monte Carlo and is comparable to pure quasirandom Monte Carlo in Model Problem 1.

Figure 2 compares errors obtained in solving Model Problem 2 with three energy states (i.e., as a 3×3 matrix problem) using the same three methods for generating random walk histories. Here we prescribed relatively low absorption probabilities (0.25) per collision in order to generate a more slowly converging Neumann series for a stricter test of our "mixed" strategy. In this case we expect the pure quasirandom vector sequence to provide the worst rate of convergence, because of the slow convergence of the Neumann series. This slower convergence is illustrated nicely in Table 2 which, like Table 1, exhibits

---

[3] For the numerical results reported in this Section, the "mixed" strategy employed used quasirandom sequences to generate the first three steps of each random walk history and pseudorandom numbers for any remaining steps.

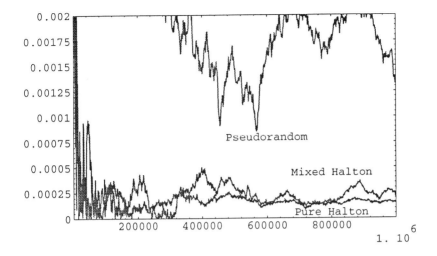

Figure 1: Model Problem 1 Combined Errors

exact and estimated probabilities p_k, $\hat{p}_k$ (Equation (4)) for Model Problem 2 executed with $P_s = 0.75$ for each collision. Using Table 2, we calculate an average number of collisions in this problem to be approximately 3.75, so the effective dimension of this problem is about 7.5 [4].

Indeed, this slower convergence of the pure quasirandom estimates is borne out, but once again the mixed strategy is superior to pseudorandom Monte Carlo by a substantial margin. The error norm used is the Euclidean norm applied to the solution vector, which has three components.

Some further experiments with this mixed strategy, in which both a smaller and a larger number of initial steps are treated quasirandomly, are reported next. In Figure 3 we graph the error from solving Model Problem 2 three separate times using the mixed strategy in which 2, 3, and 5 steps, respectively, are treated quasirandomly and all subsequent decisions are made using pseudorandom numbers. The errors are based on 10^6 histories and are compared with the error obtained when all decisions for all histories make use of pseudorandom numbers; i.e., conventional Monte Carlo. One easily sees the trend toward steady reduction of the error as more and more quasirandom sampling is used. As the number of initial steps treated quasirandomly is increased, one might expect the errors to continue falling, but they will eventually rise above those achieved using only pseudorandomly-generated steps. This is clear because the problem solved here is the same as the one shown in Figure 2. A reexamination of Figure 2 indicates

---

[4] The phase space dimension $d = 2$ in Model Problem 2 since two independent decisions are needed on each "collision" to determine whether the particle history terminates or not and, if it continues, its new energy.

Number of Steps, k	Exact Probability, p_k	Estimate of Probability, $\hat{p}_k$	Error
1	0.250000	0.249781	0.000219
2	0.187500	0.187615	0.000115
3	0.140625	0.141121	0.000496
4	0.105469	0.105724	0.000255
5	0.079102	0.078879	0.000223
6	0.059326	0.059373	0.000047
7	0.044495	0.044660	0.000165
8	0.033371	0.033278	0.000093
9	0.025028	0.024904	0.000124
10	0.018771	0.018561	0.000210
11	0.014078	0.013989	0.000089
12	0.010559	0.010381	0.000178
13	0.007919	0.007898	0.000021
14	0.005939	0.005891	0.000048
15	0.004454	0.004528	0.000074
16	0.003341	0.003409	0.000068
17	0.002506	0.002546	0.000040
18	0.001879	0.001861	0.000018
19	0.001409	0.001448	0.000039
20	0.001057	0.001010	0.000047

Table 2: Exact and Estimated Probabilities $p_k, \hat{p}_k$ for Model Problem 2.

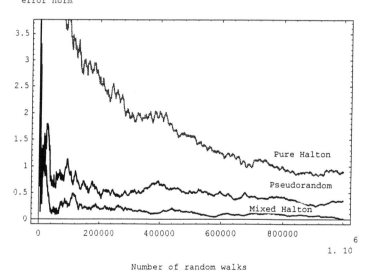

Figure 2: Model Problem 2 Combined Errors

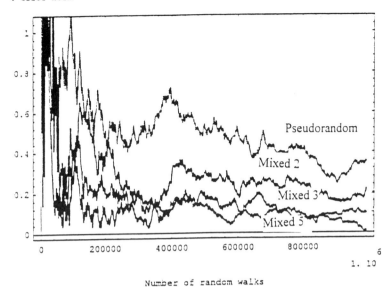

% error norm

Number of random walks

Figure 3: Model Problem 2 Comparison of Pseudorandom with Various Mixed Strategies

that the pure Halton strategy, which is what the mixed strategy approaches as the number of initial steps increases without bound, is definitely *inferior* to the pseudorandom result. This observation makes clear the possibility of optimizing convergence through a judicious choice of the way in which the quasirandom and the pseudorandom parts of the calculation are mixed and points up the need for further work of this sort.

In these initial trials of the mixed strategy, we find evidence supporting our hope for improved convergence rates, even for slowly convergent problems such as the one depicted in Figures 2 and 3. In Figure 4 we display the result of solving Model Problem 3 based on 10^5 histories[5] in a slab of material of unit cross section which is 20 mean free paths thick, and in which the scattering probability per collision $P_s = 0.9$. This input produces a transmission probability $P_T = 1.31 \times 10^{-3}$, so that only one source particle in a thousand is ever transmitted in this problem. The problem was implemented using three different sampling schemes, each of which employed variance reduction mechanisms called absorption weighting (in which absorption is prevented but the particle "weight" is steadily reduced to account for absorption), and the exponential transform [22]:

(A) pseudorandom;

(B) "pure" Halton quasirandom vector sequences;

(C) a "mixed" strategy; in which the first six collisions of every random walk history are treated using a quasirandom vector sequence[6].

A computer program developed by Chelson [6] was modified for use in solving Model

---

[5] The relatively higher cost of solving Model Problem 3 as compared with Model Problems 1 and 2 dictated the use of a smaller number of random walks here than was used in solving Model Problems 1 and 2.

[6] The sequence actually used is often referred to as the Sobol' sequence (see [1]). It is obtained from the sequences based on nets [1] by choosing the number base, b, to be 2.

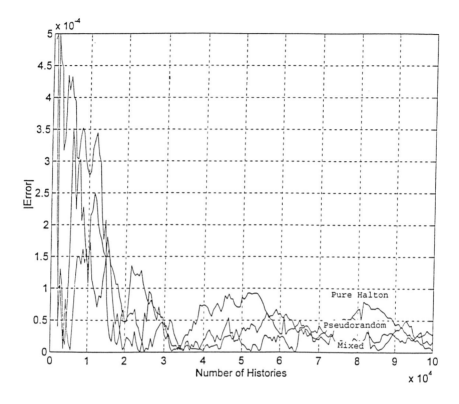

Figure 4: Model Problem 3 Combined Errors

Problem 3. The effectiveness of the mixed strategy is evident here even though the problem, while difficult, is not as difficult as the one studied next. No estimates of the effective dimension were obtained for Model Problem 3 simulations because they were implemented using various importance sampling schemes, each of which greatly distorts the distribution of the $\mu(\Lambda_k)$ of Equation (4).

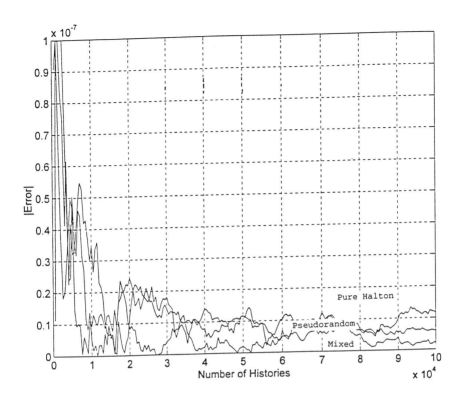

Figure 5: Model Problem 3 Combined Errors

In Figure 5 we display the result of solving Model Problem 3 with the choice $P_s = 0.5$ but all other input parameters retained as in Figure 4 (this produces the result $P_T = 7 \times 10^{-7}$; less than one source particle in a million is transmitted in this problem). This choice increases the effective dimension to . Once again, the superiority of the mixed algorithm is quite clearly demonstrated. As well, this problem exhibits dramatically that use of "pure" quasirandom vector sequences will fail to improve on pseudorandom Monte Carlo in problems in which the Neumann series produces such unfavorable convergence characteristics.

It is instructive to note that these last two results provide at least preliminary evidence that pseudorandomly implemented variance reduction techniques, such as the use of absorption weighting and the exponential transformation, which were used in solving Model Problem 3, convert quite directly to error-reducing methods when used in conjunction with quasirandom Monte Carlo implementations. Of course, this was established in the dissertations [6, 10] when "pure" quasirandom methods were employed but the results reported here constitute the first known verification of their applicability when used in connection with the hybrid strategies developed here for combining pseudorandom and quasirandom methodologies. Many open questions remain, however. For example, no theory has yet been established that would predict the calculation of *optimal* importance sampling parameters when these new quasirandom simulation techniques are employed. Our results seem to suggest that minimum *variance* choices will yield excellent results, but more work is needed here.

Next our interest shifts to tests of hybrid Strategy B, involving the scrambling of low discrepancy sequences and using these directly to replace pseudorandom sequences. Figure 6 compares errors obtained in solving Model Problem 1 using three different methods for selecting step sizes:

(A) pseudorandom;

(B) "pure" quasirandom (utilizing the Halton vector sequence); and

(C) utilizing our "scrambling" strategy, applied to the van der Corput sequence.

One can see that the pseudorandom Monte Carlo results are again worst and, in fact, that the relative comparisons are much like those shown in Figure 1. In fact, the error after 10^6 histories using Strategy B is even smaller than the error after the same number of histories using the mixed Strategy A; both have produced substantial error reductions over the pseudorandom implementation in Model Problem 1 (even though, as Figure 6 reveals, the pure Halton errors have stabilized at low levels much earlier than in the scrambled case). One might expect the scrambled strategy to eventually (i.e., for *sufficiently* large numbers of histories) outperform the pure quasirandom implementation because the Koksma-Hlawka bound for the rate of convergence of the latter is $O[(\log N)^e/N]$ (since e is the effective dimension of Model Problem 1), while the rate of convergence achieved by scrambling the van der Corput sequence should be closer to $O[(\log N)/N]$, the discrepancy of the van der Corput sequence.

Similar numerical experiments performed using Model Problem 2 with relatively small numbers of histories were equally promising for hybrid Strategy B and are not reported here for the sake of brevity. Instead, we report the first results obtained using Strategy B on the much harder class of problems described by Model Problem 3.

In this case, implementation of the scrambling algorithm entails scrambling the first three components, $\phi_2(n), \phi_3(n), \phi_5(n)$, of the Halton sequence since $d = 3$ for Model Problem 3 (1 dimension is needed to select intercollision distances, 1 to decide between

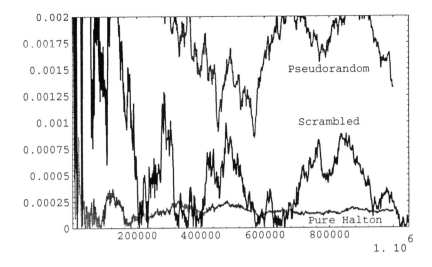

Figure 6: Model Problem 1 Pseudorandom, Pure Halton, Scrambled Errors

continuation and termination at each collision, and 1 to select a new direction). Because of the increased costs of running Model Problem 3, tests were performed using only $N = 2^{16} = 64K$ histories, comparing pseudorandom, pure Halton, and scrambled implementations. The results for the same problem depicted in Figure 4 are shown in Figure 7. The hybrid Strategy B has outperformed the other two; the results are similar to those of Figure 4. Presumably, the relative advantage of Strategy B will only increase with increased N for the same reason as was sketched above.

Finally, a comparison of the two hybrid strategies for Model Problem 3 is shown in Figure 8. While the two errors after 64K histories are quite close to each other, evidence of the superior rate of convergence of Strategy B can perhaps already be seen in Figure 8.

These initial results obtained with both the mixed and the scrambling hybrid strategies are certainly very encouraging. There is, of course, some additional cost (when compared with pseudorandom implementation) associated with each of the two hybrid algorithms, but improvements in the rates of convergence appear to more than offset these. Both techniques are actually quite easy to implement, although use of Strategy B requires more care. Both methods have the potential to provide excellent results over a very broad range of transport problems with widely varying Neumann series convergence characteristics.

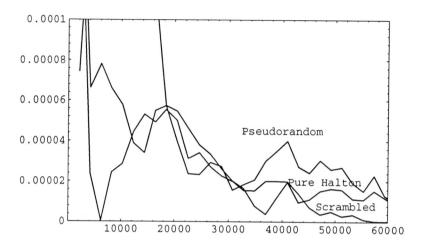

Figure 7: Model Problem 3 Pseudorandom, Pure Halton and Scrambled Errors

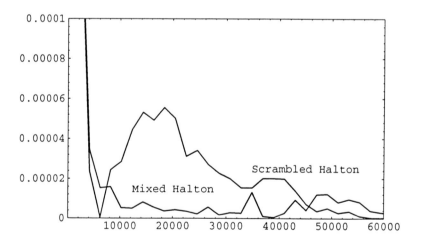

Figure 8: Model Problem 3 Mixed and Scrambled Errors

7 Convergence Rates for Model Problems

As was described earlier, a major goal of this study was to determine effective methods for replacing pseudorandom sequences in conventional Monte Carlo transport programs with appropriately selected quasirandom sequences. The expected benefit from this replacement is a gain in the *rate* of convergence of the sampling error to zero. It is therefore important to attempt to estimate these rates of convergence, at least for the model problems we have studied, for which exact answers are available.

For the hybrid Strategy A in which the initial K collisions of each random walk history are simulated using quasirandom vector components and the remaining collisions, if any, are treated pseudorandomly, one might proceed as follows: One could first guess an asymptotic rate of convergence

$$A(K, N) = O(f(K, N))$$

(for example, by utilizing the discrete probability distribution $p_k = \mu(\Lambda_k)$, Equation (4), and asymptotic rates of convergence for the individual terms, Equation (5), of the Neumann series) where N denotes the total sample size, and f designates a specific function to be tested. Then, for the test problem experimentation, one should be able to evaluate such a conjecture numerically by examining the ratio

$$\frac{|error|}{A(K, N)} \tag{25}$$

and observing whether this ratio grows, diminishes, or remains roughly constant with increases in N, for sufficiently large N.

Based on this sort of analysis, pseudorandom Monte Carlo solutions of the transport equation should produce approximately $O(N^{-\frac{1}{2}})$ errors overall since for this $K = 0$ case, this is (by the law of the iterated logarithm [1]) the average discrepancy for each term in the series, while "pure" quasi-Monte Carlo methods will produce solutions with rates of convergence that are found by averaging rates of order $O[(\log N)^s/N]$ for various integers s corresponding to the dimensionality of the individual terms in the Neumann series. Numerical experimentation of this sort, performed for Model Problems 1 and 2, are reported in [24] .These numerical results confirmed the expected $O(N^{-\frac{1}{2}})$ convergence rates for the $K = 0$ (pseudorandom) case and more rapid convergence for the $K > 0$ cases studied. We omit the details here.

From a slightly altered perspective, if one expects

$$|\delta_N| \equiv |error| = O(N^{-d}), \qquad \frac{1}{2} \le d \le 1,$$

one can estimate the exponent d from plots of $\ln |\delta_N|$ versus $\ln N$. This can be accomplished, say, by finding the best (for example, in the sense of least squares) straight line fit to a log-log plot of the data. Application of this technique has consistently produced convergence rates superior to pseudorandom for both the mixed and the scrambled sequence hybrid strategies. Once again, further details must await future publication.

Of course, model problem analysis can only provide encouragement that a hybrid strategy should outperform either pseudorandom or "pure" quasirandom methods. The issue of practical error estimation, which is treated briefly in the next section, is the more important consideration. In our opinion, this issue provides one of the outstanding unresolved questions arising in connection with the use of quasirandom or hybrid Monte Carlo methods generally.

8   Practical Error Estimation

The issue of *practical* error estimation encountered when implementing quasi-Monte Carlo or hybrid methods for transport studies is difficult for at least two reasons:

1. Error estimates based on generalizations of the Koksma-Hlawka inequality require quantitative information about the discrepancy of the initial segments of the underlying sequence, and also the (Hardy-Krause) variation of the estimator, both of which are difficult to obtain in practice.

2. Error estimates based on generalizations of the Koksma-Hlawka inequality are often very conservative.

By contrast, when conventional, pseudorandom Monte Carlo methods are used, error estimates based on the standard deviation of the estimator are both easy to compute and reasonably accurate. Of course, the Koksma-Hlawka inequality produces *strict upper bounds* for the error, so it provides a worst case analysis. On the other hand, confidence intervals based on the standard deviation for pseudorandom Monte Carlo are designed to produce *average case* error estimates, so it is not surprising that these tend to be sharper and less conservative. Put somewhat differently, if the confidence interval analysis for pseudorandom Monte Carlo were utilized for confidence levels approaching 100%, the intervals would have to increase to encompass the *entire range* of the underlying distribution of sample means. Judged in this way, it is perhaps fortunate that the Koksma-Hlawka inequality produces a *finite* interval in which the true answer *must* lie.

The most theoretically precise possibility for producing error estimates might be based on the generalized Koksma-Hlawka inequality. In the form provided in Chelson's dissertation [6], this would produce

$$|\delta_N(\xi)| \leq (b-a)HT(4mT\sqrt{mT}+1)(\mathcal{D}_N^*)^{\frac{1}{mT}} + O(\epsilon) \qquad (26)$$

where

$$\delta_N(\xi) \equiv \frac{1}{N}\sum_{i=1}^{N}\xi(\omega_i) - \int_\Omega \xi d\mu \qquad (27)$$

and where ξ is the estimator employed in the quasi-Monte Carlo process. In Equation (27), ϵ is a predetermined error criterion, m is the dimension of the physical phase space, $\xi(\Omega) \subseteq [a,b]$, T is an integer chosen so that

$$\sum_{k=T+1}^{\infty} \mu(\Lambda_k) < \epsilon$$

$H = \max_{1 \leq k \leq T} H_k = \max_{1 \leq k \leq T} \sup_t H_{k,t}$ where $H_{k,t}$ = number of components in a finitely convex decomposition of $\xi^{-1}(-\infty,t] \cap \Lambda_k$, and $\mathcal{D}_N^*$ is the discrepancy of the sequence in $[0,1)^{m \cdot T}$ that is used to make decisions about the random walk histories. Application of this result appears to be impractical because of the difficulty with obtaining estimates of the various terms appearing in it, and because it often leads to such gross overestimates of the actual error. Evidence of this latter objection is to be found in [6, 10] . Furthermore, it is known [1] that the classical Koksma-Hlawka inequality provides sharp bounds for the error in terms of the discrepancy of the underlying sequence. Thus, one knows that the exponent occurring in the inequality (22) cannot be improved beyond unity. About the only remaining remedy for the tendency to overestimate the quasirandom error bound produced by the generalized Koksma-Hlawka inequality (22) would be to find a practical way to estimate any of the expressions in Equation (21) that are equivalent to $|\delta_N(\xi)|$.

An alternative to the above-described error analysis that is available practically is to calculate the root mean square error of the estimates, as one would do for pseudorandom Monte Carlo, and use it as a figure of merit for the accuracy of the calculation. Thus, Moskowitz [21] has made effective use of such an idea, and Owen [23] has suggested that use of repeated independent randomizations of low discrepancy sequences (an idea that is similar to that of our Strategy B of Section 4) provide root mean square estimates of the error that could also yield a practical means of analyzing the error of quasirandom Monte Carlo methods.

Clearly, much more effort is needed to provide practical error estimates for quasirandom or hybrid Monte Carlo methods when applied to these difficult transport problems.

9   Acknowledgements

The author is grateful to the students of the 1993-94 CGS Chevron Mathematics Clinic team: John Sepikas (team leader), Liming Li, Giray Okten, Daniel Pick and Stephani Shi, whose enthusiasm and hard work on this project made my task as faculty supervisor so pleasant and rewarding. In particular, the students performed all of the numerical work reported here and Messrs. Li and Okten provided additional invaluable assistance with the final stages of preparation of this paper.

References

[1] H. Niederreiter, *Random Number Generation and Quasi-Monte Carlo Methods*, CBMS-SIAM, 1992.

[2] W.J. Morokoff and R.E. Caflisch, "A quasi-Monte Carlo approach to particle simulation of the heat equation", SIAM J. Num. Anal., 30, (1993), 1558-1573.

[3] J. Spanier and E.H. Maize, "Quasi-random methods for estimating integrals using relatively small samples", SIAM Rev., 36, (1994), 18-44.

[4] J. Spanier and E.M. Gelbard, *Monte Carlo Principles and Particle Transport Problems*, Addison-Wesley, 1969.

[5] J. Spanier, "Some results on transport theory and their application to Monte Carlo methods", J. Math. Anal. Appl., 17, (1967), 549- 559.

[6] P. Chelson, "Quasi-random techniques for Monte Carlo methods", Ph.D. dissertation, The Claremont Graduate School, 1976.

[7] I. Lux and L. Koblinger, *Monte Carlo Particle Transport Methods: Neutron and Photon Calculations*, CRC Press, 1990.

[8] J. Kingman and S. Taylor, *Introduction to Measure and Probability*, Cambridge Univ. Press, 1966.

[9] J. Olmstead, *Real Variables*, Appleton-Century-Crofts, 1959.

[10] E. Maize, "Contributions to the theory of error reduction in quasi-Monte Carlo Methods", Ph.D. Dissertation, The Claremont Graduate School, 1981.

[11] M.J.D. Powell and J. Swann, "Weighted uniform sampling - a Monte Carlo technique for reducing variance", J. Inst. Maths. Applics., 2, (1966), 228-236.

[12] J. Spanier, "A new family of estimators for random walk problems", J. Inst. Maths. Applics., 23, (1979), 1-31.

[13] C. Lécot, "Low discrepancy sequences for solving the Boltzmann equation", J. Comp. Appl. Math., 25, (1989), 237-249.

[14] C. Lécot, "A quasi-Monte Carlo method for the Boltzmann equation", Math. Comp., 56, (1991), 621-644.

[15] H. Babovsky, F. Gropengiesser, H. Neunzert, J. Struckmeier, and B. Wiesen, "Application of well-distributed sequences to the numerical simulation of the Boltzmann equation", J. Comp. Appl. Math., 31, (1990), 15-22.

[16] E. Braaten and G. Weller, "An improved low-discrepancy sequence for multi-dimensional integration", J. Comp. Phys., 33, (1979), 249-258.

[17] D. Knuth, "Construction of a random sequence", BIT, 5, (1965), 246-250.

[18] D. Knuth, *The Art of Computer Programming, Vol. 2: Seminumerical Algorithms*, 2nd ed., Addison-Wesley, 1981.

[19] H. Faure, Good permutations for extreme discrepancy, J. Number Theory, 1992, 47-56.

[20] H.Niederreiter, "Point sets and sequences with small discrepancy", Monatsch. Math., 104, (1987), 273-337.

[21] B.S. Moskowitz, "Application of quasi-random sequences to Monte Carlo methods", Ph.D. Dissertation, UCLA, 1993.

[22] J. Spanier, "An analytic approach to variance reduction", SIAM J. Appl. Math., 18, (1970), 172-190.

[23] A.B. Owen, "Randomly permuted $(t, m, s)-$ nets and $(t, s)-$ sequences", this volume.

[24] "Quasi-Monte Carlo methods applied to oil well logging problems", CGS Mathematics Clinic final report to Chevron Petroleum Technology Company, June, 1994.

Non-Adaptive Coverings for Optimization of Gaussian Random Fields

James M. Calvin

School of Industrial and Systems Engineering
Georgia Institute of Technology
Atlanta, GA 30332-0205

Abstract

Non-adaptive methods for global optimization of a function choose observation points independently of past observed values. We study the average performance of two simple non-adaptive algorithms for optimization of real-valued functions defined on a compact set in $\mathcal{R}^d$. One method chooses observations on a deterministic uniform grid, while the other chooses observations independently and uniformly distributed over the domain. By average performance is meant that we endow the set of objective functions with a probability measure; i.e., view the function as a sample path of a stochastic process ($d = 1$) or a random field ($d > 1$). Under the assumption of a smooth Gaussian field, we identify the limiting distributions of the (normalized) error for both methods. The deterministic grid is asymptotically 6 times as efficient in the one-dimensional case. As the dimension increases, the relative advantage of the deterministic grid decreases, and for dimensions above 6 the random grid is superior. The results on relative performance are insensitive to the particular Gaussian field assumed.

1 Introduction

We consider the problem of locating the maximum of a real-valued function defined on a compact set in $\mathcal{R}^d$ by observing the value of the function at a set of points. In this paper we consider non–adaptive algorithms that make no use of prior information in choosing the next observation site. Our primary purpose is to compare the performance of deterministic and randomized grid searches for a class of homogeneous Gaussian fields. We define the error as the difference between the global maximum and the maximum observed value. By putting a probability on the functions, the distribution of the error can be used to compare algorithms.

Here is the basic idea of the paper. Smooth Gaussian fields behave almost deterministically around a high level local maximum, looking locally like a parabola. By taking limits as the size of the domain increases (and thus the maximum increases) at an appropriate rate, we have the field, suitably normalized in a neighborhood of the global maximum, converging locally uniformly to a parabola. For non-adaptive grid searches, the distribution of the error is related to the distribution of the squared distance from the global maximizer to the nearest grid point. Assuming that the location of the maximizer is uniformly distributed, the distribution of the squared distance is calculated for different grids. This allows us to compare the performance of different grid methods.

Non-adaptive methods for global optimization (under the rubric "covering methods") are surveyed in Zhigljavsky (1991). Most studies focus on the "gaps" in the point sets generated (either the largest gap, or the expected size of the gap containing the maximizer). Anderssen and Bloomfield (1975) compared the random grid with the deterministic grid formed by the centers of a covering by cubes. They showed that for dimensions below 6, the deterministic grid is superior to the random grid, while the random grid is superior for dimensions of at least 6.

Perhaps the most useful notion of gap size for optimization is that of dispersion (see Chapter 6 of Niederreiter 1992). The dispersion of a point set is the supremum, over all points in the domain, of the distance to the nearest element of the point set. The usefulness of the concept lies in the fact that an upper bound on the error in the function approximation is given by the modulus of continuity of the objective function evaluated at the dispersion of the point set. Point sets that minimize dispersion over all point sets are known only for the one and two dimensional cases.

Instead of focusing on the gaps in the domain, we will analyze the error in terms of the objective function value. We derive limit distributions for the error under the assumption of a smooth Gaussian random field. The deterministic grid is superior for dimensions below 6 and the random grid is superior for dimensions above 6 (this is consistent with the results of Anderssen and Bloomfield, though our criterion is different). In the one-dimensional case, the deterministic grid is asymptotically six times as efficient as the random grid. These results are insensitive to the exact form of the Gaussian field.

The next section contains background and establishes the notation. Section 3 introduces the Gaussian fields that we use for our basis of comparison. The limiting error distributions, and resulting performance comparisons, are made in Section 4.

2 Background

A starting point for defining global optimization procedures is a specification of a set $\mathcal{X}$ that contains the function X to be maximized. Given a continuous real-valued function X defined on the cube $[0, T]^d$, let $X^* = \max_{t \in [0, T]^d} X(t)$ denote the global maximum of the function. Suppose we are allowed to choose n points $t_1, t_2, \ldots, t_n$ in the cube at which to observe the value of the function. Denote the maximum of the n function values by

$$M_n = \max_{1 \leq i \leq n} X(t_i). \tag{1}$$

It is of interest to choose the sites in such a way that M_n is a good approximation to X^*, where we define the approximation error by

$$\Delta_n^{\mathcal{A}}(X) = X^* - M_n \tag{2}$$

for algorithm $\mathcal{A}$ (see Figure 1). If the set $\mathcal{X}$ is small enough, algorithms can be constructed that can be analyzed in terms of worst-case performance.

An example of a worst-case result (see Törn and Žilinskas 1989) is for $\mathcal{X}$ the Lipschitz continuous functions X for which $|X(x) - X(y)| \leq L|x - y|$. In this case the uniform grid algorithm $(t_k = (k - 1)/(n - 1))$ is worst-case optimal, with

$$\sup_{X \in \mathcal{X}} \Delta_n(X) = L/n. \tag{3}$$

A more general worst-case bound is given in terms if dispersion (Niederreiter 1992). If the point set $\{t_1, t_2, \ldots, t_n\}$ has dispersion d_n given by

$$d_n = \sup_{t \in [0,T]^d} \min_{1 \leq k \leq n} \|l - l_k\|$$

and if X has modulus of continuity

$$\omega(X, s) = \sup\{|X(u) - X(v)|; u, v \in [0, T]^d, \|u - v\| \leq s\},$$

then $\Delta_n(X) \leq \omega(X, d_n)$.

Another criterion that can be used instead of the worst case error is that of average error. A probability measure P is put on $\mathcal{X}$, and we use the average error

$$E\Delta_n^{\mathcal{A}}(X) = \int_{X \in \mathcal{X}} (X^* - M_n) \, dP(X) \tag{4}$$

to compare algorithms. The average error criterion may be more appropriate if the set $\mathcal{X}$ is large (for example, if $\mathcal{X}$ is all continuous functions).

Note that the error defined in (2) depends fundamentally on the nature of the objective function X. We would expect small "gaps" in the observations, on average, to lead to small error values, but the extent of the error can not be determined without taking into account the nature of the functions. This is a basic difference between the average case and worst case approaches; in a worst case approach a single characteristic (such as modulus of continuity) suffices to describe the function behavior.

We will call the two methods we analyze the *random grid* and the *deterministic grid*. By a "random grid on $[0, T]^d$ with density n^{d}" we mean that $(nT)^d$ points are chosen independently at random, uniformly over $[0, T]^d$. By a "deterministic grid on $[0, T]^d$ with density n^{d}" we mean that observations are made at the points

$$\left(\frac{2k_1 - 1}{2n}, \frac{2k_2 - 1}{2n}, \ldots \frac{2k_d - 1}{2n} \right), \quad k_i = 1, 2, \ldots nT; \; i = 1, 2, \ldots, d. \tag{5}$$

When referring to the error for a specific method, we will write Δ_n^D for the deterministic grid and Δ_n^R for the random grid. In the following sections we derive the normalized limiting distributions of Δ_n^R and Δ_n^D for the two algorithms under the assumption that X is a sample path of a Gaussian random field.

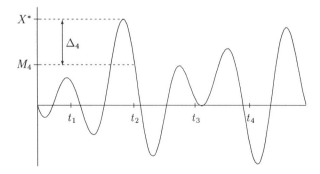

Figure 1: **One-Dimensional Sample Function and Error Variable.**

3 Gaussian Random Fields

Let $\{X(t) : t \in \mathcal{R}^d\}$ be a real, homogeneous, isotropic Gaussian field with mean 0. Thus the finite dimensional distributions of X are Gaussian, and the covariance function $R(s,t) = EX(s)X(t)$ depends only on the distance between s and t; say $R(s,t) = r(|s-t|)$, where

$$|t| = \left(\sum_{i=1}^{d} t_i^2 \right)^{1/2}.$$

We assume that r is of the form

$$r(t) = 1 - \frac{1}{2}\lambda_2 t^2 + \frac{1}{4!}\lambda_4 t^4 + o(t^4) \tag{6}$$

as $t \to 0$, for finite λ_2, λ_4. We assume that

$$|d^4 r(t)/dt^4 - \lambda_4| = O(|t|), \quad -r''(t) = \lambda_2 + O(|\log^{-a}|t||)$$

for some $a > 1$ as $t \to 0$, and also that $r(t)\log(t) \to 0$ as $t \to \infty$. These assumptions allow us to choose a version of X that has twice continuously differentiable sample functions, and we assume that X is so chosen.

Suppose that the maximum value of the field over the cube $[0, T]^d$ is M_T. The following theorem appears as Theorem 6.9.4 in Adler (1981).

Theorem 1

$$P(a_T(M_T - b_T) \le x) \to \exp(-e^{-x}) \tag{7}$$

as $T \to \infty$, where

$$a_T = \sqrt{2d\log(T)},$$

and

$$b_T = a_T + \frac{\frac{1}{2}(d-1)\log\log T + \log\left(\lambda_2^{d/2}(d/\pi)^{(d-1)/d}(2\pi)^{-1}\right)}{a_T}.$$

If X_n and X are random variables, we will use the notation $X_n \Rightarrow X$ to signify that the distribution of X_n converges weakly to the distribution of X; i.e., $Eh(X_n) \to Eh(X)$ for all bounded continuous functions h. We will also say "X_n converges in distribution to X". If X is a constant, then convergence in distribution is equivalent to convergence in probability. We will use the following fact several times in the sequel; it is a corollary of Theorem 4.4.6 in Chung (1974).

Proposition 2 *If $X_n \Rightarrow X$, $\alpha_n \Rightarrow a$, and $\beta_n \Rightarrow b$, where a and b are constants, then*

$$\alpha_n X_n + \beta_n \Rightarrow aX + b. \tag{8}$$

We will use the notation $X \overset{\mathcal{D}}{=} Y$ to signify that X and Y have the same distribution.

As a consequence of Theorem 1, $M_T/b_T \Rightarrow 1$, and so if we set $T_n = \exp(n^2/2d)$, then $M_{T_n}/n \Rightarrow 1$. That is, if the edge length of the cube increases like $\exp(n^2/2d)$, then the maximum will increase like n.

We will use the so-called *Slepian model process* (see Adler (1981) or Leadbetter et al (1983)) to analyze the sample path in a neighborhood of a high local maximum. Let $X(t)$ be a one-dimensional section of the field. Since the field is homogeneous and isotropic, X is a one-dimensional homogeneous Gaussian process with covariance function

$$E\left(X(s)X(s+t)\right) = r(t) = 1 - \frac{1}{2}\lambda_2 t^2 + \frac{1}{4!}\lambda_4 t^4 + o(t^4)$$

as $t \to 0$.

The next theorem specializes Theorem 1.2 and Lemma 3.1 of Lindgren (1972) to the present context.

Theorem 3 *Given a local maximum with height u at t^*, the conditional process $X(t^* + t)$ has the same finite–dimensional distributions as the process $\{X_u(t), t \in \mathcal{R}\}$ defined as follows:*

$$X_u(t) = ur(t) - \eta_u B(t) + \Psi(t)$$

where $\Psi(t)$ is a non-homogeneous zero-mean Gaussian process with covariance function

$$C(s,t) = r(s-t) - \frac{r'(s)r'(t)}{\lambda_2} - \frac{\lambda_4 r(s)r(t) + \lambda_2 r''(t)r(s) + \lambda_2 r''(s)r(t) + r''(s)r''(t)}{\lambda_4 - \lambda_2^2},$$

$$b(t) = \frac{\lambda_2 r(t) + r''(t)}{\lambda_4 - \lambda_2^2},$$

and η_u is a random variable that converges in distribution (as $u \to \infty$) to a normal random variable with mean 0 and variance $\lambda_4 - \lambda_2^2$.

Although we will not make use of the fact, it can be shown that

$$u\left(X_u(t^*) - X_u(t^* + t/u)\right) \to \frac{1}{2}\lambda_2 t^2$$

as $u \to \infty$, almost surely, with the convergence uniform on compact sets. This justifies the informal statement that a smooth Gaussian field looks almost deterministic near a high level local maximum.

4 Comparison of Non-Adaptive Covering Methods

We compare the deterministic uniform grid algorithm with the random algorithm based on taking observations with density n^d on $[0, T_n]^d$. We set $M_n = M_{T_n}$, where $T_n = \exp(n^2/2d)$. We then let $n \to \infty$ for our limiting comparison.

Denote by Δ_n^D and Δ_n^R the errors under the deterministic grid and random grid, respectively, based on observations with density n^d on $[0, T_n]^d$. Let t_n^* be the (first, lexicographically ordered) location where the maximum is attained. With probability one the maximizer is unique and so we can (and do) ignore the possibility of multiple maximizers in our weak convergence results. Assume that t_n^* is uniformly distributed over $[0, T_n]^d$, and let δ_n^D, δ_n^R denote the distance from t_n^* to the nearest observation site under the deterministic and random grids, respectively.

Remark: The maximizer t_n^* is not typically uniformly distributed over the domain, and this complicates the analysis of the deterministic grid since the distribution of the distance between the maximizer and the nearest observation can be complicated. One way to deal with this is to add a random "offset" to the deterministic grid; i.e., let $U_1, U_2, \ldots, U_d$ be independent uniform(0,1) random variables. Form the offset $\gamma = (U_1/2n, U_2/2n, \ldots, U_d/2n)$, and add γ to each of the points of the deterministic grid as defined at (5). Now the distribution of the distance from the maximizer to the nearest grid point has the same distribution as if the maximizer were uniformly distributed.

Another solution is to assume that the X's are periodic functions, or equivalently, view $[0, T]^d$ as the d-dimensional torus and the X's continuous. Under this assumption homogeneity implies that the maximizer is uniformly distributed.

Henceforth, we will assume that the maximizer is uniformly distributed.

With the observation density n^d per unit cube in $\mathcal{R}^d$, the distance between the maximizer and the nearest grid point is of order $1/n$. The following lemmas show that the distances δ_n^R and δ_n^D, when multiplied by n, converge in distribution and identify the limits.

Lemma 4 *Suppose that n^d observations are placed per unit cube. As $n \to \infty$,*

$$P\left(n\delta_n^R \leq x\right) \to 1 - \exp\left(-V_d x^d\right),$$

where V_d is the volume of the unit ball in $\mathcal{R}^d$:

$$V_d = \frac{\pi^{d/2}}{\Gamma(1 + d/2)}.$$

Proof: The points are uniformly distributed, so

$$P(\delta_n^R > x/n) = P(\text{no observations in a ball of radius } x/n) = \left(\frac{T_n^d - (x/n)^d V_d}{T_n^d}\right)^{n^d T_n^d}$$

$$= \left(1 - \frac{x^d V_d}{(nT_n)^d}\right)^{(nT_n)^d} \to \exp\left(-x^d V_d\right)$$

as $n \to \infty$. ∎

Lemma 5 *Suppose that n^d observations are placed per unit cube. As $n \to \infty$,*

$$n\delta_n^D \Rightarrow \frac{1}{2}D_d$$

where D_d is a random variable that has the distribution of the distance from 0 of a point uniformly distributed on the unit cube $[0,1]^d$. In particular, $E(D_d^2) = d/3$.

Proof: Since t_n^* is uniformly distributed, it is uniformly distributed over some sub-cube of side $1/n$. Thus the distance to the nearest vertex has the same distribution as the distance from 0 of a point uniformly distributed on a cube of side $1/2n$. Therefore, $2n\delta_n^D$ has the same distribution as the distance from 0 of a point uniformly distributed on a cube of side 1, which is D_d. ∎

Next we examine how the error is related to the distance of the maximizer from the nearest grid point.

Theorem 6 *Suppose that observations are chosen in such a way that $n\delta_n \Rightarrow \delta$ as $n \to \infty$, where δ_n is the distance of the maximizer from the nearest grid point. Then as $n \to \infty$,*

$$n\Delta_n \Rightarrow \frac{1}{2}\lambda_2\delta^2.$$

Proof: By Proposition 2, it suffices to show that $M_n\Delta_n \Rightarrow \frac{1}{2}\lambda_2\delta^2$, since M_n/n converges in distribution to 1. Let $X^{(n)}$ be the one-dimensional section of the field X confined to the line determined by t_n^* and the nearest grid point. Using the Slepian model process corresponding to this 1-dimensional process, we obtain

$$
\begin{aligned}
M_n\Delta_n &= M_n\left\{X^{(n)}(t^*) - X^{(n)}(t^* + \delta_n)\right\} \\
&\stackrel{\mathcal{D}}{=} M_n\left\{M_n - M_n r(\delta_n) + \eta_{M_n} b(\delta_n) - \Psi(\delta_n)\right\} \\
&= M_n\left\{M_n - M_n\left[1 - \frac{1}{2}\lambda_2(\delta_n)^2 + o(\delta_n)^2\right] + \eta_{M_n} b(\delta_n) - \Psi(\delta_n)\right\} \\
&= \frac{1}{2}\lambda_2(M_n\delta_n)^2 + M_n\, o(\delta_n)^2 + M_n\, \eta_{M_n} b(\delta_n) - M_n\Psi(\delta_n).
\end{aligned}
$$

Since $M_n\delta_n = (M_n/n)n\delta_n \Rightarrow \delta$, the first term converges in distribution to $\frac{1}{2}\lambda_2\delta^2$. We will show that the other three terms converge in probability to zero, thus completing the proof with an application of Proposition 2. By the definition of b and our assumptions on r,

$$b(t) = \frac{\lambda_2 r(t) + r''(t)}{\lambda_4 - \lambda_2^2} = \frac{1}{2}t^2 + o(t^2)$$

as $t \to 0$. Since η_u converges in distribution as $u \to \infty$, the second of the terms converges in probability to 0. It can be shown that Ψ is differentiable, with $\Psi(0) = \Psi'(0) = 0$ with probability one. This shows that the final term also converges in probability to 0. ∎

In view of this theorem, the basis for comparing the asymptotic efficiency of the algorithms is the limiting normalized squared distance of a given point from the nearest grid point. Let Z_d^R and Z_d^D be random variables such that

$$\left(n\delta_n^R\right)^2 \Rightarrow Z_d^R, \quad \left(n\delta_n^D\right)^2 \Rightarrow Z_d^D.$$

Then by Lemmas 4 and 5,

$$P(Z_d^R \leq x) = 1 - \exp\left(-V_d x^{d/2}\right),$$

and

$$Z_d^D \stackrel{\mathcal{D}}{=} \frac{1}{4}\left(U_1^2 + U_2^2 + \cdots + U_d^2\right),$$

where the U_i's are iid $\sim U(0,1)$.

Note that Z_d^D has mean $d/12$ and variance $d/180$, while Z_d^R has mean

$$V_d^{-2/d}\Gamma(1 + 2/d) = \frac{1}{\pi}\left(\Gamma(1 + d/2)\right)^{2/d}\Gamma(1 + 2/d) \sim d/(2\pi e) \approx d/17.08$$

where we have used Stirling's approximation. The variance of Z_d^R is

$$V_d^{-4/d}\left\{\Gamma(1 + 4/d) - \Gamma(1 + 2/d)^2\right\} \sim \frac{1}{6e^2}$$

as $d \to \infty$. This shows how the relative advantage of the random grid in terms of expected error grows with d, and also points out the qualitative difference that the variance of the normalized error converges to a finite constant in the random grid case, but grows linearly in the deterministic grid case.

Corollary 7 *As $n \to \infty$,*

$$\frac{E\left(\Delta_n^R\right)}{E\left(\Delta_n^D\right)} \to \frac{\frac{1}{\pi}\Gamma(! + 2/d)\left[\Gamma(1 + d/2)\right]^{2/d}}{d/12} = \frac{12}{d\pi}\Gamma(1 + 2/d)\left[\Gamma(1 + d/2)\right]^{2/d}. \tag{9}$$

It follows from Corollary 7 that for $d = 1$, the expected error is asymptotically six times as large for the random grid as for the deterministic grid. The relative advantage of the deterministic grid decreases with d, and for $d = 6$ the performance is about the same; for higher dimensions, the random grid is superior.

Notice that as $d \to \infty$,

$$\lim_{n \to \infty} nE(\Delta_n^R) \sim \frac{d}{2\pi e},$$

and so for large d, the limiting ratio of expected errors is about

$$\frac{6}{\pi e}.$$

To see this, by Stirling's formula,

$$\Gamma(1 + d/2) \sim \sqrt{2\pi}(d/2)^{(d+1)/2}e^{-d/2},$$

and so the ratio is asymptotically

$$\frac{12}{d\pi}(2\pi)^{1/d}(d/2)^{(d+1)/d}e^{-1} \sim \frac{6}{\pi e}.$$

References

[1] Adler, R. J. (1981). *The Geometry of Random Fields*. Wiley, New York.

[2] Anderssen, R. S. and P. Bloomfield (1975). Properties of the random search in global optimization. *Journal of Optimization Theory and Applications* **16** 383–398.

[3] Chung, K. L. (1974). *A Course in Probability Theory* (Second Edition). Academic Press, New York.

[4] Janson, S. (1987). Maximal spacings in several dimensions. *Ann. Probability* **15** 274–280.

[5] Leadbetter, M., G. Lindgren and H. Rootzén (1983). *Extremes and Related Properties of Random Sequences and Processes*. Springer-Verlag, New York.

[6] Lindgren, G. (1972). Local maxima of Gaussian fields. *Ark. Math.* **10** 195–218.

[7] Niederreiter, H. (1992). *Random Number Generation and Quasi-Monte Carlo Methods. CBMS-NSF Regional Conf. Series in Appl. Math.* **63**, SIAM, Philadelphia.

[8] Törn, A., and A. Žilinskas (1989). *Global Optimization*. Springer-Verlag, Berlin.

[9] Zhigljavsky, A. (1991). *Theory of Global Random Search*. Kluwer, Dordrecht.

The Method of Fundamental Solutions and the Quasi-Monte Carlo Method for Poisson's Equation

C.S. Chen
Department of Mathematical Sciences
University of Nevada, Las Vegas
Las Vegas, NV 89154, USA

Abstract

A multivariate integration method based on the quasi-Monte Carlo method which alleviates the problem of domain discretization has been implemented for computing particular solutions of Poisson's equation. Coupled with the method of fundamental solutions, a simpler and versatile technique, we have demonstrated a simple computational method which provides means of shifting the effort from analyst to computer. For illustration, we give three numerical examples for a set of standard problems and compare results of different methods.

1 Introduction

For decades the finite difference method (FDM) and finite element method (FEM) have been well established and widely used to solve numerical problems for partial differential equations (PDEs). However, the FDM and FEM require considerable amount of time and effort to discretize and index the domain elements, especially for problems in higher dimensions. In the past decade the boundary element method (BEM) has progressed rapidly and has now become an alternative technique to domain methods (FEM, FDM) for solving boundary value problems for PDEs. The major advantage of the BEM is that for homogeneous PDEs only boundary discretization is required rather than domain discretization. Efficiency is significantly improved over traditional methods (FDM and FEM). However, despite this advantage over more traditional methods, the BEM involves more sophisticated mathematics than the FEM and FDM and some difficult numerical integration of singular functions. In recent years, a new numerical technique, the method of fundamental solutions (MFS), has caught the attention of many scientists and engineers. The MFS retains the advantage of only boundary discretization, but requires very simple mathematics. Unfortunately, when the PDE has an inhomogeneous term, the main advantage of the BEM and MFS is generally lost; i.e. domain integrals are required.

Over the years, methods that deal with this problem seem to fall into three categories: (i) direct numerical integration of the domain integrals; (ii) methods such as dual and multiple reciprocity methods (DRM, MRM) [12, 13], which use radial basis functions to reduce the domain integrals to line integrals in two-dimensional problems and surface integrals in three-dimensional problems; (iii) elimination of the inhomogeneous term by the use of a particular solution.

In their paper, Golberg and Chen [6] have pointed out drawbacks of these approaches. They use a numerical integration technique, due to Atkinson [1], to eliminate the inhomogeneous term by a judicious choice of a particular solution which can be evaluated numerically without having to discretize the domain. Another approach, proposed by Gipson [5], is to evaluate the domain integral by Monte Carlo integration and then apply the BEM to solve Laplace's equation. Motivated by their work, the rapid improvement of modern computers and advanced developments in the quasi-Monte Carlo method, we propose an algorithm which extends the MFS to Poisson's equation by employing the quasi-Monte Carlo method to avoid discretization of the domain.

In recent years the Monte Carlo and quasi-Monte Carlo methods have gradually evolved into important tools in scientific computation. Indeed, Gipson [5] describes Monte Carlo integration as fast, efficient, controllable, easy to implement, reasonably accurate, applicable to a wide class of problems, not highly sensitive to dimensionality and complexity of the domain and the class of functions for which it applies. Most importantly, the Monte Carlo method requires no domain discretization. However, as pointed out by Niederreiter [11], the Monte Carlo method for numerical integration suffers the following drawbacks: (i) it provides only a probabilistic error bound; i.e. there is never any guarantee that the expected accuracy is achieved in a given calculation; (ii) the regularity of the integrand is not reflected; (iii) a true random sample is difficult to generate. Due to these drawbacks, the quasi-Monte Carlo method has become an alternative to the Monte Carlo method. It provides an explicit construction of a set of deterministic nodes and produces small error. Furthermore, the quasi-Monte Carlo method is superior to the Monte Carlo method in the following ways: the probabilistic Monte Carlo error bound for integration is $O(N^{-1/2})$, while the quasi-Monte Carlo integration yields a deterministic error bound $O(N^{-1}(\log N)^{d-1})$, where d is the dimension of the domain, for suitably chosen sets of nodes and for integrands with a relatively low degree of regularity.

Let D be a bounded set in $\mathbf{R}^d$, where d is the dimension of D, S the boundary of D with $S_1 \cup S_2 = S, S_1 \cap S_2 = \emptyset$ and $\partial u / \partial n$ is the normal derivative of u with respect to the outward pointing normal on S. We consider solving the boundary value problem

$$\Delta u = f \tag{1}$$

$$u \mid_{S_1} = g_1, \ \frac{\partial u}{\partial n} \mid_{S_2} = g_2 \tag{2}$$

where f, g_1 and g_2 are given functions.

In Section 2, we introduce a particular solution to reduce (1)-(2) to an equivalent homogeneous equation. In Section 3, we use Monte Carlo and quasi-Monte Carlo integration to approximate the particular solution mentioned in Section 2. In Section 4, we review the MFS for solving Laplace's equation. In Section 5, without loss of generality and for simplicity, we restrict our calculations to two-dimensional cases although the generalization of the idea is valid in any number of dimensions. For illustration, we give three numerical examples for a set of standard problems and compare results of different methods.

2   Basic Equations

Using Green's theorem we can reformulate (1)-(2) as the integral equation

$$
\begin{aligned}
c(P)u(P) &= \int_S G(P,Q)\frac{\partial u}{\partial n}(Q)ds(Q) - \int_S \frac{\partial G}{\partial n_Q}(P,Q)u(Q)ds(Q) \\
&\quad + \int_D G(P,Q)f(Q)dV(Q)
\end{aligned}
\tag{3}
$$

where $P \in D \cup S$, $c(P)$ is a function depending on P and $G(P,Q)$ is the fundamental solution of the Laplacian. That is,

$$
G(P,Q) = \begin{cases} \frac{1}{2\pi}\log \parallel P - Q \parallel, & P,Q \in \mathbf{R}^2, \\ \frac{-1}{4\pi\|P-Q\|^{d-2}}, & P,Q \in \mathbf{R}^d, \ d > 2 \end{cases}
$$

where $\parallel P - Q \parallel$ is the Euclidean distance between P and Q. If $f = 0$, then the domain integral on the right hand side of (3) disappears. The standard BEM or MFS can then be applied by discretizing the boundary of S. If $f \neq 0$, then the domain integral in (3) is necessary.

In this section we consider using a particular solution to transform Poisson's equation in (1)-(2) to an equivalent homogeneous equation (Laplace's equation). To be more specific, by the method of particular solutions, we write

$$
u = v + u_p
\tag{4}
$$

where u_p satisfies the inhomogeneous equation

$$
\Delta u_p = f
\tag{5}
$$

but does not necessarily satisfy the boundary conditions, and v satisfies

$$
\Delta v = 0
\tag{6}
$$

$$
v\mid_{S_1} = g_1 - u_p \mid_{S_1}, \quad \frac{\partial v}{\partial n}\mid_{S_2} = g_2 - \frac{\partial u_p}{\partial n}\mid_{S_2} .
\tag{7}
$$

If f is a simple function, then u_p in (5) may be obtained analytically. The remaining problem is to solve (6)-(7) which is a Laplace equation and can be solved by the standard MFS which will be introduced in Section 4. No domain integration is necessary in this case. However, when f is not simple, then a more general approach to find u_p in (5) is needed. The classical approach is to represent u_p as the Newtonian potential [9]

$$
u_p(P) = \int_D G(P,Q)f(Q)dV(Q)
\tag{8}
$$

and

$$
\frac{\partial u_p}{\partial n}(P) = \int_D \frac{\partial G(P,Q)}{\partial n}f(Q)dV(Q).
\tag{9}
$$

In general, we will need to evaluate the integral, (8) and (9) numerically. However, these integrations require complicated domain discretization of D, especially in the three dimensional case, and involve singular integrands $G(P,Q)$ and $\partial G(P,Q)/\partial n$. Many authors have tried to avoid these difficult domain integrations by introducing DRM, MRM with radial basis functions to approximate f in (5). Thus u_p can be approximated by solving (5) directly. In the next section, we introduce Monte Carlo and quasi-Monte Carlo integration to avoid explicit discretization of D in evaluating (8) and (9). Since (8) and (9) can be evaluated in a similar fashion, we will, for simplicity, focus on the integration of the particular solution u_p in (8).

3 Monte Carlo and Quasi-Monte Carlo Integration

3.1 Monte Carlo Integration

To evaluate the particular solution u_p in (8), we approximate u_p by the Monte Carlo estimate

$$u_p^N(P) = \frac{\lambda(D)}{N} \sum_{i=1}^{N} G(P,Q_i)f(Q_i) \tag{10}$$

with the variance given by

$$\text{Var} = \frac{\lambda(D)}{N} \sum_{i=1}^{N} [G(P,Q_i)f(Q_i)]^2 - (u_p^N(P))^2$$

where $\{Q_i\}_{i=1}^N$ is a sequence of independent, uniformly distributed random samples in D and $\lambda(D)$ the area (the volume in the three-dimensional case) of the domain D. Notice that the variance may be used to estimate the error in the calculation. From the strong law of large numbers, it is well-known that

$$u_p = \lim_{N \to \infty} u_p^N \quad \text{with probability 1.}$$

Based on the central limit theorem, it is also known that Monte Carlo integration yields a probabilistic error bound of the form $O(N^{-1/2})$. This means halving the error requires quadrupling the number of sample points.

3.2 Quasi-Monte Carlo Integration

Due to the drawbacks of the Monte Carlo method indicated in the Introduction, a great deal of research has gone into devising random sequences on computers. In recent years, the quasi-Monte Carlo method (or number-theoretic method), which is devoted to multivariate integration, has been developed as a sophisticated technique for doing numerical integrations. Technically, the quasi-Monte Carlo method is not a Monte Carlo method, since randomization is not employed. It seeks to construct a sequence of nodes $\{Q_i\}_{i=1}^N$ explicitly so that they are highly equidistributed; i.e. Q_i's are chosen in such a way that they are spread evenly and precisely throughout the domain of integration. Since the integration nodes are pre-determined, classical error analysis is possible. It gives the deterministic error bound $O(N^{-1}(\log N)^{d-1})$ [11] for suitably chosen sets of nodes. The mathematical formulation of quasi-Monte Carlo integration is similar to Monte Carlo integration in (10), except $\{Q_i\}_{i=1}^N$ are constructed non-randomly. The selection criterion of the Q_i's is based on the idea of uniformly distributed sequences and discrepancy. In his book [11], Niederreiter described the discrepancy as a quantitative measure for the deviation from a uniform distribution. Detailed error analysis has been performed in his book. He has shown that small errors are guaranteed if point sets with small discrepancy are used. For the purpose of obtaining a low-discrepancy point set, various constructions have been devised. Based on [11], we choose the following low-discrepancy point set to evaluate (10).

Let $Z_b = \{0, 1, ..., b-1\}$ where b is an integer greater than or equal 2. For every integer $n \geq 0$, there exists a unique digit expansion

$$n = \sum_{j=0}^{\infty} a_j(n)b^j$$

in base b, where $a_j(n) \in Z_b$ for all $j \geq 0$ and $a_j(n) = 0$ for all sufficiently large j. We define ϕ_b, the radical inverse function, as following:

$$\phi_b(n) = \sum_{j=0}^{\infty} a_j(n)b^{-j-1} \quad \text{for all integers } n \geq 0.$$

Definition 3.1. For any integer $b \geq 2$, the van der Corput sequence in base b is the sequence $x_0, x_1, ...$ with $x_n = \phi_b(n)$ for all $n \geq 0$.

Using the van der Corput sequence, we are able to define a d-dimensional Halton sequence which produces a low-discrepancy point set.

Definition 3.2. Let $b_1, ..., b_d$ be integers ≥ 2. Then we define the Halton sequence in the bases $b_1, ..., b_d$ as the sequence $\mathbf{x}_0, \mathbf{x}_1, ...$ with $\mathbf{x}_n = (\phi_{b_1}(n), ..., \phi_{b_d}(n))$ for all $n \geq 0$.

The star discrepancy $D^*(\mathbf{x}_1, \mathbf{x}_2, ..., \mathbf{x}_N)$ is designed to measure the irregularity of the distribution of nodes. The following theorem [11] gives an upper bound of $D^*(\mathbf{x}_1, \mathbf{x}_2, ..., \mathbf{x}_N)$.

Theorem 3.3. If $\{\mathbf{x}_1, \mathbf{x}_2, ..., \mathbf{x}_N\}$ is the d-dimensional Halton sequence in the pairwise relatively prime bases $b_1, ..., b_d$, then

$$D^*(\mathbf{x}_1, \mathbf{x}_2, ..., \mathbf{x}_N) < \frac{d}{N} + \frac{1}{N} \prod_{i=1}^{d} \left(\frac{b_i - 1}{2 \log b_i} \log N + \frac{b_i + 1}{2} \right) \text{ for all } N \geq 1.$$

According to the discrepancy bound mentioned above and the error bound of (8) which is discussed in [11], the use of Halton sequences in quasi-Monte Carlo integration results in dramatic improvement over the Monte Carlo error bound $O(N^{-1/2})$. We will illustrate further differences of these two methods by numerical examples in Section 5.

4 The Method of Fundamental Solutions

In this section we apply the MFS, which has been proved to be a highly accurate and efficient numerical technique [4, 6, 7, 8], to solve Laplace's equation (6)-(7). The MFS is based on the theory of potentials in a given region subject to given boundary conditions.

It can be viewed simply as a method for approximating the boundary data. Let Γ be an auxiliary boundary enclosing D and assume that the solution $v(P)$ at (6)-(7) has an analytic continuation defined at every point within the region enclosed by Γ. Then, by potential theory [9], it is well-known that $v(P)$ can be represented as a single layer potential

$$v(P) = \int_\Gamma \sigma(Q)G(P,Q)dS, \ P \in \overline{\Gamma}_{\text{int}} \tag{11}$$

where $\sigma(Q)$ is the strength of the source point Q (a negative value of $\sigma(Q)$ indicating a sink) and Γ_{int} denotes the interior of Γ. The purpose of moving the source points outside of the domain D is to avoid the need to evaluate singular integrals. Under the assumption that the solution $v(P)$ can be extended to the auxiliary boundary Γ, $\sigma(Q)$ can be determined by using the boundary conditions. Thus letting $P \in S$, we obtain the following integral equation

$$h(P) = \int_\Gamma \sigma(Q)G(P,Q)dS$$

where

$$h(P) = \begin{cases} g_1(P) - u_p(P), & P \in S_1, \\ g_2(P) - \frac{\partial u_p}{\partial n}(P), & P \in S_2. \end{cases}$$

To implement this procedure numerically on a computer, we have to approximate $\sigma(Q)$ in (11). The simplest way is to approximate σ by

$$\sigma_m = \sum_{k=1}^m c_k \delta(P - Q_k)$$

where $P \in S$ and $\{Q_k\}_{k=1}^m$ is a set of observation points (source or sink) on the auxiliary boundary Γ. δ is the Dirac function. Physically, in electrostatics, σ_m is a linear combination of unit point charges. Substituting the expression σ_m in the right hand side of (11) and using the properties of the delta function, it becomes

$$v_m(P) = \sum_{k=1}^m c_k G(P, Q_k), \ P \in \overline{\Gamma}_{\text{int}} \tag{12}$$

where $\{c_k\}_{k=1}^m$ have to be chosen to satisfy the boundary conditions as well as possible. One way to determine $\{c_k\}_{k=1}^m$ is to choose l points $\{P_j\}_{j=1}^l$ on S_1 and $m - l$ points $\{P_j\}_{j=l+1}^m$ on S_2 and then satisfy the boundary conditions by collocation. Thus we obtain the following $m \times m$ system of equations

$$\sum_{k=1}^m c_k G(P_j, Q_k) = h(P_j), \ j = 1, 2, \cdots, l, \ P_j \in S_1 \tag{13}$$

and

$$\sum_{k=1}^m c_k \frac{\partial G(P_j, Q_k)}{\partial n} = h(P_j), \ j = l+1, l+2, \cdots, m, \ P_j \in S_2. \tag{14}$$

This linear system can be solved via Gaussian elimination. This gives $\{c_k\}_{k=1}^m$ which in turn gives σ_m. An approximation u_m to u is then given by

$$u_m(P) = \sum_{k=1}^m c_k G(P, Q_k) + u_p^N, \ P \in D$$

where u_p^N is an approximation to u_p as described in Section 2 and 3.

Bogomolny [2] has shown that Γ could be taken as a circle with radius R and that a constant term (for two-dimensional case) added to (12) could be chosen so that v_m converges exponentially fast in m and R if v and S are analytic. Although Bolgomolny's theoretical results suggested that the constant term for the two-dimensional case should be included, his and other's numerical results showed no effect on this term. However, Golberg [8] has found that the constant term is quite sensitive to the eccentricity of the domain for the two-dimensional case. Thus, for the two-dimensional case, we modify v_m in (12) by adding a constant

$$v_m(P) = \sum_{k=1}^{m} c_k \log(P, Q_k) + c, \ P \in \overline{\Gamma}_{\text{int}}.$$

In this case, to determine $\{c_k\}_{k=1}^{m} \oplus c$, we need to choose $m + 1$ collocation points on S.

We remark that there are other ways to determine σ. For instance, Shaw, Huang and Zhao in [14] have numerically shown that better convergence can be obtained by using trigonometric approximations for σ if Γ is a circle. Another approach is to let $\{Q_k\}$ in (12) be unknown and then chosen along with $\{c_k\}$ by adaptive least squares [10].

As indicated by Bogomolny [2] and Cheng [3], the greater the radius of the auxiliary source circle, the better the approximation to be expected. The condition number of the matrix used to obtain the source densities in (13)-(14) deteriorates dramatically as the radius of the auxiliary boundary increased. However, the solution is not affected by the ill-conditioning of the system of equations (13)-(14).

5 Numerical Results

To illustrate the proposed technique, we present here three examples. Without loss of generality and for simplicity, we only consider the two-dimensional case with three different geometric domains. In Example 1, we compare the accuracy between the current method and Gipson's results in [5]. In Example 2, we consider an elliptic domain and compare our results with the DRM in [13]. Example 3 gives the results for a circular domain. For Example 1 and 3, 1000 and 3000 integration nodes were generated by using the Halton sequence indicated in Section 3 to find the particular solution. Since the solutions of all three problems are symmetric, we will only show solutions in one of the quadrants. We denote ∂D as the boundary of the domain D.

Example 1. This example involves a boundary with corners. We consider an isotropic square plate with $D = \{(x, y) \mid -6 < x < 6, -6 < y < 6\}$ subject to internal heat generation of unit value while its boundaries are held at zero temperatures. The governing equation is given by

$$\Delta u = -1, \quad \text{in } D,$$
$$u = 0, \quad \text{on } \partial D.$$

We avoid the corner points by placing nodes on the square boundary sufficiently close to corner points. 20 nodal points, which are located at $(\pm 6, \pm 5.9)$, $(\pm 5.9, \pm 6)$, $(\pm 6, \pm 3)$, $(\pm 3, \pm 6)$, $(\pm 6, 0)$ and $(0, \pm 6)$, are used for the boundary mesh. 19 uniformly distributed source points are chosen on the auxiliary circular boundary with radius 8. Two primes,

11 and 7, were chosen as bases of the Halton sequence. As shown in Table 1, the accuracy of our approach is better than Gipson's [5].

Table 1. The MFS and quasi-Monte Carlo method versus the BEM and Monte Carlo method.

x	y	BEM + MC u_{1000}	BEM + MC u_{3000}	MFS + QMC u_{1000}	MFS + QMC u_{3000}	EXACT
-2.0	2.0	8.543	8.537	8.687	8.714	8.690
-4.0	2.0	5.645	5.736	5.733	5.740	5.748
-3.0	3.0	6.362	6.477	6.537	6.543	6.522
-2.0	4.0	5.634	5.633	5.765	5.744	5.748
-4.0	4.0	3.987	3.981	3.932	3.945	3.928
-2.0	0.0	9.607	9.718	9.621	9.616	9.588
-4.0	0.0	6.161	6.234	6.240	6.283	6.286

Example 2. We consider on the elliptic domain $D = \{(x,y) \mid x^2 + 4y^2 < 4\}$ the problem

$$\Delta u = -2, \quad \text{in } D,$$
$$u = 0, \quad \text{on } \partial D.$$

The analytic solution is given by

$$u = -\frac{4}{5}(x^2 + \frac{y^2}{4} - 1).$$

We use 16 boundary nodes as in the DRM [13]. The radius of the source circle is chosen to be 8. Fifteen source points are chosen uniformly distributed on the source circle. The integration nodes for the particular solution are generated by using the Halton sequence with bases 7 and 11 in the rectangle $\{(x,y) \mid -2 \le x \le 2, -1 \le y \le 1\}$. We then discard those that do not lie in the ellipse. Table 2 shows that our results are sharper than the DRM in [13] using just 3000 integration nodes.

Table 2. The MFS and quasi-Monte Carlo methods versus the DRM.

x	y	DRM	MFS+QMC	EXACT
1.5	0.0	0.349	0.347	0.350
1.2	-0.35	0.418	0.417	0.414
0.6	-0.45	0.573	0.566	0.566
0.0	-0.45	0.646	0.637	0.638
0.9	0.0	0.643	0.635	0.638
0.3	0.0	0.789	0.784	0.782
0.0	0.0	0.807	0.798	0.800

Example 3. We consider solving the following Dirichlet problem on the unit circle $D = \{(x,y) \mid x^2 + y^2 < 1\}$:

$$\Delta u = 1, \quad \text{in } D,$$
$$u = x^2 - y^2, \quad \text{on } \partial D.$$

The analytic solution is given by

$$u = \frac{1}{4}(5x^2 - 3y^2 - 1).$$

We use 16 boundary nodes and 15 source points on a source circle with radius 8. Similar to Example 2, we generate the Halton sequence with bases 7 and 11 in the square $\{(x, y) \mid -1 \leq x \leq 1, -1 \leq y \leq 1\}$ and discard the nodes that do not lie inside the unit circle. As is expected, in Table 3 the quasi-Monte Carlo method outperformed the Monte Carlo method. For comparison, 1000 and 3000 integration nodes have been used to find the particular solution. As shown in Table 3, the results are excellent in this case.

Table 3. The potential on a circular domain.

x	y	MFS + MC		MFS + QMC		EXACT
		u_{1000}	u_{3000}	u_{1000}	u_{3000}	
0.3	0.0	-0.1357	-0.1355	-0.1377	-0.1370	-0.1375
0.5	0.0	0.0589	0.0630	0.0629	0.0629	0.0625
0.4	0.4	-0.1539	-0.1633	-0.1723	-0.1705	-0.1700
0.6	0.7	-0.1658	-0.1679	-0.1682	-0.1674	-0.1675
0.0	0.7	-0.6103	-0.6160	-0.6174	-0.6176	-0.6175
0.0	0.4	-0.3607	-0.3663	-0.3707	-0.3699	-0.3700
0.2	0.7	-0.5619	-0.5682	-0.5675	-0.5676	-0.5675

6   Conclusions

We have described how to implement quasi-Monte Carlo integration numerically to find the particular solution of Poisson's equation. Coupled with the MFS, we have demonstrated an easy way to solve Poisson's equation without domain discretization. This technique provides means of shifting the effort from analyst to computer. Even though the method will consume more computer time, it is rather easy to implement and saves considerable external modeling time. The proposed numerical method is not competitive with other methods [4, 6, 7, 8] in accuracy and efficiency. However, the virtue of this method is its simplicity. For complex domains or when there is no alternative method available, the Monte Carlo or quasi-Monte Carlo method offers an excellent approach with reasonable results. With increasing computer power, our method will be even more attractive.

Although we have restricted the investigation to Poisson's equation, our method can be extended to other linear elliptic partial differential equations with a known fundamental solution.

Acknowledgement. The author would like to thank Drs. Harald Niederreiter and Peter Shiue for their helpful comments which lead to some of the finding in this manuscript. The author also thanks the referee for his careful reading of earlier version.

References

[1] K.E. Atkinson, The Numerical Evaluation of Particular Solutions for Poisson's Equation, IMA Journal of Numerical Analysis, 5, (1985), 319-338.

[2] A. Bogomolny, Fundamental solutions method for elliptic boundary value problems, SIAM J. Numer. Anal., 22 (1985), 644-669.

[3] R.S.C. Cheng, Delta-Trigonometric and Spline - Trigonometric Methods Using the Single Layer Potential Representation, Univ. of Maryland Ph.D. Thesis, 1987.

[4] C.S. Chen, The Method of Fundamental Solutions for Nonlinear Thermal Explosions, to appear in Communication in Numerical Method in Engineering.

[5] G.S. Gipson, The Coupling Monte Carlo Integration with Boundary Integral Techniques to Solve Poisson-type Problems, Engineering Analysis with Boundary Elements, Vol. 2, (1985), 138-145.

[6] M.A. Golberg and C.S. Chen, On a Method of Atkinson for Evaluating Domain Integrals in the Boundary Element Method, Applied Mathematics and Computation, 60, (1994), 125-138.

[7] M.A. Golberg and C.S. Chen, The Theory of Radial Basis Functions Applied to the BEM for Inhomogeneous Partial Differential Equations, Boundary Elements Communications, 5, (1994), 57-61.

[8] M.A. Golberg, The Method of Fundamental Solutions for Poisson's Equation, to appear in Eng. Anal. with Boundary Elements.

[9] M.A. Jaswon and G.T. Symm, Integral Equation Methods in Potential Theory and Elastostatics, Academic Press, 1977.

[10] R.L. Johnston, G. Fairweather and A. Karageorghis, An Adaptive Indirect Boundary Element Method with Applications, Boundary Elements VIII Conference, (1986), 587-597.

[11] Harald Niederreiter, Random Number Generation and Quasi-Monte Carlo Methods, SIAM, CBMS63, Philadelphia, Pa, 1992.

[12] A.J. Nowak and C.A. Brebbia, The Multiple Reciprocity Method - A New Approach for Transforming BEM Domain Integrals to the Boundary, Eng. Analysis with Boundary Elements, 6(3), (1989), 164-167.

[13] P.W. Partridge, C.A. Brebbia and L.C. Wrobel, The Dual Reciprocity Boundary Element Method, Computational Mechanics Publications, 1992.

[14] R.P. Shaw, S.C. Huang and C.X. Zhao, The Embedding Integral Method for Potential Problems with Partitioning, Engineering Analysis with Boundary Elements, Vol. 9, (1992), 83-90.

Implementation of
A Distributed Pseudorandom Number Generator

Jian Chen and Paula Whitlock
Computer & Information Sciences Department
Brooklyn College
Brooklyn, New York 11210

Abstract

In parallel Monte Carlo simulations, it is highly desirable to have a system of pseudorandom number generators that has good statistical properties and allows reproducibility of random sequences used by the individual processes. In this work, we discuss a distributed implementation of such a system on computer networks. The system employs linear congruential pseudorandom number generators obtained from the work "Random Number Generators for MIMD Parallel Processors" [1] and is organized as a virtual, complete binary tree. Nodes of the tree are the generators and the processes using them. The relationship between a node and its subtrees reflects that of a process and its children. Any binary tree up to a million nodes can be maintained. To achieve high utilization of the generators, we implement a dynamically expanding complete binary tree. A server is maintained on a host on our computer network and application programs obtain the generators from the interface provided by the system. The system guarantees that no generator is allocated more than once to an application process and no extraneous correlations are introduced. The pseudorandom number system uses the client/server model, communicating via the Remote Procedure Calls (RPC).

Introduction

Greatly enhanced computational power can be achieved by running Monte Carlo calculations in parallel or distributed environments. In a distributed environment, the calculation is carried out by many worker processes, often on different hosts on a network. To guarantee statistical independence in the simulation, different random sequences are required for each process. That is, no random sequence may be used simultaneously in two worker processes lest the overlap biases the result. Moreover, in some applications such as nuclear physics, the Monte Carlo simulation involves the birth and death of random walks. Worker processes may terminate dynamically, and may or may not create new worker processes as needed to continue the simulation. A system of generators should be available to meet the needs of these branching, random walks. Many methods

[1-6] can be used to implement random number generators in parallel and distributed computations. Each of these methods chooses a suitable strategy to satisfy the requirement of independent random sequences for the worker processes. However, there are requirements that all such proposed parallel random number generators (prng) must satisfy, such as a long period of the sequence, minimal correlations within an individual sequence and lack of correlations between sequences.

The approach in this work to generating independent random sequences for distributed Monte Carlo calculations is to use a family of linear congruential pseudorandom number generators that are organized into a binary tree. Basic concerns of this approach include supporting process creation during a computation and guaranteeing that no generator is used by more than one worker process. In addition, any particular computation should be fully reproducible in the sense that another computation resulting from the same program will have identical numerical results provided that the process creation trees of the two computations are identical. A binary tree structure that maintains a system of prngs was first suggested by Warnock [2] and refined by Fredrickson, Hiromoto, Jordan, Smith and Warnock [3]. The basic idea was modified by Bowman and Robinson[4] and Percus and Kalos[1]. We implement a parallel random number generator system as proposed by the latter authors.

Parallel Random Number Generators

Mapping pseudorandom sequences to parallel processes which incorporated process creation was first proposed by Frederickson *et.al.* [3]. A binary tree, called a Lehmer tree, was used to manipulate two linear congruential generators as illustrated in Figure 1. At each node, the two random number generators (rng), called the left and the right, are used. The left is used to generate an initial value for a new random sequence (its seed) and the right is used for sampling.

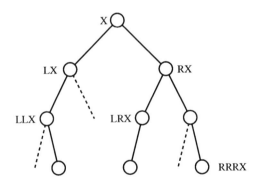

Figure 1 A pseudo-random tree

At the root of the tree (X), an initial seed is used to initiate the first pseudorandom sequence; subsequent pseudorandom numbers are obtained using the right rng (RX, RRX). When a branch is needed, corresponding to a new process creation, the left random number generator is employed to supply a new seed (LX). The newly obtained seed is used with the pseudorandom number generator on the right to continue sampling in the new process. In Figure 1, a node is labeled by the order in which the left and right random number generators have been applied. Each node is distinct in terms of its path from the root, that is, the sequence in which the generators are applied.

The generators need five parameters, a_L, c_L, a_R, c_R, m and defined as:

$$\begin{cases} X_{n+1}^L = a_L X_n^L + c_L \pmod{m}, \\ X_{n+1}^R = a_R X_n^R + c_R \pmod{m}. \end{cases} \tag{1}$$

where m is the modulus and $a_L, a_R < m$, and must be chosen to have good statistical properties. Unfortunately, the sequences produced by Eq. (1) are correlated [4,6].

Bowman and Robinson [4] proposed a modification that used two multiplicative pseudorandom number generators:

$$\begin{cases} X_{n+1}^L = a X_n^L \pmod{m}, \\ X_{n+1}^R = a^p X_n^R \pmod{m}. \end{cases} \tag{2}$$

Both generators use the same multiplier, a and modulus, m. The root of the tree starts with an initial seed and generates pseudorandom sequences using the right random number generator. As before, the left supplies a new seed every time a process creation occurs and new sequences are generated by the right rng. Two difficulties occur here. Even if a is a good generator, there is no guarantee that a^p is acceptable. Second, the number of processes to be created, p, may not be known in advance.

The method implemented in this work again employs a Lehmer tree [1]; however, a class of linear congruential generators is used to assure independence of the parallel streams of pseudorandom numbers. Each node of the tree corresponds to the use of a particular generator. When branching occurs, a *new generator* is introduced. That is, for each generator i:

$$x_{i,n+1} \equiv a x_{i,n} + b_i \pmod{m} \tag{3}$$

there are two generators on its left and right subnodes:

$$x_{L(i),n+1} \equiv a x_{L(i),n} + b_L(i) \pmod{m}$$

$$x_{R(i)=i,n+1} \equiv a x_{R(i)=i,n} + b_{R(i)=i} \pmod{m}.$$

The generator on the right subnode is the parent generator itself and the left subnode is a new generator. To use such a tree structure, we must guarantee that each node is as good a pseudorandom number sequence as any other. Percus and Kalos [1] developed a theoretical test to predict good values for the b's.

The immediate problem is how these generators are associated with worker processes. There are two plausible solutions. One involves no implicit parallelization. In this case, users of the generators have to manually call the different pseudorandom number generators in their codes to meet a computation's needs. For example, users are free to choose any generator in their code directly and in any order. As a result, it is possible that some generators employed will overlap in a large computation, especially in those applications which require long chains of process creation. The other solution is that primitive functions are built and used with a random number generator management system. The relationship between worker processes and generators is taken care of by the management system. The users need only to call the management system in their code appropriately.

We will concentrate on the latter course and introduce the Drand library [7] to support parallel computations. The functions contained within Drand can be called from a user's program which is linked with the library, libdrand.a.

Paged binary tree

Theoretically, the tree that is implied by Eq. (3) is a complete binary tree. Each of the interior nodes has left and right subtrees. When worker processes are associated with the nodes of the tree, all the processes have the same properties, that is, each process can spawn new processes and grant new generators to them. To construct a complete binary tree of 16 levels for example, 32768 generators are needed. Although such a large number of generators can be prepared, a computation can not spawn more than 16 generations of new processes, which is a severe limitation for many simulations.†
Also, a complete binary tree structure is necessary only for a computation in which the process creation tree is itself a complete binary tree.

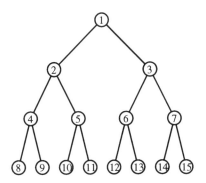
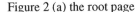
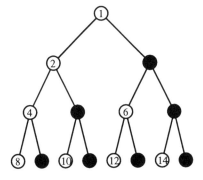

Figure 2 (a) the root page   Figure 2 (b) the reduced page

In the general case, the creation of a new process is a random event based on some probability. So, in practice, the tree that is needed is a binary tree that is not complete. Furthermore, the actual shape of the tree varies and dynamically expands. It is desirable

---

† A complete binary tree of 16 levels has $2^{16} - 1$ nodes. However, in this particular tree, only half of the nodes are distinct.

therefore, that the structure of the parallel random number generator also be capable of dynamically expanding itself. However, this introduces a significant side effect. A dynamically created binary tree of random number generators no longer has the characteristic of guaranteed reproducibility that the static, complete tree did. The structure developed must recognize this aspect and offer a solution since the reproducibility of a simulation run is highly desirable.

By using the same number of nodes as a static, complete tree of 16 levels, the dynamically expanded binary tree can be constructed to contain many hundreds of levels. The expanded binary tree is built by putting together units composed of a fixed number of nodes, called a "page", where the nodes are part of the binary tree. To increase the utilization of generators, the page size is expected to be small, but large enough to avoid unnecessary operational overhead. Consider a binary tree of 15 nodes shown in Figure 2(a). Because the right node is used by a process itself, each node and its right subnode are the same.

For example, in nodes 1, 3, 7 and 15 there is only one generator needed; so the generators in nodes 3, 7 and 15 are the same generator as in node 1, see Figure 2(b). A page of a complete subtree can be constructed using only 8 distinct generators. Each leaf of the tree will generate a subtree; however, the left subtree of a node is different from the right subtree. The root of the left subtree is a new process so it needs a new generator. The root of the right subtree is the same process and therefore inherits the generator from its parent. The configurations of left and right subtrees of the node 8 in Figure 2(b) are shown in Figure 3(a) and Figure 3(b), respectively.

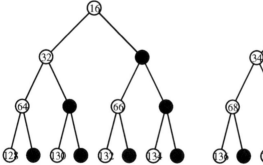 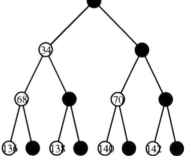

Figure 3 (a) the left expand   Figure 3 (b) the right expand

The numbers in the tree are not generator numbers. They are process identifications assigned and manipulated by the generator system. If the sequences of process creation are to be reproduced by an application program, then the processes should be assigned to the same process identification numbers.

The Drand Library

The functions implementing the expanding binary tree are written in C. A data structure called dhandle is defined to record the complete information about a generator, or a node of the tree. In our network, comprised of machines with 32-bit architecture, about 70000† generators were included and stored in a binary file where each node occupies 6 bytes. This file is managed by a network server so it supports multiple logical copies of the tree while only one copy of the physical file exists. In other words, the system supports many applications at the same time.

Three groups of user interface functions were implemented. The five fundamental primitives are: drand_create() drand_new(), drand_self(), drand_remove() and drand_random(), which supplies a pseudorandom number. In the second group are three functions provided for the user's convenience: drand_advance(), drand_branch() and drand_fork(). These functions employ the fundamental primitives to provide useful operations on the binary tree structure. Finally, there are four functions in the utility group. They are drand_copy(), drand_fprint(), drand_encode() and drand_decode().

A data structure, dhandle, is a representative of a node in the binary tree. Initially, a logical copy of the binary tree can be obtained by calling the primitive drand_create(ac, av, &handle). When the function returns, handle (of type dhandle) contains the first generator of the system, or the root node of the binary tree and process identification number 1. In addition, handle also contains the location of the generator in the page.

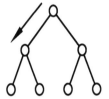 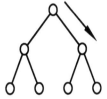

Figure 4 (a) new operation   Figure 4 (b) self operation

To build the binary tree, two primitives, drand_new() and drand_self(), are needed to manipulate the generators. Figure 4(a) shows the drand_new() operation on a subtree. Operationally, drand_new(&handle) allocates a new generator and changes the process identification number to twice that of its parent identification number. All these are updated in handle. Figure 4(b) shows the similar operation for drand_self(&handle). This primitive handles the right subtree and obtain an identification number as twice that of its parent plus 1 and no new generator is allocated.

---

† This number is limited by the period of the generator in Eq.(3)

The following code shows how these primitive are used.

```
1    main(ac, av)
2    int     ac;
3    char    **av;
4    {
5         dhandle  handle;
6         int      cid, status, ret;
7         ...
8         drand_create(ac, av, &handle);
9         if ( (cid = fork()) == 0 ) { /* child */
10             drand_new(&handle);
11             ...
12             exit(0)
13        }
14        else { /* parent */
14             drand_self(&handle);
15             ...
16        }
17   }
```

Code Segment 1

The child invokes `drand_new()`, allowing the newly created process to start using a new generator. However, the parent process has to call `drand_self()` to prepare for a possible subsequent call to `fork()`. The page expanding mechanism is transparent to the user. In the user's point of view, there are no page boundaries, but a complete binary tree, whose nodes are generators. `drand_new()` can also be called from any child process that needs different random sequences.

For convenience, three application functions are provided for alternative uses of the generators. Instead of calling `drand_self()` and then `drand_new()` to use a new generator, `drand_advance()` can be called to use the first new generator in the right subtree. Calling `drand_self()` may be necessary if the user decides that a child process is always to use the first generator on the

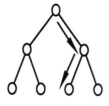 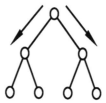 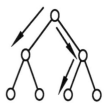

Figure 5 (a) advance   Figure 5 (b) fork   Figure 5 (c) branch

left subtree. However, the parent process can use the first generator of on the left subtree by calling `drand_new()` without having a conflict, as long as the suggested pattern given in Code Segment 1 is followed. That is, a parent process either employs the `fork()`

function to let its child process to use the left generator on the left subtree, or it calls drand_new() to grab the generator for itself, but not both. Figure 5(a) illustrates the situation when advancing to the next generator in the right subtree.

In some applications, two new individual random sequences may be required. In this case, drand_fork() or drand_branch() may be used. Pictorially, these are shown in Figure 5(b) and Figure 5(c), respectively. With either of these functions, multiple handles can be used:

```
1       #define  MAX       10  /* or whatever */
2       dhandle   handle, handle_array[MAX];
3       int       i;
4           ...
5           drand_creat(ac, av, &handle);
6           for ( i = 0; i < MAX; i++ )
7               drand_fork(&handle, &handle_array[i]);
8           ...
9           drand_remove(&handle);
10          ...
```

Code Segment 2 showing drand_fork().

drand_fork() invokes functions drand_copy() to make a duplicate copy, as well as drand_new() and drand_self() to set up new positions in the generator tree.

For programs that need to simulate branching processes, drand_branch() may be used.

```
1       #define  MAX          100   /* or whatever */
2       #define  threshold     0.5   /* or whatever */
3
4       dhandle    oldhandles[MAX], newhandles[MAX], handle;
5       int        i, j;
6           ...
7           /* assume handle and oldhandles[0..9]    */
8           /* are initialized by Code Segment 2     */
9           j = -1;
10          for ( i = 0; i < 10; i++ ) {
11              if ( drand_random(&handle) < threshold )
12                  /* newhandles[++j] and oldhandles[i] all */
13                  /* have new generators after the branch  */
14                  drand_branch(&oldhandles[i], &newhandles[++j]);
15              drand_copy(&newhandles[++j], oldhandles[i]);
16          }
17          ...
```

Code Segment 3 showing drand_branch().

The above code either copies an old handle from oldhandles[] to newhandles[] or allocates two new handles based on an old handle into newhandles[] depending on a probability generated in Line 11.

drand_random(&handle) returns a pseudorandom number using the algorithm in Eq.(3). The parameters needed are supplied in the structure, handle. The difference between drand_random() and other conventional random() functions is that handle determines the sequence, not drand_random().

The utility function drand_fprint(FILE *fp, handle *dhandle) prints the generator parameters into the file fp. drand_copy(handle *dest, handle *src) duplicates src into dest. The last two utility functions are drand_code() and drand_encode(). To support distributed message passing systems, two functions are needed. On Unix systems, remote processes are usually spawned through rexec(). A dhandle has to be transmitted via the *cmd* parameter of the rexec() call as a string. The parent process calls drand_create(ac, av, &handle) to initiate a binary tree and obtain the first generator. Then it calls drand_encode(&code, handle) to convert the handle into a string form. With concatenation of code in the parameter cmd, the handle will be delivered to a remote host through rexec(). The host on the destination side will decode the code by using drand_decode(code, &handle). Then the handle will be used in a call drand_new(&handle). The situation presented in Code Segment 1 is only for use on a single host. All processes are native to the host. Therefore, there is no need to encode/decode the drand handles. In a network, two processes, parent and child, are not related, and the handles have to be passed through rexec() in the literal string format.

System server

The Drand parallel random number generator system exists in a distributed environment. It is implemented in a RPC server/client format. The server runs on one host of a network, called drand_svc. The resources of the server can be accessed by all the other hosts, or clients, on the network. Three of the primitive functions, drand_creat(), drand_new() and drand_self(), are remote procedures. drand_remove(), drand_advance(), drand_fork() and drand_branch() are all local procedures.

The major resources that the server maintains is a file containing the additive constants given in [1] which satisfy the MIMD Spectral test introduced there. Besides the header of the file, all components are six byte long integers. The file is read-only, so the components are static and every process sees the file exactly the same.

The server maintains an array of file pointers; each of these belongs to a logical copy of the generator tree. Thus, the server can support as many simultaneous applications as the number of elements in the array. An application program registers a logical copy by calling drand_creat() and unregisters it by calling drand_remove(). Since child processes may spawn additional worker processes, the logical copy of the tree has to be kept until no further child processes are created. This means that child processes should not call drand_remove() as the child process may spawn additional worker processes. The logical copy of the tree should be kept until no further child processes are created. At that point, the parent can call drand_remove(). If the user wants to reproduce the simulation, drand_remove() should not be called.

A file pointer is advanced by seven or eight six-byte increments depending on whether it is advancing to the left subtree or the right subtree. The maximum distance that the file pointer can move is one page of generators. The file pointers, or current positions for file reading are stored in server table which can not be rewound, except when the logical copy is destroyed.

A dhandle is altered by each remote procedure call. The initial position of a page within the file is stored in the handle. In addition, the level and location of the currently used generator within the page are also stored in the handle. In the example of Figure 3, at the initial call of drand_creat(ac, av, &handle), handle will carry a generator specified by the initial page position, 1st level in the subtree and node location number 1. The level and position fields in the handle are adjusted and, therefore, pointing to a new node in the tree included in the page. When the level reaches 5 (only 4 levels of the binary tree on a page), the primitives ask the system server to adjust the initial position of the handle.

When there are multiple handles used by many processes in an application program, the only action that the server performs on the file is to advance the pointer when needed and return the new pointer to the requesting process. The information needed to specify a generator is stored in both the server and client's handle.

The remote procedures generate process identification numbers that are also stored in the parameter, `handle`, of the procedure calls. For instance, `drand_creat(ac, av, &handle)` generates an identification number 1; `drand_new(&handle)` generates a number which is twice the number it received; (the received number is its parent's identification number.) `drand_self(&handle)` generates a number which is twice the number it received plus 1. These numbers, which represent positions in the complete binary tree, can be uniquely generated for more than one simultaneously calling process. The identification numbers are not used until a new page needs to be allocated for a mapping function. When a binary tree becomes larger and larger, the identification number could be very large. The identification numbers are represented by an abstract data type which is capable of representing arbitrarily large integers.

The server also maintains other important information about applications, such as the name of the application, the client user's login id, the host from which the registration is made and a *job registration identification*.

The user's login id and the host name are used when a user queries the server to list his or her registered applications. The *job registration identification* is used to identity a registered application for the purpose of reproducing a run. These three pieces of information are not provided by the users, rather they are obtained by the `drand_creat(ac, av, &handle)` call, running on the client side. The users are notified of the *job registration identifications* when they are created in order that specific runs can be reproduced.

When a new page is allocated from the file, an ordered pair (process identification number, page offset in the file) is recorded in the server at the corresponding slot of the *systemtable*, which is an array of records. All of the records for one application are used to represent the sequence in which the remote procedures were called by the application, or the history of the application. (Note that an application may consist of many processes.) The ordered pairs stored in the table are linked together and are pointed to by a pointer with which the application was registered. The pointer is initialized when `drand_creat()` is called. Calls to `drand_new()` and `drand_self()` will update the record list whenever a new page of generators is allocated.

The system supports reproducing a result of an application based on the *job registration identification*, which is obtained by the first run of the application. If the application program is modified, the *job registration identification* can be submitted with the modified application to the server. The server guarantees to reproduce the run, that is, allocate the same generators, if the order of process creation of the application remains the same. This implies that some applications can not be reproduced. For example, if the application was not completed at the first run, the server will reproduce the result up to the place where the first run stopped in the subsequent run. Another example is if some "major" changes were made in the application program, the server will reproduce up to the point where both runs diverge. An "additional" primitive function call to the Drand library inserted somewhere in the application program during the modification is considered a "major" change. The server will not detect the change immediately until some new pages of generators are needed. If the modified application requires more pages of generators, the server will detect the change and refuse to provide generators to that modified application. An occurrence of a *page fault* indicates that the second run is not equivalent to the first run.

To reproduce a run, the drand_remove() should not be called in the program. To remove an unwanted logical copy in the server, an option -d can be used in the command line when a user program is resubmitted.

Some application issues should be made clear. Suppose that a parent and its child processes both call the following sequence after a generator tree is created by the parent: drand_new(), drand_new(), drand_new(), with no calls to drand_self() by the parent. Clearly, in this case, a random number generator may be used in the computation by two different processes, if after the sequence of drand_new() calls, both of them draw random numbers. This may be unwanted in the simulation, but it does not violate our promise that no generator is allocated twice. The reason is that 1) the server allocated a page of generators, and these generators will not be allocated a second time and 2) both processes independently move along the same path in the page. Therefore, they generate the same position numbers with respect to the entire complete binary tree, and are granted the same generators. In the server 's point of view, they are identical sequences.

However, if both processes call drand_new() one more time, then only the first call succeeds, whichever reaches the server first. The reason for this is that a new page is allocated by the server (we assume that a page has 4 levels of generators.) Only the first call survives. No generator is granted to the second call. If it did, there would be two generators allocated to one position in the generator tree. Before calling the last drand_new(), both processes have the same status in terms of position number and generator. No matter which process succeeds, the result is the same.

A companion program called drand_tool is provided to help maintain the system server. A major concern in a client/server environment supporting multiple applications is to keep the server running, even when difficulties occur. For instance, many applications may request heavy computations or the limit of reserved memory is reached. The companion program can issue a command to the server telling it to block new applications. It can delete applications that cause problems to the server, or dump the internal states of the server. drand_tool may be used to communicate with the server when the server is involved in some unexpected problem.

Examples

Usage of the Drand library is demonstrated by two sample programs. Both programs generate complete binary trees of ten levels using a recursive function, `traversal()`. The trees are representations of a Lehmer generator tree. The first sample, demo1.c, is a single process program, while the second one, demo2.c, is concurrent, and splits many processes running on a host machine. The properties of the current implementation of the distributed random number generators can be observed through the output produced by the demo programs.

The first sample program is listed in Code Segment 4 and some selected sections of the output is shown in Code Segment 5.

```
1  /*      demo1.c            */
2  #include        <stdio.h>
3  #include        "drand.h"
4  #define         MAXEXPENSION      10
5
6  traversal( handle, level, upperbd )
7  dhandle *handle;
8  int      level;
9  int      upperbd;
10 {
11         dhandle another;
12         if ( level != upperbd ) {
13                 if (drand_fork(handle, &another) < 0)
14                         return(1);
15         /* print out two subnodes */
16         (void) drand_fprint(stdout, *handle);
17         (void) drand_fprint(stdout, another);
18         return( traversal( handle, level + 1,
19                                       upperbd ) +
20                 traversal( &another, level + 1,
21                                       upperbd ));
22         }
23         return(0);
24 }
25
26 main( argc, argv )
27 int      argc;
28 char     *argv[];
29 {
30         dhandle handle;
31
32         setbuf( stdout, (char *)0 );
33         (void) drand_creat(argc, argv, &handle );
34         /* print the root of the tree */
35         (void) drand_fprint(stdout, handle);
36         (void) fprintf(stdout, "\n %s \n",
37             ( traversal(&handle, 1, MAXEXPENSION) == 0 )
38                 ? "successful" : "some failure(s)"  );
39         return(0);
40 }
41
```

Code Segment 4

In line 3 of Code Segment 4, drand.h is included. Line 4 defines a size of the tree to be generated. Lines 26 to 40 define the main program. Line 33 initializes a copy of generator tree, and line 35 prints the root generator to standard output, stdout. Then line 36 calls the recursive function traversal() by passing the root of the tree, and the initial and ending boundaries.

traversal() is listed from lines 6 to 24. Line 11 defines a temporary handle to help with recursive calls. Line 12 tests whether the desired upper bound is reached. If it does, the function returns in line 23. Line 13 calls drand_fork(handle, &another). When it returns, both handles are changed. Lines 16 and 17 print the new handles, and then the function calls itself with the lower bound incremented in line 20.

The server, drand_svc provided by Drand, can be started by typing drand_svc in a command line, as shown in Code Segment 5. demo1.c can be compiled by cc -o demo1 demo1.c -lm -ldrand, where libdrand.a is the implementation library. The program was run twice and the Unix operating system provided utility function, diff, was employed to determine the difference between the two outputs. demo1 is designed using a single thread, therefore the execution is deterministic and the outputs are the same (diff demo1.out1 demo1.out2 produces no output). The subsequent lines show the contents of demo1.out1 which is a list of the generators used, the total numbers of nodes the demo program produced, and the root generator that is propagated through the right branches.

```
$
$ export DRANDSERV=fusion
$ drand_svc > svc.out&
$
$ cc -o demo1 demo1.c -lm -ldrand
$ demo1 > demo1.out1
$ demo1 > demo1.out2
$ diff demo1.out1 demo1.out2
$
$ grep drandid demo1.out1
constb= 11863279 drandid= 1
constb= 11863279 drandid= 3
constb= 11863259 drandid= 2
constb= 11863217 drandid= 9
...
constb= 11845861 drandid= 1021
constb= 11845859 drandid= 1020
constb= 11863279 drandid= 1023
constb= 11845849 drandid= 1022
$
$ grep drandid demo1.out1 | wc
1023    4092    31629
$
$ grep drandid demo1.out1 | sort +3n
constb= 11863279 drandid= 1
constb= 11863259 drandid= 2
constb= 11863279 drandid= 3
constb= 11863217 drandid= 4
...
constb= 11845859 drandid= 1020
constb= 11845861 drandid= 1021
constb= 11845849 drandid= 1022
constb= 11863279 drandid= 1023
$
$ grep 11863279 demo1.out1
constb= 11863279 drandid= 1
```

```
constb= 11863279 drandid= 3
constb= 11863279 drandid= 7
constb= 11863279 drandid= 15
constb= 11863279 drandid= 31
constb= 11863279 drandid= 63
constb= 11863279 drandid= 127
constb= 11863279 drandid= 255
constb= 11863279 drandid= 511
constb= 11863279 drandid= 1023
```
Code Segment 5

The second program, demo2.c, is given in Code Segments 6 and 7 and its output in Code Segment 8. demo2 is initially started in the command line. It then splits itself into as many copies as the number of nodes needed in the complete binary tree.

```
 1  /*     demo2.c         */
 2  #include       <stdio.h>
 3  #include       "drand.h"
 4  #define        MAXEXPENSION    10
 5  int traversal();
 6
 7  main( argc, argv )
 8  int     argc;
 9  char    *argv[];
10  {
11      dhandle handle;
12      int     level;
13      char    path[BUFSIZ];
14
15      setbuf( stdout, (char *)0 );
16      (void) sprintf(path, "/home/tests/%s", argv0] );
17
18      if ( argc == 1 ) { /* the first process  */
19          (void) drand_creat(argc, argv, &handle );
20              level = 1;
21              drand_fprint(stdout, handle );
22      } else {
23              (void) drand_decode( &handle, argv[1] );
24              (void) drand_new( &handle );
25              (void) drand_fprint(stdout, handle);
26              level = atoi( argv[2] );
27      }
28      (void) traversal(path, argv[0], &handle, level,
29              MAXEXPENSION);
30      return 0;
31  }
32
```
Code Segment 6

In Code Segment 6, line 16 assumes that the executable, demo2, is stored in directory /home/tests so that it can be started by the program itself. demo2 differentiates the first and successor processes by the test in line 18. The first copy of the program calls drand_creat(), defines level, and prints the root generator. Then it jumps to line 28 to call the recursive function traversal() passing path, the name of the executable, the root handle, the lower and upper bound to the function.

traversal(), listed in Code Segment 7, will call the executable using its formal arguments to extract useful information. When main() is called by traversal(), its execution begins at line 23. Its parent handle and the current level are passed in the second and third arguments of main(), respectively. In line 23, its parent handle is decoded in Drand format. drand_new() is called as expected in line 24. The new handle is printed in line 20 and the current level is extracted in line 26. After these statements, it finally calls traversal(), and then exits.

```
1   traversal(path, av0, handle, level, upperbd )
2   char    *path;
3   char    *av0;
4   dhandle *handle;
5   int     level;
6   int     upperbd;
7   {
8       int             pid, status;
9       codetype        encoded;
10      char            dec[3];
11      if ( level != upperbd ) {
12          sprintf(dec, "%d", level + 1 );
13          if ( (pid = fork()) == 0 ) {
14              (void) drand_encode( &encoded, *handle );
15
16              (void) execl(path, av0, (char *) encoded,
17                              dec, (char *) 0 );
18          } else
19          if ( pid > 0 ) {
20              (void) drand_self( handle );
21              (void) drand_fprint(stdout, *handle);
22              sleep(2);
23              exit( traversal(path, av, handle,
24                      level + 1, upperbd ));
25          } else exit(1);
26      }
27      return;
28  }
```

Code Segment 7

The intent of traversal() is to launch a subtask by creating a child process which is responsible for all subtrees beneath it. Meanwhile, the parent process moves down to its right, and recursively calls traversal(). In this manner, traversal() will step through all the nodes of the complete binary tree. Each newly created process does the same: launching a new process to generate all subnodes below it and moving itself to its right subtree. The diagram in Figure 6 shows the situation when traversal() calls itself. Each triangle is a subtask and it will launch new subtasks denoted by smaller, heavier shaded triangles.

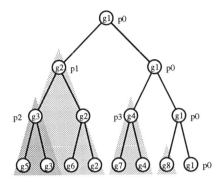

Figure 6. Generating a complete tree

Line 11 in Code Segment 7 tests whether the recursive chain should be terminated. If it is not, it increments `level` by 1 in line 12. `traversal()` splits itself into two processes. The child process executes lines 14 to 17. The handle is encoded in line 14 and executes `demo2` in line 16. When `demo2` is started, it is responsible for generating a complete binary tree rooted at a `handle` (see `main()` in Code Segment 6). The parent runs lines 20 to 24. It calls `drand_self()`, prints the new handle and then calls `sleep()` for two seconds to avoid too many concurrent processes. Then the parent generates its right branch by calling `traversal()`. The call to `exit()` is provided in case of failure.

Code Segment 8 lists some of the output after two complete runs of `demo2`. `demo2` splits into many processes and they run concurrently. The order in which the processes access the server, `drand_svc`, is non-deterministic. Therefore, the order in which the generators are associated with the complete tree differs from one run to another. However, these two consecutive runs are actually reruns of each other in fact so the relationship between generators used and node positions with respect to the tree should be static. The `grep` utility command is invoked to list the generators used in the two different runs. After sorting the two listings, the output of the two runs is the same, even though the order in which the generators were allocated was different in the two runs.

```
$ grep drandid demo2.out1
constb= 11863279 drandid= 1
constb= 11863279 drandid= 3
constb= 11863279 drandid= 7
constb= 11863259 drandid= 2
constb= 11863279 drandid= 15
...
constb= 11845741 drandid= 768
constb= 11845849 drandid= 774
constb= 11852171 drandid= 944
constb= 11849077 drandid= 624
constb= 11856653 drandid= 998
$
$ grep drandid demo2.out2
```

```
constb= 11863279 drandid= 1
constb= 11863279 drandid= 3
constb= 11863213 drandid= 6
constb= 11863279 drandid= 7
constb= 11863213 drandid= 13
...
constb= 11850907 drandid= 558
constb= 11855731 drandid= 632
constb= 11855743 drandid= 635
constb= 11855689 drandid= 634
constb= 11860031 drandid= 694
$


$ grep drandid demo2.out1 | sort +3n > out1.sort
$ grep drandid demo2.out2 | sort +3n > out2.sort
$ diff out1.sort out2.sort
$
$ cat out1.sort
constb= 11863279 drandid= 1
constb= 11863259 drandid= 2
constb= 11863279 drandid= 3
constb= 11863217 drandid= 4
...
constb= 11861879 drandid= 1020
constb= 11861887 drandid= 1021
constb= 11861873 drandid= 1022
constb= 11863279 drandid= 1023
$
```

Code Segment 8

Conclusions

Based on the theoretical work of Percus and Kalos [1], a system of linear congruential random number generators, the Drand system, has been implemented in a distributed environment. The advantages of this system are that each process uses its own sequence of pseudorandom numbers which is guaranteed to be independent from every other sequence used in the simulation and that correlations between generators taken two or three at a time have been minimized. The implementation uses a client/server model where the server keeps track of which random number generators have already been used in the computation. The user need only include a few library calls in their code in order to use the Drand system. The user has the option of rerunning simulations by using a history option in the server. The Drand system has been successfully used in simulations with a static number of processes as well as birth/death or branching random walks.

Acknowledgements

This work was supported by PSC-CUNY grant number 662488 and by National Science Foundation grant ASC-912 1428.

References

1). O.E. Percus and M. H. Kalos, "Random Number Generators for MIMD Parallel Processors," J. of Parallel and Dist. Computing, ⟨6⟩, 477-497 (1989).

2). T.T. Warnock, "Synchronization of Random Number Generators," Congressus Numeratium, ⟨37⟩, 135-144, (1983).

3). P. Fredrickson, R. Hiromoto, T.L. Jordan, B. Smith and T.T. Warnock, "Pseudo-random trees in Monte Carlo," Parallel Computing, ⟨1⟩, 175- (1984).

4). K.O. Bowman and M.T. Robinson, "Studies of random number generators for parallel processing," in Hypercube Multiprocessors, Proceedings of the Second Conference, Knoxville, Tennessee, 1986, edited by M.T. Heath (SIAM, 1987), pp. 445-453.

5). P. L'Ecuyer and S. Cote, "A Random Number Package with Splitting Facilities," ACM Trans. Math. Software, ⟨17⟩, 98-111, (1991).

6). J.H. Halton, "Pseudo-random Trees - Multiple Independent Sequence Generators for Parallel and Branching Computations," J. Comp. Phys., ⟨84⟩, 1-56, (1989).

7). J. Chen, "A Distributed Pseudorandom Number Generator Implemented on a Sun Network," Brooklyn College CIS Department Technical Report, (1994).

On the lattice test for inversive congruential pseudorandom numbers

Wun-Seng Chou and **Harald Niederreiter**

Institute of Mathematics
Academia Sinica
Nankang, Taipei 11529
Taiwan, ROC

E-mail: macws@ccvax.sinica.edu.tw

Institute for Information Processing
Austrian Academy of Sciences
Sonnenfelsgasse 19
A-1010 Vienna
Austria

E-mail: nied@qiinfo.oeaw.ac.at

Abstract. An important tool for the analysis of inversive congruential pseudorandom numbers is the lattice test. For a given prime modulus p, the optimal behavior of inversive congruential generators occurs when they pass the $(p-2)$-dimensional lattice test. We use the connection with permutation polynomials to establish several criteria for passing the $(p-2)$-dimensional lattice test. We also prove that if p is a Mersenne prime, then there exists an inversive congruential generator which has period length p and passes the $(p-2)$-dimensional lattice test.

1. Introduction

Inversive congruential pseudorandom numbers have been introduced by Eichenauer and Lehn [4] in order to overcome some of the deficiencies of the classical linear congruential method for the generation of uniform pseudorandom numbers, such as the undesirable coarse lattice structure inherent in the latter method. We refer to [12, Chapter 8] for an expository account of inversive congruential pseudorandom numbers and for a comparative analysis of the inversive congruential method and the linear congruential method; see also Eichenauer-Herrmann [7] for a recent survey of the inversive congruential method. In view of the unfavorable lattice structure produced by the linear congruential method, a key issue in the theoretical analysis of the inversive congruential method is the extent to which there is a lack of lattice structure in the pseudorandom numbers generated by this method. The appropriate technical tool for this analysis is the so-called lattice test. It is the principal aim of this paper to study conditions under which inversive congruential pseudorandom numbers show the optimal behavior under the lattice test.

Let $p \geq 5$ be a fixed prime and let $GF(p) = Z/pZ$ denote the finite field of order p, which can also be identified with the set $\{0, 1,..., p-1\}$ of integers. For $c \in GF(p)$ we define

$\bar{c} \in GF(p)$ by $\bar{c} = c^{-1}$ (= the multiplicative inverse of c in GF(p)) if $c \neq 0$ and $\bar{c} = 0$ if $c = 0$; equivalently, we may define $\bar{c} = c^{p-2}$ for all $c \in GF(p)$.

Definition 1. For $a, b \in GF(p)$ the *inversive congruential generator* $X(x_0; a, b)$ is the sequence $x_0, x_1, \ldots$ of elements of GF(p) that is obtained by choosing an initial value $x_0 \in GF(p)$ and using the recursion

$$x_{n+1} = a\bar{x}_n + b \quad \text{for } n = 0, 1, \ldots.$$

If one wants to generate pseudorandom numbers in the unit interval [0, 1), then these can be derived from $X(x_0; a, b)$ by considering $x_n/p \in [0, 1)$ for $n = 0, 1, \ldots$. However, for our purposes it is more convenient to work with the generator $X(x_0; a, b)$ itself.

The sequence $X(x_0; a, b)$ is clearly periodic and the length of its least period (or, briefly, its period length) is at most p. In practice, only those sequences with maximal period length p are of interest. Note that if $X(x_0; a, b)$ has period length p, then necessarily $ab \neq 0$. The following notion and the subsequent result due to Flahive and Niederreiter [8] are crucial. We use the standard notation $GF(p^2)$ for the extension field of GF(p) of degree 2.

Definition 2. The monic quadratic polynomial $f(x) = x^2 - bx - a$ over GF(p) with $a \neq 0$ is called an *IMP polynomial* (for "inversive maximal period polynomial") over GF(p) if the quotient of the roots of f in $GF(p^2)$ has multiplicative order $p + 1$ in $GF(p^2)$.

Lemma 3. The inversive congruential generator $X(x_0; a, b)$ has period length p if and only if the polynomial $x^2 - bx - a$ is an IMP polynomial over GF(p).

If $X(x_0; a, b)$ has period length p, then $\{x_n, x_{n+1}, \ldots, x_{n+p-1}\} = GF(p)$ for all $n \geq 0$. This implies that any inversive congruential generator $X(x_0; a, b)$ with period length p is a shifted version of some fixed generator $X(0; a, b)$, say.

Chou [2] carried out a detailed study of IMP polynomials, even over arbitrary finite fields. The following lemma is a special case of [2, Theorem 2]. We recall from [10, Definition 3.2] that a polynomial m(x) over GF(p) with $m(0) \neq 0$ has order e if e is the least positive integer such that m(x) divides $x^e - 1$ in GF(p)[x].

Lemma 4. The monic quadratic polynomial $f(x) = x^2 - bx - a$ over GF(p) with $a \neq 0$ is an IMP polynomial over GF(p) if and only if the polynomial $m_f(x) = x^2 + (b^2/a + 2)x + 1 \in GF(p)[x]$ has order $p + 1$.

An important test for the suitability of $X(x_0; a, b)$ as a pseudorandom number generator is the lattice test. Here we view $GF(p)^s$, $s \geq 1$, as an s-dimensional vector space over $GF(p)$.

Definition 5. For given $s \geq 1$, a sequence y_0, $y_1,\ldots$ of elements of $GF(p)$ passes the *s-dimensional lattice test* if the vectors $y_n - y_0$, $n = 1, 2,\ldots$, span $GF(p)^s$, where

$$y_n = (y_n, y_{n+1},\ldots, y_{n+s-1}) \in GF(p)^s \text{ for } n = 0, 1,\ldots.$$

Pseudorandom number generators should pass the lattice test for as large a dimension as possible (note that if a sequence passes the s-dimensional lattice test, then it automatically passes any lower-dimensional lattice test). A linear congruential generator y_0, $y_1,\ldots$ with prime modulus p does not pass the s-dimensional lattice test for any $s \geq 2$ (see [12, pp. 178-179]), which demonstrates the weakness of these generators with respect to lattice structure. On the other hand, inversive congruential generators $X(x_0; a, b)$ with period length p pass the lattice test for very high dimensions. Eichenauer, Grothe, and Lehn [3] showed that a generator $X(x_0; a, b)$ with period length p passes the s-dimensional lattice test for all $s \leq (p - 1)/2$, and Niederreiter [11] improved this to $s \leq (p + 1)/2$. More recently, Flahive and Niederreiter [8] proved that if $p \equiv 3 \bmod 4$, then a generator $X(x_0; a, b)$ with period length p passes the s-dimensional lattice test for all $s \leq (p + 3)/2$. Further work related to the lattice test for inversive congruential generators was carried out by Eichenauer-Herrmann [6].

It is known that a generator $X(x_0; a, b)$ with period length p cannot pass the s-dimensional lattice test for $s \geq p - 1$ (see [12, p. 180]). Therefore, as far as the lattice test is concerned, the optimal behavior of a generator $X(x_0; a, b)$ with period length p occurs when it passes the $(p - 2)$-dimensional lattice test. The focus of this paper will be on the question of the existence of inversive congruential generators $X(x_0; a, b)$ with period length p showing the optimal behavior. In Section 2 we exploit the connection with permutation polynomials and in Section 3 we establish criteria for an inversive congruential generator with period length p to pass the $(p - 2)$-dimensional lattice test. On the basis of these criteria, it is shown in Section 4 that inversive congruential generators with period length p passing the $(p - 2)$-dimensional lattice test exist in the important special case where p is a Mersenne prime. Overall, it transpires from our analysis that it is very likely that inversive congruential generators with period length p passing the $(p - 2)$-dimensional lattice test exist for all primes $p \geq 5$, but it seems to be difficult to prove this general result.

2. Associated permutation polynomials

Let $x^2 - bx - a$ be an IMP polynomial over $GF(p)$ and let $X(x_0; a, b)$ be a corresponding inversive congruential generator. Then the sequence $X(x_0; a, b)$, given by x_0, $x_1,\ldots$, has period length p by Lemma 3. We now follow Niederreiter [11] and consider the map $n \in GF(p) \mapsto x_n \in GF(p)$ which, like any self-map of a finite field, can be represented by a uniquely determined polynomial $g \in GF(p)[x]$ with $d := \deg(g) \leq p - 1$. Thus, we have

$$g(n) = x_n \text{ for all } n \in GF(p). \tag{1}$$

Since $\{x_0, x_1, \ldots, x_{p-1}\} = GF(p)$, the polynomial g permutes $GF(p)$, and so it follows from the theory of permutation polynomials that $d \leq p - 2$ (see e.g. [10, Corollary 7.5]). We may call g the *associated permutation polynomial* of the generator $X(x_0; a, b)$. The following lemma shows that the degree d of the associated permutation polynomial plays an important role in the lattice test for inversive congruential generators. This result was shown by Eichenauer, Grothe, and Lehn [3] and Niederreiter [11]; see also [12, Theorem 8.2].

Lemma 6. Let $x^2 - bx - a$ be an IMP polynomial over $GF(p)$, let $X(x_0; a, b)$ be a corresponding inversive congruential generator, and let g be the associated permutation polynomial of $X(x_0; a, b)$. Then the generator $X(x_0; a, b)$ passes the s-dimensional lattice test if and only if $s \leq d = \deg(g)$.

We now show that the degree of the associated permutation polynomial depends only on the polynomial $m_f(x) = x^2 + (b^2/a + 2)x + 1$ corresponding to $f(x) = x^2 - bx - a$ according to Lemma 4, and not on the initial value x_0 or on the specific values of a and b.

Theorem 7. Let $f_1(x) = x^2 - b_1x - a_1$ and $f_2(x) = x^2 - b_2x - a_2$ be IMP polynomials over $GF(p)$ and let $X(x_0; a_1, b_1)$ and $X(y_0; a_2, b_2)$ be inversive congruential generators with associated permutation polynomials g_1 and g_2, respectively. If $b_1^2/a_1 = b_2^2/a_2$, then there exist $e_1, e_2 \in GF(p)$ with $e_1 \neq 0$ such that $g_2(x) = e_1 g_1(x + e_2)$.

Proof. Let $\overline{g}_1$ and $\overline{g}_2$ be the associated permutation polynomials of the inversive congruential generators $X(0; a_1, b_1)$ and $X(0; a_2, b_2)$, respectively. Since the sequence $X(x_0; a_1, b_1)$ is a shifted version of $X(0; a_1, b_1)$, there is an $n_1 \in GF(p)$ satisfying $g_1(x) = \overline{g}_1(x + n_1)$. Similarly, there exists an $n_2 \in GF(p)$ such that $g_2(x) = \overline{g}_2(x + n_2)$.

Since $b_1^2/a_1 = b_2^2/a_2$, we have $m_{f_1} = m_{f_2}$. We write $m(x) = m_{f_1}(x) = m_{f_2}(x) = x^2 - cx + 1$, where $c = -b_i^2/a_i - 2$ for $i = 1, 2$. Let $e = -(c + 2)^{-1}$ and $f(x) = x^2 - x - e$, then $m_f = m$. Let g be the associated permutation polynomial of the inversive congruential generator $X(0; e, 1)$. For $i = 1, 2$ we prove by induction on n that

$$\overline{g}_i(n) = b_i g(n) \text{ for } 0 \leq n \leq p - 1. \tag{2}$$

From the definitions we have $\overline{g}_i(0) = 0 = g(0) = b_i g(0)$ and $\overline{g}_i(1) = b_i = b_i g(1)$, and so (2) holds for $n = 0, 1$. Now suppose that (2) has been shown for some $1 \leq n < p - 1$. Then from

$a_i = b_i^2 e$, the definition of inversive congruential generators, and (1), we get

$$\overline{g}_i(n + 1) = a_i/\overline{g}_i(n) + b_i = b_i^2 e/(b_i g(n)) + b_i = b_i(e/g(n) + 1) = b_i g(n + 1),$$

hence the induction is complete.

Since $\deg(\overline{g}_i) \leq p - 2$ and $\deg(g) \leq p - 2$, it follows from (2) that $\overline{g}_i = b_i g$ for $i = 1, 2$. Note that $b_i \neq 0$ for $i = 1, 2$. Combining all these results, we get

$$g_2(x) = \overline{g}_2(x + n_2) = b_2 g(x + n_2) = b_2 \overline{g}_1(x + n_2)/b_1 = b_2 g_1(x + n_2 - n_1)/b_1.$$

Take $e_1 = b_2/b_1$ and $e_2 = n_2 - n_1$. Then $g_2(x) = e_1 g_1(x + e_2)$ and $e_1 \neq 0$.

Corollary 8. If $f_1(x) = x^2 - b_1 x - a_1$ and $f_2(x) = x^2 - b_2 x - a_2$ are IMP polynomials over GF(p) with $b_1^2/a_1 = b_2^2/a_2$, then for all $x_0, y_0 \in$ GF(p) the inversive congruential generator $X(x_0; a_1, b_1)$ passes the s-dimensional lattice test if and only if $X(y_0; a_2, b_2)$ passes the s-dimensional lattice test.

Proof. This follows from Lemma 6 and Theorem 7.

Theorem 7 also leads to the following description of the family of all inversive congruential generators with period length p. We consider two generators as different if they are different as sequences.

Theorem 9. There are exactly $p(p - 1)\phi(p + 1)/2$ different inversive congruential generators $X(x_0; a, b)$ with period length p, where ϕ is Euler's totient function. Moreover, there are exactly $\phi(p + 1)/2$ different inversive congruential generators $X(0; e, 1)$ with period length p so that any other inversive congruential generator with period length p can be obtained from one of these $\phi(p + 1)/2$ generators $X(0; e, 1)$ by multiplication by a nonzero constant and shifts.

Proof. There are exactly $\phi(p + 1)/2$ polynomials $m(x) = x^2 - cx + 1$ over GF(p) of order $p + 1$ (see [2, Theorem 4] and [10, Theorem 3.5]). Consequently, there are exactly $(p - 1)\phi(p + 1)/2$ IMP polynomials $x^2 - bx - a$ over GF(p), by Lemma 4 and the relation $c = -b^2/a - 2$ (or equivalently $a = -b^2/(c + 2)$). So, in view of Lemma 3, there are exactly $(p - 1)\phi(p + 1)/2$ different inversive congruential generators $X(0; a, b)$ with period length p. Since any inversive congruential generator $X(x_0; a, b)$ with period length p is a shifted version of $X(0; a, b)$, there are exactly $p(p - 1)\phi(p + 1)/2$ different inversive congruential generators with period length p. This proves the first assertion. The second assertion follows from the proof of Theorem 7.

3. Criteria for passing the (p − 2)-dimensional lattice test

As we have pointed out in Section 1, the main interest is in inversive congruential generators $X(x_0; a, b)$ which have period length p and pass the s-dimensional lattice test for the largest possible dimension $s = p - 2$. According to Lemma 6, the generator $X(x_0; a, b)$ with period length p passes the (p − 2)-dimensional lattice test if and only if the associated permutation polynomial g has degree p − 2. A related and more explicit criterion was developed by Eichenauer and Niederreiter [5] (see also [12, p. 180]). Let $x_0, x_1,\ldots \in GF(p)$ be the terms of the generator $X(x_0; a, b)$ with period length p; then $X(x_0; a, b)$ passes the (p − 2)-dimensional lattice test if and only if

$$\sum_{n=0}^{p-1} nx_n \neq 0 \quad \text{in GF(p)}. \tag{3}$$

There are known cases where an inversive congruential generator with period length p does not pass the (p − 2)-dimensional lattice test. For instance, J. Eichenauer-Herrmann and H. Grothe have found by computer search that in the two cases

$$p = 31, a = 9, b = 1,$$
$$p = 37, a = 17, b = 1,$$

the corresponding generators $X(x_0; a, b)$ do not pass the (p − 2)-dimensional lattice test, but only the (p − 4)-dimensional lattice test.

Let $X(x_0; a, b)$ be an arbitrary inversive congruential generator with period length p. Then $f(x) = x^2 - bx - a$ is an IMP polynomial over GF(p) by Lemma 3, and the polynomial

$$m_f(x) = x^2 - cx + 1 \in GF(p)[x] \quad \text{with} \quad c = -b^2/a - 2 \tag{4}$$

has order p + 1 according to Lemma 4. If $\alpha, \beta \in GF(p^2)$ are the roots of f, then $\gamma = \alpha/\beta$ and $\gamma^{-1} = \gamma^p = \beta/\alpha$ are the roots of m_f in $GF(p^2)$, and γ has multiplicative order p + 1 in $GF(p^2)$. The criterion (3) now attains the following form.

Theorem 10. For an inversive congruential generator $X(x_0; a, b)$ with period length p let $\gamma \in GF(p^2)$ be a root of the polynomial m_f in (4). Then $X(x_0; a, b)$ passes the (p − 2)-dimensional lattice test if and only if

$$H(\gamma) := \sum_{n=1}^{p} n/(\gamma^n - 1) \neq 0 \quad \text{in GF(p}^2\text{)}.$$

Proof. It suffices to consider the case $x_0 = 0$. If $x_0 = 0$, $x_1,\ldots$ are the terms of X(0; a, b), then it follows from formulas of Flahive and Niederreiter [8, p. 76] that

$$x_n = (\alpha^{n+1} - \beta^{n+1})/(\alpha^n - \beta^n) \quad \text{for} \quad 1 \leq n \leq p.$$

With $\gamma = \alpha/\beta$ this can be written as

$$x_n = \beta(\gamma^{n+1} - 1)/(\gamma^n - 1) = \beta\gamma + \beta(\gamma - 1)/(\gamma^n - 1) \quad \text{for} \quad 1 \leq n \leq p.$$

Thus, for the sum in (3) we obtain

$$\sum_{n=0}^{p-1} n x_n = \sum_{n=1}^{p} n x_n = \beta\gamma\sum_{n=1}^{p} n + \beta(\gamma-1)H(\gamma) = \beta(\gamma-1)H(\gamma)$$

since $\sum_{n=1}^{p} n = 0$ in $GF(p^2)$. The result of the theorem follows now from the criterion (3).

We remark that it does not matter which of the two roots of m_f we consider in Theorem 10, since $H(\gamma^p) = H(\gamma)^p$. Theorem 10 implies that the question of whether $X(x_0; a, b)$ passes the $(p-2)$-dimensional lattice test just depends on the polynomial m_f, as can also be seen from Corollary 8. We now derive an alternative expression for the sum $H(\gamma)$ in Theorem 10.

Lemma 11. For an arbitrary prime p, let $\omega \neq 1$ be a $(p+1)$st root of unity in an arbitrary field of characteristic p. Then

$$(\omega - 1)^{-1} = -\sum_{i=1}^{p-1} i\omega^{-i} - 1.$$

Proof. Note that $\omega = \omega^{-p}$ since $\omega^{p+1} = 1$. Therefore

$$1 + (\omega - 1)^{-1} = \omega(\omega - 1)/(\omega - 1)^2 = \omega(\omega^{-p} - 1)/\omega^2(1 - \omega^{-1})^2 = -\omega^{-1}(1 - \omega^{-1})^{p-2}$$

$$= -\omega^{-1}\sum_{j=0}^{p-2}\binom{p-2}{j}(-1)^j\,\omega^{-j} = -\sum_{j=0}^{p-2}(j+1)\omega^{-j-1} = -\sum_{i=1}^{p-1} i\omega^{-i},$$

where we used that $\binom{p-2}{j}(-1)^j = j + 1$ for $0 \leq j \leq p - 2$ in characteristic p. The proof is complete.

Let p be a prime and let n be an integer. The integer $d_n(p)$ is defined by

$$d_n(p) = \sum_{\substack{1 \leq i,j \leq p \\ (p+1)|(ij-n)}} ij.$$

Because in our applications we work with a fixed prime $p \geq 5$, we will write d_n instead of $d_n(p)$ for simplicity. For the sum $H(\gamma)$ in Theorem 10 we now get by Lemma 11,

$$H(\gamma) = \sum_{n=1}^{p} n/(\gamma^n - 1) = \sum_{n=1}^{p} n\left(-\sum_{i=1}^{p-1} i\gamma^{-ni} - 1\right) = -\sum_{i=1}^{p}\sum_{j=1}^{p} ij\gamma^{-ij},$$

and so

$$H(\gamma) = -\sum_{n=0}^{p} d_n \gamma^n. \tag{5}$$

Theorem 12. Let $p \geq 5$ be a fixed prime. Then all inversive congruential generators with period length p fail the $(p - 2)$-dimensional lattice test if and only if the cyclotomic polynomial $Q_{p+1}(x)$ divides $D_p(x) := \sum_{n=0}^{p} d_n x^n$ in $GF(p)[x]$.

Proof. From Theorem 10 and the formula (5) we see that an inversive congruential generator $X(x_0; a, b)$ with period length p fails the $(p - 2)$-dimensional lattice test if and only if $\sum_{n=0}^{p} d_n \gamma^{-n} = 0$, i.e., if and only if γ^{-1} is a root of the polynomial $D_p(x) \in GF(p)[x]$. As we have noted after (4), γ^{-1} is also a root of the polynomial $m_f(x) = x^2 - cx + 1$ in (4), and m_f is in fact the minimal polynomial of γ^{-1} over $GF(p)$. Therefore, we have $\sum_{n=0}^{p} d_n \gamma^{-n} = 0$ if and only if $x^2 - cx + 1$ divides $D_p(x)$ in $GF(p)[x]$.

It follows from the above and from Lemma 4 that all inversive congruential generators with period length p fail the $(p - 2)$-dimensional lattice test if and only if $D_p(x)$ is divisible by all polynomials $x^2 - cx + 1$ over $GF(p)$ of order $p + 1$. The $\phi(p + 1)/2$ polynomials of the latter type are pairwise relatively prime and their product is the cyclotomic polynomial $Q_{p+1}(x)$ over $GF(p)$ by [10, Theorem 2.47]. This implies the desired result.

The proof of Theorem 12 also leads to an effective procedure for determining those inversive congruential generators $X(x_0; a, b)$ with period length p that pass the $(p - 2)$-dimensional lattice test, in the case where $Q_{p+1}(x)$ does not divide $D_p(x)$ in $GF(p)[x]$. Indeed, the first part of the proof of Theorem 12 shows that $X(x_0; a, b)$ passes the $(p - 2)$-dimensional lattice test if and only if the corresponding polynomial $x^2 - cx + 1$ in (4) does not divide $D_p(x)$. The product of all polynomials $x^2 - cx + 1$ over $GF(p)$ of order $p + 1$ that do not divide $D_p(x)$ is given by

$$T_p(x) = Q_{p+1}(x)/\gcd(Q_{p+1}(x), D_p(x)). \tag{6}$$

Therefore, $X(x_0; a, b)$ passes the $(p - 2)$-dimensional lattice test if and only if the corresponding polynomial $x^2 - cx + 1$ in (4) is a factor of $T_p(x)$ in $GF(p)[x]$. Since $\deg(Q_{p+1}) = \phi(p + 1)$, we always have $\deg(T_p) \leq \phi(p + 1)$.

An efficient method of determining quadratic factors $x^2 - cx + 1$ of the polynomial $T_p(x)$ in (6) can be based on the following argument. We define the Dickson polynomials $E_n(x) \in GF(p)[x]$ of the second kind with parameter 1 by $E_{-2}(x) = -1$, $E_{-1}(x) = 0$, and the recursion

$$E_n(x) = x E_{n-1}(x) - E_{n-2}(x) \quad \text{for} \quad n = 0, 1, \ldots.$$

These polynomials can also be explicitly given by

$$E_n(x) = \sum_{i=0}^{[n/2]} \binom{n-i}{i}(-1)^i x^{n-2i} \quad \text{for } n = 0, 1, \ldots,$$

where $[n/2]$ is the greatest integer $\leq n/2$ (see Lidl, Mullen, and Turnwald [9, Chapter 2]). For arbitrary $c \in GF(p)$ it is now shown by straightforward induction on n that

$$x^n \equiv E_{n-1}(c)x - E_{n-2}(c) \mod (x^2 - cx + 1) \quad \text{for } n = 0, 1, \ldots.$$

Thus, if we have

$$T_p(x) = \sum_{n=0}^{\phi(p+1)} t_n x^n$$

with all $t_n \in GF(p)$, then

$$T_p(x) \equiv x \sum_{n=0}^{\phi(p+1)} t_n E_{n-1}(c) - \sum_{n=0}^{\phi(p+1)} t_n E_{n-2}(c) \mod (x^2 - cx + 1).$$

Consequently, $x^2 - cx + 1$ divides $T_p(x)$ if and only if

$$\sum_{n=0}^{\phi(p+1)} t_n E_{n-1}(c) = \sum_{n=0}^{\phi(p+1)} t_n E_{n-2}(c) = 0,$$

that is, if and only if c is a common root of the polynomials

$$V_{p,1}(x) = \sum_{n=0}^{\phi(p+1)} t_n E_{n-1}(x), \quad V_{p,2}(x) = \sum_{n=0}^{\phi(p+1)} t_n E_{n-2}(x).$$

This argument can be summarized as follows.

Proposition 13. The polynomial $x^2 - cx + 1$ divides $T_p(x)$ in $GF(p)[x]$ if and only if c is a root of the polynomial $V_p = \gcd(V_{p,1}, V_{p,2})$ in $GF(p)$.

Corollary 14. The inversive congruential generator $X(x_0; a, b)$ with period length p passes the $(p - 2)$-dimensional lattice test if and only if $c = -b^2/a - 2$ is a root of the polynomial $V_p = \gcd(V_{p,1}, V_{p,2})$ in $GF(p)$.

We remark that the condition in Theorem 12, namely that $Q_{p+1}(x)$ divides $D_p(x)$ in $GF(p)[x]$, is a rather strong one, and it appears unlikely that it is satisfied. This suggests the conjecture that for all $p \geq 5$ there exists an inversive congruential generator with period length p which passes the $(p - 2)$-dimensional lattice test. Data collected by J. Eichenauer-Herrmann and H. Grothe show that this conjecture is true for all primes $p \leq 67$. In the next section we prove this conjecture in the case where p is a Mersenne prime.

4. The case of Mersenne prime moduli

For the practical implementation of inversive congruential generators, the case of Mersenne primes such as $p = 2^{31} - 1$ is particularly important (recall that p is a Mersenne prime if $p + 1$ is a power of 2). Several of the theoretical results presented in the earlier sections simplify considerably in this case. For instance, Chou [2] has shown that if p is a Mersenne prime, then a quadratic polynomial $x^2 - cx + 1$ over GF(p) has order $p + 1$ if and only if $c + 2$ and $c^2 - 4$ are nonsquares in GF(p). This condition can be checked easily by the calculation of Legendre symbols. In view of Lemma 4, this leads to a simple method for checking the property of being an IMP polynomial over GF(p). Furthermore, since the cyclotomic polynomial $Q_{p+1}(x)$ is given by $Q_{p+1}(x) = x^{(p+1)/2} + 1$ in the case of a Mersenne prime p, we can also apply the recursive procedure of Blake, Gao, and Mullin [1] for generating those values of $c \in GF(p)$ for which $x^2 - cx + 1$ is a factor of $Q_{p+1}(x)$ in GF(p)[x] (or, equivalently, for which $x^2 - cx + 1$ has order $p + 1$). It should be noted, however, that only a limited number of Mersenne primes are known.

Next we show a simpler form of the condition in Theorem 12 for Mersenne primes p which will lead to a proof of the existence of inversive congruential generators with period length p passing the $(p - 2)$-dimensional lattice test.

Lemma 15. If p is a Mersenne prime, then $Q_{p+1}(x)$ divides $D_p(x) = \sum_{n=0}^{p} d_n x^n$ in GF(p)[x] if and only if $d_{n+(p+1)/2} \equiv d_n \bmod p$ for $0 \le n \le (p - 1)/2$.

Proof. As we already noted, we have $Q_{p+1}(x) = x^{(p+1)/2} + 1$ if p is a Mersenne prime. Then

$$x^{n+(p+1)/2} \equiv - x^n \bmod Q_{p+1}(x) \text{ for all } n \ge 0,$$

and so

$$D_p(x) = \sum_{n=0}^{(p-1)/2} d_n x^n + \sum_{n=0}^{(p-1)/2} d_{n+(p+1)/2} x^{n+(p+1)/2}$$

$$\equiv \sum_{n=0}^{(p-1)/2} (d_n - d_{n+(p+1)/2}) x^n \bmod Q_{p+1}(x).$$

Therefore, $Q_{p+1}(x)$ divides $D_p(x)$ in GF(p)[x] if and only if the condition in the lemma is satisfied.

Lemma 16. If $p = 2^k - 1$, $k \ge 2$, is a Mersenne prime, then

$$d_{(p+1)/4} = (k - 1)(2^{3k-3} + 2^{2k-3}),$$

$$d_{(3p+3)/4} = (k - 1)(2^{3k-3} - 2^{2k-3}).$$

Proof. By definition,

$$d_{(p+1)/4} = \sum_{\substack{0 \le i,j \le p \\ (p+1)|(ij-(p+1)/4)}} ij.$$

Now $(p + 1)|(ij - (p + 1)/4)$ if and only if there is an integer t satisfying $ij = (p + 1)/4 + t(p + 1)$ $= (4t + 1)2^{k-2}$. So, i and j must be of the form $i = i_1 2^h$ and $j = j_1 2^{k-h-2}$ with $0 \le h \le k - 2$ and $i_1 j_1 \equiv 1 \bmod 4$. Since the solutions of the latter congruence are $i_1 \equiv j_1 \equiv 1 \bmod 4$ and $i_1 \equiv j_1 \equiv 3 \bmod 4$, we obtain

$$d_{(p+1)/4} = 2^{k-2} \sum_{h=0}^{k-2} \sum_{m=0}^{2^{k-h-2}-1} (4m + 1) \sum_{n=0}^{2^h-1} (4n + 1) + 2^{k-2} \sum_{h=0}^{k-2} \sum_{m=0}^{2^{k-h-2}-1} (4m + 3) \sum_{n=0}^{2^h-1} (4n + 3)$$

$$= 2^{k-2} \left(\sum_{h=0}^{k-2} (2^{2k-2h-3} - 2^{k-h-2})(2^{2h+1} - 2^h) + \right.$$

$$\left. + \sum_{h=0}^{k-2} (2^{2k-2h-3} + 2^{k-h-2})(2^{2h+1} + 2^h) \right)$$

$$= 2^{2k-4} \sum_{h=0}^{k-2} ((2^{k-h-1} - 1)(2^{h+1} - 1) + (2^{k-h-1} + 1)(2^{h+1} + 1))$$

$$= 2^{2k-4} \sum_{h=0}^{k-2} (2^{k+1} + 2)$$

$$= (k - 1)(2^{3k-3} + 2^{2k-3}).$$

The formula for $d_{(3p+3)/4}$ is shown by similar arguments.

Theorem 17. If $p \ge 7$ is a Mersenne prime, then there exists an inversive congruential generator which has period length p and passes the $(p - 2)$-dimensional lattice test.

Proof. By combining Theorem 12 and Lemma 15, we see that if $p \ge 7$ is a Mersenne prime, then all inversive congruential generators with period length p fail the $(p - 2)$-dimensional lattice test if and only if the condition in Lemma 15 holds. Thus, to prove the theorem it suffices to show that this condition is not satisfied for $n = (p + 1)/4$. Indeed, if $p = 2^k - 1$, then Lemma 16 shows that we have $d_{(p+1)/4} \equiv d_{(3p+3)/4} \bmod p$ if and only if $(k - 1)2^{2k-2} \equiv 0 \bmod p$. But since p is odd, this congruence is equivalent to $k - 1 \equiv 0 \bmod p$, which is obviously not valid.

The proof of Lemma 15, and thus of Theorem 17, relies on the fact that the cyclotomic polynomial Q_{p+1} has a simple form. Another case in which Q_{p+1} has a simple form is $p = 2q - 1$ with an odd prime q. Then by the standard formula for cyclotomic polynomials (see [10, Theorem 3.27]) we have $Q_{p+1}(x) = (x^q + 1)/(x + 1)$. Therefore, $Q_{p+1}(x)$ divides $D_p(x)$ in $GF(p)[x]$ if and only if $x^q + 1$ divides $(x + 1)D_p(x)$ in $GF(p)[x]$. Then, by proceeding as in the proof of Lemma 15, we get that $Q_{p+1}(x)$ divides $D_p(x)$ in $GF(p)[x]$ if and only if $d_{n+(p+1)/2} + d_{n+(p-1)/2} \equiv d_n + d_{n-1}$ mod p for $0 \le n \le (p - 1)/2$. However, because of the difficulty of providing general formulas for the integers d_n, we have not yet been able to use this condition to prove an analog of Theorem 17 for primes of the form $p = 2q - 1$.

Acknowledgments

We are grateful to J. Eichenauer-Herrmann and H. Grothe of the Technische Hochschule Darmstadt (Germany) for providing us with their computer data. The first author would like to thank the Institute for Information Processing of the Austrian Academy of Sciences in Vienna (Austria) for its hospitality and for the use of its facilities.

References

[1] I. F. Blake, S. Gao and R. C. Mullin, Explicit factorization of $x^{2^k} + 1$ over F_p with prime $p \equiv 3 \bmod 4$, *Applicable Algebra in Engrg. Comm. Comp.* 4 (1993) 89-94.

[2] W.-S. Chou, On inversive maximal period polynomials over finite fields, *Applicable Algebra in Engrg. Comm. Comp.*, to appear.

[3] J. Eichenauer, H. Grothe and J. Lehn, Marsaglia's lattice test and non-linear congruential pseudo random number generators, *Metrika* 35 (1988) 241-250.

[4] J. Eichenauer and J. Lehn, A non-linear congruential pseudo random number generator, *Statist. Papers* 27 (1986) 315-326.

[5] J. Eichenauer and H. Niederreiter, On Marsaglia's lattice test for pseudorandom numbers, *Manuscripta Math.* 62 (1988) 245-248.

[6] J. Eichenauer-Herrmann, Inversive congruential pseudorandom numbers avoid the planes, *Math. Comp.* 56 (1991) 297-301.

[7] J. Eichenauer-Herrmann, Inversive congruential pseudorandom numbers: a tutorial, *Internat. Statist. Rev.* 60 (1992) 167-176.

[8] M. Flahive and H. Niederreiter, On inversive congruential generators for pseudorandom numbers, in : G. L. Mullen and P. J.-S. Shiue, Eds., *Finite Fields, Coding Theory, and Advances in Communications and Computing* (Marcel Dekker, New York, 1993) 75-80.

[9] R. Lidl, G. L. Mullen and G. Turnwald, *Dickson Polynomials* (Longman, Harlow, Essex, 1993).

[10] R. Lidl and H. Niederreiter, *Finite Fields* (Addison-Wesley, Reading, MA, 1983).

[11] H. Niederreiter, Remarks on nonlinear congruential pseudorandom numbers, *Metrika* 35 (1988) 321-328.

[12] H. Niederreiter, *Random Number Generation and Quasi-Monte Carlo Methods* (SIAM, Philadelphia, PA, 1992).

Discrepancy lower bound in two dimensions

HENRI FAURE

C.M.I, Université de Provence
39, rue Joliot-Curie, F-13453 Marseille, Cedex 13 France
or
Laboratoire de Mathématiques Discrètes, U.P.R. 9016
163 avenue de Luminy, case 930, F-13288 Marseille Cedex 09 France

Abstract - In this communication we get, for the first time, the exact order of magnitude for the discrepancy of a two-dimensional sequence independently constructed by I.M. Sobol' and S. Srinivasan; moreover this sequence has the smallest discrepancy presently known in two dimensions.

I INTRODUCTION.

First, recall the definition of the star-discrepancy for an infinite sequence $X = (x_n)_{n \geq 1}$ in the s-dimensional unit cube $I^s = [0, 1[^s$.

A subinterval P of I^s is the product of s intervals $[a_k, b_k[$ of $[0, 1[$.

Given a set of non negative integers T, let $A(P; T; X)$ be the number of points of the sequence X with index n belonging to T such that x_n belongs to P and let $E(P; T; X)$ be the difference $A(P; T; X) - |T| |P|$ in which $|T|$ is the cardinal of T and $|P|$ the volume of P. Then the *discrepancy* $D(N, X)$ of the sequence X is defined by $D(N, X) = \sup\limits_{P \in \mathcal{P}_s} \; | E(P;]0, N]; X) |$ in which $\mathcal{P}_s$ is the set of subintervals of I^s.

If the subintervals are restricted to the family $\mathcal{P}_s^*$ for which the $a_k = 0$ one gets the *star-discrepancy* D^*. Taking in account the imprecision of knowledge in dimension $s \geq 2$, we may limit the study to D^*.

In arbitrary dimension, the only lower bound is that of K.F. Roth [8], obtained in 1954 : there exists a constant $C_s > 0$ such that, for every infinite sequence X, one has

$$\limsup_{N \to \infty} (D^*(N, X)/(Log N)^{s/2}) \geq C_s.$$

Only two improvements have been provided by W.M.Schmidt [9] in 1972 and G. Halász [4] in 1981 who, by different methods, have got the order $Log\ N$ instead of $\sqrt{Log\ N}$ in one dimension, and by J. Beck [1] in 1989 with $(Log\ N)(Log\ Log\ N)^{\frac{1}{8}-\epsilon}$ instead of $Log\ N$ in two dimensions.

A lot of sequences with very small irregularities have been constructed for numerical integration purposes; their discrepancy is bounded by a constant time $(Log\ N)^s$ in dimension s; it is conjectured that $(Log\ N)^s$ is the exact order, because the lower bound of Roth is a bound for the mean-square discrepancy.

Chronologically, the following constructions may be quoted: J.H. Halton [5] in 1960, I.M. Sobol' [10] in 1967, S. Srinivasan [11] in 1978, H. Faure [2] in 1982 and H. Niederreiter [6] [7] in 1987-1988. Note that even for these special sequences, the only lower bound is that of Roth (or Beck in two dimensions).

The aim of this communication is to announce we have got the exact order $(Log\ N)^2$ for the discrepancy of a special sequence S independently constructed by Sobol' and Srinivasan; this study is a joint work with Henri Chaix (first announcement in [3]).

In fact, this sequence S is the sequence in two dimensions with the lowest discrepancy presently known: if we set

$$d_s(X) = \lim_{N \to \infty} \sup \left(D^*(N, X)/(Log N)^s) \right),$$

we have the upper bound $d_2(S) \le \frac{1}{8(Log2)^2} = 0,26...$, obtained by Niederreiter [6] improving bounds already computed by Sobol', Srinivasan and Faure; for the best Halton sequence H we have only $d_2(H) \le \frac{1}{2 Log2\ Log3} = 0,65...$ [2].

With the notation above, our result is the following:

$$d_2(S) \ge \frac{1}{24(Log2)^2} = 0,08....$$

The numerical approach and the analogy with sequences of the same type in one dimension lead to conjecture that this lower bound is the exact value of $d_2(S)$.

In the following, we shall deal only with this special sequence S.

II DEFINITION OF THE SEQUENCE S AND THEOREM.

Van der Corput sequence in base b : given integers $b \ge 2$ and $n \ge 1$, let $n - 1 = \sum_{r=0}^{\infty} a_r(n)\ b^r$ be the expansion of $(n - 1)$ in base b ; by definition, the van der Corput sequence in base b is the sequence ϕ_b defined by $\phi_b(n) = \sum_{r=0}^{\infty} a_r(n)\ b^{-r-1}$.

Pascal matrix modulo b and associated sequence : let $C = \left(\binom{h}{l} \right)$ where $\binom{h}{l}$ is the usual binomial coefficient if $l \le h$ and $\binom{h}{l} = 0$ if $l > h$; it is easy to see that $C^k = \left(\binom{h}{l} k^{h-l} \right)$.

Then if we set $x_n^k = C^{k-1} \phi_b(n) \bmod b$, that is $x_n^k = \sum_{j=0}^{\infty} y_j^k(n)\ b^{-j-1}$

with $y_j^k(n) = \sum_{r \ge j} \binom{r}{j} (k-1)^{r-j}\ a_r(n) \bmod b$, we obtain an I^s-valued sequence $X = (x_n)$ with $x_n = (x_n^1, \ldots, x_n^s)$; this sequence has distinct coordinates as soon as $b \geq s$.

In the special case where $b = s = 2$ we get the sequence S, whose discrepancy is the aim of this study.

Remarks.

1°) Halton sequences are obtained by considering $\phi_b(n)$ with coprime basis on each coordinate. In two dimensions, the simplest is $(\phi_2(n), \phi_3(n))$; the exact behaviour of its discrepancy is still a conjecture.

2°) Both Sobol' and Srinivasan have constructed families of sequences: Sobol' by the action of primitive polynomials of $\mathbf{F}_2[X]$ on ϕ_2; Srinivasan by a step by step algorithm with two possible choices at each step; in each case, the sequence S corresponds to the simplest situation (first degree polynomials for Sobol' and the same choice at each step for Srinivasan).

THEOREM. - *Let be given the square $Q = [0, 2/3[^2$ and the sequence $N_\lambda = (16^{\lambda+1} - 1)/15$. Then $E(Q;]0, N_{2^\tau - 1}]; S) = \frac{2}{3} \, 4^\tau + \frac{17}{9} \, 2^\tau - 3^\tau - 1$, where τ is an integer greater than or equal to 1.*

In fact, we have got exact formulas for any integer λ, but they do not give an explicit final result for the present ; anyway, the numerical approach shows that the lower bound cannot be improved with arbitrary λ.

COROLLARY.

- *We have the lower bound:* $d_2(S) \geq \frac{2}{3} \, \frac{1}{(\log 16)^2} = \frac{1}{24(\log 2)^2}$.

Problems. This lower bound is a first step in the study of the general problem of lower bounds for the discrepancy of usual sequences:

What about the nearest families of Sobol' and Srinivasan ?

In dimension two, what about bases $b > 2$ and Halton sequences ?

What about dimensions greater than two ?

The more accessible question seems to concern the dimension two with bases $b > 2$.

III INDICATIONS ON THE DEMONSTRATION.

Elementary intervals and nets:

- An *elementary interval in base 2* of I^2 is an interval $[\frac{u}{2^p}, \frac{u+1}{2^p}[\times[\frac{v}{2^q}, \frac{v+1}{2^q}[$, with integers u, v, p, q verifying $0 \leq u < 2^p$ and $0 \leq v < 2^q$;

- a $(0, m, 2)$-*net in base 2* is a finite set with 2^m elements of I^2 such that every elementary interval of volume 2^{-m} contains exactly one point of the net;

- a $(0, 2)$-*sequence in base 2* is an I^2-valued sequence such that for all integers m and l, the finite set $X^l_m = \{x_{l2^m+1}, \ldots, x_{(l+1)2^m}\}$ is a $(0, m, 2)$-net in base 2.

PROPOSITION [10]. - *The sequence S is a $(0, 2)$-sequence in base 2.*

Set $E(N_\lambda) = E(Q;]0, N_\lambda]; S)$ and $E(S(T)) = E(Q; T; S)$.

LEMMA 1 *(Expression for $E(N_\lambda)$).*

- *The sequence (l_μ) being defined by $l_0 = 0$ and $l_\mu = \sum_{i=1}^{\mu} 2^{4i}$ for $\mu \geq 1$, we*

have $E(N_\lambda) = \sum_{\nu=0}^{\lambda} E(S^{l_{\lambda-\nu}}_{4\nu})$.

This lemma comes from the additivity of the remainder $E(N_\lambda)$ and from the equality $N_\lambda = \sum_{i=0}^{\lambda} 2^{4i}$.

LEMMA 2 *(Expression of $E(S^{l_{\lambda-\nu}}_{4\nu})$ for $\nu \geq 1$).*

- *Set successively :*

$$\alpha_k = \sum_{i=0}^{2\nu-k} 2^{-2i-1} \text{ for } 1 \leq k \leq \nu+1, \; \alpha'_k = \alpha_k + 2^{-4\nu+2k-1} \text{ for } 1 \leq k \leq \nu$$

and $\alpha'_{\nu+1} = \alpha'_\nu$,

$$\beta_1 = 0 \text{ and } \beta_k = \sum_{i=0}^{k-2} 2^{-2i-1} \text{ for } 2 \leq k \leq \nu, \; \beta'_k = \beta_k + 2^{-2k+1} \text{ for } 1 \leq k \leq \nu.$$

Besides set: $\pi_k = [\alpha_k, \alpha'_k[\times [\beta_k, \beta'_k[\text{ for } 1 \leq k \leq \nu, \pi_{\nu+1} = [\alpha_{\nu+1}, \alpha'_{\nu+1}[^2,$

$\pi'_k = [\beta_k, \beta'_k[\times [\alpha_k, \alpha'_k[\text{ for } 1 \leq k \leq \nu \text{ and } D_\nu = Q \cap (\bigcup_{k=1}^{\nu+1} \pi_k \cup \bigcup_{k=1}^{\nu} \pi'_k).$

Then $E(S^{l_{\lambda-\nu}}_{4\nu}) = A(D_\nu; S^{l_{\lambda-\nu}}_{4\nu}) - \frac{2}{3}(\nu + \frac{2}{3})$.

Indeed, S is a $(0, 2)$-sequence in base 2, so only the elementary intervals sitting astride Q, that is intervals π_k et π'_k, must be kept back to compute the remainder; moreover, we have $|D_\nu| = \frac{2}{3}(\nu + \frac{2}{3})2^{-4\nu}$.

Remark. For $\nu = 0$, we get directly $E(S^{l_\lambda}_0) = 1 - 4/9$.

LEMMA 3 *(Condition for $\pi_k \cap Q \cap S_{4\nu}^{l_{\lambda-\nu}} \neq \phi$).*
- For $1 \leq k \leq \nu$ and $1 \leq \nu \leq \lambda$, consider the determinant $\Delta_{\nu,k}^{\lambda}$ of order $2\nu + 1$:

$$\begin{vmatrix} 1 & \cdots & 0 & \cdots & 0 & 1 \\ 0 & & 0 & & \vdots & \vdots \\ \vdots & \ddots & 0 & \vdots & \vdots & \vdots \\ 0 & \cdots & 1 & \cdots & 0 & 1 \\ \binom{0}{0} & \cdots & \binom{2\nu-k}{0} & \cdots & \binom{2\nu-1}{0} & 1+\sum_{j=\nu+1}^{\lambda}\binom{2j}{0} \\ 0 & & & & & \\ \vdots & & \vdots & \vdots & \vdots & \vdots \\ 0 & \cdots & \binom{k-2}{k-2} & \cdots & \binom{2\nu-k}{k-2} & \cdots & \binom{2\nu-1}{k-2} & 1+\sum_{j=\nu+1}^{\lambda}\binom{2j}{k-2} \\ 0 & \cdots & 0\binom{k-1}{k-1} & \cdots & \binom{2\nu-k}{k-1} & \cdots & \binom{2\nu-1}{k-1} & \sum_{j=\nu+1}^{\lambda}\binom{2j}{k-1} \end{vmatrix}$$

Then $A(\pi_k \cap Q; S_{4\nu}^{l_{\lambda-\nu}}) = 1$ if and only if $\Delta_{\nu,k}^{\lambda} = 0$ mod 2.

This lemma is fundamental for the proof and will be detailled in the full paper to appear; the main ingredient is that the interval π_k contains exactly one point of $S_{4\nu}^{l_{\lambda-\nu}}$, which is expressed by a linear system of Cramer; moreover, the point belongs to Q and this property implies a condition of compatibility leading to the nullity of $\Delta_{\nu,k}^{\lambda}$.

Remark. The cardinal of $\pi_{\nu+1} \cap Q \cap S_{4\nu}^{l_{\lambda-\nu}}$ is always equal to one.

LEMMA 4 *(Shifting operation).*
- Given a k-tuple of integers $(l_1, \cdots, l_k)$, we define the Shift to be the operation getting the k-tuple $(l_1, l_2 + l_1, \cdots, l_{k-1} + l_{k-2}, l_k + l_{k-1})$. Then after d shifts, $d \geq k$, the last term of the k-tuple is $\sum_{h=0}^{k-1} \binom{d}{h} l_{k-h}$.

This lemma is a consequence of the properties of binomial coefficients.

LEMMA 5 *(Computation of $\Delta_{\nu,k}^{\lambda}$).*
- Given an integer ν, define the integer μ by $2^{\mu-1} \leq \nu < 2^{\mu}$. Then, for integers $2p$ and $2p + 1$ between 1 and ν, the following relations hold: :

$$\Delta_{\nu,2p}^{\lambda} = \binom{2^{\mu} - \nu + p - 1}{p} + \binom{2^{\mu} - \nu + p + \lambda}{p} \qquad \text{mod } 2$$

and $\quad \Delta_{\nu,2p+1}^{\lambda} = \binom{2^{\mu} - \nu + p - 1}{p} + \binom{2^{\mu} - \nu + p + \lambda + 1}{p+1}$ mod 2.

First we perform $(2^{\mu+1} - 2\nu + k - 1)$ shifts on the k last rows of $\Delta^\lambda_{\nu;k}$; then we add the $2\nu - k + 1$ first rows to the last one in order to yield $\Delta^\lambda_{\nu,k}$ as a product of diagonal terms equal to 1 except the last one; finally the formula we got is reduced by means of the properties of binomial coefficients, in particular the Vandermonde convolution.

For intervals π'_k in lemma 2, we get in the same way determinants $\Delta'^\lambda_{\nu,k}$ which are equal to the $\Delta^\lambda_{\nu,k}$ of lemma 3, extended to integers k running from $\nu + 1$ to 2ν as pointed out in the following lemma:

LEMMA 6 *(Computation of $\Delta'^\lambda_{\nu,k}$)*.

- *For $1 \le k \le \nu$, we have $\Delta'^\lambda_{\nu,k} = \Delta^\lambda_{\nu,2\nu-k+1}$ mod 2.*

LEMMA 7 *(particular case where $\lambda = 2^\tau - 1$)*.

- *With the notation $\Delta_{\nu,k} = \Delta^{2^\tau-1}_{\nu,k}$, we have the formulas: $\Delta_{\nu,2p} = 0$ for $1 \le p \le \nu$ and $\Delta_{\nu,2p+1} = \binom{2^\tau-\nu+p-1}{p+1}$ mod 2 for $0 \le p \le \nu - 1$.*

By means of this lemma, which expresses $\Delta_{\nu,k}$ by a single binomial coefficient, it is possible to account the points of $S^{l_{\lambda-\nu}}_{4\nu}$ belonging to D_ν in the case where $\lambda = 2^\tau - 1$.

After these lemmas, the proof of the theorem consists essentially to account the number of zeros in the sub-matrix of order 2^τ of the Pascal matrix modulo 2 restricted to the 2^τ first rows; this yields for $\lambda = 2^\tau - 1$:

$$\sum_{\nu=1}^{\lambda} A(D_\nu; S^{l_{\lambda-\nu}}_{4\nu}) = 4^\tau - 3^\tau + 2(2^\tau - 1)$$

and by subtraction of $\sum_{\nu=1}^{\lambda} 2^{4\nu} |D_\nu| = \frac{1}{3} 4^\tau + \frac{1}{9} 2^\tau - 1$, we get $E(N_\lambda) = \frac{2}{3} 4^\tau - 3^\tau + \frac{17}{9} 2^\tau - 1$, which achieves the proof.

The whole proof will be detailed in a full paper to appear later.

REFERENCES

[1] J. BECK, A two dimensional van Aardenne-Ehrenfest theorem in irregularities of distribution, *Compositio Math.*, n° 72, 1989, p. 269-339.

[2] H. FAURE, Discrépance de suites associées à un système de numération (en dimension s), *Acta Arith.*, XLI, 1982, p. 337-351.

[3] H. FAURE ET H. CHAIX, Minoration de discrépance en dimension deux, *C.R.Acad.Paris, série 1*, t.319, 1994, p.1-4.

[4] G. HALÀSZ, On Roth's method in the theory of irregularities of point distribution, *Recent Progress in Analytic Number Theory, Academic Press,* n° 2, 1981, p. 79-94.

[5] J.H. HALTON, On the efficiency of certain quasi-random points in evaluating multi-dimensional integrals, *Numer. Math.,* n° 2, 1960, p. 84-90.

[6] H. NIEDERREITER, Point sets and sequences with small discrepancy, *Monatsh. Math.,* n° 104, 1987, p. 273-337.

[7] H. NIEDERREITER, Low discrepancy and low dispersion sequences, *J. Number Theory,* n° 30-1, 1988, p. 51-70.

[8] K.F. ROTH, On irregularities of distribution, *Mathematika,* n° 1, 1954, p. 73-79.

[9] W.M. SCHMIDT, Irregularities of distribution, *Acta Arith.,* XXI, 1972, p. 45-50.

[10] I. M. SOBOL', On the distribution of points in a cube and the approximate evaluation of integrals, *USSR Comp. math. and Math. Physics,* n° 7, 1967, p. 86-112.

[11] S. SRINIVASAN, On two dimensional Hammersley sequences, *J. of Number Theory,* n° 10, 1978, p.421-429.

STABLE ESTIMATORS FOR SELF-ADJUSTING SIMULATIONS

Bennett L. Fox *   George W. Heine

Abstract

We present two estimators which converge almost surely to the nearest integer to $t = \mathrm{E}\,T$, with few restrictions on the random variable T. This works even when t is of the form $m + 1/2$ for an integer m. Applications to simulated annealing and to the design of other Monte Carlo experiments are indicated.

1 Introduction

We first briefly describe the problem and then motivate it.

The problem. At step n, we observe $T(n)$ and use $T(1), \ldots, T(n)$ to compute an estimate $H(n)$ of an unknown parameter h. If $H(n) \xrightarrow{\text{a.s.}} h$, then H is *consistent*. If there is a random time N, finite almost surely (but generally not a stopping time), such that $H(n) = h$ for all $n \geq N$, then H is *asymptotically stable*. Let $\overline{T}(n) = n^{-1} \sum_{j=1}^{n} T(j)$ and suppose that $\overline{T}(n) \xrightarrow{\text{a.s.}} t \in \mathbb{R}$. Say that we are really interested in h, the nearest integer to t (rounding up in case of ties). We give two asymptotically-stable estimators of h based on $\overline{T}(n)$.

In Section 2, we present a simple finite-precision estimator H_1, based on frequency counts, and give an elementary proof that it is asymptotically stable. In Section 3, we present another estimator H_2 which uses the sample mean and (optionally) the sample variance; the proof of its stability depends on the law of the iterated logarithm.

We give a more general and abstract statement of the problem in Section 1.1.

Remark 1: Breaking ties by rounding up is is an arbitrary choice. In Section 1.2, the problem statement implies that h is a positive integer (and that $T(n)$ is in the positive reals); so, there, we round $T(n)$ to the nearest positive integer — avoiding the possibility of estimating h as zero. Our estimators can be modified trivially to handle this situation.

Remark 2: It is important to be clear about the difference between an estimator which is asymptotically stable and one which is merely consistent. If we had wanted, say, a static point estimator, restricted to the integers, of a parameter with true value 1.5, then we would be indifferent between an estimate of 1.0 and 2.0. Asymptotic stability is needed

---

*The work of this author was partially supported by a grant from the Air Force Office of Scientific Research and the Office of Naval Research, Contract #F49620-90-C-0033.

when we are interested, not in the parameter *per se*, but in a simulation process dependent on an estimate of the parameter.

Remark 3: An example shows what can go wrong. Let $\langle x \rangle$ denote x rounded to the nearest integer, rounding upward in case of ties (i.e., $\langle x \rangle = \lfloor x + \frac{1}{2} \rfloor$). Set $t = \mathrm{E}\,T(n))$ and let $h = \langle t \rangle$. The estimator $H(n) = \langle \overline{T}(n) + \frac{1}{2} \rangle$ is not asymptotically stable when t is a half integer; for large n, it oscillates unpredictably between $\lfloor t \rfloor$ and $\lceil t \rceil$. If t is near a half integer, then damped oscillation occurs; although there is eventual stability, arguably the rate of damping is very slow — see Section 1.3. In that case, the estimators in Sections 2 and 3 also produce damped oscillation — though possibly the rate of damping is higher. Thus, our estimators have theoretical value — giving asymptotic stability even when t is a half-integer. They may also have practical value if they increase the damping rate when t is near a half-integer. We return to this point in Remark 7 close to the end of the paper.

Motivation. The problem above arises in experimental design as in Section 1.2. It also arises in simulated annealing with a noisy objective function. In the latter setup, we have two Markov chains, X and Y, both on a finite state space S, with expected cost $c(s)$ associated with state s. The cost might be unknown and estimated by simulation — for example, S might be the possible topologies of a queueing network, and $c(s)$ the expected time in the network. Let S_0 be the set of states which minimize $c(\cdot)$. While X bases its moves on a (hypothetically) known cost function $c(\cdot)$, we base move n of Y on the current estimate $\hat{c}(\cdot, n)$ of $c(\cdot)$. Suppose that $X \xrightarrow{\mathrm{pr}} S_0$. Theorem 2 of Fox and Heine (1995) says, roughly, that if $\hat{c}(\cdot)$ is (merely) consistent and if the respective move probabilities of Y approach those of X fast enough, $Y \xrightarrow{\mathrm{pr}} S_0$. With asymptotic stability, those respective probabilities coincide exactly after an almost-surely finite time. However, without it, our "fast-enough" condition is hard to check. In fact, we know of no easy-to-check condition.

1.1 Generalized setup

The parameter of interest h need only lie in some arbitrary finite or countably-infinite discrete set $\mathcal{F}$, and the observations $T(i)$ in some superset $\mathcal{G}$ of $\mathcal{F}$. Thus, $H(n)$ maps the n-fold Cartesian product of $\mathcal{G}$ to $\mathcal{F}$. Every element of $\mathcal{G}$ is represented by a single element of $\mathcal{F}$. The parameter h is the representative of some t in $\mathcal{G}$. With hardly any loss of generality, assume that h and $T(n)$ are scalars and, by rescaling if necessary, that $\mathcal{F}$ is a set of consecutive integers. The set $\mathcal{G}$ is either a subset of the real numbers (the case introduced at the outset) or a subset of the integer multiples of $1/d$, where d is an integer greater than one. In the second case, $T(n)$ is measured only to finite precision — but possibly on a finer scale than h.

Remark 4: In the simulated annealing example, suppose that the cost $c(s)$, for each state s, is a computer-representable number, hence in a finite set $\mathcal{F}$. By rescaling, we make $\mathcal{F}$ a finite set of consecutive integers — thus, in the problem format above. Fox (1995) gives a rationale for introducing $\mathcal{F}$. Ideally, it would be better to do without $\mathcal{F}$ and just use the sample mean $\overline{T}$ to estimate the cost function. This can be done — but at a price. Thus, Gelfand and Mitter (1989) give alternative hypotheses, and Yan and Mukai (1992) give an optimization scheme that is not simulated annealing; Fox (1995) criticizes both.

Remark 5: In the example of Section 1.2, the problem dictates that $\mathcal{F}$ be a set of positive consecutive integers and that $T(n)$ be in the positive reals, so there we make no approximations.

1.2 Example: Splitting

At run n, we get both $T(n)$ and an observation of the parameter ξ of direct interest. In splitting, a run branches into subruns from some (possibly random) point onwards. Here the optimal number h of branches depends on the respective expected work and output variance for each branch and on the common path up to the split; $T(n)$ is the corresponding vector of observations of these auxiliary parameters. Suppose that, for each integer parameter g, a work-normalized estimator (in the sense of Fox and Glynn (1990), for example) of ξ using the simulation strategy based on g converges weakly to a random variable $Z(g)$, as the work to compute the estimator goes to infinity. If $H(\cdot)$ is an asymptotically stable estimator of h, then Theorem 2 of Fox and Heine (1995) shows that $Z(H(n))$ converges weakly to a random variable $Z(h)$, where h is the optimal parameter in the sense that it minimizes the variance of $Z(g)$ over positive integers g.

The variance of $Z(g)$ for a positive integer g is obtained by a calculation $f(g)$ which can be applied to any positive real number. Call t the minimizer of $f(x)$ for $x \in \mathbb{R}^+$; it is obtained by simple calculus. We redefine $T(n)$ as the estimator of $\sqrt{t^2 + \frac{1}{4}}$ based on the n-th block of runs, with block size fixed. In unpublished work (circa 1983), Fox and Glynn show that h is the nearest positive integer to $\sqrt{t^2 + \frac{1}{4}}$. A sketch of their argument follows. The function to be minimized has the form $f(m) = (am + b)((c/m) + d)$ with $f(0) = f(\infty) = \infty$, and $a, b, c,$ and d positive. Since f is discretely convex, the assertion above is equivalent to $f(h) \le f(h-1)$ and $f(h) \le f(h-1)$. To make observations of the auxiliary parameters iid, we base it on two branches in each run, regardless of the number of branches actually used.

We can use a stable estimator of h in the approach of Fox and Heine (1995) to get an asymptotically-optimal work-normalized estimator of ξ.

1.3 Link to the Arc Sine Distribution

We noted above that the sequence $\overline{T}(\cdot)$ of sample means is not an asymptotically stable estimator of $h = \langle t \rangle$ when $t = \mathrm{E}\,(T)$ is a half-integer. Here we show that a naive attempt to correct the situation fails. We observe the entire sequence of sample means $\overline{T}(1), \ldots, \overline{T}(n)$ (computed to double precision, say), counting how many are above or below the half-integer line, and rounding up or down accordingly. This actually makes matters worse. For large n, the sequence of sample means closely approximates Brownian motion, and the proportion of the $\overline{T}(j)$ which lie above t has an arc sine distribution—see Feller (1968, 397 ff.). The probability is greater than 0.5 that more than 80% or fewer than 20% of the sample means will be above or below the line. When t is approximately a half-integer, this proportion still forms a beta distribution with a high probability of being near the extremes.

2 The Estimator H_1 — Frequency Counts

Let $M_k(n)$ be the number of observations equal to k up to step n; that is,

$$M_k(n) = \sum_{j=1}^{n} I\{T(j) = k\}.$$

Each observation $T(n)$ is formed by adding a random error ϵ_n to t, and then rounding to the nearest integer:

$$T(n) = \langle t + \epsilon_n \rangle$$

Assume that

(i)   the errors ϵ_n are iid;

(ii) the distribution of the ϵ_n is symmetric about zero and has a density with finite support which is strictly decreasing on the positive real numbers.

From (i), the $T(n)$ have a common discrete distribution function with mass points

$$\dots, p_{-1}, p_0, p_1, p_2, \dots$$

on the integers. From (ii), when t is not a half integer the set $\{p_k > 0 : k \in \mathbb{Z}\}$ is finite and has p_h for its unique largest element. In the exceptional case where t is a half integer, there will be maxima at both $p_{\lfloor t \rfloor}$ and $p_{\lceil t \rceil}$.

Remark 6: For large n, we expect $M_k(n)/n$ to approximately equal p_k, at least for those k not in the tail of the distribution. This is the intutive basis for our estimator.

Let $f(n)$ be a function on the positive integers which is non-decreasing in n and which satisfies both $f(n) = o(n)$ and $\log(n) = o(f(n))$. (For example, $f(n) = \sqrt{n}$.) We round up when t is a half-integer, consistent with our assumption about h in the Introduction. Below, break ties for the second-largest integer arbitrarily. At each step n, we implicitly overwrite $M_k(n-1)$ with $M_k(n)$ for $k \in \mathcal{F}$, so we suppress the index n in $M_k(n)$. At step n, the inputs are $f(n)$ and the counters M_j for $j \in \mathcal{F}$. When using arg max, if there are multiple maximizers, take the largest.

ALGORITHM STABLE1

> *For* each step n:
> > *Make* an observation $T(n)$;
> > *Set* $k \leftarrow T(n)$;
> > *Set* $M_k \leftarrow M_k + 1$;
> > *If* the set $\mathcal{N} = \{M_k : k \in \mathcal{F}\}$ has a single maximum at m,
> > > *and* $\arg\max \mathcal{N} \setminus \{m\} = m + 1$,
> > > *and* $|M_m - M_{m+1}| < f(n)$, *then*
> > > > *Set* $H_1(n) \leftarrow m + 1$.
> > *Else*
> > > *Set* $H_1(n) \leftarrow \arg\max\{M_k : k \in \mathcal{F}\}$.
> *End.*

The following two propositions show that the output $H_1(n)$ has the desired properties. We assume that the error terms ϵ_n and the function $f(n)$ satisfy the conditions stated immediately before the algorithm.

Proposition 2.1 *If t is not of the form $m + \frac{1}{2}$ for some integer m, then almost surely $H_1(n)$ equals h for all but finitely many n.*

Proposition 2.2 *If t is of the form $m + \frac{1}{2}$ for some integer m, then almost surely $H_1(n) = m + 1$ for all but finitely many n.*

Proof of Proposition 2.1: Assume w.l.o.g. that $m \le t < m + \frac{1}{2}$. By the conditions imposed on the distribution of ϵ_n, the set $\{p_k : k \in \mathbb{Z}\}$ has its unique largest value at m and its unique second-largest value at $m + 1$. The finiteness of $\{k \in \mathbb{Z} : p_k > 0\}$ and the strong law of large numbers imply that a.s. for large enough n, the set $\{M_k(n) : k \in \mathbb{Z}\}$ attains its unique maximum at $M_m(n)$ and its unique second maximum at $M_{m+1}(n)$. For arbitrary $\delta > 0$, another application of the strong law shows that for all large n, on almost all sample paths

$$|M_m(n) - M_{m+1}(n)| < n(p_m - p_{m+1}) - \delta \tag{1}$$

and the right side of (1) is bounded below by $f(n)$, for all large n. $\square$

Proof of Proposition 2.2: Here $p_m = p_{m+1}$; these two are the only maxima in the set $\{p_k : k \in \mathbb{Z}\}$. Using the strong law as in the previous proof, the two largest elements of the set $\{M_k(n) : k \in \mathbb{Z}\}$ are $M_m(n)$ and $M_{m+1}(n)$ a.s. for all large n. Let $j(0) = 0$ and

$$j(i + 1) = \min_{n > j(i)} \{T(n) = m \text{ or } T(n) = m + 1\},$$

i.e., $\{j(1), j(2), \ldots\}$ is the sequence of hitting times to the set $\{m, m + 1\}$.
   We prove first that

$$|M_m(j(i)) - M_{m+1}(j(i))| \le f(j(i)) \tag{2}$$

a.s. for all but finitely many i. Consider the Bernoulli process $\{T(j(i)) : i \in \mathbb{Z}^+\}$. It is well known and a simple consequence of the Borel-Cantelli lemmas (see, for example, Billingsley (1986, pp. 53-54)) that there exists a positive nonrandom constant K such that the events

$$\{T(j(i)) = T(j(i) + 1) = \cdots = T(j(i) + K \ln i) = m\}$$

and

$$\{T(j(i)) = T(j(i) + 1) = \cdots = T(j(i) + K \ln i) = m + 1\}$$

occur only finitely often. Let $I_0(\omega)$ be the largest i for which either of these events occurs.
   Since zero is a recurrent state in the random walk on the space of possible values of $M_m - M_{m+1}$, $M_m(j(i)) = M_{m+1}(j(i))$ for arbitrarily large values of i. Let $I_1(\omega)$ be the smallest integer larger than I_0 for which $M_m(j(I_1)) = M_{m+1}(j(I_1))$. Then for all $i \ge I_1$,

$$|M_m(j(i)) - M_{m+i}(j(i))| < K \ln i \le K \ln j(i) < f(j(i)), \tag{3}$$

where the last inequality is true for all large enough i.

Finally we show that (2) is still true when $j(i)$ is replaced by an arbitrary time n. Observing that $j(\cdot)$ is a non-decreasing function on the positive integers, we define an inverse. If $n < j(1)$, set $j^*(n) = 0$; otherwise, let $j^*(n)$ be the largest integer i such that $j(i) \leq n$. The function $j^*(\cdot)$ has the following properties:

$$j^*(j(n)) \quad = n;$$
$$j^*(n) \leq j(j^*(n)) \quad \leq n; \tag{4}$$
$$M_m(j(j^*(n))) \quad = M_m(n); \tag{5}$$
$$M_{m+1}(j(j^*(n))) \quad = M_{m+1}(n); \tag{6}$$

Thus, using (6) and (5), then (3), and finally (4),

$$
\begin{aligned}
|M_{m+1}(n) - M_m(n)| &= |M_{m+1}(j(j^*(n))) - M_m(j(j^*(n)))| \\
&\leq K \ln(j^*(n)) \\
&\leq K \ln n \\
&\leq f(n),
\end{aligned}
$$

a.s. for all large n. $\quad \square$.

3 The Estimator H_2 — Iterated Logarithms

We relax previous assumptions about the error terms ϵ_n. Instead, we assume that they are iid with a zero mean and finite variance σ^2. We compute the sample means $\overline{T}(n)$ to a higher precision d than needed for our final estimator $H_2(n)$.

Let $g(n)$ be a function which satisfies

$$\limsup \frac{\sqrt{2(\sigma^2/n) \ln \ln \sigma^2 n}}{g(n)} \leq 1 \qquad \text{and} \qquad \lim_{n \to \infty} g(n) = 0,$$

almost surely. For example, we could take $g(n) = n^{-1/2}\sqrt{2\hat{\sigma}^2 \ln \hat{\sigma}^2 n}$, where $\hat{\sigma}^2$ is the sample variance; this is valid when, in addition to our other assumptions, the sample variance is a consistent estimator of the true variance. More simply, we could take $g(n) = n^{-1/2} \ln n$. At step n, the inputs are $g(n)$ and $\overline{T}(n-1)$.

ALGORITHM STABLE2

> *For* each step n
>> *Make* an observation $T(n)$
>> *Compute* the sample mean $\overline{T}(n)$ from $T(n)$ and $\overline{T}(n-1)$
>> *Set* $J(n) \leftarrow \lfloor \overline{T}(n) \rfloor + (1/2)$
>> *If* $|\overline{T}(n) - J(n)| \leq g(n)$, *then*
>>> *Set* $H_2(n) \leftarrow \lceil \overline{T}(n) \rceil$
>> *Else*
>>> *Set* $H_2(n) \leftarrow \langle \overline{T}(n) \rangle$
> *End*

By the law of the iterated logarithm,

$$|\overline{T}_n - t| \leq \phi(n)\sqrt{2/n} \leq g(n) \tag{7}$$

for all but finitely many n, where $\phi(n) = \sqrt{\sigma^2 \ln \ln(n\sigma^2)}$. When $t = m + \frac{1}{2}$ for some integer m, then $J(n) = t$ and STABLE2 sets $H_2(n) = h$ for all large n, so that Algorithm 2 is stable by (7).

When t is not a half integer,

$$2\,g(n) < |J(n) - t|$$

for all large n; combining this with (7),

$$
\begin{aligned}
|J(n) - \overline{T}(n)| &\geq |J(n) - t| - |\overline{T}(n) - t| \\
&> 2g(n) - g(n) \\
&= g(n),
\end{aligned}
$$

for all large enough n, almost surely. Thus, once again STABLE2 sets

$$H_2(n) = h = \langle t \rangle$$

for all large n, almost surely. We have proved

Proposition 3.3 *Algorithm* STABLE2 *is stable and consistent.*

Remark 7: Algorithm STABLE2 is valid under weaker assumptions than STABLE1. When the assumptions of STABLE2 hold, we can combine the estimators of this paper: for example, round to the nearest integer unless the tests of both algorithms succeed. That combination has heuristic appeal because it interferes with what one would naturally do less than either algorithm individually, while still assuming that $h \in \mathcal{F}$. If this means we are assuming finite precision, then we could use just $\overline{T}$ up to some (large) fixed n whenever it seems empirically that the corresponding estimates fluctuate less than with the combination of STABLE2 and STABLE1 suggested above. Generally, such comparisons take far less work than the work to simulate.

Acknowledgment

We thank the referee for advice that led us to restructure the Introduction and for detailed corrections and comments.

References

Billingsley, P. (1986). *Probability and Measure* (2nd ed.). New York: John Wiley and Sons.

Feller, W. (1968). *An Introduction to Probability Theory and Its Applications* (2nd ed.), Volume II. New York: John Wiley and Sons.

Fox, B. L. (1995). Simulated annealing: folklore, facts, and directions. In this volume.

Fox, B. L. and P. W. Glynn (1990). Discrete-time conversion for simulating finite-horizon Markov processes. *SIAM J. Appl. Math. 50*, 1457–1473.

Fox, B. L. and G. W. Heine (1995). Probabilistic search with overrides. Technical report, University of Colorado at Denver.

Gelfand, S. B. and S. K. Mitter (1989). Simulated annealing with noisy or imprecise measurements. *J. Optim. Theory Appl. 69*, 49–62.

Yan, D. I. and H. Mukai (1992). Stochastic discrete optimization. *SIAM J. Control Optim. 30*, 594–612.

Authors' mailing address: Department of Mathematics, Campus Box 170, University of Colorado, P.O. Box 173364, Denver, CO 80217-3364.

Internet: bfox@castle.cudenver.colorado.edu, gheine@copper.denver.colorado.edu

A Comparison of Random and Quasirandom Points for Multidimensional Quadrature

Fred J. Hickernell [*]

Abstract

An integral over a unit volume may be approximated by the mean of the values of the integrand on some sample of points from the integration domain. A variety of sampling methods for quadrature have been proposed. These include random samples, randomized orthogonal arrays, good lattice point sets and quasirandom sequences. Recently the author has derived quadrature error bounds by using an ANOVA decomposition and reproducing kernel Hilbert spaces. The error bound coefficients are the parts of the error bound depending only on the sample and not on the integrand. This article compares the error bound coefficients for various random and quasirandom samples. The relative advantages of different sampling schemes are explored. More sophisticated samples are found to be significantly better than a simple random sample if and only if the sample size is large or the integrand is approximately a sum of parts depending on only a few variables.

1 Introduction

Many multidimensional integrals can be written in the form

$$I(f) = \int_{C^d} f(x)\, dx,$$

where $C^d = [0, 1)^d$ is the d-dimensional unit cube. One popular family of quadrature rules is the arithmetic mean of the values of the integrand on a sample, S, of N points:

$$(1) \qquad Q(f) = \frac{1}{N} \sum_{z \in S} f(z).$$

The accuracy of this quadrature rule depends on both f and S. This article explores the problem of choosing S to give accurate results for a broad class of integrands. There are many possibilities for S discussed in the literature. One choice is a simple, uniform random sample of points on the unit cube, which corresponds to the Monte Carlo method in its basic form. Extensions of the Monte Carlo Method are discussed in [1, 3]. In order to obtain greater accuracy many authors have studied deterministic or quasirandom sets that are more evenly distributed than a random sample (see [4, 10, 12, 13, 14, 18] and

[*]Department of Mathematics, Hong Kong Baptist University, Kowloon Tong, Hong Kong, E-mail: fred@hkbu.edu.hk. This research was supported by a Hong Kong UPGC-RGC grant.

the references therein). Randomized quasirandom samples have been proposed by Owen [16, 17] and Tang [20].

Historically the quadrature error analyses for random and quasirandom sampling methods have followed different approaches. For random samples one may obtain probabilistic error estimates as discussed by Davis and Rabinowitz [1] and Owen [15, 16, 17]. For quasirandom samples the most popular error bound is the Koksma-Hlawka inequality [14, Theorem 2.11]:

$$|I(f) - Q(f)| \leq D^*(S)V(f), \tag{2}$$

where $V(f)$ is the variation of f on $\bar{C}^d = [0, 1]^d$ in the sense of Hardy and Krause, and $D^*(S)$ is the discrepancy of S. The discrepancy is defined as

$$D^*(S) = \sup_{x \in C^d} \left| \frac{|S \cap [0, x)|}{N} - \text{Vol}([0, x)) \right|, \tag{3}$$

where $|S \cap [0, x)|$ denotes the number of points in $S \cap [0, x)$ and $\text{Vol}([0, x))$ denotes the volume of $[0, x)$. Although asymptotic bounds on the discrepancy of various quasirandom samples are known [14], $D^*(S)$ itself is not easy to calculate if the dimension or sample size is large. Other error bounds based on an $\mathcal{L}^2$ version of the discrepancy are also possible [12].

Recently the author [9] has derived a new worst-case error bound for quadrature rules of the form (1). The part depending on the sample only, which is analogous to the discrepancy above, is relatively simple to calculate. Furthermore, this error bound shows that the accuracy of quasirandom quadrature rules depends not on the overall dimension d, but on how well the integrand can be approximated by a sum of functions on low dimensional cubes. The derivation of this new error bound relies on two ideas: ANOVA decompositions of functions and reproducing kernel Hilbert spaces.

Statisticians [2, 16] have used ANOVA decompositions to split a function into pieces that depend only on a subset of the coordinates. Let $D = \{1, \ldots, d\}$ be the set of coordinate indices. For any index set $u \subseteq D$, let $|u|$ denote its cardinality. Let x_u denote the $|u|$-vector containing the components of x indexed by u. The quantity $dx_u = \prod_{j \in u} dx_j$ is the uniform measure on the cube $C^{|u|} = [0, 1)^{|u|}$. This cube is denoted C^u to distinguish cubes of the same dimension in different coordinate directions. Any function $f \in \mathcal{L}^2(C^d)$ may be written as the sum of its ANOVA effects, f_u, as follows:

$$f_u(x) = \int_{C^{D-u}} \left[f - \sum_{v \subset u} f_v(x) \right] dx_{D-u}, \qquad f(x) = \sum_{u \subseteq D} f_u(x).$$

The constant $f_\emptyset$ is the average value of the function, that is, the value of the integral we wish to compute. The main effect along the jth axis is $f_{\{j\}}$, while $f_{\{j,m\}}$ denotes the interaction between coordinates j and m. In general the ANOVA effect f_u depends only on the coordinates with indices in the set u. The order of f_u is $|u|$. Any function has a total of 2^d ANOVA effects, $\binom{d}{p}$ of them with order p. One may think of the pth order ANOVA effects as the parts of the function that depend only on p variables.

For example, consider the function

$$f(x) = 2x_1 + 2x_2 + 4ax_1x_2, \tag{4}$$

where a is some constant. Its ANOVA effects can be computed follows:

(5a)
$$f_\emptyset = \int_0^1 \int_0^1 f(x) dx_1 dx_2 = 2 + a,$$

(5b)
$$f_{\{1\}}(x_1) = \int_0^1 (f(x) - f_\emptyset) dx_2 = (1 + a)(2x_1 - 1),$$

(5c)
$$f_{\{2\}}(x_2) = \int_0^1 (f(x) - f_\emptyset) dx_1 = (1 + a)(2x_2 - 1),$$

(5d)
$$f_{\{1,2\}}(x_1, x_2) = f(x) - f_\emptyset - f_{\{1\}}(x_1) - f_{\{2\}}(x_2) = a(2x_1 - 1)(2x_2 - 1).$$

The quadrature error bound derived in [9] requires integrands to have a certain amount of regularity. In particular, for each ANOVA effect f_u its mixed partial derivative $\partial^{|u|} f / \partial x_u$ must be a square integrable distribution (generalized function) on C^u. This is a weaker condition than requiring $\partial^{|u|} f / \partial x_u$ to be continuous, but stronger than requiring f_u to have bounded variation. Consider, for example,

$$f_{\{1\}}(x_1) = \text{sign}(x_1 - 0.5)|x_1 - 0.5|^p, \qquad \frac{\partial f_{\{1\}}}{\partial x_1} = p|x_1 - 0.5|^{p-1}$$

The function $f_{\{1\}}$ has bounded variation on $[0,1]$ for $p \geq 0$. Its derivative is square integrable on $[0,1]$ for $p > 1/2$ but continuous only for $p > 1$. Jones [11] gives an exposition of generalized functions.

The ANOVA effects belong to the Hilbert spaces $(X_u, \langle,\rangle_u)$ defined as follows:

$$X_\emptyset = \{f : f \text{ is a constant}\}, \qquad \langle f, f' \rangle_\emptyset = f f',$$

$$X_u = \left\{ f \in \mathcal{L}^2(C^u) : \frac{\partial^{|v|} f}{\partial x_v} \in \mathcal{L}^2(C^u), \int_{C^v} f \, dx_v = 0 \quad \forall v \subseteq u \right\},$$

$$\langle f, f' \rangle_u = \int_{C^u} \frac{\partial^{|u|} f}{\partial x_u} \frac{\partial^{|u|} f'}{\partial x_u} dx_u, \quad (u \neq \emptyset).$$

The space of integrands considered, X, is the sum of all the X_u:

$$X = \{f : f_u \in X_u\}, \qquad \langle f, f' \rangle = \sum_{u \subseteq D} \gamma^{-|u|} \langle f_u, f'_u \rangle_u,$$

where f_u and f'_u are the ANOVA effects of f and f', respectively. The norm $\|f\|$ is a kind of $\mathcal{L}^2$ measure of the variation of the integrand. The arbitrary, fixed constant γ in the definition of $\langle,\rangle$ might naturally be taken to be one. However, Hickernell [9] showed that $\gamma = 4\pi^2$ leads to an error bound similar to the usual one for lattice rules [18, Equation (23)].

The Hilbert space, $(X, \langle,\rangle)$, has the property that the linear functional $\mathcal{T}_x$, defined as $\mathcal{T}_x(f) = f(x) \quad \forall f \in X$, is bounded. This implies the existence of a reproducing kernel $\eta(\bullet, x) \in X$ such that $f(x) = \mathcal{T}_x(f) = \langle \eta(\bullet, x), f \rangle \quad \forall f \in X$. The formula for η, as given in [9], is

$$\eta(x, y) = \sum_{u \subseteq D} \eta_u(x_u, y_u) = \prod_{j=1}^d [1 + \gamma \eta_1(x_j, y_j)],$$

where

$$\eta_1(x_1, y_1) = \frac{1}{2}(x_1^2 + y_1^2) + \frac{1}{3} - \max(x_1, y_1),$$

$$\eta_\emptyset = 1, \qquad \eta_u(x_u, y_u) = \gamma^{|u|} \prod_{j \in u} \eta_1(x_j, y_j).$$

The η_u are the ANOVA effects of η.

The error of quadrature rule (1), $I - Q$, is also a bounded, linear functional on $(X, \langle, \rangle)$. Its representer, ξ, corresponds to the quadrature error for the reproducing kernel. Likewise the ANOVA effects, ξ_u, are the quadrature errors for η_u:

$$I(f) - Q(f) = \langle \xi, f \rangle,$$

where

$$\xi(x) = \langle \xi, \eta(\bullet, x) \rangle = I(\eta(\bullet, x)) - Q(\eta(\bullet, x)) = 1 - \frac{1}{N} \sum_{z \in S} \eta(z, x),$$

$$\xi_\emptyset = 0, \qquad \xi_u(x) = -\frac{1}{N} \sum_{z \in S} \eta_u(z_u, x_u) \quad (u \neq \emptyset).$$

By applying the Cauchy-Schwarz inequality the following error bound was obtained in [9].

Theorem 1 *For any $f \in X$ the error of quadrature rule (1) satisfies the inequality $|I(f) - Q(f)| \leq \sum_{u \neq \emptyset} c_u \|f_u\|_u \leq c \|f\|$, where*

(6a) $$c_u^2 \equiv \gamma^{-2|u|} \|\xi_u\|_u^2 = \frac{1}{N^2} \sum_{z, z' \in S} \prod_{j \in u} \eta_1(z_j, z_j') \quad (u \neq \emptyset),$$

(6b) $$c^2 \equiv \|\xi\|^2 = \sum_{u \neq \emptyset} \gamma^{|u|} c_u^2 = -1 + \frac{1}{N^2} \sum_{z, z' \in S} \prod_{j=1}^{d} [1 + \gamma \eta_1(z_j, z_j')].$$

Equality holds when f is proportional to ξ.

The positive numbers c_u and c are called error bound coefficients. In particular, the c_u are called error bound coefficients of order $|u|$. From (6) it is clear that these coefficients depend only on the worst case integrand, ξ, which in turn depends only on the sample, S, and not on f. The error bound coefficients play a similar role to discrepancy in (2). Samples yielding smaller error bound coefficients are expected to give more accurate quadrature rules.

The error bound coefficient c_u indicates the accuracy of a quadrature rule in integrating the ANOVA effect f_u. In Sections 2 and 3 it is shown that some samples yield small c_u for low orders, while other samples yield small c_u for high orders. Although the c_u is independent of γ, c^2 is a weighted average of the c_u^2 with weights $\gamma^{|u|}$. When samples are chosen to minimize c, the value of γ can influence this choice. If one prefers to integrate high order ANOVA effects more accurately, then one should make the high order error bound coefficients as small as possible. This is done by picking a large value for γ. The converse is also true. Picking a large value for γ implies a preference for quadrature rules that accurately integrate high order ANOVA effects. However, there is a trade-off — increasing the accuracy for high order ANOVA effects typically decreases the accuracy for low order ANOVA effects, and visa versa.

The error bound in Theorem 1 has an analog for periodic integrands. Let $\mathcal{P}(\bar{C}^d)$ be the set of all continuous, periodic functions on $\mathbf{R}^d$ with period interval $\bar{C}^d$. Let $\tilde{X} =$

$X \cap \mathcal{P}(\bar{C}^d)$, so $(\tilde{X}, \langle, \rangle)$ is a subspace of $(X, \langle, \rangle)$. The reproducing kernel of this subspace, as given in [9], is

$$\tilde{\eta}(x,y) = \sum_{u \subseteq D} \tilde{\eta}_u(x_u, y_u) = \prod_{j=1}^{d} [1 + \gamma \tilde{\eta}_1(x_j, y_j)],$$

where

(7)
$$\begin{aligned}
\tilde{\eta}_1(x_1, y_1) &= \eta_1(\{x_1\}, \{y_1\}) - (\{x_1\} - 1/2)(\{y_1\} - 1/2) \\
&= \frac{1}{2} \left[\{x_1 - y_1\}(\{x_1 - y_1\} - 1) + \frac{1}{6} \right] = \frac{1}{2} B_2(\{x_1 - y_1\}),
\end{aligned}$$

$$\tilde{\eta}_\emptyset = 1, \qquad \tilde{\eta}_u(x_u, y_u) = \gamma^{|u|} \prod_{j \in u} \tilde{\eta}_1(x_j, y_j).$$

Here $\{\}$ denotes the fractional part of a number or vector, and B_2 is the quadratic Bernoulli polynomial. Note that $\tilde{\eta}$ is periodic, whereas η is not. The quadrature error bound and worst case integrand for the periodic case are analogous to those given for the general case in Theorem 1.

Theorem 2 *For any $f \in \tilde{X}$ the error of quadrature rule (1) satisfies the inequality $|I(f) - Q(f)| \leq \sum_{u \neq \emptyset} \tilde{c}_u \|f_u\|_u \leq \tilde{c} \|f\|$, where*

(8a)
$$\tilde{c}_u^2 \equiv \gamma^{-2|u|} \|\tilde{\xi}_u\|_u^2 = \frac{1}{N^2} \sum_{z,z' \in S} \prod_{j \in u} \tilde{\eta}_1(z_j, z_j') \quad (u \neq \emptyset),$$

(8b)
$$\tilde{c}^2 \equiv \|\tilde{\xi}\|^2 = \sum_{u \neq \emptyset} \gamma^{|u|} \tilde{c}_u^2 = -1 + \frac{1}{N^2} \sum_{z,z' \in S} \prod_{j=1}^{d} [1 + \gamma \tilde{\eta}_1(z_j, z_j')].$$

The remainder of this article focuses on the calculation and comparison of the error bound coefficients in equations (6) and (8) for various samples. The error bound coefficients for the simple random sample and a grid were computed in [9]. The next section gives closed form results for certain kinds of random samples. In Section 3, error bound coefficients are compared for several random and quasirandom samples. The last section discusses the relative merits of these samples.

2 Random Samples

For random samples the error bound coefficients are themselves random. However, in several interesting cases their expected values may be calculated in closed form. This section treats two kinds of randomizations. First, we consider random shifts of a deterministic set. Next, we consider the randomized orthogonal arrays studied by Owen [16, 17].

Consider a sample composed of several randomly shifted copies of a set $S_0 \subseteq C^d$. That is, given a set Δ of m independent random vectors uniformly distributed on C^d, let

(9)
$$S = \{\{z + \delta\} : z \in S_0, \delta \in \Delta\}.$$

If S_0 consists of a single point, then S is a simple random sample of m points.

Theorem 3 *Let $\tilde{c}_0$ and $\tilde{c}_{0u}$ denote the error bound coefficients defined in (8) corresponding to the quadrature rule based on $S_0 \subseteq C^d$. For S defined in (9) the expected values of the error bound coefficients are as follows:*

$$E[c_u^2] = 2^{|u|} \tilde{c}_{0u}^2/m, \qquad E[c^2] = \sum_{u \neq \emptyset} (2\gamma)^{|u|} \tilde{c}_{0u}^2/m,$$

$$E[\tilde{c}_u^2] = \tilde{c}_{0u}^2/m, \qquad E[\tilde{c}^2] = \sum_{u \neq \emptyset} \gamma^{|u|} \tilde{c}_{0u}^2/m = \tilde{c}_0^2/m.$$

Proof. Formula (6a) for c_u^2 can be written as:

$$c_u^2 = \frac{1}{N^2} \sum_{\delta, \delta' \in \Delta} \sum_{z, z' \in S_0} \prod_{j \in u} \eta_1(\{z_j + \delta_j\}, \{z_j' + \delta_j'\}).$$

When taking the expected value there are two cases to consider. If $\delta = \delta'$, then straightforward calculations give $E[\eta_1(\{z_j + \delta_j\}, \{z_j' + \delta_j'\})] = 2\tilde{\eta}_1(z_j, z_j')$. On the other hand, if $\delta \neq \delta'$, then $E[\eta_1(\{z_j + \delta_j\}, \{z_j' + \delta_j'\})] = 0$. Therefore,

$$E[c_u^2] = \frac{m}{N^2} \sum_{z, z' \in S_0} \prod_{j \in u} 2\tilde{\eta}_1(z_j, z_j') = \frac{2^{|u|}}{m} \tilde{c}_{0u}^2.$$

The formula for $E[c^2]$ follows from (6b). The formulas for $E[\tilde{c}_u^2]$ and $E[\tilde{c}^2]$ are derived in a similar fashion. The only difference is that $\tilde{\eta}_1(\{z_j + \delta_j\}, \{z_j' + \delta_j\}) = \tilde{\eta}_1(z_j, z_j')$ by (7). $\square$

The error bound coefficients associated with a simple random sample can be obtained immediately from the above theorem. We take S_0 to be a single point and note that in this case $\tilde{c}_{0u}^2 = 1/12^{|u|}$.

Corollary 4 *If S is a uniformly distributed simple random sample of N points, then the corresponding error bound coefficients are:*

$$E[c_u^2] = 1/(6^{|u|}N), \qquad E[c^2] = [(1 + \gamma/6)^d - 1]/N,$$

$$E[\tilde{c}_u^2] = 1/(12^{|u|}N), \qquad E[\tilde{c}^2] = [(1 + \gamma/12)^d - 1]/N.$$

Theorem 3 shows that for randomly shifted points the error bound coefficients for the periodic and general cases are related. This suggests that samples with small $\tilde{c}$ also have small c. For lattice rules (which include good lattice point sets as a special case) the formula for $\tilde{c}$ can be simplified from a double sum to a single sum [9]. Therefore, in Section 3 our comparison of error bound coefficients is limited to the periodic case.

Besides random shifts another type of randomization is the permutation of coordinate values of points lying on a grid. Specifically, consider the following grid with M^d points:

(10) $$G = \{(i_1, \ldots, i_d)/M : i_k = 0, \ldots, M - 1, k = 1, \ldots, d\}.$$

Let S_0 be a subset of G. This means that for every $z \in S_0$, Mz is an integer vector. For each coordinate k, let $\pi_k(0), \ldots, \pi_k(M - 1)$ denote a random permutation of the integers $0, \ldots, M - 1$. Then

(11) $$S = \{(\pi_1(Mz_1), \ldots, \pi_d(Mz_d))/M : (z_1, \ldots, z_d) \in S_0\}$$

is a random permutation of the points in S_0.

Owen [16, 17] has explored the use of randomized orthogonal arrays for computer experiments and quadrature. The notation $OA(N, d, M, t)$ denotes an orthogonal array of strength t with N points which is a subset of the grid G as defined above. "Strength t" means that when looking at any t coordinates, each of the possible M^t points occurs the same number of times. For example, the grid G above is an array of strength d, $OA(M^d, d, M, d)$. By the definition of $OA(N, d, M, t)$ there is a lower bound on the number of points, $N \geq M^t$, and this can be obtained under certain conditions on M and d. If S_0 is an orthogonal array of the form $OA(N, d, M, t)$, then S as constructed by the random permutation (11) is also an orthogonal array of the same form.

Below we compute the expected error bound coefficients of randomized strength 1 and 2 arrays with the minimum number of points. By (7) and (8a) the formula for $E[\tilde{c}_u^2]$ can be written as follows:

$$(12) \qquad E[\tilde{c}_u^2] = \frac{1}{N^2} \sum_{z,z' \in S} \prod_{j \in u} E[\tilde{\eta}_1(z_j, z_j')] = \frac{1}{2^{|u|} N^2} \sum_{z,z' \in S} \prod_{j \in u} E[B_2(\{z_j - z_j'\})].$$

If $z_j = z_j'$ then $\{z_j - z_j'\} = 0$, whereas if $z_j \neq z_j'$, then under the permutation defined in (11) $\{z_j - z_j'\}$ takes on the values $1/M, \cdots, (M-1)/M$ with equal probability. This implies

$$E[B_2(\{z_j - z_j'\})] = \begin{cases} B_2(0) = 1/6 \text{ if } z_j = z_j', \\ \frac{1}{M-1} \sum_{i=1}^{M-1} B_2(i/M) = -1/(6M) \text{ if } z_j \neq z_j'. \end{cases}$$

Thus the computation of the expected error bound coefficients is reduced to determining whether or not $z_j = z_j'$.

A strength 1 orthogonal array is also called a lattice sample. It can be generated as follows:

$$S_0 = \{i(1, \ldots, 1)/N : i = 0, \ldots, N-1\}.$$

The error bound coefficients for such an array are given by the following theorem.

Theorem 5 *Let S be a randomization of any array $OA(N, d, N, 1)$ according to (11). The expected values of the error bound coefficients for S in the periodic case are*

$$E[\tilde{c}_u^2] = \frac{1}{12^{|u|} N} \left[1 + (N-1) \left(\frac{-1}{N} \right)^{|u|} \right],$$

$$E[\tilde{c}^2] = \frac{1}{N} \left\{ \left(1 + \frac{\gamma}{12} \right)^d - 1 + (N-1) \left[\left(1 - \frac{\gamma}{12N} \right)^d - 1 \right] \right\}.$$

Proof. For this strength one array $z_j = z_j$ if and only if $z = z'$. In the double sum over S in (12), this occurs N out of N^2 times. Therefore,

$$E[\tilde{c}_u^2] = \frac{1}{2^{|u|} N^2} \left[N \left(\frac{1}{6} \right)^{|u|} + N(N-1) \left(\frac{-1}{6N} \right)^{|u|} \right] = \frac{1}{12^{|u|} N} \left[1 + (N-1) \left(\frac{-1}{N} \right)^{|u|} \right].$$

The formula for $E[\tilde{c}^2]$ follows by summing the $E[\tilde{c}_u^2]$. $\square$

Owen has recommended strength 2 arrays for computer experiments because they give an uniform grid in every pair of dimensions. Note that for $OA(M^2, d, M, 2)$ it is imperative that $d \le M + 1$. These arrays can easily be generated when M is a prime number:

$$S_0 = \{\{ (i_1, i_2, i_1 + i_2, 2i_1 + i_2, \ldots, (d-2)i_1 + i_2) / M \} : i_1, i_2 = 0, \ldots, M - 1\}.$$

The expected error bound coefficients for randomized strength 2 arrays can also be calculated in closed form.

Theorem 6 *Let S be a randomization of any array $OA(M^2, d, M, 2)$ according to (11). The expected values of the error bound coefficients for S in the periodic case are*

$$E[\tilde{c}_u^2] = \frac{1}{12^{|u|} M^2} \left[1 + (1 - |u|)(M^2 - 1) \left(\frac{-1}{M} \right)^{|u|} \right],$$

$$E[\tilde{c}^2] = \frac{1}{M^2} \left\{ \left(1 + \frac{\gamma}{12} \right)^d - 1 + (M^2 - 1) \left[\left(1 - \frac{\gamma}{12M} \right)^{d-1} \left(1 + \frac{\gamma(d-1)}{12M} \right) - 1 \right] \right\}.$$

Proof. In the double sum over S in (12) consider any fixed z'. There is one z, namely $z = z'$ for which all $z_j = z_j'$. There are $(M-1)|u|$ points z with $z_j = z_j'$ for exactly one $j \in u$. The remaining $(M-1)(M+1-|u|)$ points z have no $z_j = z_j'$ for $j \in u$. Therefore,

$$E[\tilde{c}_u^2] = \frac{1}{12^{|u|} M^2} \left[1 + (M-1)|u| \left(\frac{-1}{M} \right)^{|u|-1} + (M-1)(M+1-|u|) \left(\frac{-1}{M} \right)^{|u|} \right]$$

$$= \frac{1}{12^{|u|} M^2} \left[1 + (1 - |u|)(M^2 - 1) \left(\frac{-1}{M} \right)^{|u|} \right].$$

Again the formula for $E[\tilde{c}^2]$ follows by summing the $E[\tilde{c}_u^2]$. $\square$

The analysis above shows that randomized orthogonal arrays yield small order error bound coefficients for orders equal to their strength. However, for other orders the error bound coefficients are asymptotically the same as those for a simple random sample. Specifically, for arrays of strength $t = 1, 2$ Theorems 5 and 6 imply that

$$E[\tilde{c}_u^2] = \begin{cases} 1/(12^t N^2) & \text{for } |u| = t, \\ (1 + o(1))/(12^u N) & \text{for } |u| \ne t. \end{cases}$$

3 Numerical Comparisons

In this section the error bound coefficients are computed and compared for the following samples:

    i. simple random sample (Monte Carlo method)
    ii. randomized orthogonal arrays of strengths 1, 2 and d
    iii. good lattice point (glp) sets, and
    iv. Faure, Halton and Sobol' sequences.

Computations are carried out for dimensions up to eight and sample sizes up to 10^4.

As noted in the previous section the error bound coefficients for the general and periodic cases are similar, so only the periodic case is computed. Some samples are better for integrating low order ANOVA effects, while other samples are better for integrating high order ANOVA effects. Therefore, the $\tilde{c}_u$ are computed and compared for each order $|u|$. Since the motivation for sophisticated sampling schemes is to improve upon a simple random sample, the error bound coefficients are normalized with respect to those for the Monte Carlo method. In particular, for each sample we compute b_p and b as defined below:

$$\text{quasirandom samples:} \quad b_p^2 = \frac{12^p N}{\binom{d}{p}} \sum_{|u|=p} \tilde{c}_u^2, \quad b^2 = \frac{N\tilde{c}^2}{(1+\gamma/12)^d - 1},$$

$$\text{random samples:} \quad b_p^2 = \frac{12^p N}{\binom{d}{p}} \sum_{|u|=p} E[\tilde{c}_u^2], \quad b^2 = \frac{N\,E[\tilde{c}^2]}{(1+\gamma/12)^d - 1}.$$

For the Monte Carlo method $b_p = b = 1$ by Corollary 4. Figure 1 shows the dependence of b on N (with $\gamma = 1$) for different values of d. Figure 2 shows plots of b_p versus N for different values of p.

For the randomized orthogonal arrays of strengths 1 and 2 the normalized error bound coefficients come from Theorems 5 and 6:

$$\text{OA}(N, d, N, 1): \quad b_p^2 = 1 + (N-1)\left(\frac{-1}{N}\right)^p,$$

$$b^2 = 1 + (N-1)\frac{[1 - \gamma/(12N)]^d - 1}{(1+\gamma/12)^d - 1},$$

$$\text{OA}(N, d, N^{1/2}, 2): \quad b_p^2 = 1 + (1-p)(N-1)\left(\frac{-1}{\sqrt{N}}\right)^p,$$

$$b^2 = 1 + (N-1)\frac{[1 - \gamma/(12\sqrt{N})]^{d-1}[1 + \gamma(d-1)/(12\sqrt{N})] - 1}{(1+\gamma/12)^d - 1}.$$

The orthogonal array $\text{OA}(N, d, N^{1/d}, d)$, that is the grid G in (10), is unaffected by randomization. Its error bound coefficients are given in [9]:

$$\text{OA}(N, d, N^{1/d}, d): \quad b_p^2 = N^{1-2p/d}, \quad b^2 = N\frac{[1 - \gamma/(12N^{2/d})]^d - 1}{(1+\gamma/12)^d - 1}.$$

Note that b_p is independent of d for strength 1 and 2 arrays but not for strength d arrays. Although the above formulas are only strictly valid for certain integer values of N, they are plotted in Figures 1–2 for all N for convenience sake. The dimensions considered are $d = 1, \cdots, 8$.

Good lattice point (glp) sets are quasirandom samples of the form

$$S = \{\{ig/N\} : i = 0, \ldots, N-1\},$$

where g is a generating vector of dimension d. Following [9] a greedy algorithm is used to find good generating vectors. Specifically, for each N we set $g_1 = 1$ and then iteratively search for the $g_2, g_3, \ldots$ that minimize $\tilde{c}$ (with $\gamma = 1$). For each N there are four glp

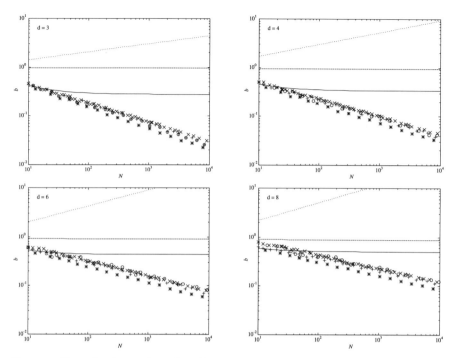

Figure 1: Normalized error bound coefficient b versus N for various dimensions d for glp sets ($*$), Faure ($\circ$), Halton ($\times$) and Sobol' ($+$) sequences, and strength 1 (solid), strength 2 (dashed) and strength d (dotted) randomized orthogonal arrays.

sets corresponding to $d = 3, 4, 6$ and 8. For any particular g and N the b_p and b can be computed in only $O(N)$ operations using the formulas in [9]. Figures 1–2 show the error bound coefficients for representative sample sizes in the range 10 to 10^4.

Error bound coefficients are computed for three d-dimensional quasirandom sequences: those of Faure [5], Halton [8] and Sobol' [19]. As opposed to glp sets and orthogonal arrays a quasirandom sequence retains its first N_1 points when the sample size is increased from N_1 to N_2. Because of this constraint the uniformity of quasirandom sequences are normally not as good as those of glp sets. To make for a fairer comparison we have increased the effective dimension of each quasirandom sequence by adding a variable with values $0, 1/N, \cdots (N-1)/N$ for each sample size N. This modification was proposed by Hammersley (see [14, Section 3.1]). Figures 1–2 show results for modified quasirandom sequences of dimension $d = 3, 4, 6$ and 8 for selected sample sizes. The prime numbers used to generate the Faure sequences are 2, 3, 5 and 7.

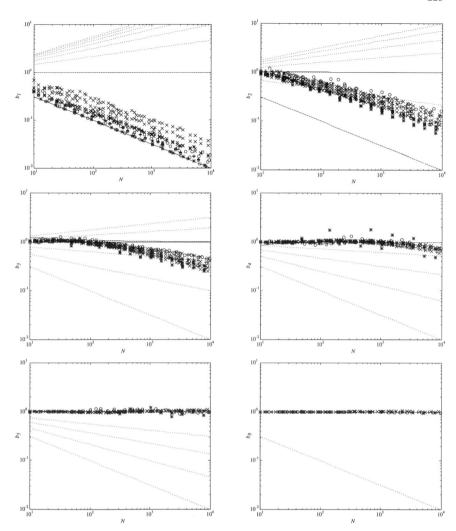

Figure 2: Normalized error bound coefficients b_p versus N for various orders p for glp sets (∗), Faure (○), Halton (×) and Sobol' (+) sequences, and strength 1 (solid), strength 2 (dashed) and strength d (dotted) randomized orthogonal arrays.

4 Discussion and Conclusion

Figure 1 indicates that the glp sets have the smallest error bound coefficients, b, followed by the quasirandom sequences and then the randomized orthogonal arrays. The grid is even worse than the Monte Carlo method for $d > 2$. The additional accuracy gained by using glp sets or quasirandom sequences instead of a simple random sample increases with the sample size since b tends to zero. However, for the strength 1 and 2 randomized orthogonal arrays b tends to a constant (less than one) for large sample sizes.

Each plot in Figure 2 displays the b_p versus N for a particular order p. Whereas b in Figure 1 is a weighted average of error bound coefficients for different orders, Figure 2 shows how well different samples integrate functions that depend purely on p variables. The glp sets and quasirandom sequences give significantly smaller b_p than those for a simple random sample when either p is small or N is large. For these quasirandom samples $b_1 < 1$ for essentially all N, $b_2 < 1$ for $N > 20$, $b_3 < 1$ for $N > 200$, and $b_4 < 1$ for $N > 2000$. On the other hand b_5 through b_8 are close to unity for N up to 10^4. Therefore, if an integrand has large high order ANOVA effects and N is not too large, the quadrature error using quasirandom samples will not be significantly better (nor significantly worse) than the error from the Monte Carlo method. In general the glp sets have slightly smaller low order b_p than quasirandom sequences, which accounts for their slightly smaller b. Among the quasirandom sequences the b_p and b are similar, although the Halton sequence seems to have a somewhat larger b_1 than the others.

Randomized orthogonal arrays of strength t are most efficient in integrating ANOVA effects of order t, but are not particularly advantageous for ANOVA effects of other orders. The sample giving the smallest high order b_p is the grid, which is an orthogonal array of strength d. However, the corresponding low order b_p for this same grid are much greater than those for the Monte Carlo method. Thus, a grid is not recommended in practice. Randomized strength 1 and 2 orthogonal arrays give small b_p for $p = 1$ and 2 respectively, but for other p the b_p are comparable to those for the Monte Carlo method.

Our assessment about the advantages of orthogonal arrays is more pessimistic than that of Owen [16, 17], who considered the variance of the quadrature estimate,

$$(13) \qquad E\{[I(f) - Q(f)]^2\}.$$

He concluded that randomized arrays $OA(N^{1/t}, d, N, t)$ are better than simple random samples because there is no contribution to the variance from ANOVA effects of order $\leq t$. Owen's analysis differs from ours in two respects. First, he replaces integration over the unit cube by summation over $N^{1/t} \times \cdots \times N^{1/t}$ equally spaced grid points. If the integrand has sufficient smoothness (more than we have assumed), this discrete approximation corresponds to a product midpoint rule with error $O(N^{-2/t})$. Thus, under Owen's assumptions a strength 2 array integrates ANOVA effects of orders 1 and 2 with error of only $O(N^{-1})$, while higher order effects are integrated with Monte Carlo accuracy. The second difference is that the integrand is fixed in (13) whereas we consider the quadrature error for the worst possible integrand:

$$E[c^2] = E\left\{\sup_{f \in X}[I(f) - Q(f)]^2 / \|f\|^2\right\}.$$

To illustrate how quadrature accuracy depends on the relative size of the ANOVA effects of different orders consider again the function (4) and its ANOVA effects (5).

Table 1: Relative quadrature error, $[I(f) - Q(f)]/I(f)$ for integrand (4)

	$a = 0$	$a = -1$	$c_{\{1\}} = c_{\{2\}}$	$c_{\{1,2\}}$
25×25 grid	0.04000	0.00160	0.02309	0.00053
	(0.04619)	(0.00213)		
glp, $N = 625, g = (1, 172)$	0.00160	0.00303	0.00092	0.00122
	(0.00185)	(0.00487)		

The norms of the ANOVA effects and the quadrature error bound from Theorem 1 are straightforward to compute:

$$\|f_\emptyset\|_\emptyset = |2 + a|, \quad \|f_{\{1\}}\|_{\{1\}} = \|f_{\{2\}}\|_{\{2\}} = 2|1 + a|, \quad \|f_{\{1,2\}}\|_{\{1,2\}} = 4|a|,$$

(14)
$$|I(f) - Q(f)| \leq 2|1 + a|(c_{\{1\}} + c_{\{2\}}) + 4|a| \, c_{\{1,2\}}.$$

If a is small in magnitude, then f is approximately a sum of functions of a single variable, and the accuracy of (1) depends primarily on the sizes of $c_{\{1\}}$ and $c_{\{2\}}$ rather than $c_{\{1,2\}}$. On the other hand, if a is close to -1, then the most important ANOVA effect is $f_{\{1,2\}}$, and the quadrature error depends mostly on the size of $c_{\{1,2\}}$. Table 1 shows the actual relative quadrature errors using a grid and a glp set with the same number of points (neither set is centered). The numbers in parentheses are the bounds on the relative quadrature error from (14). For a close to zero the glp sets are better, because they have smaller first order error bound coefficients. For a close to -1 the grid is better because it has a smaller second order error bound coefficient.

A common way to test quadrature methods is to use a suite of test integrands, such as those given in [6, 7]. The advantage is that one can see the actual performance of a method on problems of interest to practitioners. The disadvantage is that the outcome of such a test depends strongly on the types of integrands chosen, as is illustrated by the example above. Therefore, we recommend that samples for quadrature be judged on the basis of error bound coefficients — preferably by looking at different orders individually. This makes it is possible to know which samples will be more suitable for which kind of integrands.

Formulas (6) and (8) for the error bound coefficients are similar to the formula for the $\mathcal{L}^2$-discrepancy derived by Warnock [21] and studied by Morokoff and Caflisch [12] in that all involve a double sum over all points in S. However, the $\mathcal{L}^2$-discrepancy, like $\mathcal{D}^*(S)$ as defined in (3), attaches special importance to the origin. Although a reflection of the cube C^d about the plane $x_j = 1/2$ leaves the cube unchanged, the same transformation applied to S changes its discrepancy and $\mathcal{L}^2$-discrepancy. On the other hand the error bound coefficients are invariant under such a reflection since

$$\eta_1(1 - x_j, 1 - y_j) = \eta_1(x_j, y_j), \quad \tilde{\eta}_1(1 - x_j, 1 - y_j) = \tilde{\eta}_1(x_j, y_j).$$

The method for deriving the error bounds presented in this article is rather general. The Hilbert space used here was chosen for convenience sake, but others are possible. If one defines a Hilbert space and can derive its reproducing kernel, η, then the error bound $|I(f) - Q(f)| \leq \|\xi\| \|f\|$ follows immediately. The worst case integrand is $\xi(x) = I(\eta(\bullet, x)) - Q(\eta(\bullet, x))$.

In summary, the error bound coefficients are generally smallest for good lattice point sets and quasirandom sequences, larger for randomized orthogonal arrays and largest for simple random samples. The difference between more sophisticated samples and a simple random sample is more pronounced if the sample size is large or the order of the coefficients is low. For the class of integrands considered here quasirandom samples will in general give higher accuracy than simple Monte Carlo quadrature for worst case integrands. However, the gain in accuracy depends on having a large enough sample size and/or an integrand that is composed of mainly low order ANOVA effects.

Acknowledgments

The author would like to thank the organizers of the *Conference on Monte Carlo and Quasi-Monte Carlo Methods in Scientific Computing* and several participants with whom he was able to discuss this work. Special thanks go to Professors Kai-Tai Fang, Harald Niederreiter and Art Owen and an anonymous referee for their valuable input.

References

[1] P. J. DAVIS AND P. RABINOWITZ, *Methods of Numerical Integration*, Academic Press, Orlando, Florida, 1984.

[2] B. EFRON AND C. STEIN, *The jackknife estimate of variance*, Ann. Stat., 9 (1981), pp. 586–596.

[3] M. EVANS AND T. SWARTZ, *Some integration strategies for problems in statistical inference*, Computing Science and Statistics, 24 (1992), pp. 310–317.

[4] K. T. FANG AND Y. WANG, *Number Theoretic Methods in Statistics*, Chapman and Hall, New York, 1994.

[5] H. FAURE, *Discrépance de suites associées à un système de numération (en dimension s)*, Acta Arith., 61 (1982), pp. 337–351.

[6] A. GENZ, *Testing multidimensional integration routines*, in Tools, Methods and Languages for Scientific and Engineering Computation, B. Ford, J. C. Rault, and F. Thomasset, eds., Amsterdam, 1984, North-Holland, pp. 81–94.

[7] ——, *A package for testing multiple integration subroutines*, in Numerical Integration: Recent Developments, Software and Applications, P. Keast and G. Fairweather, eds., Dordrecht, 1987, D. Reidel Publishing, pp. 337–340.

[8] J. H. HALTON, *On the efficiency of certain quasi-random sequences of points in evaluating multi-dimensional integrals*, Numer. Math., 2 (1960), pp. 84–90.

[9] F. J. HICKERNELL, *Quadrature error bounds with applications to lattice rules*, SIAM J. Numer. Anal., 33 (1996).

[10] L. K. HUA AND Y. WANG, *Applications of Number Theory to Numerical Analysis*, Springer-Verlag and Science Press, Berlin and Beijing, 1981.

[11] D. S. JONES, *The Theory of Generalised Functions*, Cambridge University Press, Cambridge, second ed., 1982.

[12] W. J. MOROKOFF AND R. E. CAFLISCH, *Quasi-random sequences and their discrepancies*, SIAM J. Sci. Comput., 15 (1994), pp. 1251–1279.

[13] ———, *Quasi-Monte Carlo integration*, J. Comput. Phys., (1995).

[14] H. NIEDERREITER, *Random Number Generation and Quasi-Monte Carlo Methods*, SIAM, Philadelphia, 1992.

[15] A. B. OWEN, *A central limit theorem for Latin hypercube sampling*, J. Roy. Statist. Soc. B, 54 (1992), pp. 541–551.

[16] ———, *Orthogonal arrays for computer experiments, integration and visualization*, Statist. Sinica, 2 (1992), pp. 439–452.

[17] ———, *Lattice sampling revisited: Monte Carlo variance of means over randomized orthogonal arrays*, Ann. Stat., 22 (1994), pp. 930–945.

[18] I. H. SLOAN, *Numerical integration in high dimensions — the lattice rule approach*, in Numerical Integration: Recent Developments, Software and Applications, T. O. Espelid and A. Genz, eds., Dordrecht, 1992, Kluwer Academic Publishers, pp. 55–69.

[19] I. M. SOBOL', *The distribution of points in a cube and the approximate evaluation of integrals*, U.S.S.R. Comput. Math. and Math. Phys., 7 (1967), pp. 86–112.

[20] B. TANG, *Orthogonal array-based Latin hypercubes*, J. Amer. Statist. Assoc., 88 (1993), pp. 1392–1397.

[21] T. T. WARNOCK, *Computational investigtions of low discrepancy point sets*, in Applications of Number Theory to Numerical Analysis, S. K. Zaremba, ed., New York, 1972, Academic Press, pp. 319–343.

A Simulation Study of a Change-Point Poisson Process Based on Two Well-known Test Statistics

Chih-Hsiang Ho

Department of Mathematical Sciences
University of Nevada, Las Vegas
Las Vegas, Nevada 89154

Abstract

Methods for analysis of trends for the series of events have been extensively studied by many authors, both from a parametric and a nonparametric point of view. Most of the work on the development of the procedures for the nonhomogeneous Poisson processes are based on the fixed sample size tests. It is also desirable that in some situations, interim analyses are undertaken periodically. Suppose, for example, a repairable system is under development. A development program might consist of testing to identify deficiencies, a redesign effort to correct the deficiencies, and further testing to verify these corrections and identify new problem areas. It would be advantageous to track the reliability growth trend of the system, by means of the failure data collected during development testing, so that the program could be revised, if necessary, in order to attain the system reliability objectives. Since the data often occur naturally in a sequential fashion, it will be useful to have sequential procedures allowing for repeated significance tests to the accumulating data. In this paper, we provide side-to-side information for the following two widely used test statistics in this area: (1) the well-known Laplace test (called L test), and (2) the most powerful test for the shape parameter in the Poisson process with Weibull intensity (called Z test). Discussion of major results based on a Monte Carlo simulation study include: (1) the estimated probability of type I error at or before the nth test in sampling from a simple Poisson distribution at a constant nominal level, (2) the existing support from the well-developed sequential clinical trial designs available in the literature, (3) performance assessment for abrupt changes (increasing or decreasing in the intensity of the process), and (4) a control charting procedure which presents a visual interpretation of the trend and can be practically translated for tabular or manual use in modeling the occurrences of stochastic phenomena.

1 Introduction

Nonhomogeneous Poisson processes (NHPP's) provide models for a variety of physical phenomena. It may be reasonable to assume that the intensity, $\lambda(\cdot)$, is constant, so tests of $H_0 : \lambda(\cdot)$ is constant versus $H_A : \lambda(\cdot)$ is not constant (increasing or decreasing) are of interest. The results of such tests could indicate whether the simple homogeneous Poisson process (HPP) may be adequate or whether a more general NHPP is required in modeling the occurrences of stochastic phenomena such as earthquakes, volcanic eruptions, industrial accidents, etc.

Methods for analysis of trends for the series of events have been extensively studied by many authors, both from a parametric and a nonparametric point of view. Applications to a repairable system are discussed by Crow [7] and Lee & Lee [14]. Most of the work on the development of the procedures for the NHPP are based on the fixed sample size tests. It is also desirable that in some situations, interim analyses are undertaken periodically. Suppose, for example, a repairable system is under development. A developmental program might consist of testing to identify deficiencies, a redesign effort to correct the deficiencies, and further testing to verify these corrections and identify new problem areas. It would be advantageous to track the reliability growth trend of the system, by means of the failure data collected during development testing, so that the program could be revised, if necessary, in order to attain the system reliability objectives. Since the data often occur naturally in a sequential fashion, it will be useful to have sequential procedures allowing for repeated significance tests to the accumulating data.

In this article, we use the idea of statistical process control to distinguish between the variation inherent in the observed occurrence times and the extraordinary variation that signals a real change in the intensity of a Poisson process. The technique, which is very simple in principle and easy to apply, is described in Section 2. In Section 3 we assess the quality of the proposed procedure in the null case and provide adequate adjustments for the control limits. A power study is discussed in Section 4 for detecting a change-point Poisson process. Section 5 reports the results of a parallel study based on the well-known Laplace test statistic. Section 6 contains concluding remarks.

2 GENERAL METHOD

It should be obvious that the chronological order in which the events occur is an extremely important aspect of a stochastic process, if a significant time trend is to be objectively identified. For this reason, we formulate a general detection technique based on the statistical process control mechanism.

2.1 Statistical Process Control

Statistical control is a sophisticated concept because it recognizes that variability will be present and requires only that the pattern of variability remain the same. A variable (or process) that continues to be described by the same distribution when observed over time is said to be in statistical control, or simply in control. We are already quite advanced in the art of thinking statistically when we describe a variable as stable or in control if its distribution does not change with time. Books by Montgomery [15], and Ryan [18] review much of the work in this area.

Control charts, which were first developed in the 1920s and 1930s, provide a mechanism for recognizing whether the process is in control. A control chart will be effective if it shows a point outside the control limits almost as soon as the process goes out of control. A basic element of control charting is that data have been collected from the process of interest at a sequence of time points. Depending on the aspect of the process under investigation, some statistic is chosen. The value of this statistic is then calculated for each sample in turn. A traditional control chart then results from plotting these calculated values over time. If the points on the chart all lie between the two control limits, the process is deemed to be in control. That is, the process is believed to be operating in a stable fashion reflecting only natural random variations. An out-of-control "signal" occurs whenever a plotted point falls outside the limits. This is assumed to be attributable to a significant trend (increasing or decreasing) in the intensity of the process. We shall design the control limits so that an in-control process generates very few false alarms, whereas a process not in control quickly gives rise to a point outside the limits.

2.2 The Weibull Process

There is a strong analogy between the logic of control charting and hypothesis testing. The null hypothesis (H_0) here is that the process is in control. When an in-control process yields a point outside the control limits (an out-of-control signal), a type I error has occurred. Appropriate choice of control limits (corresponding to specifying a rejection region in hypothesis testing) will make this error probability suitably small.

Consider a nonhomogeneous Poisson process with intensity $\lambda(t) = (\beta/\theta)(t/\theta)^{\beta-1}$, called a Weibull process (WP). A goodness-of-fit test, maximum likelihood (ML) estimates of β and θ, confidence intervals, and inference procedures for this process are presented in Bain and Engelhardt [2], Bassin [5], Crow [7, 8], Finkelstein [11], and Lee and Lee [14].

A WP is appropriate for three types of processes: increasing-occurrence-rate ($\beta > 1$), decreasing-occurrence-rate ($\beta < 1$), and constant-occurrence-rate ($\beta = 1$). The generalized model can be considered a goodness-of-fit test for an exponential model ($\beta = 1$) of the occurrence times, which is equivalent to a homogeneous Poisson model of the events. Suppose we assume that the successive occurrence times of a specific process follow a single WP. Let $t_1, \cdots, t_n$ be the first n successive times of events. These times are measured from the beginning of the observation period (cumulative length of time over which the events occur), so $t_1 < t_2 < \cdots < t_n$. The following theoretical results (for development and proof see Bain and Engelhardt [3], Ch. 9) are useful for constructing the control limits:

1) The maximum likelihood estimators for β and θ at t_n are

$$\hat{\beta}_n = n / \sum_{i=1}^{n-1} \ln(t_n/t_i) \tag{1}$$

and

$$\hat{\theta}_n = t_n / n^{1/\hat{\beta}_n} \tag{2}$$

2) A size α test of $H_0 : \beta_n = \beta_0$ against $H_A : \beta_n \neq \beta_0$ is to reject H_0 if $2n\beta_0/\hat{\beta}_n \leq \chi^2_{\alpha/2}(2n-2)$ or $2n\beta_0/\hat{\beta}_n \geq \chi^2_{1-\alpha/2}(2n-2)$, where $\chi^2_{\alpha/2}(2n-2)$ is the $100\alpha/2$ percentile of a chi-square distribution with $2n - 2$ degrees of freedom.

The parameters estimated from Equations (1) and (2) provide us with quantitative values to characterize the time trend at the nth event, which is the first step toward the construction of the statistic required for plotting over time. Now suppose we wish to decide whether an exponential distribution seems appropriate (in-control signal) for the data up to the nth event. This suggests a test of $H_0 : \beta_n = 1$ against $H_A : \beta_n \neq 1$. Result 2 indicates that a chi-square test is appropriate, and the control limits are readily available from a table of the chi-square distribution.

2.3 Designing a CSLR Procedure

If the process is stable over time, the observed test statistic, $2n/\hat{\beta}_n$, should continue to be described by a chi-square distribution with $2n - 2$ degrees of freedom. We use this idea by drawing the $(1 - \alpha)100\%$ control limits at

$$
\begin{aligned}
LCL_\alpha &= \text{lower control limit} = \chi^2_{\alpha/2}(2n - 2) \\
UCL_\alpha &= \text{upper control limit} = \chi^2_{1-\alpha/2}(2n - 2).
\end{aligned}
$$

The next step in examining the process is to plot the statistic $[= 2n/\hat{\beta}_n \text{ or } 2 \sum_{i=1}^{n-1} \ln(t_n/t_i)]$ against the time order in which the measurements were recorded. Since it requires at least two occurrence times for the statistical process control, let's define cumulative sums of log ratio (CSLR) by

$$
\begin{aligned}
S_2 &= 2ln(t_2/t_1) \\
S_3 &= 2[ln(t_3/t_1) + ln(t_3/t_2)] = 2 \sum_{i=1}^{2} ln(t_3/t_i) \\
&\vdots \\
S_\ell &= 2[ln(t_\ell/t_1) + \cdots + ln(t_\ell/t_{\ell-1})] = 2 \sum_{i=1}^{\ell-1} ln(t_\ell/t_i) \\
&= S_{\ell-1} + 2(\ell - 1)ln(t_\ell/t_{\ell-1})
\end{aligned}
$$

These cumulative sums are plotted over time. That is, at time ℓ, we plot a point at height S_ℓ. At the current time point r the plotted points are $(2, S_2), (3, S_3), \cdots, (r, S_r)$. If at current time r, either $S_r \leq \chi^2_{\alpha/2}(2r - 2)$ or $S_r \geq \chi^2_{1-\alpha/2}(2r - 2)$, the process is judged to be out of control. The first inequality suggests the process has shifted to a significant increasing time trend starting at the rth event. Similarly, the second inequality suggests the process has shifted to a decreasing time trend.

3 DETERMINATION OF THE CONTROL LIMITS

The general effect of performing repeated significance tests at different stages during the accumulation of a body of data is well known. If the null hypothesis is true and if each significance test is performed at the same nominal level, the probability that at some stage or another the test criterion is significant may be substantially greater than the nominal value. The numerical results of the increased "overall significance level" or type I error for repeated tests on cumulative series of binomial, normal, and exponential observations were published by Armitage et al. [1]. To control the "null" case error rate, therefore, becomes our first priority in assessing the quality of the CSLR control charting procedure.

Table 1 contains Monte Carlo estimates of the overall error rates for the CSLR procedure. These estimates are based on 10,000 replications with sequential observations from a simple Poisson process with occurrence rate $\lambda = 1$. The CSLR procedure is applied to each generated Poisson process until an out-of-control signal is obtained or the 500th time point is reached. Control limits of 90%, 95%, and 99% are considered. The estimated error rates summarized in Table 1 are similar to those of the normal distribution and exponential distribution presented in Tables 2 and 3 by Armitage et al. [1]. Virtually all the figures agree at the first significant digit and many agree to two or even three significant digits.

Table 1. Estimated probability of type I error at or before the nth time point in sampling from a simple Poisson distribution with $\lambda = 1$ at a two-sided nominal level a.

Time Point *	$a = .1$	$a = .05$	$a = .01$
2	.095	.048	.010
3	.156	.081	.017
4	.198	.106	.024
5	.230	.126	.029
6	.257	.143	.034
7	.279	.156	.038
8	.296	.167	.042
9	.312	.177	.045
10	.324	.186	.047
20	.412	.244	.065
30	.457	.275	.076
40	.491	.301	.082
50	.516	.319	.088
100	.586	.372	.106
200	.646	.425	.127
300	.677	.456	.138
400	.697	.473	.147
500	.716	.486	.150

* Number of repeated significance tests performed at nth time point is n-1.

As can be seen from Table 1, even for large values of the nominal significance level, the asymptotic cumulative probability of an out-of-control signal (equal to unity) is approached only after a very large number of repeated investigations. For instance, the CSLR procedure using a fixed 95% control limit will require a median number of time points of 615 before an out-of-control signal, and a high probability that they will continue beyond the 1,000th time point when the process is actually in-control (also see Table 2,

Armitage et al. [1]). Alternatively, if an investigator stops at the 500th time point whether or not an out-of-control signal is reached, the average time points at which false alarms occur in such an investigation are 185, 294, and 438 respectively for a 90, 95, and 99% control limit. Table 1 also shows that the probabilities of seeing an out-of-control signal in repeated significance tests on accumulating data can be substantially above the nominal significance levels even for only moderate amounts of data. However, the probability of a type I error could be controlled by using a more stringent (nominal) significance level. Moreover, it is often more efficient and reliable to make a special effort to analyze interim results at periodic intervals, say every few time points (or months). If each analysis coincides with a meeting of program organizers then prompt action can follow. This well-known group sequential approach has been the state-of-the-art in designing clinical trials because of ethical as well as practical considerations.

Group sequential tests allow early stopping of clinical trials for detecting treatment difference or lack of it without inflating the overall significance level. There have been various proposals on how to specify group sequential boundaries for use in repeated applications of hypothesis testing; see Pocock [17], O'Brien and Fleming [16], DeMets and Ware [9, 10], Lan and DeMets [13]. In order to design a CSLR procedure with a constant control limit, tables for suitable nominal levels such as in Pocock [17] or Armitage et al. [1] have to be used. If some robustness needs to be introduced to prevent early detection of an out-of-control signal for the CSLR procedure, the boundaries proposed by O'Brien and Fleming [16] could be adopted. They suggest being very conservative early by using large critical values and decreasing the critical value during the course of the trial so that the actual level is close to the nominal level at the test scheduled at the end of the trial. For example, the stopping boundary corresponding to a five-group design with overall type I error, $\alpha = .05$ would be: .00001, .0012, .0084, .0225, and .04. In contrast to Pocock's constant nominal level ($= .0158$) for all five tests, the basic distinction between these two stopping rules are: Pocock's boundary will generally result in earlier termination and the O'Brien-Fleming boundary saves most of its rejection probability for the last test. These two methods, however, require that the total number of tests be specified in advance. Lan and DeMets [13] propose a more flexible way to construct discrete sequential boundaries based on the *use function* approach. The advantage of group sequential methods based on the use function is that the repeated analyses can be performed at arbitrary calendar times or at unequal increments of statistical information. The procedure allows unequal increments between repeated analyses, thus arbitrary numbers of repeated analyses.

As we have mentioned, the options for the adjustment are abundant. And because the type I error based on the CSLR procedure increases at the same rate as that of the repeated significance tests with normal sampling, the existing boundaries mentioned earlier can be readily carried over to the CSLR procedure for the adjustments of the control limits. On the other hand, if only informal internal quality control is concerned and for the sake of simplicity and flexibility the adjustment is not attempted, then a constant 99% control limit is recommended. In this case, one can still enjoy a relatively high power with an overall error rate of less than 0.1 for a maximum number of 80 continuous sequential tests.

4 PERFORMANCE ASSESSMENT FOR ABRUPT CHANGES

To further assess the quality of the CSLR procedure in testing for abrupt changes in equipment performance following scheduled overhauls motivates the following parallel Monte Carlo studies. In the first set of the simulation, we put a change-point at the 11th time point with $\lambda_1 = 1$ for the first 10 time points and an extremely high $\lambda_2 = 100$ for the last 11 time points. In the second set we let $\lambda_1 = 1$ for the first 10 time points and then let $\lambda_2 = .01$ for the last 11 time points. Pocock's constant nominal significance level at $\alpha' = .0075$ is used to yield a .05 overall error rate (maximum number of tests = 20, see Table 1, Pocock [17]). If this were a reasonable procedure then we would expect that a number of out-of-control signals would be declared around 11th time point. Table 2 gives the results for the simulation based on 10,000 iterations. It is interesting to note a few things about the table:

Table 2. Estimated probability of rejecting H_0 at or before the nth time point when H_A is a change-point Poisson process with $\lambda_1 = 1$ for the first segment and $\lambda_2 = 100$ and .01 for the second segment, respectively, at two-sided nominal level .0075. The location of change-point is at the 11th time point of 21 sequential observations.

Time Point	$\lambda_1 = 1, \lambda_2 = 100$	$\lambda_1 = 1, \lambda_2 = .01$
2	.008	.009
3	.013	.016
4	.020	.020
5	.023	.023
6	.025	.026
7	.028	.030
8	.031	.033
9	.032	.035
10	.035	.037
11	.041	.805
12	.054	.967
13	.079	.996
14	.116	1.000
15	.172	1.000
16	.242	1.000
17	.327	1.000
18	.412	1.000
19	.506	1.000
20	.597	1.000
21	.676	1.000

1. For the case where there is a change-point at 11 with λ changing from 1 to 100 (this is the case where the failure rate has increased), there is almost a shotgun pattern of declared change-points in the time point range of 11-21.

2. For the second case where the failure rate is decreasing from $\lambda = 1$ to $\lambda = .01$, the results are good with 76.7% and 16.2% of the change-points declared at 11 and 12, respectively.

A possible explanation for the asymmetric results is that the test statistic $S_\ell = 2 \sum_{i=1}^{\ell-1} ln(t_\ell/t_i)$ becomes stalled due to relatively small increment in t_ℓ after the change-point for the increased failure rate case. Moreover, the Monte Carlo study of Ho [12] indicates that, for the fixed sample size tests, the test statistic used for the CSLR procedure has much less power in differentiating between a constant intensity and increasing step-function intensities than between the constant and the decreasing step-function intensities for failured-truncated data. Our findings in this study are consistent with that study but are based on repeated significance tests. Of course, we have purposely chosen the extreme λ's to demonstrate the contrasts between two trends. The effects would be relatively lower (or higher) based on different combinations of occurrence rates, nominal levels, and locations of change-points.

So far, the second stage of the quality assessment for the CSLR procedure is not too discouraging, because this procedure is designed to discriminate over a broad range of alternatives. For the "smooth" intensities (e.g., Weibull, exponential, or logarithmic), Bain et al. [4] recommend the same test statistic that is used here for the CSLR procedure. However, before we try to fix the problem for the increasing failure rate case, we shall try to study a more powerful alternative pointed out by Bain et al. [4].

5 LAPLACE TEST STATISTIC AS A POSSIBLE ALTERNATIVE

Tests for a constant intensity versus an increasing alternative have been developed for other families of intensities to indicate the relative powers of all of these tests based on fixed sample size tests. In an important paper, Bain et al. [4] conclude that two tests which were derived as optimal tests for specific parameteric models also are rather powerful as tests of trend for general NHPP's. In other words, these tests are "robust" against other model assumptions. They are the well-known Laplace test (called L test), and the most powerful test for the shape parameter in a Poisson process with Weibull intensity (called Z test) as has been described in this article. The likelihood ratio test against arbitrary NHPP alternatives, two nonparametric tests for trends based on Kendall's tau and Spearman's rho, and a test based on an F statistic are also included in their study. The powers of the fixed sample size tests based on the time-truncated framework are presented by Bain et al. [4] using Monte Carlo simulation against alternatives that are increasing at an exponential rate, a power rate (Weibull intensity), and a logarithmic rate. Alternatives that are step-functions with one jump are also considered. Bain et al. [4] report that for a fixed sample size, the Z test performs quite well for the smooth alternatives (i.e. logarithmic, Weibull, and exponential intensity). In fact, for such alternatives its estimated power is at least

87% as large as that of the other five tests considered, and for the logarithmic and Weibull intensities it is the most powerful of the six. However, Bain et al. [4] also recommend that if one also wishes to guard against the step-function intensities, L could be used. The estimated smallest relative power of Z is 71% of that for L for the step-function intensities considered in their study. Therefore, it is instructive to compare the L test to the Z test in the same sequential setting, which could also provide valuable information regarding the error rates and small sample size performance.

Cox [6] discussed the use of the L test for testing $\beta = 0$ versus $\beta > 0$ in $\lambda(t) = \alpha e^{\beta t}$. A slightly different form of the test statistic is required for the control charting procedure, because the observed time period is taken up to a preassigned number of events – a failure truncated case. Consistent with the previous notation, the test statistic in examining the process at the current time point r is

$$L_r = \left[\sum_{i=1}^{r-1} t_i/t_r - (r-1)/2 \right] / \left[(r-1)/12 \right]^{1/2},$$

which approaches normality quite rapidly as r increases. Thus, based on L test, the $(1-\alpha)$ 100% control limits are

$LCL_\alpha \;\; = z_{\alpha/2}$, the $100\alpha/2$ percentile of the standard normal distribution
$UCL_\alpha \;\; = z_{1-\alpha/2} = -z_{\alpha/2}$.

If at current time r, either $L_r \leq z_{\alpha/2}$ or $L_r \geq z_{1-\alpha/2}$, then the process is judged to be out of control. The performance of the L test is also evaluated by Monte Carlo simulation. Attention is first focused on tests of small fixed sample sizes. Although, for L, the calculable minimum sample size is 2, the minimum sample sizes that make sense for a nominal level of .1, .05 and .01 are 2, 3 and 4 respectively. The empirical results show that the large sample approximation for the L test for small sample sizes yield true significance levels smaller than the nominal levels. For instance, in the fixed sample size case with n = 2 and nominal level $\alpha' = .1$, the estimated true significance level is only .055, which jumps quickly to approximately .1 if n is 3. However, for $\alpha' = .01$, the estimated true significance level is .0035 for the required minimum sample size, 4, and increases steadily to the nominal level (= .01) at approximately $n = 11$. Obviously, this test has to be used very cautiously for small values of n and α'. Due to the same starting values of $n = 2$, we choose the nominal significance level .1 to compare the rates of type I error between the Z and L tests for the control charting procedure. Interestingly enough, our simulation results show that the L test yields a considerably larger overall error rate than the Z test. The error rate for the L test hits .5 at the 13th time point (versus 43 for the Z test) and reaches .9391 (versus .7163 for the Z test) at the 500th time point. The overall performance of L in detecting an abrupt change in a process as described in Section 4 is better than the Z test using the same nominal level. The L test picks up the out-of-control signals faster than the Z test in both cases of trend at the expense of higher overall type I errors. Unfortunately, the shotgun pattern still exists in the increasing failure rate case.

6   CONCLUSIONS

Although the outcomes of the extended search for a possible better alternative are not successful (i.e., the L-test) for the control charting procedure, some of the results are interesting. They provide side-to-side information for these two well-known and widely used test statistics for trend analysis. We now conclude this section with a few comments and point to some further work.

1. The Z test has an asymmetric performance in detecting an alternative which is a step-function intensity in both the fixed sample size case [12] and the CSLR sequential procedure developed in this simulation study. Moreover, the Z test is the UMPU test of $\beta = 1$ versus $\beta \neq 1$ for the family of intensities, $\lambda(t) = (\beta/\theta) (t/\theta)^{\beta-1}$, is also recommended for other smooth alternatives [4].

2. A trend in the sequence $\{t_i /t_n\}$ for $1 \leq i \leq n - 1$ is reflected as the reverse trend in the sequence $\{1 - t_i /t_n\}$ for $1 \leq i \leq n - 1$. Ho [12] developed a backward Z test and recommended that parallel tests (backward and forward) be performed on the same data set to guard against the step-function intensities of both kinds (increasing and decreasing). The former study [12] was based on a fixed sample size design. A logical extension of the current study is to add the backward test at each time point to guard against the change-point Poisson process with an increasing failure rate. The out-of-control signal sent by either test, however, would further inflate the overall type I error. Additional adjustment of the significance levels to account for the unified dual process control needs to be investigated. Our efforts for future studies will be devoted to this goal and to some quality assessments (i.e., power study) of the suggested modified procedure.

7 Acknowledgements

The author thanks David Lerman and Andy Tsang for help with computations. This work was supported by the Nevada Nuclear Waste Project Office.

References

[1] Armitage, P., McPherson, C.K., and Rowe, B.C., Repeated Significance Tests on Accumulating Data. Journal of the Royal Statistical Society, Ser. A, 132 (1969) 235-244.

[2] Bain, L.J., and Engelhardt, M., Inferences on the Parameters and Current System Reliability for a Time Truncated Weibull Process, Technometrics, 22 (1980) 421-426.

[3] Bain, L.J., and Engelhardt, M., Statistical Analysis of Reliability and Life-Testing Models - Theory and Methods, (2nd ed.), (1991) New York: Marcel Dekker.

[4] Bain, L.J., Engelhardt, M., and Wright, F.T., Tests for an Increasing Trend in the Intensity of a Poisson Process: A Power Study, Journal of the American Statistical Association, 80 (1985) 419-422.

[5] Bassin, W.M., Increasing Hazard Functions and Overhaul Policy, Proceedings of the 1969 Annual Symposium on Reliability, Chicago: IEEE, (1969) 173-178.

[6] Cox, D.R., Some Statistical Methods Connected with Series of Events, Journal of the Royal Statistical Society, Ser. B, 17 (1955) 129-164.

[7] Crow, L.H., Reliability Analysis for Complex Repairable Systems, Reliability and Biometry, eds. F. Proschan and R.J. Serfling, Philadelphia: SIAM: (1974) pp. 379-410.

[8] Crow, L.H., Confidence Interval Procedures for the Weibull Process with Applications to Reliability Growth. Technometrics, 24 (1982) 67-72.

[9] DeMets, D.L., and Ware, J.H., Group Sequential Methods for Clinical Trials with a One-Sided Hypothesis, Biometrika, 67 (1980) 651-60.

[10] DeMets, D.L., and Ware, J.H., Asymmetric Group Sequential Boundaries for Monitoring Clinical Trials, Biometrika, 69 (1982) 661-663.

[11] Finkelstein, J.M., Confidence Bounds on the Parameters of the Weibull Process. Technometrics, 18 (1976) 115-117.

[12] Ho, C-H, Forward and Backward Tests for an Abrupt Change in the Intensity of a Poisson Process, Journal of Statistical Computation and Simulation, 48 (1993) 245-252.

[13] Lan, K.K.G., and DeMets, D.L., Discrete Sequential Boundaries for Clinical Trials, Biometrika, 70 (1983) 659-663.

[14] Lee, L., and Lee, S.K., Some Results on Inferences for the Weibull Process, Technometrics, 20 (1978) 41-45.

[15] Montgomery, D.C., Introduction to Statistical Quality Control, (1985) New York: John Wiley.

[16] O'Brien, P.C., and Fleming, T.R., A Multiple Testing Procedure for Clinical Trials, Biometrics, 35 (1979) 549-556.

[17] Pocock, S.J., Group Sequential Methods in the Design and Analysis of Clinical Trials, Biometrika, 64 (1977) 191-199.

[18] Ryan, T.P., Statistical Methods for Quality Improvement, (1989) New York: John Wiley.

A Quasi-Monte Carlo Algorithm for the Global Illumination Problem in the Radiosity Setting

Alexander Keller

Fachbereich Informatik
AG Numerische Algorithmen
Universität Kaiserslautern
Postfach 3049, D-67653 Kaiserslautern
e-mail: keller@informatik.uni-kl.de

Abstract

One of the main problems in computer graphics is to solve the global illumination problem, which is given by a Fredholm integral equation of the second kind, called the radiance equation (REQ). In order to achieve realistic images, a very complex kernel of the integral equation, modelling all physical effects of light, must be considered. Due to this complexity Monte Carlo methods seem to be an appropriate approach to solve the REQ approximately. We show that replacing Monte Carlo by quasi-Monte Carlo in some steps of the algorithm results in a faster convergence.

1 Introduction

Realistic image synthesis [1] mainly consists of two steps: First the radiance equation (REQ) has to be solved to determine the global illumination. Then the solution of the REQ is sampled by simulating a lens in order to compute the radiance value for each pixel of the image matrix.

In our case the scene to be illuminated is composed of disjoint triangles A_k, $1 \leq k \leq K$, where K depends on the detail of modelling and is in the range of $10^3 \ldots 10^6$ for practical scenes. The union S of all triangles is the surface of the scene. We assume that the scene is finite and closed and that each triangle has a local coordinate system and is not degenerate. To solve the REQ means to know the radiance $L(x, \omega)$ in any location $x \in \cup_{k=1}^{K} A_k = S$ for any direction $\omega \in \Omega$, where Ω is the hemisphere in normal direction in x. The radiance L is measured in $[\frac{W}{m^2\,sr}]$ and as such is intensity per solid angle. The radiance equation (REQ), a Fredholm integral equation of the second kind, is given by

$$L(x, \omega) = L_0(x, \omega) + \int_{\Omega'} L(h(x, \omega'), -\omega')\, f_r(-\omega', x, \omega)\, \cos \theta(x, \omega')\, d\omega' \qquad (1)$$

---

[1] For a profound introduction into realistic image synthesis see [CW93], for a survey on Monte Carlo methods in computer graphics see [MP93].

where $L(x,\omega)$ is the radiance emerging from x in direction ω. L is the sum of the source radiance L_0 and the radiance from all over the hemisphere Ω' (over surface location x) reflected at x into direction ω. $h(x,\omega')$ is the first point hit from location x in direction ω' and $\theta(x,\omega')$ is the angle between the surface normal in x and the direction ω'. The f_r term is a bidirectional reflection distribution function (BRDF) describing the surface properties in a surface location x for light coming from direction $-\omega'$ and being reflected into direction ω. In this way the BRDF characterizes color, gloss, etc. of the surface. Due to the Helmholtz principle we have

$$f_r(-\omega', x, \omega) = f_r(-\omega, x, \omega')$$

which means that all light paths can be reversed. This is an important property and the basis of ray tracing. The algorithm which will be derived in the next section is designed for the so-called radiosity setting [2]. In this setting all surfaces are diffuse reflectors, i.e. we only allow constant BRDFs, describing a uniform reflection independent of direction:

$$f_r(x) = \frac{\rho_d(x)}{\pi}$$

The REQ is a convenient formulation of a radiative transfer problem. Usually the radiance L is a vector of components representing the color basis red, green and blue. So all equations would be replicated for each component. For the sake of simplicity we now restrict to only one wavelength.

We rewrite equation (1) as operator equation:

$$L = L_0 + T_f L \qquad (2)$$

The solution is accessible via the Neumann series. Since the operator norm is less than 1 in the diffuse environment, the series is converging and can be cut off at a certain level M with little loss of accuracy [3]:

$$L = \sum_{i=0}^{\infty} T_f^i L_0 \approx \sum_{i=0}^{M} T_f^i L_0 \qquad (3)$$

Since the operator T_f is positive, the truncation of the series causes a certain underestimation.

The radiance value of one pixel P is the mean value integral over its area:

$$L_P = \frac{1}{|P|} \int_P L(h(Eye, \mathrm{dir}(Eye, x)), -\mathrm{dir}(Eye, x)) \, dx \qquad (4)$$

where Eye is the position of the observer's eye and $\mathrm{dir}(Eye, x)$ is the direction from the eye through location x on the screen into the scene.

---

[2] See the conclusion for elusion of this restriction.
[3] In practical diffuse environments $M \approx 10$ is sufficient.

2   Algorithm

The main idea of our algorithm is to calculate average local solutions of the REQ for the triangles A_k, as proposed in [HK94b], and then to reconstruct the image from these average informations. The average local solutions will be computed by quasi-Monte Carlo integration (see [Nie92a]). The idea of applying low discrepancy particle methods to integral equations goes back to the work of [Hla62], [NW73], and [SP87].

In order to solve the REQ we insert (2) into itself and get

$$
\begin{aligned}
L &= L_0 + T_f\, L_0 + T_f\, T_f\, L \\
&= L_d + L_i
\end{aligned}
\tag{5}
$$

where

$$
\begin{aligned}
L_d &:= L_0 + T_f\, L_0 \quad \text{is the direct illumination and} \\
L_i &:= T_f\, T_f\, L \quad\ \ \text{is the indirect illumination.}
\end{aligned}
$$

The direct illumination L_d is calculated separately, since there exist fast techniques for its computation. For the Monte Carlo integration, we have to integrate over all solid angles of triangles with $L_0 \neq 0$:

$$
\begin{aligned}
& L_d(x,\omega) \\
&= L_0(x,\omega) + \int_{\Omega'} L_0(h(x,\omega'),-\omega')\, f_r(-\omega',x,\omega)\, \cos\theta(x,\omega')\, d\omega' \\
&= L_0(x,\omega) + \sum_{k=1}^{LS} \int_{\Omega'_k} L_0(h(x,\omega'),-\omega')\, f_r(-\omega',x,\omega)\, \cos\theta(x,\omega')\, \chi_{A_k}(h(x,\omega'))\, d\omega' \\
&\approx L_0(x,\omega) + \sum_{k=1}^{LS} \frac{|\Omega'_k|}{N} \sum_{i=0}^{N-1} L_0(h(x,\omega'_i),-\omega'_i)\, f_r(-\omega'_i,x,\omega)\, \cos\theta(x,\omega'_i)\, \chi_{A_k}(h(x,\omega'_i))
\end{aligned}
\tag{6}
$$

where LS is the number of lightsources, Ω'_k is the solid angle of triangle A_k seen from location x without obstruction. The χ_{A_k} function is the visibility function:

$$
\chi_{A_k}(h(x,\omega')) ::= \begin{cases} 1 & h(x,\omega') \in A_k \\ 0 & \text{else} \end{cases}
$$

For the determination of the indirect contribution L_i, we first apply a functional Ψ_{A_k} to the solution of the REQ:

$$
\langle L, \Psi_{A_k} \rangle = \int_S \int_\Omega L(x,\omega)\, \cos\theta(x,\omega)\, \chi_{A_k}(h(x,\omega))\, d\omega\, dx
\tag{7}
$$

The functional $\Psi_{A_k} = \cos\theta(x,\omega)\, \chi_{A_k}(h(x,\omega))$ is the "detector"-function for the total incoming power on surface A_k, which is given by the inner product above. So the mean diffuse radiance reflected by this triangle is

$$
\overline{L_{A_k}}(x) = \frac{\rho_d(x)}{\pi\, |A_k|}\, \langle L, \Psi_{A_k} \rangle
\tag{8}
$$

Supposed we have an approximation for $\overline{L_{A_k}}(x)$, the approximation for the indirect illumination looks like:

$$
\begin{aligned}
L_i(x,\omega) &= (T_f\,T_f\,L)(x,\omega) \\
&\approx \int_{\Omega'} \left(\sum_{k=1}^{K} \overline{L_{A_k}}(h(x,\omega'))\, \chi_{A_k}(h(x,\omega')) \right)\, f_r(-\omega',x,\omega)\,\cos\theta(x,\omega')\,d\omega' \quad (9)
\end{aligned}
$$

To finally generate the image, we have to resample the approximation of L for each pixel. For one pixel we have

$$
L_P \approx \frac{1}{OS} \sum_{i=0}^{OS-1} L(h(Eye,\mathrm{dir}(Eye,x_i)), -\mathrm{dir}(Eye,x_i)) \quad (10)
$$

where OS is the oversampling rate per pixel, i.e. the number of rays shot from the eye through the pixel into the scene.

3   Quasi-Monte Carlo Integration

The quasi-Monte Carlo integration uses the same formula (11) as the (pseudo-) Monte Carlo method, but replaces the (pseudo-) random numbers by low discrepancy sequences.

$$
\int_{[0,1]^s} f(u)\,du \approx \frac{1}{N} \sum_{i=0}^{N-1} f(u_i) \quad (11)
$$

The (u_i) then are a deterministic pattern on $\bar{I}^s = [0,1]^s$ specifically designed for integration. That means that quasi-random numbers may fail several statistical tests applied for pseudo-random numbers; they solely keep the property of uniform distribution. Since we can only use a finite number of samples, the pattern has a deviation from the uniform distribution measured as discrepancy. The discrepancy of a given (deterministic) point set $P_N = \{u_0, \ldots, u_{N-1}\}$ with regard to a family of subsets $\mathcal{F}$ is defined as

$$
D(\mathcal{F}, P_N) = \sup_{A \in \mathcal{F}} \left| \lambda_s(A) - \frac{1}{N} \sum_{i=0}^{N-1} \chi_A(u_i) \right|, \quad (12)
$$

where the supremum is taken over all subsets $A \subseteq \bar{I}^s$ of the family $\mathcal{F}$ and λ_s is the Lebesgue measure on $\bar{I}^s$. The discrepancy can be seen as maximum integration error for the characteristic functions of such a family of subsets.

For $\mathcal{J}^* = \{A = \prod_{i=1}^{s}[0,a_i] \mid 0 \le a_i \le 1\}$ we have the star-discrepancy

$$
D^*(P_N) ::= D(\mathcal{J}^*, P_N) .
$$

The inequality of Koksma and Hlawka

$$
\left| \int_{\bar{I}^s} f(u)\,du - \frac{1}{N} \sum_{i=0}^{N-1} f(u_i) \right| \le V(f) \cdot D^*(P_N) \quad (13)
$$

is a separation of upper error bound into the star-discrepancy, depending only on the sampling pattern, and the total variation $V(f)$ of the integrand. This decomposition makes sense if the total variation $V(f)$ in the sense of Hardy and Krause (for further details see [Nic92a]) can be bounded by a finite constant.

Since L, L_0 and f_r are not continuous, this estimation is not directly applicable. In [HK94a] we showed an upper error bound for the pixel oversampling (10), whereas an upper error bound for algorithms for L and $\overline{L_{A_k}}$ is an open problem in the setting of computer graphics.

3.1 Low Discrepancy Sequences

Let us now recall two standard types of low discrepancy sequences. First, we define Φ as the radial inverse function:

$$\Phi_b(i) = \sum_{j=0}^{\infty} a_j(i)\, b^{-j-1} \quad \text{when } i = \sum_{j=0}^{\infty} a_j(i)\, b^j \tag{14}$$

where the natural number $b > 1$ is the base and $i \in I\!N$. The values Φ_b always are in the unit interval $[0, 1)$. For illustration: $\Phi_2(i)$ simply is the binary representation of i mirrored at the decimal point [4].

The Halton and the Hammersley sequence are s-dimensional vectors built from radial inverse functions in relatively prime bases b_j, that is, we choose subsequent primes for the b_j. For $0 \le i < N$ we have:

$$\text{Halton points:} \quad u_i = (\Phi_{b_1}(i), \dots, \Phi_{b_s}(i))$$

$$\text{Hammersley points:} \quad u_i = (\frac{i}{N}, \Phi_{b_1}(i), \dots, \Phi_{b_{s-1}}(i))$$

The discrepancies $D^*(P_N)$ for both sets have the following order of magnitude (see [Nie92a]):

$$D^*_{Halton} \in O\left(\frac{\log^s N}{N}\right)$$

$$D^*_{Hammersley} \in O\left(\frac{\log^{s-1} N}{N}\right).$$

The Halton sequence is an incremental pattern, meaning that increasing the number of samples is possible without discarding the samples already drawn. In contrast to this the Hammersley sequence is not incremental since increasing the sampling rate results in discarding all samples computed so far. In consequence this pattern is not useful for adaptive sampling, although it has an asymptotically smaller discrepancy. No values of either sets need to be precomputed, since all values are directly accessible.

Using a permutation σ of the set $\{0, 1, \dots, b - 1\}$ for the generation of the radial inverse functions is called scrambling:

$$\Phi_b(i, \sigma) = \sum_{j=0}^{\infty} \sigma(a_j(i))\, b^{-j-1} \quad \text{when } i = \sum_{j=0}^{\infty} a_j(i)\, b^j$$

---

[4]For fast algorithms for computing Φ we refer to [Str93] or [HW64].

Scrambling is used for breaking the dependencies of the radical inverse function for large bases in high dimensions. Scrambling, which is useful in most applications of quasi-Monte Carlo integration, in computer graphics has no big effect. That fact arises from the complex scene geometry, where an ϵ-change of one ray direction can change a random walk path completely.

There is a variety of more low discrepancy sequences as the Faure-, the Sobol-, and further scrambled sequences. All these sequences differ a little bit in quality for the environment of computer graphics, but have about the same order of magnitude in discrepancy.

4 Implementation

The implementation is split into a preprocessing step for the evaluation of (7) and the rendering step for calculating the mean pixel radiances of (10).

We rewrite equation (7) by substituting the truncated Neumann series (3) for the solution L:

$$
\langle L, \Psi_{A_k} \rangle
$$

$$
\approx \left\langle \sum_{j=0}^{M} T_f^j L_0, \Psi_{A_k} \right\rangle
$$

$$
= \sum_{j=0}^{M} \langle T_f^j L_0, \Psi_{A_k} \rangle
$$

$$
= \sum_{j=0}^{M} \int_S \int_\Omega (T_f^j L_0)(x,\omega) \, \cos\theta(x,\omega) \, \chi_{A_k}(h(x,\omega)) \; d\omega \, dx
$$

$$
= \sum_{j=0}^{M} \int_{\Omega^{j+1}} \int_{S_0} \prod_{l=1}^{j} f_r(-\omega_l, x_l, \omega_{l+1}) \prod_{l=1}^{j+1} \cos\theta(x_{l-1}, \omega_l)
$$
$$
L_0(x_0, \omega_1) \, \chi_{A_k}(x_{j+1}) \, dx_0 \, d\omega_1 \cdots d\omega_{j+1}
$$

$$
= \sum_{j=0}^{M} \int_{Q^{j+1}} \int_{S_0} \prod_{l=1}^{j} f_r(-\omega_l, x_l, \omega_{l+1}) \prod_{l=1}^{j+1} \frac{\sin 2\theta_l}{2}
$$
$$
L_0(x_0, \omega_1) \, \chi_{A_k}(x_{j+1}) \, dx_0 \, d\theta_1 \, d\phi_1 \cdots d\theta_{j+1} \, d\phi_{j+1}
$$

$$
= \sum_{j=0}^{M} |S_0| \, \pi^{2j+2} \int_{I^{2j+4}} \prod_{l=1}^{j} f_r(-\omega_l, x_l, \omega_{l+1}) \prod_{l=1}^{j+1} \frac{\sin \pi u_{2l}}{2}
$$
$$
L_0(x_0, \omega_1) \, \chi_{A_k}(x_{j+1}) \, du_0 \cdots du_{2j+3} \tag{15}
$$

where

$$
x_l = h(x_{l-1}, \omega_l) \text{ for } l > 0 \text{ and}
$$
$$
Q = [0, \frac{\pi}{2}] \times [0, 2\pi]
$$

S_0 is the surface of the lightsources where $L_0 > 0$. For the Monte Carlo evaluation of (15) we use a $(2M + 4)$-dimensional low discrepancy sequence (u_i) for all decisions of a random walk simulation. Using the first two components $(u_{i,0}, u_{i,1})$ we choose a starting point $x_0 \in S_0$ for the path on a light source. Then the particle is shot into direction $\omega_1 = (\theta_1, \phi_1) = (2\pi u_{i,2}, \frac{\pi}{2} u_{i,3})$. The triangle which is hit first in this direction receives the incoming power. Subsequently the particle is attenuated due to the BRDF in the hitpoint and traced into the next direction given by $(2\pi u_{i,4}, \frac{\pi}{2} u_{i,5})$. This procedure is extended to a path length of $M + 1$ lines. By this we do the evaluation of (7) for all triangles A_k simultaneously by only one low discrepancy sequence.

As mentioned in section 3.1, for adaptive termination of the preprocessing step we have to apply infinite low discrepancy sequences like the Faure-, Halton-, or Sobol-sequence. The termination criterion is similar to [Pas94]. For two numbers N_1 and N_2 of iterated paths, we determine an error by

$$\triangle E(N_1, N_2) = \frac{1}{\sum_{k=1}^{K} L_{0,k} |A_k|} \sqrt{\frac{\sum_{k=1}^{K} d(\overline{L_{A_k}(N_1)}, \overline{L_{A_k}(N_2)})^2 |A_k|}{\sum_{k=1}^{K} |A_k|}}$$

where the $\overline{L_{A_k}(N_i)}$ is the approximation of (7) by N_i samples and L_0 is the radiance of the different light sources. Further we select an interval $\triangle N$ for the measurements. The distance d is the Euclidean distance between the two color vectors $\overline{L_{A_k}(N_i)}$. The process is terminated if for a fixed T and the smallest $n \in I\!N$

$$\triangle E((n + t) \triangle N, (n + t + 1) \triangle N) < \epsilon \text{ for } 0 \le t < T .$$

Since the error is weighted by the size of the triangles, after termination the bigger areas A_k are integrated more exactly than the smaller areas. This makes sense, because in the resampling step (9) the bigger areas are hit more often than the smaller ones.

The image generation is done by selecting positions in a pixel and sending rays from the eye through these positions into the scene. As shown in [HK94a], the Halton sequence is well suited for adaptive oversampling in this case.

Having determined the hitpoints in the scene, the radiance emerging from them into the direction of the eye has to be calculated by (5) for averaging (10). The direct light L_d is calculated by (6). It would be very expensive to calculate the contribution of all lights for all hitpoints of the oversampling process. So for the OS hitpoints, we calculate the light contribution of $\frac{\text{number of lights}}{OS}$ lights selected by a random permutation (which changes from pixel to pixel so as to avoid aliasing), so that altogether every light is used exactly once for one pixel.

For each first hitpoint from the eye into the scene, we resample (7) by jittering rays over the hemisphere. From (9) we derive

$$
\begin{aligned}
L_i(x, \omega) &\approx \int_{\Omega'} \left(\sum_{k=1}^{K} \overline{L_{A_k}}(h(x, \omega')) \, \chi_{A_k}(h(x, \omega')) \right) f_r(-\omega', x, \omega) \, \cos\theta(x, \omega') \, d\omega' \\
&= \int_{Q} \left(\sum_{k=1}^{K} \overline{L_{A_k}}(h(x, \omega')) \, \chi_{A_k}(h(x, \omega')) \right) f_r(-\omega', x, \omega) \, \frac{\sin 2\theta'}{2} \, d\theta' \, d\phi' \\
&= \int_{Q} \left(\sum_{k=1}^{K} \overline{L_{A_k}}(h(x, \omega')) \, \chi_{A_k}(h(x, \omega')) \right) f_r(-\omega', x, \omega) \, \frac{d \sin^2 \theta'}{2} \, d\phi'
\end{aligned}
$$

$$\approx \frac{\pi}{SR} \sum_{i=0}^{SR-1} \left(\sum_{k=1}^{K} \overline{L_{A_k}}(h(x,\omega')) \, \chi_{A_k}(h(x,\omega'_i)) \right) f_r(-\omega'_i, x, \omega)$$

where $x = h(Eye, -\omega)$ is the point first hit by a ray from the Eye into direction $-\omega$. SR is the number of scattered rays used for the evaluation of the integral. For reasons of efficiency a kind of importance sampling is applied. The rays are not uniformly distributed over the hemisphere Ω, but distributed with respect to the $\sin^2 \theta'$-term.

This sampling method applies to all BRDFs. In order to reduce the variance, it would be better to use a basis decomposition of the BRDF and to integrate over the basis vector separately (e.g. see [War92]). But since here we only allow the evaluation of diffuse, i.e. constant BRDFs no decomposition is necessary.

In addition to the algorithm explained above, we implemented two enhancements: First we included mirrors. A perfect mirror can be modelled by a δ-function. So whenever a ray hits a singular surface, it is reflected without changing its radiance. The second enhancement concentrates on the fact that the $h(x,\omega)$ operation is the most expensive operation in computer graphics. Therefore we implicitly increase the number of samples for the indirect light $\overline{L_{A_k}}$ of one pixel by using the samples of the eight neighbouring pixels. In order to assure a small error for this averaging, we have to guarantee a small deviation of normals. Therefore we only take those indirect light rays, which hit the same surfaces as considered in the current pixel. Since we exclusively used triangles to model our scenes, the deviation then is equal to zero. This kind of increment of samples acts as a low pass filter (since Monte Carlo integration simply is averaging) and so smoothes the indirect light contribution. Note that this method does not smear sharp contours, because it is not an image space method.

This method acts conveniently to the observer's eye: If we see areas which extend over big areas of the image, we tend to perceive high levels of noise. Exactly in this case the averaging smoothes this effect. If on the other hand we have many small objects in one pixel, the eye cannot distinguish and such does not perceive the high variance.

4.1 Measurements

Due to the complexity of the kernel of the REQ, for realistic scene description no analytic solution is accessible to serve as benchmark for the various rendering methods. So there are two ways left to test the performance of illumination algorithms. The first is to compute a solution with a very high number of samples and to take this solution as reference for calculations with less samples (see [HK94b]).

The second method is to simplify the kernel in such a way that an analytical solution becomes accessible. Let $\rho_d = \frac{1}{2}$, $f_r = \frac{\rho_d}{\pi}$, and $L_0 = \frac{1}{4}$ for all triangles A_k. Then the solution

$$L = L_0 + \int_{\Omega} \frac{\rho_d}{\pi} L \cos \theta(\omega) \, d\omega$$

$$= L_0 + T_f L = \sum_{i=0}^{\infty} T_f^i L_0$$

$$= \sum_{i=0}^{\infty} \left(\frac{1}{2\pi}\right)^i \frac{1}{4} T^i = \frac{1}{4} \sum_{i=0}^{\infty} \frac{1}{2^i} = \frac{1}{2}$$

where

$$T = \int_\Omega \cos\theta \, d\omega = \int_0^{2\pi} \int_0^{\frac{\pi}{2}} \cos\theta(\omega) \, \sin\theta \, d\theta \, d\phi = \pi$$

simply is the projection of the unit-hemisphere onto the plane. By simplifying the kernel this way, only the complexity of the construction of the scene remains. This analytic solution can be used for finding obvious errors in the simulation. But more importantly it can serve as a benchmark test for illumination algorithms by calculating the distance to the analytic solution.

For the experiments we chose four different scenes. All scenes were modelled in meters, that is in real measure. The scenes are an empty cube ($K = 12$ triangles), a living room ($K = 3604$ triangles), an office ($K = 276$ triangles) and the computer graphics teapot (*Utah Teapot*) in a box ($K = 1572$ triangles). Images of the last three scenes can be seen in figure 1.

The algorithm described in section 2 will always have a small systematic error due to the truncation of the Neumann series. In the measurements we used $M = 29$ for the experiments with the analytic solution in order to keep the truncation error very small and $M = 6$ for the realistic experiments.

For the analytic solution experiments (all surfaces were grey with $f_r = \frac{1}{2\pi}$ and emitted white light with $L_0 = \frac{1}{4}$) we calculated the mean square deviation $\Delta_a(N)$ and the weighted mean square deviation $\Delta_{aw}(N)$:

$$\Delta_a(N) = \sqrt{\frac{\sum_{k=1}^{K}(\overline{L_{A_k}(N)} - \frac{1}{2})^2}{K}}$$

$$\Delta_{aw}(N) = \sqrt{\frac{\sum_{k=1}^{K}(\overline{L_{A_k}(N)} - \frac{1}{2})^2 |A_k|}{K \sum_{k=1}^{K} |A_k|}}$$

The comparison of different sampling patterns for various numbers of samples is illustrated in tables 2 and 3. It can be observed that the low discrepancy sequences acquire the same level of error faster than the (pseudo-) random sequences (for a description of the inversive generator, see [Nie92b]). Note that inside the cube there are not any obstructions, and so the algorithm converges very fast due to the simple, symmetric geometry.

For the realistic setting (i.e. real surface textures and light sources, except for the cube, for which we still used the analytic setting), table 1 shows the behaviour of adaptive termination as described above. For a given accuracy ϵ of the preprocessing step, the number N of samples needed for termination ($\Delta N = 1000$, $T = 2$) is printed. From these tables it can be seen that the complexity of construction of the scene nearly replaces scrambling of the low discrepancy sequences. It also can be seen that the low discrepancy sequences are superior to the (pseudo-) random sequences, i.e. they terminate faster.

5   Conclusion and Further Work

We proposed an algorithm to approximately solve the global illumination problem for scenes consisting of diffuse and pure specular objects. The quasi-random, i.e. deterministic preprocessing step is applicable to any BRDF. In order to make the resampling step apply to not only diffuse BRDF and especially to gather caustics, the incoming radiance in equation (7) has to be stored for any incoming direction (for example by spherical harmonics).

By our experiments we proved that the quasi-random approach for the preprocessing step is slightly superior to the pseudo-random algorithm.

6 Acknowledgement

The author would like to thank Stefan Heinrich for helpful discussions and suggestions.

References

[CW93] M.Cohen, J.Wallace: Radiosity and Realistic Image Synthesis, Academic Press Professional, Cambridge 1993.

[HW64] J.H.Halton, G.Weller: Algorithm 247: Radical-inverse quasi-random point sequence [G5], Comm. ACM, No.12, 7(1964), pp. 701-702.

[HK94a] S.Heinrich, A.Keller: Quasi-Monte Carlo methods in computer graphics, Part I: The QMC-Buffer, Technical Report 242/94, University of Kaiserslautern, 1994.

[HK94b] S.Heinrich, A.Keller: Quasi-Monte Carlo methods in computer graphics, Part II: The Radiance Equation, Technical Report 243/94, University of Kaiserslautern, 1994.

[Hla62] E.Hlawka: Lösung von Integralgleichungen mittels zahlentheoretischer Methoden I, Sitzungsber., Abt. II, Österr. Akad. Wiss., Math.-Naturwiss. Kl., 171(1962), pp. 103-123.

[MP93] S.P.Mudur, S.N.Pattanaik: Monte Carlo methods for computer graphics, in State of the Art Reports, Eurographics, 1993.

[NW73] H.Neunzert, J.Wick: Die Theorie der asymptotischen Verteilung und die numerische Lösung von Integrodifferentialgleichungen, Numer. Math., 21(1973), pp. 234-243.

[Nie92a] H.Niederreiter: Random Number Generation and Quasi-Monte Carlo Methods, SIAM Philadelphia, Pennsylvania 1992.

[Nie92b] H.Niederreiter: New Methods for Pseudorandom Number and Pseudorandom Vector Generation, Proc. 1992 Winter Simulation Conference (Arlington, VA, 1992), IEEE Press, Piscataway, NJ, 1992, pp. 264-269.

[Pas94] S.Paskov: Termination Criteria for Linear Problems, Technical Report, to appear, Columbia University, 1994.

[SP87] P.K.Sarkar, M.A.Prasad: A comparative study of pseudo and quasi random sequences for the solution of integral equations, J. Comp. Physics, 68(1987), pp. 66-88.

[Str93]   J.Struckmeier: Fast generation of low-discrepancy sequences, Berichte der Arbeitsgruppe Technomathematik, Nr.93, Fachbereich Mathematik, Universität Kaiserslautern.

[War92]   G.Ward: Measuring and Modeling Anisotropic Reflection, Computer Graphics, No.2, 26(1992), pp. 265-272.

Sequence	N per scene			
	cube	living room	office	teapot
$\epsilon = 10^{-3}$				
Lin. Congr.	42000	36000	48000	31000
Inversive	35000	37000	37000	36000
Halton	33000	36000	28000	29000
Faure	32000	29000	30000	30000
Sobol	30000	35000	30000	24000
$\epsilon = 3.3 \cdot 10^{-4}$				
Lin. Congr.	100000	99000	149000	102000
Inversive	101000	115000	126000	106000
Halton	92000	100000	72000	85000
Faure	99000	95000	72000	77000
Sobol	95000	87000	78000	77000
$\epsilon = 10^{-4}$				
Lin. Congr.	330000	343000	439000	334000
Inversive	337000	314000	397000	316000
Halton	266000	305000	235000	238000
Faure	276000	283000	276000	219000
Sobol	310000	298000	253000	228000

Table 1: Samples needed for a termination by ϵ

Samples	Sequence				
N	Lin. Congr.	Inversive	Halton	Faure	Sobol
empty cube					
10^3	0.0647699	0.0469468	0.0459244	0.0375936	0.0458288
10^4	0.0138841	0.0150312	0.015322	0.0101492	0.0154578
10^5	0.00416856	0.00426497	0.00347108	0.00388516	0.0139228
living room					
10^3	5.68352	3.89019	2.0911	1.46717	5.49831
10^4	1.87948	1.5128	2.15754	1.94902	1.51938
10^5	0.64136	0.550875	0.524932	0.727295	0.573517
office					
10^3	0.46748	0.456453	0.399271	0.280669	0.431063
10^4	0.259767	0.202392	0.331709	0.475231	0.234959
10^5	0.136946	0.127394	0.136938	0.136222	0.14049
teapot					
10^3	3.55052	24.4236	5.28716	3.58575	4.63836
10^4	1.36052	2.67153	0.962619	1.01024	0.920303
10^5	0.460942	0.438209	0.434956	0.412762	0.422177

Table 2: Mean square deviation from analytical solution

Samples	Sequence				
N	Lin. Congr.	Inversive	Halton	Faure	Sobol
empty cube					
10^3	0.00589147	0.00485275	0.00514532	0.00398542	0.00500996
10^4	0.00171686	0.00181801	0.00147316	0.00128062	0.00184561
10^5	0.000478878	0.000538219	0.000402798	0.000395118	0.00152823
living room					
10^3	0.0056218	0.00591197	0.00478811	0.00455163	0.00582923
10^4	0.00325591	0.00320881	0.00322973	0.00330551	0.00322495
10^5	0.0028332	0.00282448	0.00282852	0.00282892	0.00284094
office					
10^3	0.00861589	0.00845805	0.00825801	0.0070116	0.00775931
10^4	0.00672887	0.00668762	0.00670869	0.00680193	0.00695615
10^5	0.00661765	0.00654544	0.00655064	0.00657714	0.00685
teapot					
10^3	0.00624067	0.0116751	0.00809856	0.00610271	0.00580429
10^4	0.00183252	0.00199039	0.00156816	0.00158062	0.00159681
10^5	0.000581887	0.000552315	0.000543979	0.000565605	0.000613286

Table 3: Weighted mean square deviation from analytical solution

Figure 1: Images of the living room, the office and the teapot

Multivariate Walsh series, digital nets and quasi-Monte Carlo integration

Gerhard Larcher   Wolfgang Ch. Schmid*

Salzburg

1   Introduction

Quite recently in a series of papers ([6], [11], [9], [10], [8]) a theory of the numerical integration of multivariate Walsh series by means of (t, m, s)-nets was developed.

It is the aim of this paper on the one hand to give a short introduction to the, for applications, most important method developed in the theory, and on the other hand to give some new facts on multivariate Walsh series and digital nets and to give concrete calculations thereby completing some aspects of the theory and illustrating its usefulness in concrete applications.

In section 2 we recall some facts on multivariate Walsh series in a base b and introduce in this context the concept of "base b finite precision functions". The integration method in question especially is of interest for the integration of rapidly converging Walsh series belonging to certain classes $_b\overline{E}_s^\alpha(c)$.

In section 3 we give connections between these classes of functions and b-adic derivatives of functions.

In section 4 we give a summary on the main integration-error estimates and recall a method of Niederreiter for the construction of digital (t, m, s)-nets of high quality constructed over $\mathbb{Z}_b$. Moreover we give a lower bound for the quality parameter t of digital nets which in the light of recent developments is essentially best possible.

Finally in section 5 in two examples we compare the method of integrating base b Walsh series by digital nets over $\mathbb{Z}_b$ with quasi-Monte Carlo methods using alternative point sets like good lattice point sets or Halton sequences, and in a third example we consider integration results obtained by different point sets for a multivariate function without special Walsh representation properties.

2   Multivariate Walsh series

We use the Paley enumeration of the one-dimensional Walsh system in a base b. This enumeration was introduced by Paley in [16] for base $b = 2$ and adapted by Chrestenson [2] for arbitrary base $b \geq 2$.

---

*The second author was supported by the CEI Project PACT, WP5.1.2.1.3

Definition 1 *For an integer $b \geq 2$ let the function $_b\phi$ on $\mathbb{R}$ be defined by $_b\phi(x) := w^k$ if $\frac{k}{b} \leq x < \frac{k+1}{b}$ for an integer k, where $w = e^{\frac{2\pi i}{b}}$.*

For a non-negative integer n with base b representation $n = \sum_{k=0}^{m-1} a_k b^k$ the n-th Walsh function in base b is defined on $[0, 1)$ by

$$_b wal_n(x) = {_b\phi^{a_0}}(x) \cdot {_b\phi^{a_1}}(bx) \cdots {_b\phi^{a_{m-1}}}(b^{m-1}x) .$$

The system of Walsh functions in a base b is a complete orthonormal system of functions in $L^2([0, 1))$. (See [17] for $b = 2$ and [2] for arbitrary b.)

We will use the following simple principle for evaluating the value of $_b wal_n$ at a certain point x:

$$\text{if } n = \sum_{k=0}^{m-1} a_k b^k \text{ and } x = \sum_{k=1}^{\infty} c_k b^{-k}$$

are the base b representations of n respectively of x in base b, then

$$_b wal_n(x) := w^{a_0 c_1 + a_1 c_2 + \cdots + a_{m-1} c_m} . \tag{1}$$

Definition 2 *For dimension $s \geq 2$ and non-negative integers $k_1, \ldots, k_s$ we define*

$$_b wal_{k_1, \ldots, k_s}(x_1, \ldots, x_s) := {_b wal_{k_1}}(x_1) \cdots {_b wal_{k_s}}(x_s) .$$

The system of s-dimensional Walsh functions in a base b is a complete orthonormal system of functions in $L^2([0, 1)^s)$. See [17] for $s = b = 2$. The proof there easily can be extended to the general case.

Definition 3 *By an s-dimensional Walsh series in base b we denote a function f on $[0, 1)^s$ of the form*

$$f(x_1, \ldots, x_s) = \sum_{h_1, \ldots, h_s = 0}^{\infty} \hat{W}_f(h_1, \ldots, h_s) \cdot {_b wal_{h_1, \ldots, h_s}}(x_1, \ldots, x_s)$$

with certain $\hat{W}_f(h_1, \ldots, h_s) \in \mathbb{C}$. $\hat{W}_f(h_1, \ldots, h_s)$ is called the $(h_1, \ldots, h_s)$-Walsh-Fourier coefficient of f.

A Walsh polynomial is a finite Walsh series. By the degree of a Walsh polynomial f we understand the maximal d for which there is an i, $1 \leq i \leq s$, and integers $d_1, \ldots, d_{i-1}, d_{i+1}, \ldots, d_s \geq 0$ with $\hat{W}_f(d_1, \ldots, d_{i-1}, d, d_{i+1}, \ldots, d_s) \neq 0$.

As indicated in the introduction we mainly are interested in the numerical integration of rapidly converging multivariate Walsh series. What we mean by "rapidly converging" will become clear in the next section. In any case a Walsh polynomial is a rapidly converging Walsh series.

The class of Walsh polynomials now contains a class of functions which is of great importance in concrete calculations.

Definition 4 *For integers $b \geq 2$ and $s, n \geq 1$ a function $f : [0, 1)^s \longrightarrow \mathbb{C}$ is called finite precision function in base b of precision n if f is constant on all intervals of the form*

$$\prod_{i=1}^{s} \left[\frac{e_i}{b^n}, \frac{e_i + 1}{b^n} \right) , \quad 0 \leq e_i < b^n, \ e_i \in \mathbb{Z} .$$

In concrete calculations because of finite computer precision in fact we always are dealing with such finite precision functions. In most cases here we are dealing with base $b = 2$ or 10 and certain degree of precision depending on the machine precision. Each finite precision function in base b is representable pointwise by a Walsh polynomial in base b. We have

Proposition 1 *Each s-dimensional finite precision function f in base b of precision n is pointwise represented by an s-dimensional Walsh polynomial in base b of degree at most $b^n - 1$.*

Proof: For $s = 1$ and $b = 2$ this is a Lemma of Paley (see [17]).

For arbitrary s and b it suffices to show that the characteristic function of each interval $\prod_{i=1}^{s} \left[\frac{e_i}{b^n}, \frac{e_i+1}{b^n}\right)$ is an s-dimensional Walsh polynomial in base b of degree $\leq b^n - 1$. Since the product of s one-dimensional Walsh polynomials, each of degree $\leq N$, in the s different variables $x_1, \ldots, x_s$ is an s-dimensional Walsh polynomial in the variables $x_1, \ldots, x_s$ of degree $\leq N$, it suffices to show the above for $s = 1$.

Now it is easily checked by using the principle (1) that for each integer e with representation $e = \sum_{k=0}^{n-1} c_k b^k$ in base b, the characteristic function of the interval $\left[\frac{e_i}{b^n}, \frac{e_i+1}{b^n}\right)$ is given by $\sum_{k=0}^{b^n-1} \gamma(e. k) \cdot {}_b\mathrm{wal}_k$ where $\gamma(e, k) = \frac{1}{b^n} w^{-(c_{n-1}a_0 + \cdots + c_0 a_{n-1})}$ when $k = \sum_{j=0}^{n-1} a_j b^j$.

$\square$

Conversely, of course, each s-dimensional Walsh polynomial in base b of degree $\leq n$ is a finite precision function in base b of precision $\leq m$ where m is such that $b^{m-1} \leq n < b^m$. So in concrete applications we always are dealing with rapidly converging Walsh series.

3 Rapidly converging Walsh series and b-adic derivatives

In analogy to the theory of good lattice points for the numerical integration of Fourier series in [11] we defined the following classes of rapidly converging Walsh series.

Definition 5 *For base $b \geq 2$, dimension $s \geq 1$ and $\alpha > 1, c > 0 (b, s \in \mathbb{N}; c, \alpha \in \mathbb{R})$ let ${}_b\overline{E}_s^{\alpha}(c)$ be the class of all functions $f : [0, 1)^s \longrightarrow \mathbb{C}$ which are representable pointwise by an absolutely converging Walsh series in base b with Walsh coefficients $\hat{W}_f(h_1, \ldots, h_s)$ satisfying*

$$|\hat{W}_f(h_1, \ldots, h_s)| \leq \frac{c}{(\overline{h}_1 \cdots \overline{h}_s)^{\alpha}} \text{ for all } (h_1, \ldots, h_s) \neq (0, \ldots, 0) \ .$$

(Here $\overline{h} := \max(1, |h|)$.)

In analogy to connections between the "smoothness" of a function and the speed of convergence of its Fourier series (see [4]) we have connections between b-adic differentiability of a function and the speed of convergence of its Walsh series.

To give this connection we extend the concept of the dyadic derivative (this concept was introduced by Butzer and Wagner in [1], see also [17, section 1.7]) to the concept of the b-adic derivative ($b \geq 2$) of a function. (Compare this extension with the concept of Onneweer [15].)

Definition 6 *For an integer base* $b \geq 2$ *and reals* x *and* y *with representation* $x = \sum_{j=w}^{\infty} x_j b^{-j}$ *and* $y = \sum_{j=w}^{\infty} y_j b^{-j}$ *in base* b *(*$w \in \mathbb{Z}$*), the* b*-adic sum* $x \oplus_b y$ *of* x *and* y *is defined by*

$$x \oplus_b y = \sum_{j=w}^{\infty} z_j b^{-j} \quad \text{where } z_j = x_j + y_j \pmod{b} .$$

Definition 7 *For an integer base* $b \geq 2$*, a real valued function* f*, a real* $x \in [0,1)$ *and a non-negative integer* n *let*

$$d_n f(x) := \sum_{j=0}^{n} b^{j-1} \sum_{k=0}^{b-1} k \sum_{l=0}^{b-1} w^{-lk} \cdot f\left(x \oplus_b \frac{l}{b^{j+1}}\right)$$

(Here again $w = e^{\frac{2\pi i}{b}}$*.)*

f is b-adically differentiable at x if $f^{[1]}(x) := \lim_{n \to \infty} d_n f(x)$ exists and is finite; $f^{[1]}$ is the dyadic derivative of f at x. Higher order b-adic derivatives of f at x and partial b-adic derivatives of multivariate functions in a point $(x_1, \ldots, x_s)$ are defined in the natural way.

A function $f \in L^p([0,1))$, $1 \leq p < \infty$ is called strongly b-adically differentiable in $L^p([0,1))$ if there is a $g \in L^p([0,1))$ such that $d_n f \longrightarrow g$ in L^p-norm for $n \to \infty$. g then is called strong b-adic derivative of f in $L^p([0,1))$ and is denoted by $\mathbf{d}f$. Higher order strong derivatives $\mathbf{d}^{[r]}f$ and strong derivatives and strong partial derivatives of multivariate functions are defined in the natural way.

Analogously to Theorem 1 in [15] we have for each $n \geq 0$ and each $x \in [0,1)$:

$$_b\mathrm{wal}_n^{[1]}(x) = n \cdot {_b\mathrm{wal}_n(x)} .$$

Further quite analogously to Theorem 13 in [17] we have: if $f : [0,1) \longrightarrow \mathbb{C}$ is r times strongly differentiable in $L^1([0,1))$ for some $r \in \mathbb{N}$ then

$$\hat{W}_{\mathbf{d}^{[r]}f}(k) = k^r \cdot \hat{W}_f(k) \tag{2}$$

for all $k \geq 0$.

Then we have

Proposition 2 *For a base* $b \geq 2$ *and a dimension* $s \geq 1$ *let* $\alpha > 1$ *be such that* αs *is an integer, and let* $\alpha_1, \ldots, \alpha_s \in \{0,1,\ldots,\alpha s\}$ *be such that* $\alpha_1 + \cdots + \alpha_s = \alpha s$*. Let* $f \in L^1([0,1)^s)$ *be such that for all permutations* $v_1, \ldots, v_s$ *of* $\alpha_1, \ldots, \alpha_s$ *the strong* b*-adic derivatives* $\frac{d^{[\alpha s]}f}{dx_1^{[v_1]} \cdots dx_s^{[v_s]}}$ *of* f *in* $L^1([0,1)^s)$ *exist, then* $f \in {_b\overline{E}_s^{\alpha}}(c)$ *with*

$$c = \max_{v_1, \ldots, v_s} \int_0^1 \cdots \int_0^1 \left| \frac{d^{[\alpha s]}f}{dx_1^{[v_1]} \cdots dx_s^{[v_s]}}(x_1, \ldots, x_s) \right| dx_1 \cdots dx_s .$$

Proof: For $v_1, \ldots, v_s$ like above and $m_1, \ldots, m_s \geq 0$ not all zero and without restriction of generality with $|m_1| \geq \cdots \geq |m_k| \geq 1$ and $m_{k+1} = \cdots m_s = 0$ we have by (2)

$$\hat{W}_{\frac{d^{v_1 + \cdots v_k}f}{dx_1^{[v_1]} \cdots dx_s^{[v_s]}}}(m_1, \ldots, m_s) = m_1^{v_1} \cdots m_k^{v_k} \cdot \hat{W}_f(m_1, \ldots, m_s)$$

$$\Rightarrow \left| \hat{W}_f(m_1, \ldots, m_s) \right| \leq \frac{c}{|m_1|^{v_1} \cdots |m_k|^{v_k}} \ .$$

If $v_1, \ldots, v_s$ are such that $v_1 \geq \cdots \geq v_s$ then

$$\frac{1}{|m_1|^{v_1} \cdots |m_k|^{v_k}} \leq \frac{1}{(|m_1| \cdots |m_k|)^{\beta_k}}$$

$$\left(\text{where } \beta_k := \frac{1}{k}(v_1 + \cdots + v_k) \right)$$

$$\leq \frac{1}{(|m_1| \cdots |m_k|)^{\alpha}}$$

and the result follows.

$\square$

So we may interprete functions in a class $_b \overline{E}_s^{\alpha}(c)$ for some $\alpha > 1$ as functions which are "smooth" with respect to b-adic derivation.

4 Digital nets constructed over $\mathbb{Z}_b$ and numerical integration of Walsh series

The concept of (t, m, s)-nets and of digital (t, m, s)-nets constructed over a ring R was systematically introduced and studied by Niederreiter in [12] and [13] (see [14] and further references given there). First examples of nets were already studied by Sobol [18] and Faure [3].

Definition 8 *Let $b \geq 2$, $s \geq 1$, and $0 \leq t \leq m$ be integers. Then a point set $P = \{x_0 \ldots, x_{N-1}\}$ consisting of $N = b^m$ points of $[0, 1)^s$ forms a (t, m, s)-net in base b if the number of n with $0 \leq n \leq N - 1$, for which x_n is in the subinterval J of $[0, 1)^s$, is b^t for all $J = \prod_{i=1}^{s} \left[a_i b^{-d_i}, (a_i + 1) b^{-d_i} \right)$ with integers $d_i \geq 0$ and $0 \leq a_i < b^{d_i}$ for $1 \leq i \leq s$, and with s-dimensional volume b^{t-m}.*

Definition 9 *Let $b \geq 2$, $s \geq 1$, and $0 \leq t \leq m$ be integers. Let $c_{jr}^{(i)}$ for $1 \leq i \leq s$, $1 \leq j \leq m$ and $0 \leq r \leq m - 1$ be integers. For $n \in \{0, \ldots, b^m - 1\}$ let $n = \sum_{r=0}^{m-1} a_r(n) b^r$ be the representation of n in base b. Let*

$$x_n^{(i)} = \sum_{j=1}^{m} y_{nj}^{(i)} b^{-j} \quad \in [0, 1)$$

for $0 \leq n < b^m$ and $1 \leq i \leq s$, where

$$y_{nj}^{(i)} = \sum_{r=0}^{m-1} c_{jr}^{(i)} a_r(n) \pmod{b}$$

for $0 \leq n < b^m$, $1 \leq i \leq s$ and $1 \leq j \leq m$.
If for some integer t with $0 \leq t \leq m$ the point set

$$x_n = \left(x_n^{(1)}, \ldots, x_n^{(s)} \right) \quad \in [0, 1)^s \ ; \quad n = 0, \ldots, b^m - 1$$

is a (t, m, s)-net in base b, then it is called a digital (t, m, s)-net constructed over $\mathbb{Z}_b$.

Then we have the following integration-error estimate (see [10] and [8]):

Theorem 1 *Let $b \geq 2$ and let $\mathbf{x}_0, \ldots, \mathbf{x}_{N-1}$ be a digital (t, m, s)-net constructed over $\mathbb{Z}_b$, then with a positive constant $K(s, \alpha, c, b)$ depending only on s, α, c and b, we have*

$$\left| \int_{[0,1)^s} f(\mathbf{x})\, d\mathbf{x} - \frac{1}{N} \sum_{k=1}^{N} f(\mathbf{x}_k) \right| \leq K(s, \alpha, c, b) \cdot b^{t\alpha} \cdot \frac{(\log N)^{s-1}}{N^\alpha}$$

for all $f \in {}_b\overline{E}_s^\alpha(c)$.

This result is, up to logarithmic factors, best possible in the sense that there is a constant $K'(s, \alpha, c, b) > 0$ such that for all N and all point set $\mathbf{y}_0, \ldots, \mathbf{y}_{N-1}$ in $[0, 1)^s$ there is a $f \in {}_b\overline{E}_s^\alpha(c)$ with

$$\left| \int_{[0,1)^s} f(\mathbf{x})\, d\mathbf{x} - \frac{1}{N} \sum_{k=1}^{N} f(\mathbf{y}_k) \right| \geq K'(s, \alpha, c, b) \cdot \frac{(\log N)^{s-1}}{N^\alpha} \quad .$$

(See Theorem 4 in [11].)

Efficient methods for constructing digital (t, m, s)-nets over $\mathbb{Z}_b$ of high quality were given for b prime by Niederreiter in [12] and in [13], and for arbitrary b in [8].

The latter method is based on the method given in [13] and provides for all bases b, all dimensions s and all m, digital $(\tau_b(s), m, s)$-nets constructed over $\mathbb{Z}_b$ where

$$\tau_b(s) = \max_{1 \leq v \leq h} T_{p_v}(s-1) \quad ,$$

$p_1 \leq \cdots \leq p_h$ are the different prime divisors of b and where T_p is defined like in [14, formula 4.69]. By Theorem 4.54 in [14] we have

$$\tau_b(s) < (s-1) \cdot \left(\log_p(s-1) + \log_p \log_p(s-1) + 1 \right) \quad ,$$

where p is the smallest prime divisor of b and where $\log_p$ denotes the logarithm in base p.

We will show in the following that up to constant and logarithmic factors Niederreiter's method provides digital (t, m, s)-nets over $\mathbb{Z}_b$ of optimal quality. Quite recently (still unpublished) Niederreiter and Xing developed a new construction method which is based on the theory of algebraic curves over finite fields. It seems that this method beats the quality of all former explicit construction methods. So it provides for any base b, for all m and all s digital $(t(s, b), m, s)$-nets with $t(s, b) = O(s)$ (the O-constant depends on b only), which by the following proposition is the best possible order of magnitude.

Proposition 3 *Let $s \geq 1$ and $b \geq 2$ be given. If $T \geq 1$ is such that for all m there exists a digital (T, m, s)-net constructed over $\mathbb{Z}_b$ then $T > s \cdot \frac{p-1}{p^3}$, where p is the smallest prime factor of b.*

Remark: Note that by Corollary 2 in [8] a digital $(0, m, s)$-net over $\mathbb{Z}_b$ exists for all m if and only if $0 > s \cdot \frac{1}{p+2} - 1$.

Proof: Assume that for some $t \leq s \cdot \frac{p-1}{p^3}$ there is a digital $(t, 3t, s)$-net constructed over $\mathbb{Z}_b$. Let

$$C = \left\{ c_{jr}^{(i)} \,\middle|\, 1 \leq i \leq s, 1 \leq j \leq 3t, 0 \leq r \leq 3t - 1 \right\}$$

be the set of integers defining this net according to Definition 9. By Lemma 4 and Lemma 7 in [8] C then also defines a $(t, 3t, s)$-net over $\mathbb{Z}_p$. For $i = 1, \ldots, s$ let $c^{(i)}$ be the vector $\left(c_{1,0}^{(i)}, \ldots, c_{1,3t-1}^{(i)} \right) \in \mathbb{Z}^{3t}$. If we choose arbitraryly $2t$ of these vectors, say $d^{(1)}, \ldots, d^{(2t)}$ and form the matrix $D := \begin{pmatrix} d^{(1)} \\ \vdots \\ d^{(2t)} \end{pmatrix}$ then by the definition of a digital net constructed over $\mathbb{Z}_p$ for each choice of $f = \begin{pmatrix} f_1 \\ \vdots \\ f_{2t} \end{pmatrix} \in \mathbb{Z}_p^{2t}$ the system $D \cdot x \equiv f$ (mod p) has exactly p^t solutions. Therefore $d^{(1)}, \ldots, d^{(2t)}$ have to be linearly independent over $\mathbb{Z}_p$. Consequently for two different choices $\left(d^{(1)}, \ldots, d^{(t)} \right)$ and $\left(e^{(1)}, \ldots, e^{(t)} \right)$ out of $c^{(1)}, \ldots, c^{(s)}$ we have $\lambda_1 d^{(1)} + \cdots + \lambda_t d^{(t)} \not\equiv \mu_1 e^{(1)} + \cdots + \mu_t e^{(t)} \not\equiv 0 \pmod{p}$ for all $\lambda_1, \ldots, \lambda_t, \mu_1, \ldots, \mu_t \in \{1, \ldots, p-1\}$. Therefore we have the condition

$$(p-1)^t \cdot \binom{s}{t} < p^{3t} \Longrightarrow$$

$$\left(\frac{s}{t} \right)^t \leq \binom{s}{t} < \left(\frac{p^3}{p-1} \right)^t \Longrightarrow$$

$$t > \frac{p-1}{p^3} \cdot s, \qquad \text{a contradiction.}$$

5 Three computational examples

We give three concrete examples for the integration with digital nets constructed over $\mathbb{Z}_b$ and compare the results with the results we get by using other point sets for the quasi-Monte Carlo integration.

The first two examples are certain Walsh series namely a finite precision function (cf. Definition 4), respectively a rapidly converging Walsh series out of the class $_2\overline{E}_s^4(c)$ (cf. Definition 5). The third example is a function without a special Walsh series representation. In each of the three examples we compare results we get by using certain digital (t, m, s)-nets with results we achieve by using as well good lattice point sets as Halton sequences. The tabulated values for the good lattice point sets were taken from [19]. For more details on the s-dimensional Halton sequence see [5].

The following tables are just a small representative part of a large number of numerical experiments which were carried out by the authors as part of the CEI Project PACT. These results will appear as a Technical Report of the Austrian Center for Parallel Computation.

Example 1. We consider the following finite precision function f in base b of precision n: For $i = 1, \ldots, s$ let $x_i \in [0, 1)$ be represented in base b by $x_i = \sum_{j=1}^{\infty} \frac{x_i(j)}{b^j}$, then

$$f(x_1, \ldots, x_s) := \sum_{i=1}^{s} \sum_{j=1}^{n} x_i(j) \pmod{b} .$$

For all b, s, n we have

$$\int_{[0,1)^s} f(\mathbf{x})dx = \frac{b-1}{2}.$$

In the following we give results for the special cases: base $b = 5$, precision $2 \leq n \leq 6$, and dimension $7 \leq s \leq 10$. We integrate these functions first with base 5 digital $(1,8,7)-$, $(2,8,8)-$, $(3,8,9)-$ and $(4,8,10)$-nets that is with 5^8 sample points. We compare these results with the results we get by using first base 2 digital $(8,19,7)-$, $(11,19,8)-$, $(14,19,9)-$ and $(18,19,10)$-nets, then good lattice point sets with 2^{19} points and finally the first 2^{19} elements of the Halton sequence. Although in the first case we only used $5^8 = 390625$ points instead of $2^{19} = 524288$ points in all the other cases, the errors obtained by base 5 digital nets for the most parameters are significantly smaller than the errors obtained by the other point sets.

Table 1

s	n	good lattice 2^{19} points	Halton 2^{19} points	(t_s, m, s), base 2 2^{19} points	(t_s, m, s), base 5 5^8 points
7	2	1.0204315e-03	4.5776367e-05	1.3923645e-03	7.6800000e-06
7	3	3.8146973e-06	7.9917908e-04	1.4171600e-03	3.8400000e-05
7	4	6.3133240e-04	4.8427582e-03	1.5716553e-03	9.4720000e-05
7	5	2.0580292e-03	2.3269653e-03	1.7776489e-03	1.5360000e-05
7	6	1.9226074e-03	2.6988983e-03	3.8528442e-04	2.3808000e-04
8	2	3.8146973e-06	1.8100739e-03	2.1362305e-03	2.0480000e-05
8	3	7.2479248e-04	4.5776367e-05	1.9245148e-03	1.2800000e-04
8	4	9.8609924e-04	2.1419525e-03	1.8329620e-03	7.6800000e-06
8	5	3.5934448e-03	1.5125275e-03	3.2806396e-04	1.3824000e-04
8	6	1.7566681e-03	9.1743469e-04	2.0561218e-03	7.1680000e-05
9	2	6.8664551e-05	1.6784668e-04	1.4591217e-03	1.0240000e-05
9	3	2.2945404e-03	1.4972687e-03	2.3269653e-03	1.5360000e-04
9	4	3.8146973e-06	1.2168884e-03	1.3656616e-03	9.9840000e-05
9	5	1.1367798e-03	2.9487610e-03	6.8473816e-04	3.5072000e-04
9	6	3.8146973e-04	8.9263916e-04	1.1100769e-03	1.5360000e-05
10	2	1.3332367e-03	2.8610229e-04	1.3790131e-03	1.2800000e-05
10	3	1.1177063e-03	7.8201294e-05	1.1863708e-03	5.1200000e-05
10	4	1.0757446e-03	5.2585602e-03	6.8473816e-04	1.5360000e-04
10	5	1.9264221e-04	1.5869141e-03	1.1920929e-03	4.7360000e-04
10	6	2.2144318e-03	4.2724609e-04	1.5068054e-03	2.9440000e-04

Example 2. Here we consider the rapidly converging Walsh series f_β with parameter $\beta = 3$, described in detail in [9, Section 4]. This function is contained in $_2\overline{E}_s^4(c)$ with a constant $c = \left(8 \cdot \frac{14}{15}\right)^s + \frac{16^s}{2} \cdot \frac{16^s - 2}{16^2 - 1}$.

In the following we give the results for the special cases: dimension $7 \leq s \leq 10$, by using 2^{21}, 2^{22} and 2^{23} sample points. We compare the results we get by using (t,m,s)-nets in base 2 with the values of t given in Table 2 (these nets are provided by the tables of optimal polynomials in [7]) with the results obtained by using good lattice point sets and Halton sequences.

For all parameters the digital base 2 nets give essentially smaller errors than good lattice point sets and Halton sequences.

Table 2

s	number of nodes	good lattice error	Halton error	(t, m, s)-net, base 2	
				t	error
7	2097152	1.7515935e-05	5.2438885e-05	7	6.5064620e-13
7	4194304	4.7665038e-06	7.5918397e-05	7	1.3405943e-13
7	8388608	1.0427638e-05	4.4513814e-05	8	1.0880185e-14
8	2097152	4.8566452e-05	2.4882064e-04	8	2.7386704e-11
8	4194304	1.2603024e-04	1.6392513e-04	7	1.1632894e-10
8	8388608	1.1412178e-04	8.5103615e-05	8	2.1038726e-14
9	2097152	2.1427557e-05	1.5685086e-04	10	3.8510394e-10
9	4194304	4.0961074e-06	4.7410372e-05	10	7.2016281e-11
9	8388608	8.4314002e-07	1.4556830e-05	10	5.2745030e-12
10	2097152	3.0443267e-04	3.5434624e-04	12	7.8417594e-10
10	4194304	2.7096240e-04	1.2353423e-04	11	5.5515592e-12
10	8388608	2.3489566e-04	4.1152916e-05	11	7.5073447e-11

Example 3. Here we consider the function $f(x_1, \ldots, x_s) = \sqrt{x_1 + \cdots + x_s}$. We again give the special cases: dimension $7 \leq s \leq 10$, by using 2^{21}, 2^{22} and 2^{23} sample points. We compare the results we get by using the same (t, m, s)-nets in base 2 as in example 2 with the results obtained by using good lattice point sets and Halton sequences.

This function shows no special behaviour in its Walsh representation. The integration results for good lattice point sets and for base 2 digital nets are of about the same quality and they are, on average, by a factor 10 better than the results obtained by using the Halton sequence.

Table 3

s	number of nodes	good lattice error	Halton error	(t, m, s)-net, base 2	
				t	error
7	2097152	1.4147531e-06	4.7537729e-06	7	7.2870240e-07
7	4194304	7.6427836e-07	2.4123119e-06	7	3.1239455e-07
7	8388608	4.1313934e-07	1.4135486e-06	8	1.8215721e-07
8	2097152	1.2091152e-06	5.5669197e-06	8	6.6352108e-07
8	4194304	1.1383449e-07	2.8861645e-06	7	3.3873157e-07
8	8388608	5.4233034e-07	1.6104916e-06	8	1.9254427e-07
9	2097152	7.0977311e-07	6.8404553e-06	10	3.6024590e-06
9	4194304	4.5424279e-07	3.2042023e-06	10	9.3517849e-07
9	8388608	2.4682497e-07	1.8020614e-06	10	3.4818688e-07
10	2097152	1.1358496e-06	7.4891350e-06	12	1.9625840e-06
10	4194304	2.8406080e-07	3.5343143e-06	11	1.4699121e-06
10	8388608	4.7217250e-07	2.0742485e-06	11	4.4534013e-07

References

[1] P.L. Butzer and H.J. Wagner. Walsh–Fourier series and the concept of a derivative. *Applicable Analysis*, **3**:29–46, 1973.

[2] H.E. Chrestenson. A class of generalized Walsh functions. *Pacific J. Math.*, **5**:17–31, 1955.

[3] H. Faure. Discrépance de suites associées à un système de numération (en dimension *s*). *Acta Arith.*, **41**:337–351, 1982.

[4] N.M. Korobov. *Number-Theoretic Methods in Approximate Analysis*. Fizmatgiz, Moscow, 1963. (Russian).

[5] L. Kuipers and H. Niederreiter. *Uniform Distribution of Sequences*. John Wiley, New York, 1974.

[6] G. Larcher. A class of low-discrepancy point-sets and its application to numerical integration by number-theoretical methods. In F.Halter-Koch and R.Tichy, editors, *Österreichisch-Ungarisch-Slowakisches Kolloquium über Zahlentheorie*, volume 318 of *Grazer Math. Ber.*, pages 69–80, Graz, 1993. Karl-Franzens-Univ. Graz.

[7] G. Larcher, A. Lauß, H. Niederreiter, and W.Ch. Schmid. Optimal polynomials for (t, m, s)-nets and numerical integration of multivariate Walsh series. To appear in: SIAM J. Numer. Analysis, 1995.

[8] G. Larcher, H. Niederreiter, and W.Ch. Schmid. Digital nets and sequences constructed over finite rings and their application to quasi-Monte Carlo integration. Submitted to: Monatsh. Math., 1995.

[9] G. Larcher, W.Ch. Schmid, and R. Wolf. Representation of functions as Walsh series to different bases and an application to the numerical integration of high-dimensional Walsh series. *Math. Comp.*, **63**:701–716, 1994.

[10] G. Larcher, W.Ch. Schmid, and R. Wolf. Quasi-Monte Carlo methods for the numerical integration of multivariate Walsh series. To appear in: Math. Comp. Modelling, 1995.

[11] G. Larcher and C. Traunfellner. On the numerical integration of Walsh series by number-theoretic methods. *Math. Comp.*, **63**:277–291, 1994.

[12] H. Niederreiter. Point sets and sequences with small discrepancy. *Monatsh. Math.*, **104**:273–337, 1987.

[13] H. Niederreiter. Low-discrepancy and low-dispersion sequences. *J. Number Theory*, **30**:51–70, 1988.

[14] H. Niederreiter. *Random Number Generation and Quasi-Monte Carlo Methods*. Number **63** in CBMS–NSF Series in Applied Mathematics. SIAM, Philadelphia, 1992.

[15] C.W. Onneweer. Differentiability for Rademacher series on groups. *Acta Sci. Math.*, **39**:121–128, 1977.

[16] R.E.A.C. Paley. A remarkable system of orthogonal functions. *Proc. Lond. Math. Soc.*, **34**:241–279, 1932.

[17] F. Schipp, W.R. Wade, and P. Simon. *Walsh Series. An Introduction to Dyadic Harmonic Analysis.* Adam Hilger, Bristol and New York, 1990.

[18] I.M. Sobol'. The distribution of points in a cube and the approximate evaluation of integrals. *Ž. Vyčisl. Mat. i Mat Fiz.*, **7**:784–802, 1967. (Russian).

[19] Gerhard Wesp. A new and extensive table of good lattice points. Deliverable D5Z-1 of the CEI Project PACT, 1994.

Address of the authors:

Gerhard Larcher and Wolfgang Ch. Schmid
Institut für Mathematik
Universität Salzburg
Hellbrunnerstraße 34
A–5020 Salzburg
Austria
e-mail address: schmidw@edvz.sbg.ac.at

PARALLEL PSEUDORANDOM NUMBER GENERATION USING ADDITIVE LAGGED-FIBONACCI RECURSIONS

MICHAEL MASCAGNI
M. L. ROBINSON
DANIEL V. PRYOR
STEVEN A. CUCCARO

Supercomputing Research Center, I.D.A.

ABSTRACT. We study the suitability of the additive lagged-Fibonacci pseudorandom number generator for parallel computation. This generator has a relatively short period with respect to the size of its seed. However, the short period is more than made up for with the huge number of full-period cycles it contains. We call these different full-period cycles equivalence classes. We show how to enumerate the equivalence classes and how to compute seeds to select a given equivalence class. The use of these equivalence classes gives an explicit parallelization suitable for a fully reproducible asynchronous MIMD implementation. To explore such an implementation we introduce an exponential sum measure of quality for the additive lagged-Fibonacci generators used in serial or parallel. We then prove the first non-trivial results we are aware of on this measure of quality.

1. Introduction.

In Knuth's well known exposition on pseudorandom number generation [5], several methods of generation are considered. Among these is the additive lagged-Fibonacci pseudorandom number generator:

$$(1) \qquad x_n = x_{n-k} + x_{n-\ell} \pmod{M}, \quad \ell > k.$$

This generator is defined by the modulus, M, the register length, ℓ, and the lag, k. When M is prime, periods as large as $M^\ell - 1$ are possible. However, it is more common to consider lagged-Fibonacci generators with $M = 2^m$, for some m. These generators with power-of-two moduli are considerably easier to implement than general prime moduli; however, their periods are much smaller than in the prime-modulus case.

In Marsaglia's empirical study of pseudorandom number generators [9], the additive lagged-Fibonacci generator with power-of-two modulus was one among those that were tested. Overall, this generator did well on all of Marsaglia's "stringent" tests, save the "non-overlapping birthday spacing test." However, Marsaglia noted that by choosing a

1991 *Mathematics Subject Classification.* 11L07, 65C10, 65Y05.

Key words and phrases. random number generation, parallel computation, Fibonacci generator, MIMD, SIMD, reproducible, exponential sums.

generator with a large register length, ℓ, improvements are seen in the "non-overlapping birthday spacing test."

There are several other compelling reasons to study this generator, [11]. This generator is used by Thinking Machines Corporation in their "Connection Machine Scientific Subroutine Library" (CMSSL) as a parallel pseudorandom number generator.[1] In addition, Brent has recently added to the understanding of this generator in both theory and practice, [1, 2]. Aside from clarifying the conditions for obtaining the maximum possible period, Brent carefully analyzed the use of the additive lagged-Fibonacci integer generator and its floating-point counterpart. In most Monte Carlo applications, uniformly distributed floating-point numbers, not integers, are desired. The floating-point counterpart of equation (1) is $\omega_n = \omega_{n-k} + \omega_{n-\ell} \pmod 1$. Here the $\omega_n \in [0, 1)$ are floating-point numbers. Besides being able to compute directly floating-point pseudorandom numbers, this formulation and the integer counterpart in equation (1) are amenable to efficient vectorization, [1]. Thus we see that this generator in both the integer and floating-point versions is versatile, and that there are several good reasons for exploring the additive lagged-Fibonacci generator with power-of-two modulus to find an effective parallel implementation.

First, we must understand some of the properties that are desirable in a parallel pseudorandom number generator. Besides efficiency and pseudorandomness, which are properties of the generator when used in serial, we require that:

1. The generator must be easy to parallelize (this question is the primary concern of this paper).
2. The generator must be reproducible in a "strong" sense.
3. The streams generated in parallel must seem "independent."

Property 2 is very important to computational scientists. When doing Monte Carlo calculations on new machines, exact agreement with previous and trusted calculations is essential. This is not an easy task, as many sophisticated Monte Carlo calculations can be quite complicated. By reproducibility in the "strong" sense we require reproducibility both on the same machine with a different partitioning of the processing resources and between different machines. This demanding definition of reproducibility ensures the portability of a parallel generator to any parallel machine. This is a rather lofty goal, but one that we show is accessible to the additive lagged-Fibonacci generator.

Property 3 has not been adequately addressed in a general sense. One partial solution is to use exponential sums as theoretical measures of quality for both the serial and the parallel use of pseudorandom number generators. With additive lagged-Fibonacci generators modulo a power-of-two, this leads to the consideration of full-period exponential sums that have not been previously studied in the literature. In addition, empirical measures of quality are important in our understanding of property 3. Such empirical measures of the parallel use of these generators has already lead to improvements in their implementation, [3].

The plan of the paper is as follows. In §2 we review the conditions on equation (1) and the seed for obtaining the maximal-period for these generators. We next introduce an equivalence relationship on the set of seeds that are in some maximal-period cycle. The

---

[1]It is the careful study of Thinking Machines' (lack of a) seeding algorithm that was the prime motivation for this work.

large number of equivalence classes (ECs) that result is the basis for our parallelization. We then describe an enumeration of all of the ECs. This leads to an algorithm for the computation of a seed in a given EC. We conclude §2 with a discussion on the quality of this generator in terms of exponential sums. §3 is devoted to proving some results for the full-period exponential sums measures of quality. These include an exact value for a collection of full-period sums and a non-trivial upper bound for the general case of interest. This last bound is weak in general, but with additional empirical data, it can be used to get much better bounds for specific generators. In addition, we use an enumeration of the ECs as the basis for parallelization and analyze the parent-child generators in terms of their exponential sum quality. Finally, in §4 we summarize the results and propose directions for further study.

2. Properties of the Generator.

2.1 Cycle Structure.

Let us begin with computing the period of these generators. If M is prime, then a period of $M^\ell - 1$ is possible provided that the characteristic polynomial $f(x) = x^\ell - x^k - 1$ is primitive modulo M. In the case of interest, when $M = 2^m$, the maximum possible period is $(2^\ell - 1)2^{m-1}$. In general, a linear recurrence modulo 2^m has period $(2^\ell - 1)2^{m-1}$ if and only if the following three conditions hold:

(I) modulo 2 the sequence has period $(2^\ell - 1)$,
(II) modulo 2^2 the sequence has period $(2^\ell - 1)2$,
(III) modulo 2^3 the sequence has period $(2^\ell - 1)2^2$.

For a proof see [12] or exercise 11 of §3.2.2 in [7]. We are only interested in working with simple additive lagged-Fibonacci generators. **For the remainder of this paper, we consider only linear recurring sequences that satisfy the following three properties:**

(i) **Equation (1) holds for some $\ell > 2$ and for some M, a power of two.**
(ii) **The characteristic polynomial $f(x) = x^\ell - x^k - 1$ is primitive mod 2.**
(iii) **The seed is not all even.**

Brent recently proved, [2], that if a linear recurring sequence satisfies properties (i) - (iii), then it has the maximal possible period of $(2^\ell - 1)2^{m-1}$.

There is a conceptual benefit from working modulo a power-of-two, instead of modulo a prime: the cycle structure present can be exploited. Taken modulo 2, equation (1) defines a shift-register sequence. With $f(x)$ primitive modulo 2, we obtain a maximal-period shift-register sequence of period $2^\ell - 1$, [4, 17]. It is well known that this sequence cycles over all possible nonzero contents of its l-bit state. Next consider equation (1) taken modulo 4. The least-significant bits of the register are just the maximal-period shift-register sequence from the modulo 2 case. The most-significant bits are the superposition of two sequences: (a) the maximal-period shift-register sequence from the initial values of the most-significant bits, and (b) the impulse responses due to the carries from the least-significant bits.[2] Sequence (a) cycles with period $2^\ell - 1$ if it is nonzero and adds bit-wise to sequence (b). Thus, without loss of generality we can assume (a) is zero and concentrate on (b) to analyze the cycle structure. Sequence (b)

---

[2] Recall that the impulse response in a shift-register is the sequence obtained by starting from the "unit" fill $[1, 0, 0, \ldots, 0]$, e.g. see [4].

is the superposition of impulse responses forced by carries from the least-significant bit. The impulse response has period $2^\ell - 1$. Since we have the maximum-possible period, the carries must also have period $2^\ell - 1$. Thus a particular carry will force a period $2^\ell - 1$ impulse response, after which the periodic repeat of the first carry will zero the impulse response. Thus each carry will produce a period $(2^\ell - 1)2$ response made up of a $2^\ell - 1$ length maximal-period sequence followed by $2^\ell - 1$ zeros. Because we obtain the maximum possible period, the superposition of these sequences also has period $(2^\ell - 1)2$. This explains the doubling of the period when a new most-significant bit is added and gives an understanding of the cycle structure of the additive lagged-Fibonacci generator.

We now understand how each new most-significant bit doubles the period of this generator; however, this adds ℓ bits of seed to the generator, not just one. The maximum possible period of these generators is extremely short, given the size of the seed. In the prime modulus case, the maximum possible period is equal to the number of nonzero fills in the register. With $M = 2^m$, the maximum possible period of $(2^\ell - 1)2^{m-1}$ is considerably smaller than the number of nonzero fills, $2^{\ell m} - 1$. Where has all this state gone?

The answer to this question comes by considering the condition on the seed for obtaining the maximum possible period. Unlike many pseudorandom number generators, the additive lagged-Fibonacci generator's seed is an ℓ-tuple of residues modulo 2^m. Since we see that the lesser significant bits of the ℓ-tuple perturb the more significant bits through period-doubling carries, starting the generator with an all zero least-significant bit must reduce the period. In fact, the only condition to obtain the maximum possible period is that the ℓ-tuple seed must not contain only zeros in the least-significant bit, [10, 2]. In terms of residues modulo 2^m, this means that the elements of the seed cannot all be even. It is easy to calculate that the number of seeds that give the maximum possible period is $(2^\ell - 1)2^{\ell(m-1)}$. Since each of these seeds is in a maximum possible period cycle, there must be

$$(2) \qquad\qquad E = \frac{(2^\ell - 1)2^{\ell(m-1)}}{(2^\ell - 1)2^{m-1}} = 2^{(\ell-1)(m-1)}$$

cycles with maximum possible period. If we define an equivalence relationship among seeds as being in the same cycle, then we see that these generators have E distinct ECs.

2.2 Equivalence Class Canonical Form.

The use of these ECs will be the key to parallelizing this generator. Thus we must be able to enumerate the ECs and to calculate a seed from each of the ECs given this enumeration. To derive an explicit enumeration, we must decide on one seed from the full period to serve as the representative for the given EC. We call this representative seed the EC's canonical form. Since the least-significant bit of this generator is a maximal-period shift-register sequence, we can choose some given nonzero fill for the least-significant bits of the canonical form. To transform an arbitrary seed into a seed with the canonical form's least-significant bits, one need only advance the seed at most $2^\ell - 2$ times to match the least-significant bits.[3]

---

[3] A table of size j of least-significant bits spaced equally around the cycle will reduce this to no more than $\lceil \frac{2^\ell - 2}{j} \rceil$ steps.

Now that we have placed the least-significant bits of a seed in canonical form, we must decide what to do with the most-significant bits. Given that we want simultaneously to fix the least-significant bits and remain in the same EC, we must leap ahead in the generator's cycle some multiple of $2^\ell - 1$, the period of the least-significant bits.

At this point we must define some notation to simplify the subsequent discussion. Let us first recast equation (1) into a matrix recursion modulo 2^m. First we write $\mathbf{x}_n = [x_n, x_{n-1}, \ldots x_{n-\ell+1}]^T$ for the contents of the register at the nth step. We may then write equation (1) as $\mathbf{x}_n = \mathbf{A}\mathbf{x}_{n-1} \pmod{2^m}$ with the $\ell \times \ell$ matrix $\mathbf{A}$ defined by:

$$(3) \qquad \mathbf{A} = \begin{pmatrix}
0 & 0 & 0 & \ldots & 0 & 1 & 0 & \ldots & 0 & 0 & 1 \\
1 & 0 & 0 & \ldots & 0 & 0 & 0 & \ldots & 0 & 0 & 0 \\
0 & 1 & 0 & \ldots & 0 & 0 & 0 & \ldots & 0 & 0 & 0 \\
0 & 0 & 1 & \ldots & 0 & 0 & 0 & \ldots & 0 & 0 & 0 \\
\vdots & \vdots & \vdots & \ddots & \vdots & \vdots & \vdots & & \vdots & \vdots & \vdots \\
0 & 0 & 0 & \ldots & 1 & 0 & 0 & \ldots & 0 & 0 & 0 \\
0 & 0 & 0 & \ldots & 0 & 1 & 0 & \ldots & 0 & 0 & 0 \\
0 & 0 & 0 & \ldots & 0 & 0 & 1 & \ldots & 0 & 0 & 0 \\
\vdots & \vdots & \vdots & & \vdots & \vdots & \vdots & \ddots & \vdots & \vdots & \vdots \\
0 & 0 & 0 & \ldots & 0 & 0 & 0 & \ldots & 1 & 0 & 0 \\
0 & 0 & 0 & \ldots & 0 & 0 & 0 & \ldots & 0 & 1 & 0
\end{pmatrix}$$

with k and ℓ labeling columns.

Recall that $\mathbf{x}_n$ is column vector of integers modulo 2^m. Let us also define an alternative notation for $\mathbf{x}_n$ so that we can refer to a particular set of bits down its ℓ elements. Thus we write $\mathbf{x}_n$ as a row vector of m-bit column vectors of length ℓ as $\mathbf{x}_n = [b_{m-1}^{(n)}, b_{m-2}^{(n)}, \ldots, b_0^{(n)}]$. Here $b_0^{(n)}$ is the column bit-vector holding $\mathbf{x}_n$'s least-significant bits while $b_{m-1}^{(n)}$ hold $\mathbf{x}_n$'s most-significant bits.[4]

With the above notation we can say that since $\mathbf{A}$ is a recursion matrix for a maximum possible period additive lagged-Fibonacci generator, $\mathbf{A}^{2^\ell-1} \equiv \mathbf{I} \pmod 2$. Let us call $\mathbf{J} = \mathbf{A}^{2^\ell-1}$, so we may write $\mathbf{J} \equiv \mathbf{I} \pmod 2$ and $\mathbf{J}^2 \equiv \mathbf{I} \pmod 4$. Thus applying $\mathbf{J}$ to $\mathbf{x}_n$ leaves the least-significant bits fixed, while $\mathbf{J}^2$ leaves the two least-significant bits fixed. If we assume that we have changed a given seed's least-significant bits into the canonical form, application of $\mathbf{J}$ yields two possible b_1's, call them b_1 and $b_1^{\mathbf{J}}$. We choose one of these to be the b_1 for our EC representative. An unambiguous choice is to select the smallest of b_1 and $b_1^{\mathbf{J}}$ viewed as ℓ-bit integers. Here we define the least-significant bit of b_1 as the corresponding bit in x_0. However; this procedure still produces an unambiguous choice if this bit is viewed as the most-significant bit. Note that b_1 and $b_1^{\mathbf{J}}$ can never be equal, as that would contradict achieving the maximum possible period. Next we use $\mathbf{J}^2$ in a similar manner to choose between b_2 and $b_2^{\mathbf{J}^2}$. This procedure continues until we have our EC representative. By construction, this algorithm produces the same seed for an EC when given any seed in the full period. Additionally, application of this procedure to seeds from different ECs will produce different canonical form seeds.

---

[4] We will usually not use the superscript in subsequent discussion as we rarely need to refer to the bits from one time step to another.

Using this algorithm, we can produce a single seed that is the representative for its EC and is in a canonical form. How do we now enumerate the different ECs? The number of bits in the seed is $\ell \times m$, while the number of ECs is $E = 2^{(\ell-1)(m-1)}$. Thus a set of $(\ell-1)(m-1)$ bits specifies a unique EC. Our canonical form has already specified the ℓ least-significant bits, so it could be hoped that the canonical form gives the following explicit enumeration:

(4)

	m.s.b			l.s.b.	
b_{m-1}	b_{m-2}	$\ldots$	b_1	b_0	
$\square$	$\square$	$\ldots$	$\square$	$b_{0\,\ell-1}$	$x_{\ell-1}$
$\square$	$\square$	$\ldots$	$\square$	$b_{0\,\ell-2}$	$x_{\ell-2}$
$\vdots$	$\vdots$	$\vdots$	$\vdots$	$\vdots$	$\vdots$
$\square$	$\square$	$\ldots$	$\square$	$b_{0\,1}$	x_1
0	0	$\ldots$	0	$b_{0\,0}$	x_0

This enumeration leaves exactly $(\ell-1)(m-1)$ bits to be specified in the canonical form and yields exactly $E = 2^{(\ell-1)(m-1)}$ different possibilities. The remarkable thing is that this is the case! To prove it we first will have to understand the relationship between the bit vectors b_1 and $b_1^{\mathbf{J}}$ from our explicit construction of the EC representative. Recall that the bit vector b_1 is modified into $b_1^{\mathbf{J}}$ by the application of $\mathbf{J}$. The mechanism for this modification is the superposition of the carries from b_0 and the evolution of b_1 viewed as a shift-register. However, since $\mathbf{J} = \mathbf{A}^{2^\ell-1}$ and the period of the shift-register is $2^\ell - 1$, $b_1^{\mathbf{J}}$ is only a function of b_0 though the carries. Thus it follows that $b_1^{\mathbf{J}} = b_1 \oplus C_1(b_0)$, where $C_1(b_0)$ is a bit-vector valued function that when vectorially added modulo 2 to b_1 transforms it to $b_1^{\mathbf{J}}$ via the cumulative superposition of the carries. Such a transformation exists for any linear functional on bit-vectors. One consequence of this representation is that $C_1(b_0)$ must be the value of $b_1^{\mathbf{J}}$ when b_1 is all zeros. Since b_1 and $b_1^{\mathbf{J}}$ cannot be equal, it must be that $C_1(b_0)$ is nonzero and hence has a most-significant one bit, when viewing $C_1(b_0)$ as an ℓ-bit integer. This most-significant one indicates that its bit position changes from b_1 to $b_1^{\mathbf{J}}$. Above we chose the smallest of b_1 and $b_1^{\mathbf{J}}$ viewed as ℓ-bit integers in part of our canonical form. By choosing this position to be a zero, we ensure that choice among all b_1, $b_1^{\mathbf{J}}$ pairs and are free to fill in the remaining $\ell-1$ as we choose.

This procedure can be repeated for b_2 and $b_2^{\mathbf{J}^2}$ via the calculation of $C_2(\cdot)$, and we can continue to repeat this procedure for each successive b_i to produce the following EC tableau:

(5)

	m.s.b			l.s.b.	
b_{m-1}	b_{m-2}	$\ldots$	b_1	b_0	
$\square$	$\square$	$\ldots$	0	$b_{0\,\ell-1}$	$x_{\ell-1}$
0	$\square$	$\ldots$	$\square$	$b_{0\,\ell-2}$	$x_{\ell-2}$
$\vdots$	$\vdots$	$\vdots$	$\vdots$	$\vdots$	$\vdots$
$\square$	0	$\ldots$	$\square$	$b_{0\,1}$	x_1
$\square$	$\square$	$\ldots$	$\square$	$b_{0\,0}$	x_0

The location of the zeros in each x_n is the location of the most-significant one in the $C_i(\cdot)$ bit-vector. It turns out this bit-vector is only a function of the least-significant bit, b_0. A simple proof of this fact can be found in [12].

The fact that $C_i(\cdot)$ is a function only of b_0 has a profound effect in computing a seed in a given EC. Once we have settled on the b_0 for our EC canonical form we can precompute all of the $C_i(b_0)$. This lets us precompute the location of the fixed zeros in the EC representative. Thus we have reduced the problem of producing a seed in a given equivalence class to some precomputation and the translation of a $(\ell-1)(m-1)$ bit equivalence class number into $(\ell-1)(m-1)$ open bit locations in the seed tableau!

Another consequence of $C_i(\cdot)$'s dependence on b_0 alone is that we may try several b_0's to find one that gives an EC canonical form like equation (4). We have implemented a seeding scheme based on these explicit EC computations for the case of the additive lagged-Fibonacci used in Thinking Machines' CMSSL. Here the recurrence is:

$$(6) \qquad x_n = x_{x-5} + x_{n-17} \pmod{2^{32}}.$$

With very little work a particular b_0 was found so that the tableau in (4) could be used. In fact, for all primitive trinomials of degree up to 255, a special b_0 was found that gave a canonical form as in equation (4).

2.3 Equivalence Classes for Parallelism.

With this huge number of ECs, parallel implementation is easy. The key is to associate each independent parallel process in the computation with a unique parallel process identifier, K. This K is then used to select the Kth EC for this process.[5] This procedure works without difficulty provided the parameters for the generator are chosen so that no K is required in the computation that exceeds $2^{(\ell-1)(m-1)} - 1$.

One of the most demanding applications for pseudorandom generators is transport Monte Carlo, [16]. Here the path of a particle is followed and modified via sampling. Particles are emitted, absorbed, and created along a trajectory based on the parameters of the problem and the outcome of the pseudorandom number generator. The overall solution is then the average over many different particle trajectories. The partitioning of this problem among different processors is conceptually trivial via the independent trajectories. However, ensuring the reproducibility of this computation on an asynchronous MIMD machine is not nearly as easy. Since a particle may create new particles, this leads to new trajectories to be followed. It is common practice to place the information from a particle creation into a computational queue. The queued particles are then processed when a free processor becomes available. Reproducibility requires that particles are queued with information sufficient for the pseudorandom number generator to produce the same stream of pseudorandom numbers regardless of what processor or in what order the particle is processed. When using the additive lagged-Fibonacci generator, one need only provide a unique K for each particle to ensure reproducibility in this very general sense.

In general, if one is computing on an arbitrary asynchronous MIMD machine, it is desirable to be able to produce a child process identifier, K, that is guaranteed to be

---

[5]In practical implementations, the Kth parallel process obtains the $f(K)$th EC, where $f(\cdot)$ is an appropriately chosen function.

distinct from others created elsewhere in the computation. In addition, the assignment of K should be a local computation based only on the parent's process identifier. This is easily accomplished by associating all possible parallel processes with a binary tree. When the process for node K is required to create n children, it does so by assigning the n nodes closest and below it on the binary tree. This assures a local computation. In particular, if the process assigned to node K has two children, they receive nodes $2K + 1$ and $4K + 2$.

This procedure does not totally solve the problem of the reuse of ECs in an asynchronous MIMD computation. It is possible to have each particle in a computation create one child particle and hence rapidly descend down the $(l-1)(m-1) + 1$ levels of the process binary tree. However, this seems to be a very unlikely situation, as even the relatively small CMSSL generator has 497 levels!

2.4 Quality Issues.

An important issue in pseudorandom number generation is the quality of the numbers produced by a given recursion. There are many desirable randomness properties that a sequence should possess, and it is important that it does well on empirical tests of statistical randomness. However, empirical testing has practical as well as theoretical limitations, [5]. Thus the inclusion of qualitative theoretical results that impact on pseudorandomness is always important.

A very powerful tool for the theoretical exploration of the quality of pseudorandomness is the exponential sum. Most importantly, the exponential sum is related to the discrepancy through upper and lower bounds, [8, 7, 14]. In turn, the discrepancy appears explicitly in the Koksma-Hlawka bound on numerical integration error. This is very important since numerical integration is **the** fundamental Monte Carlo application.

The exponential sum auto-correlation for a modulo M sequence, $\{x_n\}$, is defined as:

$$(7) \qquad A(i,j) = \sum_{n=0}^{i-1} e^{\frac{2\pi\sqrt{-1}}{M}(x_n - x_{n-j})}.$$

When $i = \text{Per}(x_n)$, the period length, these are called full-period exponential sums, otherwise they are called partial-period exponential sums. The manipulation of these types of sums in order to calculate or bound them is fundamental to many areas of number theory, [6].

An important use of exponential sums in the case of parallel pseudorandom number generation is to use them as a measure of the exponential sum cross-correlation among different parallel pseudorandom number sequences. Suppose we have two modulo M pseudorandom number sequences, $\{x_n\}$ and $\{y_n\}$. Their exponential sum cross-correlation is given by:

$$(8) \qquad C(i,j) = \sum_{n=0}^{i-1} e^{\frac{2\pi\sqrt{-1}}{M}(x_n - y_{n-j})}.$$

In our case we are interested in sequences that are both generated by (1) and have seeds in different ECs. Since $C(i,j)$ is a sum over the difference of sequences at a fixed offset, j, we can compute it by considering it as an exponential sum of the same recurrence in a potentially different EC. This is because the difference of two sequences obeying a given recursion will itself obey the recursion. Thus equation (8) gives us a qualitative tool to explore relationships between related ECs. This approach is discussed in §3.

3. Some Properties of Full-Period Sums.

To analyze these additive lagged-Fibonacci recursions using full-period exponential sums from (7) we must consider such sums with $M = 2^m$. Exponential sums over finite fields have yielded quite remarkable results on the quality of many types of pseudorandom number generators, [8, 13, 14]. The best results for these sums when the recursions are modulo a power-of-two, however, are considerably less informative. For example, the general bound for a ℓ-term recursion modulo 2^m is $O(2^{\frac{\ell m}{2}})$, [13]. Given that there are $\mathrm{Per}(x_n) = (2^\ell - 1)2^{m-1}$ terms of magnitude one in this sum, a trivial upper bound is $\mathrm{Per}(x_n) << O(2^{\frac{\ell m}{2}})$. To our knowledge, no bounds on these types of exponential sums exist that take into account the relatively short period of these sequences. In §3.1 we calculate the sum of all of the full-period exponential sums over all ECs. In §3.2 we show how the exponential sum cross-correlations among different ECs depends on the similarity of the two ECs' representative and leads to sums related to full-period sums with smaller powers-of-two. Finally, in §3.3 we prove the first non-trivial upper bounds we are aware of for these exponential sums. The order of magnitude of these sums is substantially inferior to the analogous sums over finite fields. We hope these humble results encourage experts to improve upon them.

3.1 Full-Period Exponential Sums over All Equivalence Classes.

One exact calculation that can be done relating to the full-period exponential sum measure of quality given in (7) is the exact determination of all of the full-period sums over all of the ECs. When $m = 1$ the sequences are just maximal-period shift-register sequences and there is only $E = 2^{(\ell-1)(m-1)} = 1$ equivalence class. It is well known that in this case $C(\mathrm{Per}(x_n), \cdot) = -1$, [4].

To compute these sums in general we will use a well-known trick. Consider the sum:

$$(9) \qquad \mathcal{S}_m = \sum_{k \in \mathcal{K}_m^\ell} e^{\frac{2\pi\sqrt{-1}}{2^m}k_0}.$$

Here $\mathcal{K}_m^\ell$ is the set of all nonzero ℓ-tuples of residues modulo 2^m, and k_0 is the 0th element of the ℓ-tuple, k. Thus this is a sum over the integers in all nonzero ℓ-tuples modulo 2^m with the appropriate multiplicity. This sum has exactly $2^{\ell m} - 1$ terms. Thus $\mathcal{S}_m = -1$ since k_0 takes on the value of each residue modulo 2^m exactly 2^ℓ times, except for 0, which is taken on only $2^\ell - 1$ times. We will now show how to rewrite $\mathcal{S}_m$ as a sum over full-periods over all ECs.

First a simple example, $m = 2$. We can rearrange $\mathcal{S}_2$ into a sum over all full-period cycles modulo 4 plus the rest. Recall that a cycle is full-period if and only if at least one of its elements is odd. For a general m there are $E_m \times \mathrm{Per}_m(x_n) = 2^{(\ell-1)(m-1)} \times (2^\ell - 1)2^{m-1} = (2^\ell - 1)2^{\ell(m-1)}$ such elements in these full-period cycles. In particular, for $m = 2$ there are $E_2 = 2^{\ell-1}$ ECs of length $\mathrm{Per}_2(x_n) = (2^\ell - 1)2$. We can characterize the rest of the elements as those that do not have at least one odd element, i.e., all elements of the cycle have zero as their least-significant-bit. There are $2^\ell - 1$ of these accounting for $E_2 \times \mathrm{Per}_2(x_n) + 2^\ell - 1 = 2^{2\ell} - 1$ terms, which is all of the nonzero 2-tuples. Thus:

$$(10) \qquad \mathcal{S}_2 = -1 = \sum_{E_2} \sum_{\mathrm{Per}_2(x_n)} e^{\frac{2\pi\sqrt{-1}}{4}x_n} + \sum_{E_1} \sum_{\mathrm{Per}_1(x_n)} e^{\frac{2\pi\sqrt{-1}}{4}x_n}.$$

The last sum is over all ECs modulo 2, $E_1 = 1$, over the full period, $Q_1(x_n) = 2^\ell - 1$, and over a sequence $x_n = 2 \times y_n$ where y_n is a maximal-period shift-register sequence modulo 2. This is easily recognized as $C(\text{Per}(x_n), \cdot) = -1$, and thus $\sum_{E_2} \sum_{\text{Per}_2(x_n)} e^{\frac{2\pi\sqrt{-1}}{4}x_n} = 0$. Let us denote a sum over all ECs and over all full-period cycles modulo 2^m as $S(m)$. Thus the first term in (10) is $S(2)$ and the second term is $S(1)$.

Thus we see a pattern: $\mathcal{S}_m = \sum_{k=1}^{m} S(k) = S(m) + \mathcal{S}_{m-1}, m > 1$. We will formally prove this statement below. We have already proven that $\mathcal{S}_m = \mathcal{S}_1 = S(1) = -1$. If we assume that the formula $\mathcal{S}_m = \sum_{k=1}^{m} S(k) = S(m) + \mathcal{S}_{m-1}, m > 1$ is true, then for $m > 1$ $S(m) = \mathcal{S}_m - \mathcal{S}_{m-1} = -1 - (-1) = 0$. Thus we have

$$S(m) = \begin{cases} 0, & \text{for } m > 1 \\ -1, & \text{for } m = 1. \end{cases}$$

This means that for the exponential sum over all full-period ECs of the additive lagged-Fibonacci generator yields zero except when $m = 1$. In this case we have the sum equal to one, which follows from more classical results of maximal-period shift-register sequences, [4].

Now let us prove the main result:

Theorem 1. $\mathcal{S}_m = \sum_{k=1}^{m} S(k) = S(m) + \mathcal{S}_{m-1}, m > 1$.

Proof. We prove by induction. The case of $m = 2$ was proven above. Assume the formula true for $m - 1$. It suffices to prove that $\mathcal{S}_m - \mathcal{S}_{m-1} = S(m)$. The sum, $\mathcal{S}_m$, is taken over residues modulo 2^m and has $2^{\ell m} - 1$ terms while $\mathcal{S}_{m-1}$ is over residues modulo 2^{m-1} with $2^{\ell(m-1)} - 1$. Let us rewrite

$$(11) \qquad \mathcal{S}_{m-1} = \sum_{k \in \mathcal{K}_{m-1}^\ell} e^{\frac{2\pi\sqrt{-1}}{2^{m-1}}k_0} = \sum_{k \in \mathcal{K}_{m-1}^\ell} e^{\frac{2\pi\sqrt{-1}}{2^m}(2 \times k_0)}.$$

Thus $\mathcal{S}_{m-1}$ can be viewed as a sum over $2^{\ell(m-1)} - 1$ even residues modulo 2^m. In fact, this reformulation casts $\mathcal{S}_{m-1}$ as the sum over all the modulo 2^m ℓ-tuple that have only even elements. Thus the difference $\mathcal{S}_m - \mathcal{S}_{m-1}$ is the sum over all ℓ-tuples of residues modulo 2^m that have at least one odd element. These ℓ-tuples are exactly those that belong to a full-period cycle. We have called the sum over all modulo 2^m ℓ-tuples in a full-period cycle $S(m)$, so that $\mathcal{S}_m - \mathcal{S}_{m-1} = S(m)$, as was desired. $\square$

3.2 Equivalence Classes and Exponential Sums.

We now turn to the analysis of exponential sum cross-correlations among ECs. If $\{x_n\}$ and $\{y_n\}$ come from different ECs, we know we can find an offset, j, so that $\mathbf{x}_n$ and $\mathbf{y}_{n+j}$ agree in their least-significant bit. In fact, if $\mathbf{x}_n$ and $\mathbf{y}_n$ are their respective EC representatives, they already agree in their least-significant bit. Now assume that we are working with recurrences from equation (1), i.e., $M = 2^m$, with a fixed recursion (i.e., ℓ and k given) and have the full-period exponential sum as a function of m, i.e., $C(\text{Per}(x_n), \cdot) = F(m)$. In addition, assume that $\mathbf{x}_n$ and $\mathbf{y}_{n+j}$ agree only in their least-significant bit. We notice that the difference, $\tilde{z}_n = x_n - y_{n-j}$, is even and may be written as $\tilde{z}_n = 2z_n$, where z_n is a maximum period additive lagged-Fibonacci sequence modulo 2^{m-1}. Thus

$$\sum_{n=0}^{\mathrm{Per}(x_n)-1} e^{\frac{2\pi i}{2^m}\tilde{z}_n} = \sum_{n=0}^{\mathrm{Per}(x_n)-1} e^{\frac{2\pi i}{2^m}2z_n} =$$

(11)
$$\sum_{n=0}^{\mathrm{Per}(x_n)-1} e^{\frac{2\pi i}{2^{m-1}}z_n} = \sum_{n=0}^{2\mathrm{Per}(z_n)-1} e^{\frac{2\pi i}{2^{m-1}}z_n} =$$

$$2 \sum_{n=0}^{\mathrm{Per}(z_n)-1} e^{\frac{2\pi i}{2^{m-1}}z_n} = 2F(m-1).$$

In general, if there is an offset, j, that makes $\mathbf{x}_n$ and $\mathbf{y}_{n+j}$ agree in their r least-significant bits, we have that the sum in (11) will equal $2^r \times F(m-r)$. This gives us a clear understanding of how different ECs are related, since one way to see how many least-significant bits of overlap there are between $\{x_n\}$ and $\{y_n\}$ is to place them in canonical form. Because of the periods of the different bits, $\{x_n\}$ and $\{y_n\}$ can be made to agree in their r least-significant bits if and only if their EC representatives agree in their r least-significant bits. This means that a local scheme of computing child processor numbers based on the mapping of parallel processes onto the binary tree will always produce children with cross-correlations as small as possible. In fact, this analysis allows us to modify the assignment of process numbers in appropriate ways to avoid large cross-correlations if a particular computation chooses related ECs poorly in this respect. In our implementation of this generator, we did not use this assignment procedure for child processes, for the reason of correlations in lesser-significant bits in sequences all started from the EC representatives. One solution is to apply some pseudorandom power of $\mathbf{A}$ to a given seed, but in our implementation [15] we believe we have a more elegant solution.

3.3 Bounds on Full-Period Exponential Sums of Maximal Period.

Throughout this subsection, the notation $\max_{\{x_n\}}$ means that the maximum is taken over all maximal period sequences $\{x_n\}$ satisfying a recursion relation of the form

$$x_n = x_{n-k} + x_{n-\ell} \pmod{2^m}, \quad \ell > k,$$

where ℓ and k are fixed. Our goal in this subsection is to obtain nontrivial upper bounds in certain cases for $|U(m)|$ where

$$U(m) = \max_{\{x_n\}} \sum_{j=0}^{\mathrm{Per}(x_n)} e^{\frac{2\pi\sqrt{-1}}{2^m}x_j}.$$

It is convenient to introduce the quantity

$$Z(m) = \max_{\{x_n\}} \#\{j \,|\, x_j = 0 \text{ and } 0 \le j < \mathrm{Per}(x_n)\},$$

the largest number of zeros in the full-period cycles of all ECs modulo 2^m. Our first result bounds $|U(m)|$ in terms of $Z(m/2)$ and suggests the possibility of a descent.

Proposition 1. If $1 < m \in \mathbb{Z}^+$, then

$$|U(m)| \leq 2^{\lceil m/2 \rceil} Z(\lfloor m/2 \rfloor)$$

Proof. Let $\{x_n\}$ be a linear recurrent sequence modulo 2^m that produces the full-period exponential sum $U(m)$. Note that if $\{x_n\}$ is viewed modulo $2^{\lceil m/2 \rceil}$, then it has period $H = (2^\ell - 1)2^{\lceil m/2 \rceil - 1}$. It follows that

$$x_{n+H} - x_n = 2^{\lceil m/2 \rceil} y_n,$$

where $\{y_n\}$ is a maximal period sequence defined modulo $2^{\lfloor m/2 \rfloor}$ that satisfies the same recurrence as $\{x_n\}$.

Now observe if μ is a non-negative integer, then

$$x_{n+(\mu+1)H} = x_{n+\mu H} + 2^{\lceil m/2 \rceil} y_{n+\mu H} = x_{n+\mu H} + 2^{\lceil m/2 \rceil} y_n,$$

by the periodicity of $\{y_n\}$. It follows by induction that

$$x_{n+\mu H} = x_n + \mu 2^{\lceil m/2 \rceil} y_n.$$

Recall that each non-negative integer j with $0 \leq j < (2^\ell - 1)2^{m-1}$ can be written uniquely as $j = N + \mu H$, where $0 \leq N < H$, and $0 \leq \mu < 2^{\lfloor m/2 \rfloor}$. It follows at once that

$$\sum_{j=0}^{\text{Per}(x_n)} e^{\frac{2\pi\sqrt{-1}}{2^m} x_j} = \sum_{N=0}^{H-1} \sum_{\mu=0}^{2^{\lfloor m/2 \rfloor}-1} e^{\frac{2\pi\sqrt{-1}}{2^m} x_{N+\mu H}}.$$

Hence,

$$U(m) = \sum_{N=0}^{H-1} \sum_{\mu=0}^{2^{\lfloor m/2 \rfloor}-1} e^{\frac{2\pi\sqrt{-1}}{2^m}(x_N + \mu 2^{\lceil m/2 \rceil} y_N)}.$$

Further algebraic manipulation yields

$$U(m) = \sum_{N=0}^{H-1} e^{\frac{2\pi\sqrt{-1}}{2^m} x_N} \sum_{\mu=0}^{2^{\lfloor m/2 \rfloor}-1} \left(e^{\frac{2\pi\sqrt{-1}}{2^{\lfloor m/2 \rfloor}} y_N} \right)^\mu.$$

Now the innermost sum is a geometric series that vanishes unless y_N is zero, in which case it equals $2^{\lfloor m/2 \rfloor}$. Thus

$$U(m) = 2^{\lfloor m/2 \rfloor} \sum_{\substack{N=0 \\ y_N=0}}^{H-1} e^{\frac{2\pi\sqrt{-1}}{2^m}(x_N)}.$$

Estimating the sum trivially, we have that

$$|U(m)| \leq 2^{\lfloor m/2 \rfloor} \#\{N | y_N = 0 \text{ and } 0 \leq N < H\}.$$

When m is even, H is the period of $\{y_n\}$. In this case,

$$|U(m)| \leq 2^{m/2} Z(m/2),$$

and the proposition follows for even m. When m is odd, H is twice the period of $\{y_n\}$. In this case,

$$|U(m)| \leq 2^{\lfloor m/2 \rfloor} 2 Z(\lfloor m/2 \rfloor),$$

proving the proposition when m is odd. $\quad\square$

Note that the trivial upper bound for $|U(m)|$ is $(2^\ell - 1)2^{m-1}$, the common period of the linear recurrent sequences under consideration. Note that if the values of a maximal period $\{x_n\}$ were approximately uniformly distributed over the congruence classes mod $2^{m/2}$, then Proposition 1 would produce a decidedly nontrivial bound for $|U(m)|$. Indeed, if $|Z(\lfloor m/2 \rfloor)|$ is well-approximated by the expected $(2^\ell - 1)/2$, then Proposition 1 would yield an estimate of the form $|U(m)| = O(2^{m/2+\ell})$. This is better that the trivial bound $|U(m)| = O(2^{m+\ell})$ by a factor of $2^{m/2}$, although it is still a factor of $2^{\ell/2}$ greater than the empirically conjectured $O(\sqrt{\text{Per}(x_n)})$, [12].

Currently, however, we have only the following estimate for $|Z(m)|$.

Proposition 2. *For any* $m \in \mathbb{Z}^+$, *we have*

$$|Z(m)| \leq \frac{2^\ell - 1}{2} + \frac{(|U(m)| + \cdots + |U(1)|)}{2}.$$

Proof. Let $\delta(x)$ be the usual Kronecker function which takes the value 1 at the origin and vanishes elsewhere. Let $\{x_n\}$ be the maximal period linear recurrent sequence modulo 2^m that vanishes $Z(m)$ times over a full period. Then

$$Z(m) = \sum_{j=0}^{\text{Per}(x_n)-1} \delta(x_j) = \sum_{j=0}^{\text{Per}(x_n)-1} \frac{1}{2^m} \sum_{h=0}^{2^m-1} e^{\frac{2\pi\sqrt{-1}}{2^m}h x_j}.$$

Changing the order of summation yields

$$Z(m) = \frac{1}{2^m} \sum_{h=0}^{2^m-1} \sum_{j=0}^{\text{Per}(x_n)-1} e^{\frac{2\pi\sqrt{-1}}{2^m}h x_j}.$$

The contribution from the $h = 0$ term is simply $\text{Per}(x_n)/2^m$, which equals $(2^\ell - 1)/2$. When h is odd, $\{hx_n\}$ is a maximal period sequence modulo 2^m that satisfies the same recurrence as $\{x_n\}$, and there are precisely 2^{m-1} odd values of h. Thus the terms for which h is odd contribute at most $|U(m)|/2$. When h is precisely divisible by 2^k, then the innermost sum can be viewed as 2^k copies of a full-period sum modulo 2^{m-k}, and there are precisely 2^{m-k-1} such values of h. Such terms contribute at most $|U(m-k)|/2$. The proposition is proved. $\quad\square$

Note that applying Propositions 1 and 2 recursively yields the estimate

$$|U(m)| \leq 2^{\lceil m/2 \rceil} \left(\frac{2^{\ell} - 1}{2} + 1/2 \sum_{j=0}^{\lfloor m/2 \rfloor - 1} |U(\lfloor m/2 \rfloor - j)| \right).$$

Iterating this approach down to the case where $m = 1$ does not in general improve on the trivial bound mentioned earlier. However, suppose the actual value of $Z(m)$ is obtained by brute force on a computer. Then one could apply Proposition 1 and obtain nontrivial estimates for $|U(2m)|$, even when the size of $2m$ is so large that a brute force attack on $U(m)$ itself is not feasible. For example, we have computed $|Z(2)| \leq 65790$ in the case where $\ell = 17$ and $k = 5$. Applying Proposition 1 then yields the nontrivial result

$$|U(4)| \leq (\mathrm{Per}(x_n))^{.9003}$$

in this special case.

4. Discussion and Conclusions.

We have provided the theoretical background for the use of a two-term additive lagged-Fibonacci pseudorandom number generator in the most general parallel setting. The algorithms are based on the realization that equation (1) produces a vast number of full-period cycles. These cycles can be explicitly chosen through the calculation of an appropriate seed. We have also provided a simple and general local computation to produce child ECs from parents. In addition, we have analyzed the theoretical quality of these sequences and have understood the exponential sum cross-correlations among different ECs.

The bounds we have provided for these full-period exponential sums are surely far from the best possible. In fact, if we consider a fixed recursion modulo 2^m, an exhaustive computation of all of the exponential sums can be undertaken. This provides a concrete number to place in this cascade of inequalities that empirically improves what can be proven for a particular generator. For example, in the case where $\ell = 17$ and $k = 5$, the CMSSL generator, we have computed exhaustively that $Z(2) \leq 65790$. This motivates further research into both computational and theoretical improvements in the determination of upper bounds for these full-period exponential sums.

In addition to this work, we have provided an implementation of these ideas, [15]. This implementation is very similar to the ideas presented here, except that the seeding method has been improved. This implementation is based directly on the integer recursion in (1) and produces $[0, 1)$ random variables by $u_n = x_n/M$. A more direct approach is to work directly with floating-point numbers and take equation (1) modulo 1. This can be done while still ensuring absolute EC integrity by providing zero valued guard bits in the mantissa. For our simple three term recursions with ± 1 coefficients, a mantissa of length s bits can hold floating-point values in the top $s - 1$ bits without risking an operation that changes the EC.

Acknowledgments

The authors would like to thank R. G. E. Pinch of Queen's College at Cambridge University, England, for suggesting the induction used in the main proof of §3.3.

References

1. R. P. Brent, *Uniform Random Number Generators for Supercomputers*, Proceedings Fifth Australian Supercomputer Conference, SASC Organizing Committee, 1992, pp. 95–104.

2. R. P. Brent, *On the periods of generalized Fibonacci recurrences*, Math. Comput. **63** (1994), 389–401.

3. S. A. Cuccaro, M. Mascagni and D. V. Pryor, *Techniques for testing the quality of parallel pseudorandom number generators*, Proceedings of the Seventh SIAM Conference on Parallel Processing for Scientific Computing, pp. 279–284.

4. S. W. Golomb, *Shift Register Sequences*, Revised Edition, Aegean Park Press, Laguna Hills, California, 1982.

5. D. E. Knuth, *The Art of Computer Programming, Vol. 2: Seminumerical Algorithms, Second edition*, Addison-Wesley, Reading, Massachusetts, 1981.

6. N. M. Korobov, *Exponential sums and their applications*, Kluwer Academic Publishers, Dordrecht, The Netherlands, 1992.

7. L. Kuipers and H. Niederreiter, *Uniform distribution of sequences*, John Wiley and Sons, New York, 1974.

8. R. Lidl and H. Niederreiter, *Introduction to finite fields and their applications*, Cambridge University Press, Cambridge, London, New York, 1986.

9. G. Marsaglia, *A current view of random number generators*, Computing Science and Statistics: Proceedings of the XVIth Symposium on the Interface, 1985, pp. 3–10.

10. G. Marsaglia and L.-H. Tsay, *Matrices and the structure of random number sequences*, Linear Alg. and Applic. **67** (1985), 147–156.

11. M. Mascagni, S. A. Cuccaro, D. V. Pryor and M. L. Robinson, *Recent Developments in Parallel Pseudorandom Number Generation*, Volume II, Proceedings of the Sixth SIAM Conference on Parallel Processing for Scientific Computing (D. E. Keyes, M. R. Leuze, L. R. Petzold, D. A. Reed, ed.), SIAM, Philadelphia, Pennsylvania, 1993, pp. 524–529.

12. M. Mascagni, S. A. Cuccaro, D. V. Pryor and M. L. Robinson, *A fast, high-quality, and reproducible lagged-Fibonacci pseudorandom number generator*, in the press, Comput. Physics (1995).

13. H. Niederreiter, *Quasi-Monte Carlo methods and pseudo-random numbers*, Bull. Amer. Math. Soc. **84** (1978), 957–1041.

14. H. Niederreiter, *Random number generation and quasi-Monte Carlo methods*, SIAM, Philadelphia, Pennsylvania, 1992.

15. D. V. Pryor, S. A. Cuccaro, M. Mascagni and M. L. Robinson, *Implementation and usage of a portable and reproducible parallel pseudorandom number generator*, in Proceedings of Supercomputing '94, 1994, pp. 311–319.

16. J. Spanier and E. M. Gelbard, *Monte Carlo Principles and Neutron Transport Problems*, Addison-Wesley, Reading, Massachusetts, 1969.

17. R. C. Tausworthe, *Random numbers generated by linear recurrence modulo two*, Math. Comput. **19** (1965), 201–209.

SUPERCOMPUTING RESEARCH CENTER; INSTITUTE FOR DEFENSE ANALYSES; 17100 SCIENCE DRIVE; BOWIE, MARYLAND 20715-4300 USA

E-mail address: `mascagni@super.org`, `robinson@super.org`, `pryor@super.org`, `cuccaro@super.org`

Quasirandom Diffusion Monte Carlo

Bradley Moskowitz *

Mathematics Department. UCLA

February 5. 1995

Abstract

Diffusion Monte Carlo is a common method for estimating the properties of quantum mechanical systems by computing averages over sets of random walk simulations. We have found that by using quasirandom sequences of points in place of random or pseudorandom points in generating the simulation paths, we are able to obtain improved convergence rates and consequently reduced Monte Carlo errors for Diffusion Monte Carlo. Computational results are presented for a three dimensional harmonic oscillator and the Helium atom.

A key element in successfully applying the quasirandom sequence is a recently developed technique involving renumbering the simulations paths after each time step, which allows a lower dimensional quasirandom sequence of points to be used.

1 Introduction

Diffusion Monte Carlo (DMC) [5, 9, 29, 33] is a common method for estimating the properties of quantum mechanical systems by computing averages over sets of random walk simulations. It is closely related to Green's Function Monte Carlo (GFMC) [4, 14, 18].

Quasirandom sequences [10, 25, 28] are sequences of points designed to produce more accurate integration estimates, called quasi-Monte Carlo, than standard Monte Carlo with random or pseudorandom points. This is accomplished by relaxing the requirement that successive points be independent, and instead seeking a set of points which spreads out over the domain as uniformly as possible. In one dimension, the best such sequence is a uniform grid, but in higher dimensions more complicated sequences are preferable [7, 25]. The particular quasirandom sequence which we have used is called the base-2 Niederreiter sequence [3].

---

*At Westinghouse Bettis Atomic Power Laboratory, West Mifflin, PA, as of 11/94. This work was done while at UCLA. Research supported in part by the Air Force Office of Scientific Research under grant number F49620-94-1-0091.

Although originally used for integration estimates, quasirandom sequences can be used for simulations as well. There have been several examples of such quasi-Monte Carlo simulations [11, 17, 26, 30], however they have often been limited to cases in which the number of time steps or iterations was small.

We have found that for Diffusion Monte Carlo, which often involves hundreds or thousands of time steps, we can use quasirandom sequences by making use of a recently developed technique involving renumbering the simulations after each time step [21, 23]. Without this technique either the dimensions of the quasirandom sequence would need to be too high to be effective (quasirandom sequences lose their effectiveness as the dimensions increase [2, 19, 22, 24, 28]), or else correlation errors between time steps would lead to erroneous results. With quasirandom sequences and renumbering, Diffusion Monte Carlo results can be significantly improved in terms of expected error versus either number of simulations or total cpu time, as will be demonstrated in the examples below.

Note: The renumbering technique mentioned above is still in its infancy and has not yet been extensively tested nor has its convergence been proven in all cases. However, computational results using this technique so far have been strongly encouraging [23]. The results presented here represent a further confirmation that renumbering can be effective, and that it has a potentially wide variety of useful applications.

2   Diffusion Monte Carlo

In this section a brief description of Diffusion Monte Carlo will be given. As mentioned above, Diffusion Monte Carlo is a method for estimating the properties of quantum mechanical systems. For simplicity, the property which we shall deal with will be the ground state energy. Diffusion Monte Carlo involves converting the time-dependent Schrödinger Equation into a diffusion equation by changing the time variable to imaginary time, as shown below. The equation is then multiplied by a 'trial function,' which is a rough estimate of the exact solution, in order to reduce the expected errors. (Note: This is a type of variance reduction technique [9].) Monte Carlo estimates are computed by following a set of random 'walkers' as they drift and diffuse through the system according to the diffusion equation above and then averaging their trial energy levels.

2.1   Schrödinger Equation

For a quantum mechanical system of M bodies with a time-independent potential function, the time-dependent Schrödinger Equation is the following:

$$-\sum_{k=1}^{M} \frac{\hbar^2}{2m_k} \nabla_k^2 \psi(\boldsymbol{y}, t) + v(\boldsymbol{y})\, \psi(\boldsymbol{y}, t) = i\hbar\, \frac{\partial}{\partial t} \psi(\boldsymbol{y}, t) \tag{1}$$

where m_k is the mass of the k'th body, $\boldsymbol{y} = (\boldsymbol{x}_1, \ldots, \boldsymbol{x}_M)$ is a $3M$-dimensional vector representing the position of all of the bodies, ∇_k^2 is the Laplacian operator for the k'th body, $v(\boldsymbol{y})$ is the potential function, and $\psi(\boldsymbol{y}, t)$ is the wavefunction or probability amplitude for the system.

Defining the Hamiltonian operator, $H = -\sum_{k=1}^{M} \frac{\hbar^2}{2m_k} \nabla_k^2 + v(\boldsymbol{y})$, (1) can be rewritten,

$$H\psi(\boldsymbol{y},t) = i\hbar \frac{\partial}{\partial t}\psi(\boldsymbol{y},t)$$

A formal solution to (1) is the following:

$$\psi(\boldsymbol{y},t) = \sum_{j=0}^{\infty} a_j \, e^{-i\hbar^{-1} E_j t} \psi_j(\boldsymbol{y})$$

when v is continuous. $\psi_j(\boldsymbol{y})$, $j = 0, \ldots$. are a complete orthogonal set of eigenfunctions of H with $E_0 < E_1 < \cdots$, and a_j is the $\psi_j(\boldsymbol{y})$ component of $\psi(\boldsymbol{y},0)$, $a_j = \int_{\mathbb{R}^{3M}} \psi(\boldsymbol{y},0)\,\psi_j(\boldsymbol{y})\,d\boldsymbol{y}$.

To transform to imaginary time. let $\phi(\boldsymbol{y},\tau) = \psi(\boldsymbol{y},it)$, then

$$H\phi(\boldsymbol{y},\tau) = -\hbar \frac{\partial}{\partial \tau}\phi(\boldsymbol{y},\tau) \tag{2}$$

This has formal solution,

$$\phi(\boldsymbol{y},\tau) = \sum_{i=0}^{\infty} a_i \, e^{-\hbar^{-1} E_i \tau} \psi_i(\boldsymbol{y})$$

where the eigenfunctions and eigenvalues are unchanged since H is a purely spatial operator. (2) is a diffusion equation which, unlike (1), converges exponentially fast to the ground state, ψ_0. The only exception is when the wavefunction is specifically restricted to be orthogonal to ψ_0, as when higher energy states are being sought.

2.2 Trial Function

A trial function, which is a rough estimate of the exact solution, is multiplied into the diffusion equation. (2), in order to reduce the variance and eliminate discontinuities.

Rewriting (2) as follows:

$$\frac{\partial}{\partial \tau}\phi(\boldsymbol{y},\tau) = \sum_{k=1}^{M} \frac{\hbar}{2m_k} \nabla_k^2 \phi(\boldsymbol{y},\tau) - \frac{1}{\hbar} v(\boldsymbol{y})\,\phi(\boldsymbol{y},\tau) \tag{3}$$

we see a partial differential equation corresponding to diffusion and growth.

The potential, $v(\boldsymbol{y})$, acts in the growth term. When it is highly variable or singular (for example a Coulomb potential) it leads to high variance Monte Carlo estimates. Multiplying by a well-chosen trial function, ψ_T, can eliminate discontinuities and reduce the variance of the Monte Carlo estimates.

Let $f(\boldsymbol{y},\tau) = \psi_T(\boldsymbol{y})\,\phi(\boldsymbol{y},\tau)$, then multiplying (3) by ψ_T and rearranging terms gives us the following:

$$\frac{\partial}{\partial \tau}f(\boldsymbol{y},\tau) = \sum_{k=1}^{M} \frac{\hbar}{2m_k} \nabla_k^2 f(\boldsymbol{y},\tau) - \sum_{k=1}^{M} \nabla_k \cdot [a_k(\boldsymbol{y})\,f(\boldsymbol{y},\tau)] - b(\boldsymbol{y})\,f(\boldsymbol{y},\tau) \tag{4}$$

where

$$b(\boldsymbol{y}) \;=\; \frac{H\psi_T(\boldsymbol{y})}{\psi_T(\boldsymbol{y})} \;\;, \text{ the trial energy function.}$$

$$\boldsymbol{a}_k(\boldsymbol{y}) \;=\; \frac{\hbar}{m_k}\cdot\frac{\nabla_k\psi_T(\boldsymbol{y})}{\psi_T(\boldsymbol{y})} \;, \; k=1,\dots,M \;\;. \text{ the drift functions.}$$

The new equation (4) is similar to (3) but with an extra set of terms with 1st derivatives of f, which add a drift to the diffusion process. Importantly, b, the trial energy, replaces v, the potential, in the growth term. When ψ_T has been well chosen, the trial energy will be relatively constant, leading to lower variance Monte Carlo estimates of E_0. Note: Hypothetically, if one used exactly $\psi_T = \psi_0$ then b would be a constant, equal to E_0, and trivial zero-variance Monte Carlo estimates would result.

In practice the ground state energy, E_0, is typically estimated using what is called the variational estimate. Let $\psi_*(\boldsymbol{y})$ be an arbitrary wavefunction with a non-zero ground state component. Then using orthogonality properties,

$$E_0 = \frac{\int_{\mathbb{R}^{3M}} H\psi_*(\boldsymbol{y})\,\psi_0(\boldsymbol{y})\,d\boldsymbol{y}}{\int_{\mathbb{R}^{3M}} \psi_*(\boldsymbol{y})\,\psi_0(\boldsymbol{y})\,d\boldsymbol{y}}$$

This is estimated by computing,

$$E(\tau) = \frac{\int_{\mathbb{R}^{3M}} H\psi_*(\boldsymbol{y})\,\phi(\boldsymbol{y},\tau)\,d\boldsymbol{y}}{\int_{\mathbb{R}^{3M}} \psi_*(\boldsymbol{y})\,\phi(\boldsymbol{y},\tau)\,d\boldsymbol{y}} \tag{5}$$

and then letting $E(\tau) \to E_0$ as $\tau \to \infty$ and $\phi(\boldsymbol{y},\tau) \to \psi_0(\boldsymbol{y})$.

It is natural then to take $\psi_* = \psi_T$. Then we have,

$$E(\tau) \;=\; \frac{\int_{\mathbb{R}^{3M}} H\psi_T(\boldsymbol{y})\,\phi(\boldsymbol{y},\tau)\,d\boldsymbol{y}}{\int_{\mathbb{R}^{3M}} \psi_T(\boldsymbol{y})\,\phi(\boldsymbol{y},\tau)\,d\boldsymbol{y}} \;=\; \frac{\int_{\mathbb{R}^{3M}} b(\boldsymbol{y})\,f(\boldsymbol{y},\tau)\,d\boldsymbol{y}}{\int_{\mathbb{R}^{3M}} f(\boldsymbol{y},\tau)\,d\boldsymbol{y}} \tag{6}$$

In which case we observe that $E(\tau)$ can be computed as the expected value of $b(\boldsymbol{y})$ when $\boldsymbol{y}$ is distributed according to $f(\boldsymbol{y},\tau)$ as a probability density.

Diffusion Monte Carlo consists of estimating $E(\tau)$ by using the diffusion, drift, and growth rates from Equation 4 to generate a large set of simulation paths, or random walkers, distributed in space according to $f(\boldsymbol{y},\tau)$, and then averaging b over the set of paths to compute a Monte Carlo estimate of $E(\tau)$. As long as the simulations have traveled for a long enough imaginary time, τ, so that $E(\tau)$ has converged to E_0, we have our desired estimate of E_0.

One important assumption above is that $f(\boldsymbol{y},\tau)$ is strictly positive, so that it can function as a probability density. In many systems with fermions, however, the wavefunction must be negative in certain regions, leading f to be negative as well. As a result, further approximations such as the fixed node approximation [4] are necessary. For simplicity, we will only consider examples in which f is strictly positive so this problem does not arise. For more details on this active area of research see [1, 4, 34].

3 Stochastic Simulation

In this section, a few of the details of simulating the random walks for Diffusion Monte Carlo will be discussed. The stochastic process which corresponds to (4) in the previous section is governed by the following system of stochastic differential equations:

$$\begin{aligned} dZ_k(\tau) &= a_k(Z(\tau))\,d\tau + \sigma_k\,dW \ , \quad k = 1,\dots,M \\ dY(\tau) &= -b(Z(\tau))\,d\tau \end{aligned}$$

where $Z = (Z_1,\dots,Z_M)$, $Z(0) \sim f(y,0)$ (initial distribution), $Y(0) = 0$, $\sigma_k = \sqrt{h/m_k}$.

For a random walker, $Z(\tau)$ indicates its position in $3M$-dimensional space. $Y(\tau)$ is related to the relative survival probability of the walker, which is usually handled by associating a weight factor with each walker, but can also be handled by actually allowing some walkers to be killed off while others split into multiple walkers [29].

Using Ito's Formula [27], one can prove the following important result:

$$E(\tau) = \frac{E\left[b(Z(\tau))e^{Y(\tau)}\right]}{E\left[e^{Y(\tau)}\right]} \ , \quad \text{over stochastic paths } (Z,Y). \tag{7}$$

where the quantities on the right hand side are expectations over the set of all possible stochastic paths followed by a random walker.

In order to generate simulation paths for $(Z(\tau), Y(\tau))$ it is necessary to discretize the imaginary time. This is commonly done using the *Euler-Maruyama* scheme:

$$\begin{aligned} Z_k^{(j+1)} &= Z_k^{(j)} + \sqrt{h}\,\sigma_k W_k^{(j)} + h\,a_k(Z^{(j)}) \ , \quad k = 1,\dots,M \\ Y^{(j+1)} &= Y^{(j)} - h\,b(Z^{(j)}) \end{aligned}$$

where $Z^{(0)} \sim \psi_T^2(y)$ to approximate $\psi_T(y)\psi_0(y)$, $Y^{(0)} = 0$, h is the time step size and j the time step counter so that $\tau = hj$, and $W_k^{(j)}$ is a $3D$ vector of Gaussian, $\mathcal{N}(0,1)$, random variables. For estimates of $E(\tau)$ as an expected value, this scheme is first order accurate in h (Kloeden and Platen [15]).

In order to take larger time steps, without large discretization errors, one can use higher order discretization schemes. In particular, the following scheme is third order accurate (for estimates of $E(\tau)$). It is related to Runge-Kutta methods for ODE's and was derived by Helfand and Greenside ([8, 12]):

$$\begin{aligned} g_{1,k} &= a_k(Z^{(j)} + \sqrt{h}\,\zeta_1) \ , \quad k = 1,\dots,M \\ g_{2,k} &= a_k(Z^{(j)} + h\,.516719\,g_1 + \sqrt{h}\,\zeta_2) \\ g_{3,k} &= a_k(Z^{(j)} + h(-.397300\,g_1 + .427690\,g_2) + \sqrt{h}\,\zeta_3) \\ g_{4,k} &= a_k(Z^{(j)} + h(-1.587731\,g_1 + 1.417263\,g_2 + 1.170469\,g_3) + \sqrt{h}\,\zeta_4) \\ Z^{(j+1)} &= Z^{(j)} + h(.644468\,g_2 + .194450\,g_3 + .161082\,g_4) + \sqrt{h}\,\zeta_0 \\ Y^{(j+1)} &= Y^{(j)} - .5\,h\left(b(Z^{(j)}) + b(Z^{(j+1)})\right) \end{aligned}$$

where $Z^{(0)} \sim \psi_T^2(y)$, $Y^{(0)} = 0$, $\zeta_0 = W_k^{(j)}$, $\zeta_1 = .271608\,V_k^{(j)}$, $\zeta_2 = .516719\,W_k^{(j)} + .499720\,V_k^{(j)}$, $\zeta_3 = .030390\,W_k^{(j)} - .171658\,V_k^{(j)}$, $\zeta_4 = V_k^{(j)}$, and $W_k^{(j)}$ and $V_k^{(j)}$ are two independent $3D$ vectors of Gaussian, $\mathcal{N}(0,1)$, random variables. This scheme involves more computation than Euler-Maruyama and twice the number of Gaussian random variables per time step. However, convergence in imaginary time, τ, is reached in far fewer time steps by taking much larger time steps.

Note: To be precise, the weights, Y, are only simulated to second order accuracy in terms of h above. Also see [6] for further discussion of the application of higher order approximation schemes to Diffusion Monte Carlo.

4 Standard Pseudorandom Diffusion Monte Carlo

Standard pseudorandom Diffusion Monte Carlo consists of using the simulation techniques of the last section to compute an estimate of E_0 using (7), as follows:

$$\hat{E}_N = \frac{\sum_{i=1}^{N} b(Z_i^{(m)}) e^{Y_i^{(m)}}}{\sum_{i=1}^{N} e^{Y_i^{(m)}}} \tag{8}$$

where $(Z_i^{(m)}, Y_i^{(m)})$ represents the position and weight of the i'th stochastic simulation path at time $\tau = mh$. Simulation paths are computed using pseudorandom numbers which are transformed into Gaussian random variables using any one of a number of standard techniques [20].

The error in this estimate can be broken down as follows:

$$
\begin{aligned}
\hat{E}_N - E_0 &= [\hat{E}_N - E^{(h)}(m)] + [E^{(h)}(m) - E(\tau)] + [E(\tau) - E_0] \\
&= \epsilon_{mc} + \epsilon_h + \epsilon_\tau \\
&= O(N^{-1/2}) + O(h^\alpha) + O(e^{-\tau})
\end{aligned}
$$

where $E^{(h)}(m)$ is defined as the expected value of $\hat{E}_N$ as $N \to \infty$, and α is the order of accuracy of the time discretization method.

In the equation above, ϵ_{mc} is the statistical or 'Monte Carlo' error, ϵ_h is the approximation error from the time discretization, and ϵ_τ is the convergence in imaginary time error.

4.1 Ensemble-Based Sampling

In practice, the estimate (8) above is seldom used precisely as written. Instead of generating N independent stochastic simulation paths through time τ, it is much more common to follow a smaller set, or ensemble, of R 'random walkers' as they travel forward in imaginary time [16, 29]. After a start-up period of J time steps, which reduces the time convergence error, ϵ_τ, to acceptably small levels, the positions and weights of the walkers are used after every j'th time step in computing the estimate of E_0, where j is taken sufficiently large so that the new positions are (nearly) independent of the previous positions used. For example, for a total sample size of $N = 10000$, an ensemble size of 50 walkers would be followed for $J + 200j$ time steps in order to accumulate the full sample. J and j are determined empirically.

Ensemble-based sampling saves memory. since R can be much smaller than N. and time, since j can be much smaller than J. As long as h is sufficiently small so that ϵ_h is negligible, and J is sufficiently large so that ϵ_r is negligible, the primary source of error left is the Monte Carlo error, ϵ_{mc}.

5   Quasirandom Diffusion Monte Carlo

In this section we examine how quasirandom sequences can be used to reduce the Monte Carlo error for Diffusion Monte Carlo estimates.

Quasirandom sequences are designed to produce more accurate integration estimates than standard Monte Carlo when used in place of random or pseudorandom points. By spreading out as uniformly as possible over the integration domain, quasirandom sequences can produce Monte Carlo errors which converge at rates as fast as $O(N^{-1})$ instead of the usual $O(N^{-1/2})$ for Monte Carlo [25].

However, quasirandom Monte Carlo estimates, called quasi-Monte Carlo, lose their advantage over standard Monte Carlo as the number of dimensions increases. This is a well established result, theoretically and computationally [2, 19, 22, 24, 28]. While there is no specific maximum number of dimensions for which quasi-Monte Carlo is useful, it appears that for many applications once the number of dimension reaches levels near 30 or 40 quasi-Monte Carlo may not be any better than standard pseudorandom Monte Carlo, and at higher dimensions quasi-Monte Carlo can often produce results which are worse than standard Monte Carlo. This limitation on the number of dimensions has important implications for quasirandom Diffusion Monte Carlo, as we shall see below.

5.1   Quasirandom Simulation

Although originally designed for integration estimates, quasirandom sequences have been used for simulations as well. This is accomplished by treating a simulation problem as being equivalent to the evaluation of an integral, where the integral is the expected value of the simulation's output over the space of all possible simulation paths [23, 31]. The integrand is then a function of all the random inputs at every time step of the simulation path since ultimately the output is a function of all of these inputs. Therefore, the total number of dimensions is equal to the sum of the dimensions, or number of inputs, at each time step. For a simulation path consisting of m time steps in d-dimensions then, we would have md dimensions. Hence, the number of dimensions can easily become very high whenever many time steps are involved, as is typically the case for Diffusion Monte Carlo. As discussed above, this means that quasi-Monte Carlo may not be effective in such instances since it loses its effectiveness as the number of dimensions increases. As a result, most previous applications of quasi-Monte Carlo estimates to simulations have been limited primarily to simulations involving only a few time steps or iterations.

Recently a way to overcome the limitation on time steps has been found which involves renumbering, or scrambling, the order of a set of simulation paths after each time step [17, 21, 23]. As discussed in [23], this technique can effectively reduce the number of dimensions from md to simply d (or $d + 1$) by separating each time step apart from those preceding and following it.

5.2 Application to Diffusion Monte Carlo

For Diffusion Monte Carlo, the number of random input variables for each time step is equal to $3M$, where M is the number of bodies in the quantum mechanical system, for the Euler-Maruyama scheme. For the higher order Helfand-Greenside scheme this doubles to $6M$ per time step. Therefore, regardless of any renumbering, there is an essential limitation to the size of the quantum mechanical systems for which quasirandom Diffusion Monte Carlo may be useful. The examples discussed below include a three dimensional harmonic oscillator ($M = 1$) and the Helium atom ($M = 2$, fixed nucleus). It is anticipated that similar results should be obtainable in the near future for systems of 3 or 4 bodies as well. For larger systems of say 10 or more bodies, it is not anticipated that quasirandom Diffusion Monte Carlo, as presented here, will be useful.

The use of renumbering is essential since Diffusion Monte Carlo typically involves hundreds or thousands of time steps, m, which without renumbering would lead to dimensions, $3Mm$ or $6Mm$, far too high for quasi-Monte Carlo to be effective. Renumbering allows us to use just $3M$ or $6M$ dimensional quasirandom points regardless of the number of time steps, as long as renumbering is done following each time step. This makes quasirandom Diffusion Monte Carlo feasible.

5.3 Renumbering

Renumbering, described in detail in [23], involves generating a full set of simulation paths together, and, after each time step, scrambling up the order of the numbering of the simulation paths. The method of renumbering can be either random – using a random permutation of $\{1, \ldots, N\}$ to renumber N paths, or a more sophisticated technique known as binary renumbering, in which paths are renumbered according to where they are located in space after each step.

For binary renumbering, which is used in the examples below, after each time step, simulation paths are grouped according to which subregion of space they each fall within. Subregions are formed by dividing each dimension into L levels where L is a power of two (for convenience), creating a total of L^d subregions in d-dimensions. When every path's subregion has been determined, the paths are renumbered by taking all of the paths in one subregion followed by all the points in the next subregion, and so on through all the subregions. The order within each subregion is simply random. The order of the subregions themselves is fixed in a way that minimizes the distance between successive subregions. This particular arrangement is used because it reduces the average distance between successively numbered paths, which (as discussed in [23]) leads to improved quasi-Monte Carlo results by reducing a quantity known as the variation.

It is important to mention at this point that without renumbering, if we used just $3M$ or $6M$ dimensional quasirandom points for Diffusion Monte Carlo, correlations between the quasirandom points used in successive time steps would lead to biased and incorrect results. The other alternative is to use extremely high dimensional quasirandom points, which as discussed previously, is not recommended because quasirandom points tend to lose their effectiveness as the number of dimensions increases.

5.4 Continuation Method

Given the very high number of time steps required in general for Diffusion Monte Carlo, since τ must be large enough so that the walkers have converged in imaginary time to the ground state (so that e_τ is sufficiently small), while h must be small enough so that discretization error, ϵ_h, is sufficiently small as well, it is impractical to use quasirandom sequences throughout the simulations. Renumbering allows us to use $3M$ or $6M$ dimensional quasirandom points regardless of the number of time steps, but going through the renumbering process after every one of a large number of time steps would add a great deal of extra work, which it would be preferable to avoid. A further difficulty is ensemble-based sampling, described in Section 4.1, to which there is no clear way to apply quasirandom sequences effectively.

A solution to the two difficulties posed above is to combine standard pseudorandom ensemble-based sampling with quasirandom sampling in what is termed a 'continuation' method. For continuation, we first use standard Monte Carlo with ensemble-based sampling to generate a set of N pseudorandom simulation path 'endpoints' and weights, $\{Z_i, Y_i\}_{i=1}^N$. (Note: Endpoints is in quotations because with ensemble-based sampling, the positions and weights used are actually intermediate values for a smaller set of random walkers.) Then, quasirandom points are used to 'continue' the paths from each (Z_i, Y_i) for an additional fixed number of time steps, using renumbering after each of these steps, after which a Monte Carlo estimate of E_0 is computed as the weighted average of b at the set of new 'continued' endpoints. (Note: For standard Monte Carlo, the estimate would be computed as a weighted average of b at the original 'non-continued' endpoints.)

The basic idea motivating this procedure is that we are first generating a sample, $\{Z_i, Y_i\}_{i=1}^N$, which is relatively free of errors ϵ_h and ϵ_τ, and dominated by error ϵ_{mc}. Then, the additional quasirandom steps are used to reduce this Monte Carlo error from $O(N^{-1/2})$ levels to accelerated levels closer to $O(N^{-1})$, which are characteristic of quasi-Monte Carlo.

One drawback of continuation is that by requiring us to store the entire set of N pseudorandom endpoints before continuing them, this technique negates the memory savings of ensemble-based sampling. This could be a significant drawback when N is large. The same problem arises, however, if an entirely quasirandom simulation is done without ensemble-based sampling and without continuation. Variations in which the samples are continued in smaller blocks to reduce the memory requirements might be possible [32].

On the other hand, continuation can produce a large improvement in cpu time, since the number of renumberings can be limited to just the number of continuation steps rather than the total number of time steps. Then, as long as the number of quasirandom continuation steps is relatively small, the additional cpu time required will not be excessive.

Since increasing the number of continuation steps allows the continued paths to travel further away from the original pseudorandom endpoints. resulting in a better opportunity for Monte Carlo errors to be significantly reduced, but also costing extra cpu time, there is a trade-off between decreased error and increased cpu time as the number of continuation steps increases. In practice, the optimal number of continuation steps can be determined empirically by observing the resultant estimated errors and average cpu times.

It is at this point that using a higher order accurate discretization method can be beneficial. since it allows larger time steps to be used. With larger time steps, fewer continuation steps and renumberings are needed in order to move sufficiently far away from the original pseudorandom endpoints to see an appreciable improvement.

The examples below will demonstrate the effectiveness of quasirandom Diffusion Monte Carlo using the continuation method with renumbering.

6 Quantum Mechanical Examples

We examine two quantum mechanical examples to study the effectiveness of quasirandom Diffusion Monte Carlo using continuation and renumbering, as discussed above, when compared with standard pseudorandom Diffusion Monte Carlo. In all cases, the high-order accurate Helfand-Greenside discretization scheme is used. This allows larger time steps to be taken than if a less accurate method, such as Euler-Maruyama, were used instead. The two examples are a three dimensional harmonic oscillator and the Helium atom.

6.1 Error Measurement

With quasi-Monte Carlo estimates, the issue of error measurement is always an important one since no truly satisfactory standard procedure has yet been established. Let N be the number of simulations or equivalently the 'sample size.' Then, for each of the examples, each different estimate of E_0, pseudorandom or quasirandom, is repeatedly computed R times at each of several different fixed values of N. This allows the expected error as a function of sample size to be observed.

Ideally, each repetition above should be independent. However, even the best pseudorandom sequences are ultimately deterministic and so the independence of successive trials is in doubt. In the case of quasirandom sequences, this is a particular concern because successive points can be highly correlated. In the present examples, each set of R quasirandom estimates for a given sample size, N, is made using R successive blocks of points in the quasirandom sequence (the base-2 Niederreiter sequence [3]). However, in order to try to ensure that these estimates are not too highly correlated, a random number of points between 1 and N is skipped between each block of points.

It is not inconceivable that given just the right scenario, successive quasirandom estimates could be correlated, thereby skewing the error estimates made below, particularly the second one when the exact solution is unknown. Although no evidence of any •ich problem was found in the examples below, in which the exact solutions, or highly accurate estimates, were known in advance, the problem of reliable error estimation remains a critical area for further research in quasi-Monte Carlo because without adequate error estimation techniques these methods will not be widely adopted for use.

Assuming the independence discussed above, the root mean square error of each estimate of E_0 at sample size N can itself be estimated as follows:

$$\hat{\epsilon}_{rms}(N) = \sqrt{\frac{1}{R} \sum_{r=1}^{R} (\hat{E}_N^{(r)} - E_0)^2}$$

where $\hat{E}_N^{(r)}$ is the r'th estimate at sample size N and E_0 is the exact ground state energy.

An estimate of the error which does not rely on the exact solution (presumably unknown in most practical situations) is also needed. In the absence of any standard reliable error estimate of this form for quasi-Monte Carlo, a natural possibility which is consistent with the use of the root mean square error above is the sample variance of the set of R repeated trials as follows:

$$\hat{\sigma}(N) = \sqrt{\frac{1}{R-1} \sum_{r=1}^{R} (\hat{E}_N^{(r)} - \bar{E})^2}$$

where $\bar{E}$ is the average of all R estimates. This estimate will be accurate for unbiased estimators as long as the assumption of independent trials is valid. However, if there is doubt about the independence of the trials, as may well be the case for quasi-Monte Carlo, extreme caution is appropriate. In the examples below, both estimates, $\hat{\epsilon}_{rms}(N)$ and $\hat{\sigma}(N)$ were computed, with similar results in either case. However, as indicated above, the issue of accurate error estimation for quasi-Monte Carlo remains a critical unsolved problem.

6.2 Harmonic Oscillator

This is a simple example which readily demonstrates the effectiveness of the quasirandom techniques.

Example 1 *The ground state energy, E_0, is estimated for a three dimensional harmonic oscillator. The potential energy function is the following:*

$$v(x,y,z) = \frac{1}{2}k\,r^2$$

where $k > 0$ is the oscillator strength, and $r^2 = x^2 + y^2 + z^2$ is the squared distance from the oscillator's center, which is taken to be the origin.

In atomic units, with $\hbar = 1$, we consider a harmonic oscillator of mass $m = 1$ and strength $k = 1$, which has an exact ground state energy of $E_0 = 1.5$. This exact quantity is used in computations of the estimated root mean square error, $\hat{\epsilon}_{rms}(N)$. Note: In atomic units, energy is measured in units of $4.36 \cdot 10^{-18}$ Joules (Hartrees), length in units of $5.3 \cdot 10^{-11}$ meters (Bohrs), and time in units of $2.4 \cdot 10^{-17}$ seconds.

The exact ground state wave function, $\psi_0(x,y,z)$, is equal to $(4\pi)^{-1}e^{-r^2/2}$, leading us to use as our trial function (as described in Section 2.2):

$$\psi_T(x,y,z) = (3.92157\pi)^{-1}e^{-.51r^2}$$

where we deliberately take a function slightly different from the exact ground state in order to avoid the trivial case of exact zero-variance results.

The resultant trial energy function is then the following:

$$b(\boldsymbol{y}) = \frac{H v_T(\boldsymbol{y})}{v_T(\boldsymbol{y})} = 1.53 - 0.202r^2$$

While the vector drift function is the following:

$$\boldsymbol{a}(\boldsymbol{y}) = (-1.02x, -1.02y, -1.02z)$$

For the simulations, the time steps are of size $h = 0.15$ using the Helfand-Greenside discretization scheme (see Section 3). This time step size was determined to be small enough so that discretization errors were insignificant relative to Monte Carlo errors. An ensemble size of $R = 50$ random walkers is used, with a delay of $J = 70$ time steps before positions are stored, and a gap of $j = 12$ time steps between successively stored positions (see Section 5.4).

Standard pseudorandom Diffusion Monte Carlo is compared with quasirandom Diffusion Monte Carlo using continuation (with 2, 4, and 12 continuation steps) and renumbering (with each dimension divided into 64 equally spaced levels between -5 and +5 for a total of 262144 subregions in three dimensions).

Figure 1 is a plot of the error, $\hat{e}_{rms}(N)$, computed over $R = 60$ repetitions, as a function of N on a log-log scale. The log-log scale allows the plot of an error of the form $\epsilon = b N^a$ to appear as a straight line with slope a. We expect a convergence rate of $a = -0.5$ for standard Monte Carlo, and a lower (or faster) rate for quasi-Monte Carlo.

Figure 2 is a similar plot of the error against average cpu time instead of N. The graph is intended to provide a more realistic comparison of the various estimates of E_0.

6.2.1 Discussion of Results: Example 1

We see in Figures 1 and 2 that quasirandom Diffusion Monte Carlo clearly outperforms standard Diffusion Monte Carlo for this example. The convergence rate of the error improves from close to $O(N^{-1/2})$ for the standard method to $O(N^{-0.641})$ for the quasirandom method, with this rate steadily improving as the number of continuation steps is increased.

Figure 2 makes it clear that the extra cpu time required for renumbering and continuation is more than made up for by the reduced errors. For an error level of about 10^{-4}, we see a reduction in cpu time by a factor of about 5 for quasirandom Diffusion Monte Carlo with either 4 or 12 continuation steps.

We also observe that taking any more than 12 continuation steps would probably not be worthwhile since in Figure 2 the results using 12 steps, in terms of cpu time, are comparable to the results using just 4 steps.

6.3 Helium Atom

For the Helium atom, a number of additional features are incorporated into the quasirandom simulations. First of all, in order to reduce storage requirements by nearly half,

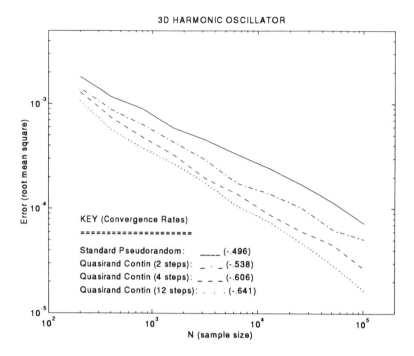

Figure 1: Log-Log Plot for Example 1.

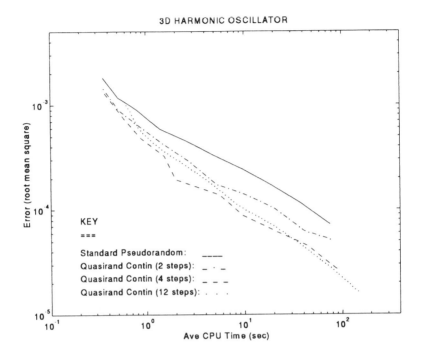

Figure 2: Log-Log Plot (Timing) for Example 1.

some symmetries are exploited. Given configuration. $\boldsymbol{y} = (x_1, y_1, z_1, x_2, y_2, z_2)$. the potential function and therefore the wavefunction solution is only dependent on r_1, r_2, and r_{12}. the distance from the first electron to the nucleus. the second electron to the nucleus, and between the electrons, respectively. Therefore, when the pseudorandom endpoints are stored to be reused for continuation. only the triplet r_1, r_2, r_{12} is saved rather than all six dimensions. The same thing is done for each of the continuation steps as well.

During the continuation steps. for each path, the missing angles are sampled quasirandomly at each time step after retrieving the three distances from storage and before taking the next time step. This reduces the storage required by nearly a half, and it also, importantly, simplifies the renumbering process because instead of requiring subregions over $6D$ space, subregions over only $3D$ space are necessary. This use of symmetry does not introduce any new approximation errors into the results, and should be applicable to many quantum mechanical systems with various symmetries.

A second feature of the Helium atom simulations is that in the renumbering process, an adaptive method is used to select the location of subregions. Initially the $3D$ space is divided into 4096 subregions by dividing each dimension into 16 levels. Then. depending on where most of the pseudorandom endpoints fall, some of these subregions are further subdivided into 512 smaller subregions (8 more sublevels in each dimension). This method gives us fine subdivisions in regions of greater interest where more of the simulation paths are, and coarser subdivisions in lower interest regions where fewer simulation paths appear.

Example 2 *The ground state energy, E_0, of the Helium atom is estimated using Diffusion Monte Carlo. Experimental results [6] have produced the following estimated value: $E_O = -2.903724$, in atomic units, which shall be considered the 'exact' solution for the purpose of computing errors here. We assume a fixed nucleus located at the origin so that the system is just a two body system ($M = 2$).*

The potential energy function. in atomic units, is the following:

$$v(\boldsymbol{y}) = -\frac{2}{r_1} - \frac{2}{r_2} + \frac{1}{r_{12}}$$

where $\boldsymbol{y} = (x_1, y_1, z_1, x_2, y_2, z_2)$, r_1 is the distance from the first electron to the nucleus, r_2 the distance from the second electron to the nucleus, and r_{12} the distance between the electrons.

The trial function chosen is of the form:

$$\psi_T(\boldsymbol{y}) = \exp(-2r_1)\exp(-2r_2)\exp\left(\frac{a\,r_{12}}{1 + b\,r_{12}}\right)$$

with $a = 0.5$ and $b = 0.2$, which ensures that trial energy, $b(\boldsymbol{y})$, remains finite at $r_1 = 0$, $r_2 = 0$, and $r_{12} = 0$ [13]. The resultant trial energy function is then the following:

$$b(\boldsymbol{y}) = -4 + \frac{1}{r_{12}} - \frac{1}{r_{12}(1 + b\,r_{12})^3} - \frac{1}{4(1 + b\,r_{12})^4} + \frac{(r_1 + r_2)(1 - r_1 \cdot r_2/r_1 r_2)}{r_{12}(1 + b\,r_{12})^2}$$

While the vector drift function is the following:

$$\boldsymbol{a}(\boldsymbol{y}) = \frac{\boldsymbol{y} - \boldsymbol{y}^*}{2r_{12}(1 + b\,r_{12})^2} - \frac{2\boldsymbol{y}}{r_1}$$

where $\boldsymbol{y}^* = (x_2, y_2, z_2, x_1, y_1, z_1)$.

For the simulations, the time steps are of size $h = 0.03$ *and the Helfand-Greenside discretization scheme is used. As in Example 1, the time step size was chosen to be small enough so that discretization errors were insignificant relative to Monte Carlo errors. An ensemble size of* $R = 100$ *random walkers is used, with a delay of* $J = 500$ *time steps before positions are stored, and a gap of* $j = 50$ *time steps between successively stored positions.*

Standard pseudorandom Diffusion Monte Carlo is compared with quasirandom Diffusion Monte Carlo using continuation (with 10 and 20 continuation steps) and renumbering (using the adaptive technique described above).

Figure 3 is a plot of the error, $e_{rms}(N)$*, computed over* $R = 60$ *repetitions, as a function of N on a log-log scale. As in Example 1, the log-log scale allows the plot of an error of the form* $e = bN^a$ *to appear as a straight line with slope a. We expect a to be near* -0.5 *for standard Monte Carlo and lower for quasi-Monte Carlo. Figure 4 is a similar plot of the error against average cpu time instead of N.*

6.3.1 Discussion of Results: Example 2

The results for Example 2 are similar to those for the first example. We see in Figures 3 and 4 that quasirandom Diffusion Monte Carlo still clearly outperforms standard Diffusion Monte Carlo, although not to as large an extent as in Example 1. This is not surprising since this system is of size $M = 2$ while the first was $M = 1$, as discussed in Section 5.2. The convergence rate of the error improves from close to $O(N^{-1/2})$ for the standard method, as expected, to $O(N^{-0.596})$ for the quasirandom method with 20 continuation steps. The level of improvement is somewhat disappointing given the hope for rates closer to $O(N^{-1})$ based on discrepancy theory for quasirandom sequences [25]. However the problems being dealt with here are far more complicated than simple integration estimates, and discontinuities are introduced by reordering and continuation which tend to lower the convergence rate [23]. This may be looked upon as an indication of the difficulties involved in translating theoretical predictions into practical results. As the techniques introduced here are improved upon one would expect the convergence rate to move closer to $O(N^{-1})$ with consequent further reductions in the error as a function of cpu time. It is encouraging that even given the modest gain in convergence rate for this example, there is still a significant gain in terms of cpu time as indicated below.

As in Example 1, the plot against cpu time shows that the extra cpu time required for renumbering and continuation is more than made up for by the reduced errors. For the Helium atom, at an error level of about 10^{-3} we see a reduction in cpu time by a factor of almost 3 for quasirandom Diffusion Monte Carlo with 20 continuation steps.

It was found that taking more than 20 steps was not profitable in terms of error reduction versus extra cpu time for this example.

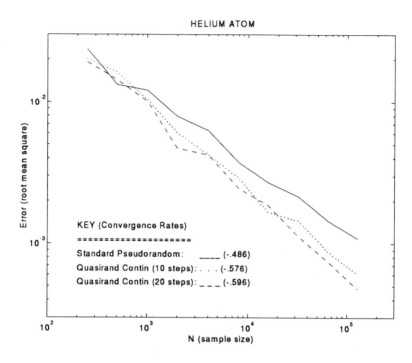

Figure 3: Log-Log Plot for Example 2.

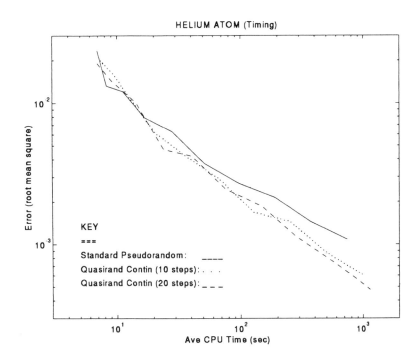

Figure 4: Log-Log Plot (Timing) for Example 2.

7 Conclusions

The results of the two examples above demonstrate that quasirandom Diffusion Monte Carlo is not only feasible, but can actually produce more accurate results in a given amount of cpu time than standard Diffusion Monte Carlo. It is the two techniques discussed above, renumbering and continuation, which make this possible.

There are many possible directions for further work in this area. One would be to better understand how and why renumbering works, and to prove conclusively that it does work for some well defined class of problems. Another would be to apply the methods described here to more complicated quantum mechanical systems involving perhaps 3 or 4 bodies in motion. A third direction would be to adapt the methods here to fermion problems in which the wavefunction changes sign, introducing new difficulties, as mentioned in Section 2.2. In addition, the problem of quasirandom error measurement needs to be further addressed, as mentioned in Section 6.1.

Finally, the results presented here should demonstrate that quasirandom sequences can perhaps be beneficial for a far wider range of Monte Carlo applications than presently thought possible.

References

[1] J. B. Anderson and C. A. Traynor. Quantum chemistry by random walk: Exact treatment of many-electron systems. Research Report, May 1991.

[2] M. Berblinger and C. Schlier. Monte Carlo integration with quasi-random numbers: Some experience. *Computer Physics Communications*, 66:157–166, 1991.

[3] P. Bratley, B. L. Fox, and H. Niederreiter. Implementation and tests of low-discrepancy sequences. *ACM Transactions on Modeling and Computer Simulation*, 2(3):195–213, July 1992.

[4] D. Ceperly and B. Alder. Quantum Monte Carlo for molecules: Green's function and nodal release. *Journal of Chemical Physics*, 81(12):5833–5844, 1984.

[5] D. Ceperly and B. Alder. Quantum Monte Carlo. *Science*, 231:555–560, Feb. 1986.

[6] S. A. Chin. Quadratic diffusion Monte Carlo algorithms for solving atomic many-body problems. *Physical Review A*, 42(12):6991–7005, Dec. 1990.

[7] B. L. Fox. Implementation and relative efficiency of quasirandom sequence generators. *ACM Transactions of Mathematical Software*, 12(4):362–376, Dec. 1986.

[8] H. Greenside and E. Helfand. Numerical integration of stochastic differential equations - II. *The Bell System Technical Journal*, pages 1927–1941, Oct. 1981.

[9] R. Grimm and R. Storer. Monte Carlo solution of Schrödinger's equation. *Journal of Computational Physics*, 7:134–156, 1971.

[10] J. Halton. On the efficiency of certain quasi-random sequences of points in evaluating multi-dimensional integrals. *Numerische Mathematik*, 2:84–90, 1960.

[11] S. Heinrich and A. Keller. Quasi-monte carlo methods in computer graphics, part II: The radiance equation. Technical Report 243/94, Fachbereich Informatik, AG Numerische Algorithmen, Universität Kasierslautern, 1994.

[12] E. Helfand. Numerical integration of stochastic differential equations. *The Bell System Technical Journal*, pages 2289–2299, Dec. 1979.

[13] C. Joslin and S. Goldman. Quantum Monte Carlo studies of two-electron atoms constrained in spherical boxes. *Journal of Physics B - Atomic, Molecular, and Optical Physics*, 25(9):1965–1975, 1992.

[14] M. H. Kalos and P. A. Whitlock. *Monte Carlo Methods, Volume I: Basics*. J. Wiley and Sons, New York, 1986.

[15] P. E. Kloeden and E. Platen. *Numerical Solution of Stochastic Differential Equations*. Springer-Verlag, Berlin; New York, 1992.

[16] S. E. Koonin. *Computational Physics*. Benjamin/Cummings, Menlo Park, California, 1986.

[17] C. Lecot. A quasi-Monte Carlo method for the Boltzmann equation. *Mathematics of Computation*, 56(194):621–644, Apr. 1991.

[18] M. E. Lee and K. E. Schmidt. Green's Function Monte Carlo. *Computers in Physics*, pages 192–197, March/April 1992.

[19] Y. L. Levitan, N. Markovich, S. Rozin, and I. Sobol´. Short communications on quasirandom sequences for numerical computations. *Zh. vychisl. Mat. mat. Fiz.*, 28(5):755–759, 1988.

[20] G. Marsaglia. Normal (Gaussian) random variables for supercomputers. *The Journal of Supercomputing*, 5:49–55, 1991.

[21] W. Morokoff and R. Caflisch. A quasi-Monte Carlo approach to particle simulation of the heat equation. *SIAM Journal on Numerical Analysis*, to appear, 1993.

[22] W. J. Morokoff. *Quasi-Monte Carlo Methods for Numerical Integration and Simulation*. PhD thesis, New York University, May 1990.

[23] B. Moskowitz. Improved stochastic simulation using quasi-Monte Carlo: A computational study. *Mathematical and Computer Modelling*, submitted, Aug. 1994.

[24] B. Moskowitz and R. E. Caflisch. Smoothness and dimension reduction in quasi-Monte Carlo methods. *Mathematical and Computer Modelling*, submitted, Mar. 1994.

[25] H. Niederreiter. *Random Number Generation and Quasi-Monte Carlo Methods*. SIAM, Philadelphia, 1992.

[26] D. O'Brien. Accelerated quasi Monte Carlo integration of the radiative transfer equation. *J. Quant. Spectrosc. Radiat. Transfer*, 48(1):41–59, 1992.

[27] B. Oksendal. *Stochastic Differential Equations*. Springer-Verlag, New York, 1992.

[28] W. H. Press and S. A. Teukolsky. Quasi- (that is, sub-) random numbers. *Computers in Physics*, pages 76–79, Nov/Dec 1988.

[29] P. J. Reynolds, J. Tobochnik, and H. Gould. Diffusion quantum Monte Carlo. *Computers in Physics*, pages 662–668, Nov/Dec 1990.

[30] B. Shuhman. Application of quasirandom points for simulation of gamma radiation transfer. *Progress in Nuclear Energy*, 24:89–95, 1990.

[31] I. Sobol´. Quasi-Monte Carlo methods. *Progress in Nuclear Energy*, 24:55–61, 1990.

[32] J. Spanier, June 1994. presentation at Las Vegas Conference on Monte Carlo and Quasi-Monte Carlo Methods in Scientific Computing.

[33] C. A. Traynor, J. B. Anderson, and B. M. Boghosian. A quantum Monte Carlo calculation of the ground state energy of the hydrogen molecule. *Journal of Chemical Physics*, 94(5):3657–3664, Mar. 1991.

[34] S. Zhang and M. Kalos. Bilinear quantum Monte-Carlo - expectations and energy differences. *Journal of Statistical Physics*, 70(3-4):515–533, Feb. 1993.

Randomly Permuted (t, m, s)-Nets and (t, s)-Sequences

Art B. Owen

Dept. of Statistics, Sequoia Hall
Stanford University
Stanford, CA 94305, U.S.A.

E-mail: owen@playfair.stanford.edu

Abstract: This article presents a hybrid of Monte Carlo and Quasi-Monte Carlo methods. In this hybrid, certain low discrepancy point sets and sequences due to Faure, Niederreiter and Sobol' are obtained and their digits are randomly permuted. Since this randomization preserves the equidistribution properties of the points it also preserves the proven bounds on their quadrature errors. The accuracy of an estimated integrand can be assessed by replication, consisting of independent re-randomizations.

The hybrid method is applied to two published sets of test integrands. The results on the larger set of test integrands suggests that the randomized sequences tend to give more accurate integral estimates than the original unpermuted points.

The method presented here is a further step in a sequence of methods from the statistical literature, that starts with Latin hypercube sampling and develops through randomized orthogonal arrays and orthogonal array based Latin hypercube sampling.

1 Introduction

The problem at hand is to calculate an estimate of $I = \int_{[0,1)^s} f(X) dX$ where f is an integrable function on the s dimensional unit cube $[0, 1)^s$. When s is small and f is smooth, special techniques described in Davis and Rabinowitz (1984) can achieve great accuracy with a small number of function evaluations. As s increases, Monte Carlo and Quasi-Monte Carlo methods become competitive, and ultimately dominant. Davis and Rabinowitz (1984, Chap 5.10) consider $s > 15$ to be a high enough dimensionality that "sampling or equidistribution methods are indicated".

1.1 Monte Carlo versus Quasi-Monte Carlo

Quasi-Monte Carlo methods are often more accurate than Monte Carlo methods, as evidenced by some asymptotics and by some numerical results. Unfortunately, it can be difficult to assess the accuracy attained by a Quasi-Monte Carlo approximation. Monte Carlo methods are often less accurate than Quasi-Monte Carlo methods, but it is ordinarily easy to assess the accuracy of a Monte Carlo approximation. One therefore has to trade off conflicting goals: getting an accurate estimate and knowing how accurate the estimate is likely to be. This paper presents a hybrid method which randomizes the equidistribution methods known as (t, m, s)-nets and (t, s)-sequences. Repeated independent randomizations provide a basis for estimating the accuracy of the integral.

Let $U[0, 1)^s$ denote the uniform distribution on $[0, 1)^s$. The simplest Monte Carlo method uses n independent random s vectors $X_1, \ldots, X_n$ from $U[0, 1)^s$. The estimate is $\hat{I}_{IND} = n^{-1} \sum_{i=1}^{n} f(X_i)$. By the strong law of large numbers, $\hat{I}_{IND} \to I$ as $n \to \infty$ with probability 1. If also $\int_{[0,1)^s} f(X)^2 dX < \infty$, then by the central limit theorem $n^{1/2}(\hat{I}_{IND} - I) \to N(0, \sigma^2)$ in distribution as $n \to \infty$, where $\sigma^2 = \int_{[0,1)^s} (f(X) - I)^2 dX$.

These facts underlie the main features of Monte Carlo methods. The error of simple Monte Carlo integration is of order $n^{-1/2}$ in probability. The asymptotic magnitude of the error does not depend explicitly on the dimension s of the integrand. It depends instead on σ^2. Moreover, it is easy to estimate the accuracy of Monte Carlo methods, using a sample based estimate of σ^2, usually $\hat{\sigma}^2 = (n-1)^{-1} \sum_{i=1}^{n} (f(X_i) - \hat{I}_{IND})^2$.

More sophisticated Monte Carlo methods can be used to reduce the variance of the estimated integral. Some examples are importance sampling, antithetic variables and control variates. Ripley (1987) gives a broader discussion of Monte Carlo methods. Ripley (1987, Chap. 5) considers variance reduction techniques.

To describe equidistribution methods, let $\hat{F}_n$ be the probability measure which assigns equal probability n^{-1} to each of $X_1, \ldots, X_n \in [0, 1)^s$. Then we may write

$$\hat{I} = \int_{[0,1)^s} f(X) d\hat{F}_n$$

and $\hat{I}$ will be close to I if the discrete uniform distribution $\hat{F}_n$ is close to the continuous uniform distribution $U[0, 1)^s$, and f is reasonably well behaved. This heuristic is quantified by the Koksma-Hlawka inequality

$$|\hat{I} - I| \leq V_{HK}(f) D_n^*(F, \hat{F}_n)$$

where D_n^* is the star discrepancy of $X_1, \ldots, X_n$ and $V_{HK}(f)$ is the total variation of f in the sense of Hardy and Krause. D_n^* is the greatest absolute difference between the continuous uniform probability F and the discrete uniform probability $\hat{F}_n$, taken over all subcubes of $[0, 1)^s$ containing the origin. It is an s dimensional generalization of the Kolmogorov-Smirnov distance between two distribution functions. See Niederreiter (1992, Chap. 2) for the definition of D_n^* and other discrepancy measures and for the definition of V_{HK}.

The integration error is thus bounded by the product of a quantity measuring how uniform $X_1, \ldots, X_n$ are and a quantity measuring how rough f is.

It is possible to construct point sets $X_1, \ldots, X_n$ such that $D_n^*(F, \hat{F}_n) = O((\log n)^{s-1}/n)$ and infinite sequences $(X_i)_{i \geq 1}$ such that for the first n points of the sequence $D_n^*(F, \hat{F}_n) = O((\log n)^s/n)$. Thus functions of bounded variation in the sense of Hardy and Krause (BVHK functions), can be integrated over the unit cube with errors of order $(\log n)^{s-1}/n$ or smaller, which is a superior rate to $n^{-1/2}$.

Note that $\log n$ raised to a moderately large power does not become negligible compared to n until n is very large. For example $(\log n)^9/n > n^{-1/2}$ for $n = 10^{34}$. Despite this, some empirical investigations, especially in problems stemming from physics, have found equidistribution methods to be superior to Monte Carlo methods for certain test functions in moderate dimensions. That is, equidistribution methods appear to be more accurate than the Koksma-Hlawka bound. The numerical results in this paper also bear this out.

A disadvantage of equidistribution methods is that it can be difficult to get an estimate or an upper bound for $V_{HK}(f)$.

1.2 Hybrid Methods

The idea of combining Monte Carlo and Quasi-Monte Carlo has been used before. Cranley and Patterson (1976) randomize the origin in the number theoretic method. Braaten and Weller (1979) randomly scrambled the Halton sequence. Joe (1990) randomizes some lattice rules.

Recent statistical work stemming from computer experiments may also be viewed as combining Monte Carlo and Quasi-Monte Carlo methods. McKay, Conover and Beckman (1979) introduce Latin hypercube sampling, a form of Monte Carlo that is stratified along each of the s coordinate axes in $[0,1)^s$. Owen (1992b, 1994a) extends this to randomized orthogonal arrays. These are point sets that appear as regular grids in any r dimensional coordinate projection but are otherwise randomized. Here r, usually 2 or 3, is the strength of the orthogonal array used. Independently, Tang (1993) presents orthogonal array based Latin hypercube samples that combine univariate stratification of Latin hypercube samples with r variate stratification of orthogonal arrays. This work presents a further step in this direction. When using $(0, m, s)$-nets and $(0, s)$-sequences in base b, it allows simultaneous stratification on all r variate coordinate projections, for all $r \geq 1$ for which n is a multiple of b^r.

1.3 Outline and Further Developments

Section 2 gives basic definitions and notation for this article, and describes a sequence of Monte Carlo by Quasi-Monte Carlo hybrid methods arising from the statistical theory of computer experiments. Section 3 describes the randomized (t, m, s)-nets and (t, s)-sequences, proves some elementary properties of them (uniformity and equidistribution) and discusses some practical issues in their computation. Section 4 presents some numerical examples using test functions of Roos and Arnold (1963) and Genz (1984) in 10 dimensions. Independent replication of the randomization provides a reasonable assessment of uncertainty on the Roos and Arnold integrands. On the more comprehensive Genz integrands, it even appears that randomization improves the accuracy of the low discrepancy points used.

The Monte Carlo variance of this hybrid method is investigated in Owen (1994b). A square integrable integrand f is expressed in a multidimensional multiresolution (wavelet) expansion, in which the coarsest terms are integrated exactly by a randomized (t, m, s)-net, while finer terms contribute to the sampling variance. It is shown there that for any f with $\sigma^2 < \infty$, the sampling variance of $\hat{I}$ is $o(n^{-1})$ along the sample size sequence, $n = \lambda b^k$, $1 \leq \lambda < b$, $k \geq 1$. Moreover for a scrambled $(0, m, s)$-net in base $b \geq \max(s, 2)$, the sampling variance of $\hat{I}$ is not larger than $\sigma^2 (b/(b-1))^{\min(s-1,m)} n^{-1}$, with this holding uniformly in f.

2 Notation and Definitions

This section gives a brief outline of several background areas used in this paper, in order to make a self contained presentation. References are provided for readers who desire more comprehensive coverage.

We use f to denote the integrand, and $[0,1)^s$ is the domain of f, where s is a positive integer. Integrals written without an explicit domain of integration are assumed to be over $[0,1)^s$. The domains $(0,1)^s$ and $[0,1]^s$ could equally well have been used. The convention of using intervals closed on the left and open on the right makes it easier to partition a cube like $[0,1)^s$ into subcubes.

A generic point in $[0,1)^s$ will be denoted by X. The components of X are denoted by superscripts, that is $X = (X^1, \ldots, X^s)$. Integration rules considered here use n points $X_1, \ldots, X_n$ from $[0,1)^s$. We write $X_i = (X_i^1, \ldots, X_i^s)$. The s-dimensional (Lebesgue) volume of a (measurable) subset $E \subseteq [0,1)^s$ will be denoted by $\lambda_s(E)$.

All of the integration rules considered here use $\hat{I} = n^{-1} \sum_{i=1}^n f(X_i)$ to approximate $I = \int f(X)dX$. The rules differ only in how the X_i are chosen, and this is indicated by subscripting $\hat{I}$.

A sequence of points in $[0,1)^s$, finite or infinite, will be denoted by (X_i). Notation such as $(X_i)_{i \geq 1}$ or $(X_i)_{i=1}^n$ will also be used.

2.1 ANOVA Decomposition of $L^2 : [0,1)^s$

If the integrand f is measurable and $\int f(X)^2 dX < \infty$ then we may define the analysis of variance (ANOVA) decomposition of f given below. See Efron and Stein (1981) or Owen (1992b, 1994a), or Stein (1987) for more details. The article by Hickernell in this volume also discusses the ANOVA decomposition of functions.

Let $u \subseteq \{1, 2, \ldots, s\}$ be a subset of the coordinates of $[0,1)^s$, let $|u|$ be the cardinality of u, and let $\bar{u}$ denote the complement $\{1, 2, \ldots, s\} - u$. For $X \in [0,1)^s$, let X^u denote those coordinates of X indexed by elements of u and write $[0,1)^u$ for the domain of X^u. This allows us to distinguish among coordinate projections of $[0,1)^s$ having the same dimension $|u|$. The ANOVA decomposition of f takes the form

$$f(X) = \sum_{u \subseteq \{1,2,\ldots,s\}} \alpha_u(X)$$

where the sum is over all 2^s subsets of coordinates of $[0,1)^s$. The terms in this decomposition are defined recursively starting with

$$\alpha_\emptyset(X) = \int f(Z)dZ = I$$

and using the rule

$$\alpha_u(X) = \int_{Z^u = X^u, Z^{\bar{u}} \in [0,1)^{\bar{u}}} \left(f(Z) - \sum_{v \subset u} \alpha_v(Z) \right) \prod_{j \in \bar{u}} dZ^j \tag{1}$$

The sum in (1) is over proper subsets $v \neq u$. Each α_u gives the joint effect on f of the variables X^u, averaged over the possible values of $X^{\bar{u}}$, after removing all effects due to proper subsets of the variables in X^u.

The function α_u only depends on X through X^u. It is convenient to abuse notation somewhat and consider α_u to be defined on either $[0,1)^s$ or $[0,1)^u$ depending on circumstances. We note some properties of this decomposition: $j \in u$ implies $\int_0^1 \alpha_u(X)dX^j = 0$ for any fixed values of $X^k, k \neq j$, $\int \alpha_u(X)\alpha_v(X)dX = 0$ for $u \neq v$, $\int f(X)^2 dX = \sum_u \int \alpha_u(X)^2 dX$ and

$$\sigma^2 = \int (f(X) - \alpha_\emptyset)^2 dX = \sum_{|u|>0} \int \alpha_u(X)^2 dX. \tag{2}$$

Expression (2) is an analogue of the ANOVA decomposition from the statistical theory of experimental design. The variation in f is partitioned among sources u which have effects α_u with mean square $\int \alpha_u(X)^2 dX$.

The notation α_j will be used as a shortcut for $\alpha_{\{j\}}$. The function $f_{\text{add}} = \alpha_\emptyset + \sum_{j=1}^s \alpha_j(X^j)$ is called the additive part of f.

2.2 (t, m, s)-nets and (t, s)-sequences

This subsection gives a brief outline of elementary intervals, (t, m, s)-nets and (t, s)-sequences. For more details see Niederreiter (1992, Chap. 4).

Let $s \geq 1$ and $b \geq 2$ be integers. An elementary interval in base b is a subinterval of $[0,1)^s$ of the form

$$E = \prod_{j=1}^s \left[\frac{a_j}{b^{l_j}}, \frac{a_j+1}{b^{l_j}} \right)$$

for integers l_j, a_j with $l_j \geq 0$ and $0 \leq a_j < b^{l_j}$. Clearly E has volume $\lambda_s(E) = b^{-\sum_{j=1}^s l_j}$.

Let $m \geq 0$ be an integer. A finite sequence of b^m points from $[0,1)^s$ is a $(0, m, s)$-net in base b if every elementary interval in base b of volume b^{-m} contains exactly one point of the sequence. In other words, every elementary interval that "should" have one point of the sequence does have one point of the sequence.

Let $t \leq m$ be a nonnegative integer. A finite sequence of b^m points from $[0,1)^s$ is a (t, m, s)-net in base b if every elementary interval in base b of volume b^{t-m} contains exactly b^t points of the sequence. Elementary intervals that "should" have b^t points do have b^t points, though when $t > 0$ elementary intervals of volume b^{-m} don't necessarily contain exactly 1 point from the sequence. Smaller values of t imply greater equidistribution.

For $t \geq 0$, an infinite sequence of points $X_1, X_2, \ldots$ is a (t, s)-sequence in base b if for all $k \geq 0$ and $m \geq t$ the finite sequence $X_{kb^m+1}, \ldots, X_{(k+1)b^m}$ is a (t, m, s)-net in base b. As with (t, m, s)-nets, smaller values of t imply greater equidistribution. In both cases, taking $m = t$ corresponds to the trivial statement that all the points of (X_i) are in $[0,1)^s$.

If $(X_i)_{i=1}^n$ is a (t, m, s)-net in base b then $D_n^* = O(\log(n)^{s-1}/n)$, and if $(X_i)_{i\geq 1}$ is a (t, s)-net in base b then $D_n^* = O(\log(n)^s/n)$. Niederreiter (1992, Theorems 4.10 and 4.17) gives more precise statements of these facts.

Constructions of (t, m, s)-nets and (t, s)-sequences in base 2 were given by Sobol' (1967). These are described by Niederreiter (1978). Faure (1982) constructs $(0, m, s)$-nets and $(0, s)$-sequences in prime bases $p \geq s$. Niederreiter (1987) provides $(0, m, s)$-nets and $(0, s)$-sequences in prime power bases $p^r \geq s$, and more generally, for bases that are products of prime powers, with each prime power $\geq s$. Niederreiter (1988) provides some constructions of (t, m, s)-nets and (t, s)-sequences for $t \geq 0$.

2.3   Latin Hypercubes

Latin hypercube sampling and extensions described here, originated in the context of computer experiments. This area of research considers how one might choose good sets of points for exploring the input space of a computer program. For an introduction to the design and analysis of computer experiments, see Sacks, Mitchell, Welch and Wynn (1989).

In a Latin hypercube sample,

$$X_i^j = \frac{\pi_j(i) + U_i^j}{n}, \tag{3}$$

where π_j is a random permutation of $\{0, \ldots, n-1\}$ uniformly distributed over the $n!$ possible permutations, U_i^j is a $U[0,1)$ random variable and all the π_j and U_i^j are mutually independent. A common variant of Latin hypercube sampling takes

$$X_i^j = \frac{\pi_j(i) + 1/2}{n}. \tag{4}$$

Version (3) was introduced by McKay, Conover and Beckman (1979) for computer experiments. In this version each X_i^j has the $U[0,1)$ distribution. Under version (4), which appears in Patterson (1954), each X_i^j has the discrete uniform distribution on $1/2n, 3/2n, \ldots, (2n-1)/2n$.

Latin hypercube samples have a mild form of equidistribution. Consider an interval of the form $E = [a/n, (a+1)/n)$ where a is an integer and $0 \le a < n$. Then for each $j = 1, \ldots, s$, the interval E contains exactly one of the points $X_1^j, \ldots, X_n^j$. Under (4), the point is at the center of E, while under (3), the point is uniformly distributed within E.

This equidistribution reduces the Monte Carlo variance of a Latin hypercube sample. Let $\hat{I}_{LHR}$ denote $n^{-1} \sum_{i=1}^{n} f(X_i)$ for X_i generated by (3), use $\hat{I}_{LHC}$ for X_i generated by (4) and use $\hat{I}_{LH}$ when it is not necessary to distinguish the methods. Using the ANOVA decomposition

$$\hat{I}_{LH} = I + \sum_{j=1}^{s} \frac{1}{n} \sum_{i=1}^{n} \alpha_j(X_i^j) + \sum_{|u|>1} \frac{1}{n} \sum_{i=1}^{n} \alpha_j(X_i).$$

Under LHC, $n^{-1} \sum_{i=1}^{n} \alpha_j(X_i^j)$ is an n point midpoint rule estimate of $\int_0^1 \alpha_j(x)dxj = 0$. Such a rule has an error of order $O(n^{-2})$ if its integrand has a bounded second derivative and an error of order $O(n^{-1})$ with a bounded first derivative. Under LHR, $n^{-1} \sum_{i=1}^{n} \alpha_j(X_i^j)$ is a one dimensional Riemann integral estimate of $\int_0^1 \alpha_j(x)dx = 0$. In either case, the error in integrating the additive part of f is of smaller order than $n^{-1/2}$. Stein (1987) shows that $\text{Var}(\hat{I}_{LHR}) = n^{-1} \left(\sigma^2 - \sum_{j=1}^{s} \int \alpha_j(X)^2 dX \right) + o(n^{-1})$. Owen (1992a) shows that $n^{1/2}(\hat{I}_{LHR} - I)$ has an asymptotic normal distribution as $n \to \infty$, when f is bounded.

2.4   Orthogonal Arrays

Working independently, Owen (1992b,1994a) and Tang (1993) have extended the variance reduction property of Latin hypercube sampling to higher order terms of the ANOVA decomposition of f. Both used orthogonal arrays.

An orthogonal array is an n by s matrix A of integers between 0 and $b-1$ inclusive. The array A is said to be of strength $r \leq s$ if in every n by r submatrix of A, each of the b^r possible rows appears the same number, $\lambda \geq 1$, of times. Of course $n = \lambda b^r$. The array is denoted by $OA(n, s, b, r)$. See Raghavarao (1971) for more details on orthogonal arrays.

Owen (1992b) proposed the use of randomized orthogonal arrays for computer experiments and numerical integration, taking

$$X_i^j = \frac{\pi_j(A_i^j) + U_i^j}{b}$$

or

$$X_i^j = \frac{\pi_j(A_i^j) + 1/2}{b}$$

where the π_j are uniform random permutations of $0, \ldots, b-1$, the U_i^j are $U[0, 1)$ random variables and all π_j, U_i^j are independent. Call the corresponding integral estimates $\hat{I}_{OAR}$ and $\hat{I}_{OAC}$. Latin hypercube samples correspond to using $OA(n, s, n, 1)$.

The points of a randomized orthogonal array have an r dimensional equidistribution property. Let u be a subset of $\{1, \ldots, s\}$ with $|u| = r$. Every interval of the form $\prod_{j \in u}[a_j b^{-1}, (a_j + 1)b^{-1}) \prod_{j \notin u}[0, 1)$, where the a_j are integers with $0 \leq a_j < b$, contains exactly λ points of the randomized orthogonal array.

Owen (1992b) shows that $\hat{I}_{OAC}$ integrates terms α_u, $|u| \leq r$ with the accuracy of a b level product midpoint rule. For smooth α_u, and α_u may be smoother than f, this yields errors of size $O(b^{-2}) = O(n^{-2/r})$, as $b \to \infty$ through a family of orthogonal arrays with constant λ. Thus for $r < 4$ these terms are integrated with better than Monte Carlo accuracy. Owen (1992b) also considers mapping the levels of the orthogonal array onto a certain multipanel Gauss rule, which for smooth α_u leads to errors of size $O(b^{-3}) = O(n^{-3/r})$.

Owen (1994a) shows that $\text{Var}(\hat{I}_{OAC}) \doteq n^{-1} \sum_{|u|>r} \int \alpha_u^2 dX$ for square integrable f and many common families of orthogonal arrays.

Tang (1993) proposed orthogonal array based Latin hypercube sampling. An OA-based Latin hypercube is one for which $\lfloor bX_i^j \rfloor$ form an orthogonal array $OA(n, s, b, r)$, where $\lfloor x \rfloor$ is the greatest integer less than or equal to x. It thus enjoys the r dimensional equidistribution property of orthogonal arrays, and the one dimensional equidistribution property of Latin hypercube samples. Tang shows that for continuous f and $r = 2$ that $\text{Var}(\hat{I}_{OALH}) = \left(\sigma^2 - \sum_{j=1}^{s} \int \alpha_j(X)^2 dX - \sum_{j<k} \int \alpha_{\{jk\}}(X)^2 dX\right) + o(n^{-1})$.

Latin hypercube and orthogonal array based schemes remove some terms from the Monte Carlo variance, but still produce an error of order $n^{-1/2}$. They may be replicated independently to estimate this reduced variance. But adding replicates does not improve the order of accuracy. That is fixed by the choice of t, the strength of the array, and b, the number of levels used.

By contrast (t, s)-sequences provide estimates with better convergence rates than $n^{-1/2}$, and have equidistribution properties that grow stronger as the number of evaluations increases through natural sample sizes.

If $(X_i)_{i=1}^{n}$ is a (t, m, s)-net in base b, then for positive integers $q \leq m - t$, $\lfloor b^q X_i^j \rfloor$ form an orthogonal array $OA(n, s, b^q, \lfloor (m-t)/q \rfloor)$.

3 Randomized (t, m, s)-nets and (t, s)-sequences

Let (A_i) be a (t, m, s)-net or a (t, s)-sequence in base b. The i'th term in the sequence is denoted $A_i = (A_i^1, \ldots, A_i^s)$. We may write the components of A_i in their base b expansion

$$A_i^j = \sum_{k=1}^{\infty} a_{ijk} b^{-k},$$

where $0 \leq a_{ijk} < b$ for all i, j, k.

A randomized version of (A_i) is a sequence (X_i) with elements $X_i = (X_i^1, \ldots, X_i^s)$ written as

$$X_i^j = \sum_{k=1}^{\infty} x_{ijk} b^{-k},$$

with x_{ijk} defined in terms of random permutations of the a_{ijk}

$$
\begin{aligned}
x_{ij1} &= \pi_j(a_{ij1}) \\
x_{ij2} &= \pi_{ja_{ij1}}(a_{ij2}) \\
x_{ij3} &= \pi_{ja_{ij1}a_{ij2}}(a_{ij3}) \\
&\vdots \\
x_{ijk} &= \pi_{ja_{ij1}a_{ij2}\cdots a_{ijk-1}}(a_{ijk}).
\end{aligned}
$$

Each permutation π is uniformly distributed over the $b!$ permutations of $\{0, 1, \ldots, b-1\}$, and the permutations are mutually independent. Here π_j permutes the first digit in the base b expansion of A_i^j for all i. The second digit is permuted by $\pi_{ja_{ij1}}$. The permutation applied to the second digit a_{ij2} depends on value of the first digit a_{ij1}. Similarly, the permutation applied to the k'th digit a_{ijk} depends on values of the first $k - 1$ digits a_{ij1} through a_{ijk-1}. Let π^{-1} denote the permutation which is the inverse of π.

Let $\hat{I}_{RS} = n^{-1} \sum_{i=1}^{n} f(X_i)$ where X_i are the first n points of a (t, s)-sequence in base b and let $\hat{I}_{RN}$ be similarly defined with X_i the points of a (t, m, s)-net in base b. We show below that this randomization preserves the equidistribution properties of (t, m, s)-sequences and (t, s)-nets. Therefore if $f \in BVHK$

$$|\hat{I}_{RS} - I| = O((\log n)^s/n), \quad n \to \infty. \tag{5}$$

We also show below that X_i is uniformly distributed on $[0, 1)^s$, and so for any integrable f, the integral estimates are unbiased:

$$E(\hat{I}_{RS}) = E(\hat{I}_{RN}) = I. \tag{6}$$

For $f \in BVHK$ of bounded variation the Monte Carlo variance of $\hat{I}_{RS}$ is $O((\log n)^{2s}/n^2)$ by (5), which is a better asymptotic order than the usual $O(1/n)$.

We recommend using an initial subsequence of a randomly permuted (t, s)-sequence. This allows one to take additional points in such a way as to preserve equidistribution, if the need arises. The natural sample sizes to consider are of the form $n = kb^m$ for $1 \leq k < b$, $m \geq 0$. As k increases from 1 through $b - 1$ the number of points in an elementary interval of volume b^{t-m} is kb^{t-m}. That is one can increase n by less than a factor of b while maintaining equidistribution. The step from $k = b-1$, $m = m_0$ to $k = 1$, $m = m_0 + 1$ maintains equidistribution over all elementary intervals of volume b^{t-m_0} while achieving equidistribution over all elementary intervals of volume b^{t-m_0-1}.

The $(0, s)$-sequences have the strongest equidistribution properties, as assessed by elementary intervals, and these are available in base b for any prime power $b = p^r \geq s$. Ordinarily the smallest such prime power is a good choice, since it minimizes b^m, the smallest sample size for which some equidistribution property appears in every m dimensional coordinate projection. It should be noted that $(0, s)$-sequences in the smallest prime power base $p^r \geq s$ don't always achieve the smallest rates for the discrepancy D_n^* for a given s. See Niederreiter (1992, Chap. 4).

3.1   Equidistribution

Here we show that the randomization described above preserves the defining properties of (t, m, s)-nets and (t, s)-sequences.

Proposition 1.   If (A_i) is a (t, m, s)-net in base b then (X_i) is a (t, m, s)-net in base b with probability 1. If (A_i) is a (t, s)-sequence in base b then (X_i) is a (t, s)-sequence in base b with probability 1.

Proof: Consider an elementary interval in base b

$$E = \prod_{j=1}^{s} \left[\frac{a_j}{b^{l_j}}, \frac{a_j + 1}{b^{l_j}} \right)$$

where $0 \leq a_j < b^{l_j}, j = 1, \ldots, s$. Write $a_j = \sum_{k=1}^{l_j} c_{jk} b^{l_j - k}$ where $0 \leq c_{jk} < b$. Then $X_i \in E$ if and only if

$$0 \leq \sum_{k=1}^{l_j} (x_{ijk} - c_{jk}) b^{l_j - k} + \sum_{k=l_j+1}^{\infty} x_{ijk} b^{l_j - k} < 1, \quad j = 1, \ldots, s. \tag{7}$$

Assume for the present that $\sum_{k=l_j+1}^{\infty} x_{ijk} b^{l_j-k} < 1$ always holds in equation (7). Because $\sum_{k=1}^{l_j} (x_{ijk} - c_{jk}) b^{l_j-k}$ is an integer and $0 \leq \sum_{k=l_j+1}^{\infty} x_{ijk} b^{l_j-k} < 1$, $X_i \in E$ if and only if $x_{ijk} = c_{jk}$ for $k = 1, \ldots, l_j$, $j = 1, \ldots, s$.

Thus $X_i \in E$ if and only if $a_{ijk} = \pi_{ja_{ij1} \cdots a_{ijk-1}}^{-1}(c_{jk})$. Let $\tilde{a}_j = \sum_{k=1}^{l_j} \pi_{ja_{ij1} \cdots a_{ijk-1}}^{-1}(c_{jk}) b^{l_j-k}$, so that $0 \leq c_j < b^{l_j}$. Define the elementary rectangle

$$\tilde{E} = \prod_{j=1}^{s} \left[\frac{\tilde{a}_j}{b^{l_j}}, \frac{\tilde{a}_j + 1}{b^{l_j}} \right).$$

Now $X_i \in E$ if and only if $A_i \in \tilde{E}$. Therefore the number of elements of $(X_i)_{i=1}^n$ in an elementary rectangle E is the same as the number of elements on $(A_i)_{i=1}^n$ in the related elementary rectangle $\tilde{E}$. This equivalence establishes both claims of the proposition, except for the issue of whether $\sum_{k=l_j+1}^{\infty} x_{ijk} b^{l_j-k} < 1$, to be addressed next.

If (A_i) is a (t, m, s)-net then so is (X_i) unless $\sum_{k=l+1}^{\infty} x_{ijk} b^{l-k} = 1$ for some $i = 1, \ldots, b^m$, $j = 1, \ldots, s$, $l = 0, \ldots, m - t$. But $\sum_{k=l+1}^{\infty} x_{ijk} b^{l-k} = 1$ if and only if $x_{ijk} = b - 1$ for all $k \geq l + 1$. For any given i, j, l, the x_{ijk} for $k = l + 1, \ldots, \infty$ are independent random variables taking the values $0, 1, \ldots, b-1$ each with probability $1/b$. Therefore $P(\sum_{k=l+1}^{\infty} x_{ijk} b^{l-k} = 1) = 0$ for any i, j, l. This establishes that if (A_i) is a (t, m, s)-net, then so is (X_i) with probability one, because the union of a finite number of zero probability events has probability zero.

Finally if (A_i) is a (t, s)-sequence, then so is (X_i) unless $\sum_{k=l+1}^{\infty} x_{ijk} b^{l-k} = 1$ for some $i \geq 1$, $j = 1, \ldots, s$, $l \geq 0$. Therefore if (A_i) is a (t, s)-sequence then so is (X_i) with probability one, because the union of countably many zero probability events has probability zero. $\square$

3.2 Uniformity

Here we show that each X_i has the uniform distribution on $[0, 1)^s$. This is a property of the scrambling and holds whether or not A_i is a member of a (t, m, s)-net or a (t, s)-sequence.
Proposition 2. *Let $A \in [0, 1)^s$ and let X be the result of the random scrambling described above applied to A. Then X has the uniform distribution on $[0, 1)^s$. That is for any Lebesgue measurable $G \subseteq [0, 1)^s$, $P(X \in G) = \lambda_s(G)$, the s-dimensional Lebesgue measure of G.*
Proof: Consider an elementary subinterval of $[0, 1)^1$ in base b

$$E = \left[\frac{a}{b^l}, \frac{a+1}{b^l} \right),$$

where $0 \leq a < b^l$. Write $A^j = \sum_{k=1}^{\infty} a_{jk} b^{-k}$, $X^j = \sum_{k=1}^{\infty} x_{jk} b^{-k}$ and $a = \sum_{k=1}^{l} c_k b^{l-k}$ where $0 \leq c_k < b$.

As was shown in the proof of Proposition 1, there is zero probability that the base b expansion of any X^j ends with an infinite sequence of $b-1$'s. Therefore, apart from an event with probability zero, $X^j \in E$ if and only if

$$\pi_{j a_{j1} \cdots a_{jk-1}}(a_{jk}) = c_k, \quad k = 1, \ldots, l.$$

Because the permutations are uniform and independent we have that $X^j \in E$ with probability b^{-l}. That is $P(X^j \in E) = \lambda_1(E)$ on all elementary rectangles in $[0, 1)^1$.

Standard measure extension techniques lead to the desired result. First, it follows that $P(X^j \in E) = \lambda_1(E)$ for any interval E of the form $[ub^{-l}, vb^{-l})$ for integers l, u, v with $l \geq 0$ and $0 \leq u \leq v \leq b^l$. The endpoints of these intervals are dense in $[0, 1)$. Second, a Corollary of Chung (1974, P. 28), extends the result $P(X^j \in B) = \lambda_1(B)$ to Borel measurable $B \subseteq [0, 1)$. (Chung actually uses $(-\infty, \infty)$, but this is not an essential difference.) Third, the equivalence $P(X^j \in B) = \lambda_1(B)$ extends to Lebesgue measurable sets B, because subsets of sets of measure zero, have probability zero of containing X^j. Finally, let B_j, $j = 1, \ldots, s$ be measurable subsets of $[0, 1)$. Because the components X^j, $j = 1, \ldots, s$ are independent,

$$P(X^j \in B_j, j = 1, \ldots, s) = \prod_{j=1}^{s} \lambda_1(B_j) \tag{8}$$

Finally, λ_s is the unique measure on $[0, 1)^s$ which satisfies (8). $\square$

3.3   Computational Issues

For practical computation one must truncate the expansion of X_j^i at some finite $K_{\max}$. One choice is to take $K_{\max}$ large enough that $b^{-K_{\max}}$ is small compared to the error committed in approximating real values on the computer. Then one may take

$$X_i^j = \sum_{k=1}^{K_{\max}} x_{ijk} b^{-k}.$$ (9)

Another choice is to take $K_{\max} = M$ if one is sure that at most b^M points will ever be used. The truncation effect in (9) might not be negligible with $K_{\max} = M$, so one might take

$$X_i^j = \sum_{k=1}^{K_{\max}} x_{ijk} b^{-k} + b^{-K_{\max}}/2$$ (10)

or

$$X_i^j = \sum_{k=1}^{K_{\max}} x_{ijk} b^{-k} + b^{-K_{\max}} U_i^j$$ (11)

where for $i = 1, \ldots, n$, $j = 1, \ldots, s$, the U_i^j are independent $U[0,1)$ random variables. Using (10) gives $E(X_i^j) = 1/2$ compared to $(1 - b^{-K_{\max}})/2$ under (9) and using (11) gives X_i^j the $U[0,1)$ distribution, simulating the effects of $k > K_{\max}$.

The next practical issue is that the definition of (X_i) requires $s * (1 + b + b^2 + \ldots + b^{K_{\max}-1}) = s(b^{K_{\max}} - 1)/(b - 1)$ permutations. If b storage locations are used to store a permutation the result is that about $sb^{K_{\max}}$ storage locations are required.

If $K_{\max}$ is chosen to make $b^{-K_{\max}}$ small compared to the machine precision, then most of this storage is unnecessary. In practice one uses a (t, m, s)-net, which may be an initial subsequence of a (t, s)-sequence. If there are n points in the net, each with $K_{\max}$ digits then at most $nsK_{\max}$ permutations are required. This puts the storage requirement on the order of $nsbK_{\max}$ integers. A further space reduction is possible for large $K_{\max}$ because most of the permutations used are only used on one value of a. Therefore only the corresponding $\pi(a)$ need be computed and stored, and the storage demands are on the order of $nsK_{\max}$ integers, plus some extra for bookkeeping. This amount of storage is comparable to the ns floating point locations required to store the X_i^j.

4   Numerical Examples

The author has written code to construct randomized $(0, s)$-nets in prime power bases $b = p^r \geq s$. The construction used is due to Faure (1982), for prime bases b, and to Niederreiter (1987) for prime power bases.

4.1   Roos-Arnold Test Functions

Davis and Rabinowitz (1984, p. 405) present some examples drawn from Roos and Arnold (1963). The integrands considered are

$$f_1(X) = \frac{1}{s} \sum_{j=1}^{s} |4X^j - 2|,$$

$$f_2(X) = \prod_{j=1}^{s} |4X^j - 2|, \text{ and}$$

$$f_3(X) = \prod_{j=1}^{s} \frac{\pi}{2} \sin(\pi X^j).$$

The true values are $\int f_i(X)dX = 1$, $i = 1, 2, 3$, $s \geq 1$. Integrand f_1 is not a hard enough test function, because it is a sum of univariate functions. Any point set $X_1, \ldots, X_n$ for which $X_1^j, \ldots, X_n^j$ is a good univariate quadrature rule, $j = 1, \ldots, s$ will work well on f_1.

The comparisons in Davis and Rabinowitz use $s = 20$ and $s = 25$. Here we consider $s = 10$, and three rules: simple Monte Carlo with X_i independent and uniformly distributed on $[0, 1)^s$, a $(0, 10)$-sequence in base 11 and a randomized $(0, 10)$-sequence in base 11. Let the estimates be called $\hat{I}_{IID}$, $\hat{I}_S$ and $\hat{I}_{RS}$ respectively. For each method $n = 1331 = 11^3$ points were used. The random methods were repeated 500 times, to provide some insight into their sampling distribution.

For f_2 the estimate from the first 1331 points of the $(0, s)$-sequence was $\hat{I}_S = 1.65$ so the error was .65. The random methods have 500 errors and these may be summarized by root mean square errors:

$$\left(\frac{1}{500} \sum_{r=1}^{500} (\hat{I}_{IID,r} - 1)^2 \right)^{1/2} = 0.094$$

and similarly the RMSE for $\hat{I}_{RS}$ was 0.046. Thus the randomized sequences are roughly twice as accurate on this integrand as is simple Monte Carlo and both are much more accurate than the original unrandomized sequence. This large error reflects some folklore well known among researchers on quadrature: $(0, s)$-sequences tend to get off to a bad start, and in practice it may pay to use a tail sequence like $(X_i)_{i \geq b^M + 1}$ where b^M is large compared to the number of points to be used in quadrature.

Now suppose that one generated the first 11 replications of each of the randomized methods. Over the first 11 replications, the mean of $\hat{I}_{RS}$ was 1.029. An approximate 95% confidence interval based on these 11 replications is $1.029 \pm 2.228 \hat{s}_{RS}/11^{1/2} = 1.029 \pm 0.031$. (The constant 2.228 comes from the t distribution with 10 degrees of freedom.) The true value, $I = 1$, lies just barely inside the approximate 95% confidence interval. Over the first 11 replications, the mean of $\hat{I}_{IID}$ was 1.026. A confidence interval around this value could be constructed from the 14641 evaluations used, had they been saved. Such an interval would have been roughly $(1.96/2.228) * (0.094/0.046) = 1.8$ times as wide as the one around $\hat{I}_{RS}$. The first factor is a ratio of t critical points and the second is the estimated ratio of standard deviations. By comparison, based on the first 14641 function evaluations $\hat{I}_S = 1.0085$. This is better than both randomized methods, but it is hard to estimate its error from the data.

For f_3 the error in $\hat{I}_S$ on 1331 evaluations was .0072. The RMSEs for the random methods are: 0.071 for $\hat{I}_{IID}$ and 0.030 for $\hat{I}_{RS}$. Again the randomized sequences are roughly twice as accurate as simple Monte Carlo but now both are less accurate than the original unrandomized sequence.

Over the first 11 replications, the mean of $\hat{I}_{RS}$ was 0.996. The approximate 95% confidence interval based on these 11 replications is $0.996 \pm .027$. The true value, $I = 1$, lies well within the approximate 95% confidence interval. Over the first 11 replications, the mean of $\hat{I}_{IID}$ was .976. By comparison, based on the first 14641 function evaluations $\hat{I}_S = 0.980$. This is slightly better than simple Monte Carlo, not as good as the randomized sequences, and *worse* than the result of $\hat{I}_S$ on the first 1331 evaluations.

The confidence intervals constructed around $\hat{I}_{RS}$ are based on the assumption that each $\hat{I}_{RS}$ is approximately normally distributed with mean I and some variance. It was to assess whether this is realistic, that 500 evaluations of $\hat{I}_{RS}$ were calculated. Figure 1 shows a histogram of these 500 values with a scaled normal density superimposed.

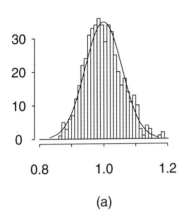
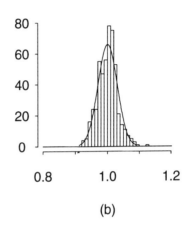

(a) (b)

Figure 1: Histograms of 500 replications of $\hat{I}_{RS}$. Part (a) is for Roos-Arnold integrand f_2, part (b) is for Roos-Arnold integrand f_3.

The sampling distribution of $\hat{I}_{RS}$ for these integrands is nearly normal, so that confidence intervals are in fact reasonably reliable. When there is doubt about the applicability of the normal distribution, or when greater accuracy of confidence intervals is an issue, one can use more accurate methods than $\hat{I} \pm \hat{\sigma} t_{(r-1)}^{.95} r^{-1/2}$. For example, bootstrap confidence intervals (Efron and Tibshirani (1993), Hall (1992)) may be used. Owen (1990) reports some simulations in which the bootstrap-t method achieves confidence levels very close to 95% for means of small non-normal samples.

In summary, replicated versions of $\hat{I}_{RS}$ provide a reliable way to form confidence intervals, at least on integrands like these. Sometimes $\hat{I}_{RS}$ was more accurate than $\hat{I}_S$ on these Roos-Arnold integrands and sometimes it was not. A more thorough comparison of accuracy is given in Section 4.2.

Owen (1994b) considers forming confidence intervals based on viewing a single realization of $\hat{I}_{RS}$ as b internal replications of a smaller net with n/b evaluations. It is possible to construct schemes in which a variance estimate based on such internal replicates has an expectation no smaller than the true variance of $\hat{I}_{RS}$.

4.2 Genz's Test Functions

A more comprehensive set of test functions was proposed by Genz (1984):

$$
\begin{aligned}
f_1(X) &= \cos\left(2\pi u_1 + \sum_j a_j X^j\right) && \text{(Oscillatory)} \\
f_2(X) &= \prod_j \left(a_j^{-2} + (X^j - u_j)^2\right) && \text{(Product Peak)} \\
f_3(X) &= \left(1 + \sum_j a_j X^j\right)^{-s-1} && \text{(Corner Peak)} \\
f_4(X) &= \exp\left(-\sum_j a_j^2 (X^j - u^j)^2\right) && \text{(Gaussian)} \\
f_5(X) &= \exp\left(-\sum_j a_j |X^j - u^j|\right) && (C_0) \\
f_6(X) &= \exp\left(-\sum_j a_j X^j\right) 1_{X^1 > u_1} 1_{X^2 > u_2} && \text{(Discontinuous)}
\end{aligned}
$$

The u_j are taken to be uniformly distributed on $[0,1]$. They are not thought to affect the difficulty of the integration problem, though they do break up some of the unnatural symmetry found in the Roos-Arnold test functions. The a_j are thought to greatly affect the difficulty of the integration problem, with larger values of a_j making the integrands harder. The a_j used were generated by taking independent $U[0,1]$ random variables and scaling them so that $\sum_{j=1}^{s} a_j = h_k s^{-e_k}$, where $k = 1, \ldots, 6$ indexes the test functions. Genz suggests $e = (1.5, 2, 2, 1, 2, 2)$. The values $h = (110, 600, 600, 100, 150, 100)$ were chosen by trial and error so that for $s = 10$, the simple Monte Carlo estimates $\hat{I}_{IID}$ gets I correct to roughly 2 digits of relative accuracy (in a RMSE sense).

Note that integrand $f_6 \notin BVHK$ because of its discontinuities.

As in the Section 4.1, we use $s = 10$. For each of the functions above, 30 realizations of that function (differing in a_j and u_j) were generated. Each of these was integrated using $n = 14641 = 11^4$ points and the same three rules $\hat{I}_{IID}$, $\hat{I}_S$ and $\hat{I}_{RS}$ as were used for the Roos-Arnold integrands. It should be noted that one set of independent random variables X_i^j was generated for the simple Monte Carlo method and it was used for all 30 replicates of all 6 integrands. It is conceivable that a lucky (or unlucky) set of $ns = 146410$ random X_i^j was obtained, but this seems a remote possibility, since 180 independent random integrands were generated. Similarly one randomized $(0, 4, 10)$-net in base 11 was used.

Figure 2 shows the RMS relative error of the three methods, applied to the oscillatory integrand f_1 of Genz. The sample sizes plotted are integer multiples of powers of 11, from $n = 11$ through $n = 14641$. These are chosen from the natural sample sizes discussed above.

For large enough samples, Figure 2 shows that simple Monte Carlo is less effective than the equidistribution methods. Figure 2 also shows that the randomly permuted sequence attains better accuracy than does the original $(0, 10)$-sequence. Not only does permuting the sequence allow one to assess its accuracy, it also appears to improve the accuracy. For small n this is not surprising, as it is well known that these sequences take a while to "warm up". It is noteworthy that this extra accuracy extends to larger n.

Another feature evident in Figure 2 is a wavy pattern in the RMS error for the original $(0, 10)$-sequence. As n increases the RMS error improves slowly, if at all, until n begins to approach a new power of b. Then the error rapidly declines. This suggests that for $(0, s)$-sequences in base b there is a big advantage to using $n = b^m$. By contrast the curve for the randomly permuted $(0, 10)$-sequence shows a steadier improvement as n increases from b^{m-1} to b^m in steps of b^{m-1}.

Integrand 1 : Oscillatory

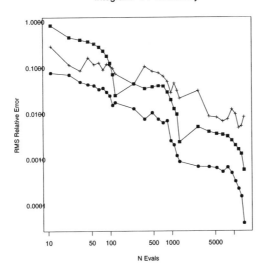

Figure 2: RMS relative error versus number of integrand evaluations for estimating Genz's Oscillatory integrand f_1 in 10 dimensions. 30 independently generated integrands were used. Simple Monte Carlo (+), a $(0, 10)$-sequence in base 11 (■), a randomly permuted $(0, 10)$-sequence in base 11 (●).

Figure 3 shows that very similar conclusions may be drawn for Genz's integrands f_2 through f_5. It is clear in these plots that the equidistribution method and randomized equidistribution method are both achieving a better rate than the pure Monte Carlo method. The randomized $(0, 10)$-sequence is usually, but not always, providing better accuracy than the original sequence.

Figure 4 shows the accuracy attained for Genz's Discontinuous integrand f_6. This integrand is not of bounded variation, so the Koksma-Hlawka inequality does not apply. Figure 4 suggests that $\hat{I}_S$ and $\hat{I}_{RS}$ are somewhat more accurate than $\hat{I}_{IID}$. They do not show the same evidence of achieving a better rate of accuracy than $\hat{I}_{IID}$.

Acknowledgements

I thank Fred Hickernell for many stimulating discussions while this work was in progress. Thanks also to the organizers of the conference "Monte Carlo and Quasi-Monte Carlo Methods in Scientific Computing" held in Las Vegas in June 1994. Attending that very pleasant meeting led to significant changes in the way I think about this material.

References

Braaten, E. & Weller, G. (1979). An Improved Low-Discrepancy Sequence for Multidimensional Quasi-Monte Carlo Integration. *J. Comp. Physics* **33**, 249–258.

Chung, K.L. (1974). *A Course in Probability Theory*. Academic Press, New York.

Cranley, R. & Patterson, T.N.L. (1976). Randomization of Number Theoretic Methods for Multiple Integration. *SIAM J. Numer. Anal.* **13**, 904–914.

Davis, P.J. & Rabinowitz, P. (1984). *Methods of Numerical Integration, 2nd Ed.* Academic Press Inc, Orlando FL.

Efron, B. & Stein, C. (1981). The Jackknife Estimate of Variance. *Ann. Stat.* **9**, pp 586–596.

Efron, B. & Tibshirani, R. (1993). *An Introduction to the Bootstrap.* Chapman and Hall, London.

Faure, H. (1982). Discrépance de Suites Associées à un système de Numération (en Dimension *s*). *Acta Arith.* **41**, 337–351.

Genz, A. (1984). Testing Multidimensional Integration Routines. *in Tools, Methods and Languages for Scientific and Engineering Computation, B. Ford, J. C. Rault, & F. Thomasset, eds., 81–94* North-Holland, Amsterdam.

Hall, P. (1992). *The Bootstrap and Edgeworth Expansion*. Springer, New York.

Joe, S. (1990). Randomization of Lattice Rules for Numerical Multiple Integration. *J. Comp. Appl. Math.* **31**, 299-304.

McKay, M.D., Conover, W.J. & Beckman, R.J. (1979). A Comparison of Three Methods for Selecting Values of Input Variables in the Analysis of Output From a Computer Code. *Technometrics* **21**, 239–245.

Niederreiter, H. (1978). Quasi-Monte Carlo Methods and Pseudo-Random Numbers. *Bull. Amer. Math. Soc.* **84** 957–1041.

Niederreiter, H. (1987). Point Sets and Sequences with Small Discrepancy. *Monatsh. Math.* **104**, 273–337.

Niederreiter, H. (1988). Low Discrepancy and Low-Dispersion Sequences. *J. Number Theory* **30**, 51–70.

Niederreiter, H. (1992). *Random Number Generation and Quasi-Monte Carlo Methods.* CBMS-NSF, SIAM, Philadelphia.

Owen, A. B. (1990). Empirical Likelihood and Small Samples. *Computing Science and Statistics, Proceedings of the 22nd Symposium on the Interface* A.S.A., Alexandria VA.

Owen, A. B. (1992a). A Central Limit Theorem for Latin Hypercube Sampling. *Journal of the Royal Statistical Society, Ser. B* .

Owen, A.B. (1992b). Orthogonal Arrays for Computer Experiments, Integration and Visualization. *Stat. Sinica* **2**, 439–452.

Owen, A.B. (1994a). Lattice Sampling Revisited: Monte Carlo Variance of Means Over Randomized Orthogonal Arrays. *Ann. Stat.* **22** pp. 930–945.

Owen, A.B. (1994b). "Monte Carlo Variance of Scrambled Equidistribution Quadrature". Dept. of Statistics Technical Report Number 466, Stanford University.

Patterson, H.D. (1954). The Errors of Lattice Sampling. *J.R.S.S. B* **16**, 140–149.

Raghavarao, D. (1971). *Constructions and Combinatorial Problems in Design of Experiments.* Wiley, New York.

Ripley, B.D (1987). *Stochastic Simulation.* J. Wiley, New York.

Sacks, J., Welch, W.J., Mitchell, T.J. & Wynn, H.P. (1989). Design and Analysis of Computer Experiments. *Statistical Science* **4**, 409–435.

Roos, P. & Arnold, L. (1963). Numerische Experimente zur Mehrdimensionalen Quadratur. *Österreich. Akad. Wiss. Math.-Natur. Kl. S.-B. II* **172**, 271–286.

Sobol', I.M. (1967). The Distribution of Points in a Cube and the Accurate Evaluation of Integrals (in Russian). *Zh. Vychisl. Mat. i Mat. Phys.* **7**, 784–802.

Stein, M. (1987). Large Sample Properties of Simulations Using Latin Hypercube Sampling. *Technometrics* **29**, 143–151.

Tang, B. (1993). Orthogonal Array-Based Latin Hypercubes. *J.A.S.A.* **88**, 1392–1397.

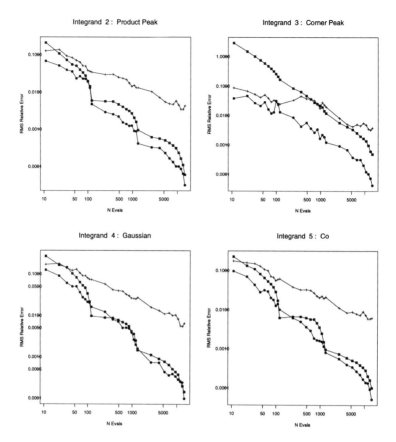

Figure 3: RMS relative error versus number of integrand evaluations for estimating Genz's Product Peak integrand f_2, Corner Peak integrand f_3, Gaussian integrand f_4, and C_0 integrand f_5. All integrands are over 10 dimensions. 30 independently generated integrands were used from each family. Simple Monte Carlo ($+$), a $(0, 10)$-sequence in base 11 ($\blacksquare$), a randomly permuted $(0, 10)$-sequence in base 11 ($\bullet$).

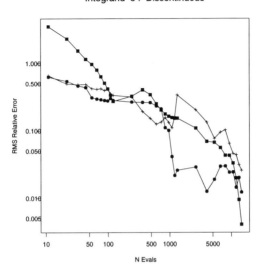

Integrand 6 : Discontinuous

Figure 4: RMS relative error versus number of integrand evaluations for estimating Genz's Discontinuous integrand f_6 in 10 dimensions. 30 independently generated integrands were used. Simple Monte Carlo (+), a $(0,10)$-sequence in base 11 (■), a randomly permuted $(0,10)$-sequence in base 11 (●).

Quantum Monte Carlo Simulation: Algorithm and Applications

Tao Pang [1]

ABSTRACT   The algorithm of the diffusion quantum Monte Carlo simulation method and some of its applications are presented. The emphasis is on the actual calculations of electronic structures of small systems, including a hydrogen molecule confined in a spheroidal box and the ionic hydrogen clusters H_{2n+1}^{+} with n=1, 2, 3, 4. The Monte Carlo calculations on the molecular energies and structures of these systems reported here are the best theoretical calculations to date. Equilibrium geometric structures of some clusters are searched through optimizing their molecular energies and good agreement is found with the previous configuration interaction calculations. The electronic structures of the H_{2n+1}^{+} clusters are analyzed based on the bonding formation in the systems. Estimate on the dissociation energy of each cluster is made and compared with the available experimental measurements.

1   Introduction

Monte Carlo techniques now are widely used in many areas of scientific computing, ranging from quantum transport on the surface, to structural phase transitions of various materials [1]. Quantum Monte Carlo simulation techniques have been developed in studying electronic structure of atoms, molecules and small clusters and they are found to be extremely promising in accurate evaluation of the ground-state energies, geometric structures, and vibrational spectra of small systems [2, 3].

In this article, we will present some of our recent quantum Monte Carlo simulation results on the study of hydrogen molecules under quantum confinement [4] and electronic structure of small ionic hydrogen clusters H_{2n+1}^{+} with $n = 1, 2, 3, 4$ [5]. We will first discuss the algorithm of the diffusion quantum Monte Carlo method which we have adopted in calculating electronic structure of small systems. Then we will report the applications of this scheme, specifically in the study of the electronic structure of a hydrogen molecule under quantum confinement of a spheroidal box and the study of the geometric and electronic structures of the ionic hydrogen clusters H_{2n+1}^{+} with $n = 1, 2, 3, 4$.

The properties of hydrogen molecules under high pressures have attracted a lot of attention recently due to the advances in the high-pressure technology and the unusual behavior of materials under high pressures [6, 7, 8]. For example, solid hydrogen experiences metal-insulator transitions [6, 7, 9, 10] and structural transitions [11, 12] when the

---

[1] Department of Physics, University of Nevada, Las Vegas, Nevada 89154-4002, USA

pressure is above 100 GPa. Recent experiments reveal that the properties of hydrogen molecules, confined in neon, helium and argon matrices under high pressures [13] have a very strong pressure dependence. Qualitatively, it is noted that the pressure dependence of the bond length and the pressure dependence of the vibrational frequency of the molecule under the high pressures [13] is similar to the theoretical calculations from a variational method [14] for a hydrogen molecule in an infinite potential box of the shape of a spheroid. Since this previous calculation is variational in nature, one could not know for sure that the difference between the experiment and the theory is purely due to the deficiency of the model. The quantum Monte Carlo simulations, which we have carried out, clarify this issue, since the Monte Carlo simulations are in principle exact, that is, the errors are statistical and controllable.

Ionic hydrogen clusters are among the simplest clusters which can be produced experimentally. There have been many interesting studies of the properties of the ionic hydrogen clusters from both experiment [15, 16, 17, 18] and theory [19, 20, 21, 22, 23, 24]. The rich vibrational spectra and unusual energetic behavior of these clusters have been studied by the infrared spectroscopy [15, 16] and fragmentation and collision [17, 18] experiments. The interplay of the quantum zero-point motion of the protons and the strong electron-electron correlation effects of the electrons in the clusters has made it a challenge for obtaining an accurate account of their properties both experimentally and theoretically. All typical methods in quantum chemistry, such as, the Hartree-Fock methods [21] and configuration interaction methods [19, 20, 22, 23, 24], have been applied in studying these clusters and the theoretical results have formed a basic understanding of the ionic clusters, including the ones with higher ionization, more electrons, and atomic core structures.

2   Diffusion Quantum Monte Carlo Scheme

The goal of many-body theory is to understand the properties of a many-body Hamiltonian in given conditions. Typically a many-body system is described by the Hamiltonian

$$H = -\frac{\hbar^2}{2m}\sum_{i=1}^{N}\nabla_i^2 + \sum_{i<j}^{N}V(\mathbf{r}_i, \mathbf{r}_j) + \sum_{i=1}^{N}U_0(\mathbf{r}_i), \tag{1}$$

where $\hbar$ is the Planck constant, m is the mass of a particle, $V(\mathbf{r}_i, \mathbf{r}_j)$ is the interaction between the ith and jth particles, and $U_0(\mathbf{r}_i)$ is the external potential on the ith particle.

Usually it is impossible to solve the above Hamiltonian exactly and approximations have to be tolerated in obtaining the properties of the system. The diffusion quantum Monte Carlo (DQMC) scheme was deviced to gain the information of the ground state of the system by solving an imaginary time Schrödinger equation. Formally one can transform the time-dependent Schrödinger equation with an imaginary time into a diffusion equation [25, 26]. Here we adopt the version introduced in Ref. [27], which converts the Schrödinger equation into a diffusion equation of a time-dependent function in the configuration space with a drifting term and a branching term [27, 2]. In the DQMC approach for an N-electron system, one takes a variational wavefunction $\Phi(\mathbf{R})$ with $\mathbf{R} = (\mathbf{r}_1, \mathbf{r}_2, ..., \mathbf{r}_N)$ as a guiding wavefunction for the ground state and then generates another wavefunction $\Psi(\mathbf{R}, t)$, which is the result of the propagation of $\Phi(\mathbf{R})$ in the imaginary time t:

$$\Psi(\mathbf{R}, t) = e^{\int_0^t E_n(t')dt' - Ht}\Phi(\mathbf{R}), \tag{2}$$

where $E_n(t')$ is an adjustable constant, which is used during the simulation to ensure that the projection of $\Psi(\mathbf{R}, t)$ to the true ground state is of the order of 1. One can easily show that as long as $\Phi(\mathbf{R})$ is not orthogonal to the exact ground state of H, $\Psi(\mathbf{R}, t)$ will approach the true ground state of H when t goes to infinity. In fact, if one takes a time derivative of Eq. (2), one obtains the imaginary time Schrödinger equation

$$\frac{\partial \Psi(\mathbf{R}, t)}{\partial t} = (E_n - H)\Psi(\mathbf{R}, t). \tag{3}$$

One can now introduce a time-dependent function $F(\mathbf{R}, t)$ with

$$F(\mathbf{R}, t) = \Psi(\mathbf{R}, t)\Phi(\mathbf{R}). \tag{4}$$

After multiplying the imaginary time Schrödinger equation of Eq. (3) with the guiding wavefunction $\Phi(\mathbf{R})$, one has

$$\frac{\partial F}{\partial t} = \frac{1}{2}\nabla^2 F - \nabla \cdot F\mathbf{U} - [E_n - E(\mathbf{R})]F, \tag{5}$$

with

$$\mathbf{U} = \nabla \ln \Phi(\mathbf{R}), \tag{6}$$

as the drifting velocity and

$$E(\mathbf{R}) = \Phi^{-1}(\mathbf{R})H\Phi(\mathbf{R}), \tag{7}$$

as the local energy for a given configuration $\mathbf{R}$. If one introduces a time-dependent expectation value,

$$E(t) = \frac{\langle \Phi(\mathbf{R})|H|\Psi(\mathbf{R}, t)\rangle}{\langle \Phi(\mathbf{R})|\Psi(\mathbf{R}, t)\rangle} = \frac{\int d\mathbf{R}\, F(\mathbf{R}, t)E(\mathbf{R})}{\int d\mathbf{R}\, F(\mathbf{R}, t)}, \tag{8}$$

at time t, one can obtain the true ground-state energy,

$$E_0 = \lim_{t \to \infty} E(t). \tag{9}$$

The multi-dimensional integral in Eq. (8) can be carried out by the Monte Carlo quadrature with $F(\mathbf{R}, t)$ treated as a time-dependent distribution function if it is positively defined. In practice, the simulation is done by rewriting the diffusion equation for $F(\mathbf{R}, t)$ into an integral form:

$$F(\mathbf{R}', t + \tau) = \int d\mathbf{R}\, F(\mathbf{R}, t)G(\mathbf{R}', \mathbf{R}; \tau), \tag{10}$$

where $G(\mathbf{R}', \mathbf{R}; \tau)$ is the Green function of the diffusion equation. If τ is very small, the Green function can be approximated as

$$G(\mathbf{R}', \mathbf{R}; \tau) \simeq W(\mathbf{R}', \mathbf{R}; \tau)G_0(\mathbf{R}', \mathbf{R}; \tau), \tag{11}$$

where

$$G_0(\mathbf{R}', \mathbf{R}; \tau) = (\frac{1}{2\pi\tau})^{3N/2}e^{-[\mathbf{R}'-\mathbf{R}-\mathbf{U}\tau]^2/2\tau}, \tag{12}$$

is a propagator due to the drifting, and

$$W(\mathbf{R'}, \mathbf{R}; \tau) = e^{-\{[E(\mathbf{R}) + E(\mathbf{R'})]/2 - E_n(t)\}\tau}, \qquad (13)$$

is a branching factor.

In order to treat $F(\mathbf{R}, t)$ as a distribution function or a probability, we have to have it positively defined. A common practice is to apply a fixed-node approximation, which forces the function $F(\mathbf{R}, t)$ to be zero in case it becomes negative [27]. This fixed-node approximation still provides an upper bound for the evaluation of the ground-state energy [27] and gives good molecular energies for small systems if the guiding wavefunction is properly chosen [2]. The fixed-node approximation can be removed by releasing the fixed nodes but it will require additional computing efforts [28]. Here is a summary of the major steps in the actual simulations:

(a) One first performs variational quantum Monte Carlo (VQMC) simulations to optimize the variational parameters in the guiding wavefunction.

(b) Then one can generate an initial ensemble with many independent configurations from the VQMC simulations.

(c) Each configuration is updated by a drifting and a Gaussian random walk, with the new coordinate,

$$\mathbf{R'} = \mathbf{R} + \mathbf{U}\tau + \chi, \qquad (14)$$

where χ is a $3N$-dimensional Gaussian random number generated to have a variance of τ on each dimension.

(d) Not every step is accepted and a probability,

$$P = \min[1, A(\mathbf{R'}, \mathbf{R}; \tau)], \qquad (15)$$

is used to judge if the updating move should be accepted or not. Here

$$A(\mathbf{R'}, \mathbf{R}, \tau) = \frac{\Phi(\mathbf{R'})^2 G(\mathbf{R}, \mathbf{R'}; \tau)}{\Phi(\mathbf{R})^2 G(\mathbf{R'}, \mathbf{R}; \tau)}, \qquad (16)$$

which is necessary in order to have a detailed balance between points $\mathbf{R'}$ and $\mathbf{R}$. When the move crosses a node, the move is rejected under the the fixed-node approximation.

(e) A new ensemble is then created with branching: $M = [W(\mathbf{R'}, \mathbf{R}; \tau_a) + \xi]$ copies of the configuration $\mathbf{R'}$ are put in the new ensemble. Here ξ is a uniform random number between [0,1] which is used to make sure that the fraction of W is properly taken care of. τ_a is the effective diffusion time which is proportional to τ with a coefficient as the ratio of the mean square distance of the accepted moves to the mean square distance of the attempted moves.

(f) The average local energy at each time step $\bar{E}(\mathbf{R'})$ is then evaluated from the summation of $E(\mathbf{R'})$ of each configuration and weighted by the corresponding probability $W(\mathbf{R'}, \mathbf{R}; \tau)$. The time dependent energy $E_n(t)$ is updated at every step with

$$E_n(t) = \frac{\bar{E}(\mathbf{R}) + \bar{E}(\mathbf{R'})}{2}, \qquad (17)$$

to ensure a smooth convergence.

Before one can take the data for the calculations, one needs to run the above steps long enough so the error is dominated by statistics. The data itself is taken with an interval of typical ten steps. The exact size of the interval can be determined from the autocorrelation function of the physical quantities evaluated in the simulations. One can then group and average the data to a desired accuracy. Several independent runs can be carried out for a better average.

In the next couple of sections, I will demonstrate the application of the DQMC method outlined above in the study of the electronic structure of small systems.

3 Hydrogen Molecule under Confinement

In the experiments to study the hydrogen molecules in an inert element matrix, such as a neon matrix under high pressures, the molecules behave qualitatively the same as they are confined inside quantum cavities. A significant question is how sensitive that the properties, such as the molecular energy, bond length, and vibrational frequency are, to the shape and confinement of the cavities. Since all the previous calculations are variational, it is impossible to determine the quantitative difference between the experimental measurements and the calculations is purely due to the deficiency of the model or the inaccuracy of the variational calculations. This is the major motive for us to study the system by the diffusion quantum Monte Carlo simulations, which, in principle, provide the exact information of the ground state of the model.

The spheroidal box which provides the potential barrier for the electrons in the hydrogen molecule can be written as

$$\frac{x^2}{b^2} + \frac{y^2}{b^2} + \frac{z^2}{a^2} = 1, \tag{18}$$

with a the semimajor axis and b the semiminor axis. The Hamiltonian of the system, that is, two electrons and two protons located at the foci of the spheroidal box, is given by

$$H = H(1) + H(2) + \frac{e^2}{r_{12}} + \frac{e^2}{R_{ab}}, \tag{19}$$

with

$$H(i) = -\frac{\hbar^2}{2m}\nabla_i^2 - \frac{e^2}{r_{ia}} - \frac{e^2}{r_{ia}}, \tag{20}$$

where m and e are the mass and charge of an electron, r_{ia} and r_{ib} are the distances between the ith electron and two protons, and r_{12} is the distance between two electrons. R_{ab} is the separation of two protons. Since two protons are located at the foci of the spheroidal box at the equilibrium, we have $R_{ab}^2 = 4(a^2 - b^2)$. So for a given a, one can adjust R_{ab} and equally b to search for the equilibrium state of the system with the molecular energy of the system optimized.

To simplify the calculations, the Born-Oppenheimer approximation has been adopted to the system so the proton wavefunction and electron wavefunction are treated separately, that is, the electron wave equation is solved with a fixed and given proton configuration. This is also why the Hamiltonian of Eq. (19) did not include the kinetic energy of the protons. The external potential now is given from both the Coulomb potential of the

Table 1: The optimized DQMC ground-state energy E_m and optimized bond length R_{ab} E_m^v and R_{ab}^v are the corresponding variational calculations of Ref. [14]. and E_m are the results of the DQMC calculations of the present work. All quantities are given in atomic units.

a	R_{ab}^v	E_m^v	R_{ab}	E_m
∞	1.403	-1.1716	1.4010	$-1.1746(5)$
5	1.395	-1.1638	1.3895	$-1.1702(3)$
4	1.355	-1.1440	1.3503	$-1.1533(5)$
3	1.208	-1.0441	1.1771	$-1.0523(9)$
2	0.893	-0.4749	0.8949	$-0.4790(10)$
1	0.455	4.5947	0.4493	4.5944(30)

protons and the spheroidal box, that is, an infinite potential barrier defined by Eq. (18). The total molecular energy or the potential energy surface of the protons can be written as

$$E_m = E_0 + \frac{e^2}{R_{ab}}, \tag{21}$$

with E_0 the ground-state energy of electrons for a fixed proton positions. We will adopt the atomic units (a.u.) throughout this article, which is equivalent to setting $m = e = \hbar = c = 1$. Then the energy is given in hartrees and the length is in terms of the Bohr radius.

We have chosen a relatively simple wavefunction for the two electrons in the system as the guiding wavefunction for DQMC simulations,

$$\Phi(\mathbf{R}) = J(r_{12})\psi(\mathbf{r}_1)\psi(\mathbf{r}_2), \tag{22}$$

where a two-body Jastrow correlation factor $J(r)$ is introduced to account for the electron-electron correlation effects,

$$J(r) = \exp[\frac{r}{2(1 + \beta r)}]. \tag{23}$$

Here β is a variational parameter which is determined in the VQMC simulations before running the DQMC simulations. $\psi(\mathbf{r}_i)$ is a single particle orbital with a selected form of

$$\psi(\mathbf{r}) = [e^{-\alpha r_a} + e^{-\alpha r_b}][\Gamma(\mathbf{r}) - 1], \tag{24}$$

and the function $\Gamma(\mathbf{r})$ is chosen as

$$\Gamma(\mathbf{r}) = \frac{x^2}{b^2} + \frac{y^2}{b^2} + \frac{z^2}{a^2}, \tag{25}$$

to ensure that the wavefunction goes to zero when one of the electrons approaches the spheroidal box. α can be determined from $\alpha = 1/[1 + \exp(-R_{ab}/\alpha)]$ which ensures the cusp condition between the electron and two protons. The wavefunction outside the spheroidal box is set to zero always and the DQMC simulations are started after a search for an optimized β by the VQMC simulations as initial runs.

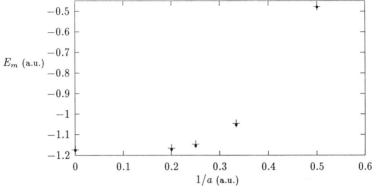

Figure 1: The optimized DQMC ground-state energy (dots) versus the inverse of the semimajor axis of the spheroidal box together with the previous variational results of Ref. [14] (crosses).

The quantum Monte Carlo simulations for a hydrogen molecule confined inside a spheroidal box have been carried out on RISC/6000 workstation and on a Cray-YMP single processor. The ground-state energy of a molecule in a spheroidal box with a fixed semimajor axis is calculated. The equilibrium bond length and minor axis of the spheroidal box are searched by optimizing the ground-state energy of the molecular system. In Table 1, we show the optimized ground state energy and the optimized bond length with a given semimajor axis. The variational calculations [14] done on the same systems are listed, too, for comparison. As one can see from the table, the variational calculations on both energy and bond length are quite accurate and are improved considerably over the previous variational calculations [14].

The DQMC results of the molecular energy of a hydrogen molecule versus the inverse of the bond length of the molecule inside a spheroidal box are plotted together with the previous variational calculations in Fig. 1. Clearly the trend of the molecular energy versus the inverse of the semimajor axis of the spheroidal box, the measure of the confinement, is well described by the variational calculations of LeSar and Herschbach [14]. This is an indication that the calculations on the pressure dependence of the bond length and vibrational frequency from the variational method [14] may have given a good description of the system. The variational calculations of the pressure dependence of the vibrational frequency and the equilibrium bond length of a hydrogen molecule [14] are, in fact, in qualitative agreement with the recent experimental measurements of the pressure dependence of the vibrational frequency and bond length of hydrogen molecules in neon and other matrices under high pressures [13]. It will be extremely interesting to verify this point by a direct evaluation of the vibrational frequencies of the system by the quantum Monte Carlo method [3] for the protons with the potential energy surface given by the DQMC simulations of the electrons. Another possible future direction is to simulate the electrons and protons together and the motion of the electrons can become comparable to the motion of the protons when the localization of the electrons due to the quantum confinement becomes significant.

4 Ionic Hydrogen Clusters

There are $2n$ electrons, interacting with each other, in the ionic hydrogen cluster H_{2n+1}^+. In order to simplify the calculations, the Born-Oppenheimer approximation is applied to the system so we can treat the proton motion and electron motion separately. The properties of the system are then dominated by the electron wavefunction with the proton configuration adjusted adiabatically. The role of the protons now is in the external potential of the electrons,

$$U_0(\mathbf{r}_i) = -\sum_{\alpha=1}^{2n+1} \frac{e^2}{r_{i\alpha}}, \tag{26}$$

with $r_{i\alpha}$ defined as the distance between the ith electron and the αth proton. The interaction between two electrons is still given by the Coulomb potential

$$V(\mathbf{r}_i, \mathbf{r}_j) = \frac{e^2}{|\mathbf{r}_i - \mathbf{r}_j|}. \tag{27}$$

The molecular energy of the system, or the potential energy surface of the protons, can be written as

$$E_m = E_0 + \sum_{\beta > \alpha}^{2n+1} \frac{e^2}{r_{\alpha\beta}}, \tag{28}$$

with E_0 the total ground-state energy of the electronic Hamiltonian for a fixed proton configuration, that is, with $r_{\alpha\beta}$, the distance between the αth and βth protons given.

Based on the understanding of the electronic structure of the ionic hydrogen clusters H_{2n+1}^+, the following wavefunction is chosen as the guiding wavefunction for the DQMC simulations:

$$\Phi(\mathbf{R}) = \det |\mathbf{D}^\uparrow| \det |\mathbf{D}^\downarrow| e^{-\Lambda(\mathbf{R})}, \tag{29}$$

where $\mathbf{D}^\sigma$ is a matrix which is constructed from linear combinations of the Gaussian states located around each proton.

$$D_{ki}^\sigma = \sum_{\alpha=1}^{2n+1} C_{k\alpha} e^{-r_{i\alpha}^2/\omega_\alpha^2}, \tag{30}$$

is for the ith electron in the kth orbital in the Slater determinant $\det |\mathbf{D}^\sigma|$ with spin σ. The linear combination coefficients $C_{k\alpha}$ and the variances ω_α are variational parameters of the guiding wavefunction and are determined in the initial VQMC simulations. The exponent $\Lambda(\mathbf{R})$ is the Jastrow correlation factor which accounts for the short-range correlation effects between two electrons or an electron and a proton. $\Lambda(\mathbf{R})$ is parameterized to satisfy all the cusp conditions with

$$\Lambda(\mathbf{R}) = -\frac{1}{2} \sum_{j>i} \frac{\gamma_{ij} r_{ij}}{1 + \eta r_{ij}} + \sum_{i,\alpha} \frac{r_{i\alpha}}{1 + \eta r_{i\alpha}}, \tag{31}$$

where γ_{ij} varies with the specific particles involved. For two electrons with the same spin orientation, $\gamma_{ij} = 1$ and for two electrons with opposite spin orientations, $\gamma_{ij} = 1/2$. η is another variational parameter which is also optimized by the VQMC simulations.

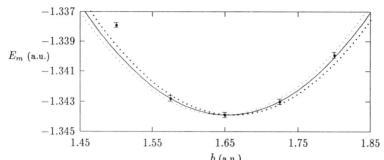

Figure 2: DQMC molecular energy of H_3^+ with different bond length b (dots with error bars). The quadratic fit is also shown (solid line) together with possible variations due to the error bars in the two points adjacent to the central point (dotted lines).

The VQMC and DQMC simulations for H_{2n+1}^+ with $n = 1, 2, 3, 4$ have been carried out on a Cray-YMP single processor and several RISC/6000 workstations. For the H_3^+ and H_5^+ clusters, we have also searched for their most stable geometric structures by optimizing the molecular energies of the systems with each individual bond varied. A quadratic dependence on each individual length is assumed for the change of the molecular energy:

$$E_m^i = E_m + \mu(b_i - b)^2, \tag{32}$$

where b_i is the bond length to be varied around its optimized value b, E_m^i is the corresponding molecular energy, and μ is a parameter which is determined by the actual data in the fitting. The optimization is confined in the same symmetry of the lowest geometric structure from the best CI calculations.

From the best CI calculations, we know that the stable geometric structure of H_3^+ is an equal-bond-length triangle [29]. The molecular energy of H_3^+ is calculated with several different bond lengths around the stable geometry. In order to obtain the equilibrium geometry, the three evaluated points around the equilibrium point are used for the quadratic fit. In Fig. 2, the DQMC simulations of five different bond lengths are shown together with the quadratic curves obtained from fitting. The equilibrium bond length of H_3^+ is estimated as 1.654 ± 0.008 a.u., which is in excellent agreement with the result from the best CI calculation [29], 1.650 a.u. The DQMC error bar of the equilibrium bond length is estimated by altering the fitting curve with the possible change, due to the statistic error in the data, in the two points next to the equilibrium point.

Based on the best CI calculations [20], the lowest energy configuration of H_5^+ is in C_{2v} symmetry as illustrated in Fig. 3. We have performed DQMC simulations to search the equilibrium structure for H_5^+ with each individual bond length varied but with the symmetry of the cluster maintained. For each individual bond length, we vary it around the optimized value and obtain three evaluations. All other lengths are kept fixed during the optimization. We then apply the quadratic fit to obtain the optimized bond lengths. The best CI results from Ref. [20] are used as references. The DQMC evaluations of the optimized bond lengths are summarized in Table 2, together with the calculations from

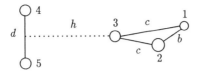

Figure 3: The equilibrium geometric structure of H_5^+ in C_{2v} symmetry with all relevant lengths and the proton sites indicated.

Table 2: The DQMC result of the stable geometric structure of H_5^+ together with the HF result of Ref. [21] and the CI result of Ref. [20]. All quantities are given in atomic units. The bond lengths are as shown in Fig. 3.

Bond Length	HF	CI	DQMC
b	1.577	1.544	1.550
c	1.704	1.815	1.841
h	2.970	2.530	2.483
d	1.413	1.444	1.472

the HF method [21] and the best CI work [19, 20]. It seems that the electron-electron correlation effects have made distance between the H_2 and H_3^+ segments in H_5^+ shorter.

From the geometric structures of the H_{2n+1}^+ clusters, one would conclude that they are just simple combinations of $n-1$ H_2 segments attached to an H_3^+ core. However, based on the optimized linear combination coefficients obtained from the VQMC simulations performed for H_5^+, the electrons do not simply stay on the states of H_3^+ and H_2. In fact the single-particle states mainly located around H_2 penetrate into the H_3^+ region and form local itinerant orbitals between the H_3^+ core and H_2 segments. The positive charge of the cluster now is delocalized and distributed among all the segments. This is important in order to understand the formation of all the H_{2n+1}^+ clusters and other similar clusters. The bond formed between the H_3^+ core and each H_2 segment in the H_{2n+1}^+ clusters is clearly different from the bonding behavior in solid hydrogen. In fact, there is no real bonding formation between one H_2 segment and its neighbors in solid hydrogen and the system is kept together by external pressure. Electrons in solid hydrogen become itinerant in the whole system when the pressure is high enough to force the H_2 states to overlap. Based on the VQMC results, we can conclude that the electronic structure of the H_{2n+1}^+ clusters is very unique and the electrons are in some types of local itinerant states formed between each H_2 segment and the H_3^+ core. This is certainly the main reason to have the stable H_{2n+1}^+ clusters.

The molecular energies of H_{2n+1}^+ with $n = 1, 2, 3, 4$ are calculated by the DQMC simulations described. After we have completed the simulations for the H_3^+ and H_5^+

Table 3: DQMC molecular energies of H_{2n+1}^+ with $n = 1, 2, 3, 4$ together with the HF results of Ref. [21] and the CI results of Refs. [19], [20], [22], and [29]. All quantities are given in atomic units and the error bars of the DQMC results are given in parentheses.

System	HF	CI	DQMC
H_3^+	−1.3001	−1.3438 [a]	−1.3438(2)
H_5^+	−2.4427	−2.5232 [b]	−2.5316(5)
H_7^+	−3.5824	−3.6909 [c]	−3.7133(11)
H_9^+	−4.7206	−4.8640 [c]	−4.8942(21)

[a] From Ref. [29]. [b] From Refs. [19, 20]. [c] From Ref. [22].

clusters, we have also studied the H_7^+ and H_9^+ clusters. We did not search for equilibrium structures of H_7^+ and H_9^+ but instead, we calculated their molecular energies with the geometric structures obtained from the best CI calculations [19, 20, 29]. All the molecular energies for the H_{2n+1}^+ clusters obtained from the DQMC simulations are listed in Table 3, together with the results of the HF [21] and the CI calculations [19, 20, 29, 22]. The improvement over the best CI results varies. It seems that the improvement of the DQMC results over the best CI calculations increases with the number of the electrons involved.

5 Conclusions

We have presented the algorithm of the diffusion quantum Monte Carlo simulation method for the calculations of the electronic structures of small systems and the calculations of the equilibrium ground-state energy and bond length of a hydrogen molecule confined in a spheroidal box with a given semimajor axis and the molecular energies of the ionic hydrogen clusters H_{2n+1}^+.

The algorithm is quite efficient in the study of the small systems and as shown from our examples, the scheme adopted here provides the evaluations of the ground-state energies and structures of small systems with very high accuracy.

The DQMC calculations of the ground-state properties of a hydrogen molecule inside a spheroidal box with different bond lengths are extremely accurate and the DQMC molecular energies, which we have obtained, are mostly about 1% lower than the previous variational calculations of Ref. [14]. The DQMC equilibrium bond lengths are systematically shorter than those of the variational calculations for the given semimajor axes. However, the qualitative behavior of the bond length dependence of the molecular energy is in excellent agreement with the variational calculations. The implication is that the bond length dependence obtained from the previous variational calculations reflects precisely the behavior of the model. The experimental data on the pressure dependence of the vibrational frequencies of the hydrogen molecule in all inert element matrices under high pressures is in qualitative agreement with the variational results. The detailed quantitative difference should have come from the deficiency of the model in describing the systems, that is, the shape of the potential energy surfaces and finite potential barrier of the matrices. Even though the DQMC simulations in principle provide the exact ground-

Table 4: The DQMC estimates of the dissociation energies of H_{2n+1}^+ together with the HF and CI results from Refs. [19], [20], and [22] and the experimental data from Refs. [31] and [32]. The selected CI calculations of zero-point motion corrections, 3.0 kcal/mol for H_5^+ [20], 2.6 kcal/mol for H_7^+ [19], and 2.1 kcal/mol for H_9^+ [22], are used in deducing the DQMC dissociation energies. All quantities are given in kcal/mol.

System	HF	CI	DQMC	Exp.
H_5^+	1.88	5.45 [a]	5.3	5.8±1.2 [c]
				6.6±0.3 [d]
H_7^+	0.72	0.42 [b]	1.9	3.40 [d]
H_9^+	0.94	1.24 [b]	1.9	3.16 [d]

[a] From Refs. [19, 20]. [b] From Ref. [22]. [c] From Refs. [31]. [d] From Ref. [32].

state energy of the system within statistical errors, one important aspect, which has not been investigated yet, is the effects of the image charges of the electrons and protons due to the local inhomogeneity of the dielectric constant. It will be a very interesting future direction to explore the shape dependence of the molecular energy and to calculate the vibrational frequencies directly from the DQMC simulations.

The calculated DQMC molecular energies of the H_{2n+1}^+ clusters allow us to analyze the dissociation energies of these clusters. The chemical process which we are interested in here is

$$H_{2n+1}^+ + \Sigma_{2n+1}^0 \rightarrow H_{2n-1}^+ + H_2, \tag{33}$$

where Σ_{2n+1}^0 is the heat absorbed during the process, or the dissociation energy of the reaction. One can estimate Σ_{2n+1}^0 if the molecular energies of the three systems H_{2n+1}^+, H_{2n-1}^+, and $+ H_2$, and the corrections due to quantum zero-point motion are given,

$$\Sigma_{2n+1}^0 = \Sigma_{2n+1}^e - \Sigma_{2n+1}^c, \tag{34}$$

where Σ_{2n+1}^c is the corrections of the quantum zero-point motion and Σ_{2n+1}^e is the electronic binding energy,

$$\Sigma_{2n+1}^e = E_{2n-1}^+ + E_2 - E_{2n+1}^+, \tag{35}$$

with E_{2n-1}^+, E_2 and E_{2n+1}^+ the molecular energies of H_{2n-1}^+, H_2, and H_{2n+1}^+ separately. We have summarized the estimated DQMC dissociation energies for the H_{2n+1}^+ clusters in Table 4, together with the results from the HF calculations [22], CI calculations [19, 20, 22], and selected experimental measurements [31, 32]. We have taken the corrections of the quantum zero-point motion from the CI calculations [19, 20, 22] to estimate the DQMC dissociation energies for all the H_{2n+1}^+ clusters. As a matter of fact, the zero-point motion corrections are not sensitive to the accuracy of the theory [19, 20] itself, but the difference of all molecular energies needed. The ground state energy of the hydrogen molecule E_2 is taken from the DQMC result of Ref. [4]. The DQMC dissociation energies for all H_{2n+1}^+ are considerably improved over the the best CI calculations in comparison with the selected experimental results.

Further improvement in the evaluation of the molecular and dissociation energies is possible if the guiding wavefunctions are further improved or the nodes in the guiding wavefunctions are released. An extremely interesting problem in the future is to calculate the vibrational frequencies of these clusters in two steps: First one can calculate the electronic structure and then obtain the molecular energy or potential energy surface of the system; Then one can apply the quantum Monte Carlo simulation techniques to the protons with the potential surface given and then extract the vibrational frequencies of the systems as done in Ref. [3].

Acknowledgements:
This work was supported in part by the NSF under Cooperative Agreement OSR-9353227, the Supercomputer Center for Energy and the Environment at the University of Nevada, Las Vegas (UNLV), the University Fellowships and Grants Committee of UNLV, and the Army High Performance Computing Research Center at the University of Minnesota.

References

[1] K. Binder (Ed.). *The Monte Carlo Method in Condensed Matter Physics*, Springer-Verlag, Berlin, 1992.

[2] W. A. Lester, Jr. and B. L. Hammond. Quantum Monte Carlo for the electronic structure of atoms and molecules. *Annual Review of Physical Chemistry*, 41:283-311, 1990.

[3] M. A. Suhm and R. O. Watts. Quantum Monte Carlo studies of vibrational states in molecules and clusters. *Physics Reports*, 204:293-329, 1991.

[4] T. Pang. Hydrogen molecule under confinement: Exact results. *Physical Review A*, 49:1709-1713, 1994.

[5] T. Pang. Properties of ionic hydrogen clusters: a quantum Monte Carlo study. *Chemical Physics Letters*, 228:555-561, 1994.

[6] H. K. Mao and R. J. Hemley. Optical studies of hydrogen above 200 Gigapascals: Evidence for metallization by band overlap. *Science*, 244:1462-1465, 1989.

[7] R. J. Hemley and H. K. Mao. Critical behavior in the hydrogen insulator-metal transition. *Science*, 249:391-393, 1990.

[8] D. M. Ceperley and B.J. Alder. Ground state of solid hydrogen at high pressures. *Physical Review B*, 36:2092-2106, 1987.

[9] H. Chacham, S. G. Louie. Metallization of solid hydrogen at megabar pressures: A first-principles quasiparticle study. *Physical Review Letters*, 66:64-67, 1991.

[10] M. P. Surh, T. W. Barbee III, and C. Mailhiot. Zero-point motion and the insulator-metal transition in solid molecular hydrogen. *Physical Review Letters*, 70:4090-4093, 1993.

[11] H. E. Lorenzana, I. F. Silvera, and K. A. Goettel. Orientational phase transition in hydrogen at megabar pressures. *Physical Review Letters*, **64**:1939-i1942, 1990.

[12] K. J. Runge, M. P. Surh, C. Mailhiot, and E. L. Pollock. Path integral Monte Carlo calculations of orientational order in compressed H_2. *Physical Review Letters*, **69**:3527-3530, 1992; **70**:2974 (E), 1993.

[13] P. Loubeyre, R. LeToullec and J.C. Pinceaux, Properties of H_2 under strong compression in a Ne matrix. *Physical Review Letters*, **67**:3271-3274, 1991; Raman measurements of the vibrational properties of H_2 as a guest molecule in dense helium, neon, argon, and deuterium systems up to 40 Gpa. *Physical Review B*, **45**:12844-12853, 1992.

[14] R. LeSar and D.R. Herschbach. Electronic and vibrational properties of molecules at high pressures. Hydrogen molecule in a rigid spheroidal box. *Journal of Chemical Physics*, **85**:2798-2804, 1981.

[15] M. Okumura, L. I. Yeh, and Y. T. Lee. The vibrational predissociation spectroscopy of hydrogen cluster ions. *Journal of Chemical Physics*, **83**:3705-3706, 1985.

[16] M. Okumura, L. I. Yeh, and Y. T. Lee. Infrared spectroscopy of the cluster ions $H_3^+ \cdot (H_2)_n$. *Journal of Chemical Physics*, **88**: 79-91, 1988.

[17] B. Farizon, M. Farizon, M. J. Gaillard, E. Gerlic and S. Ouaskit. Fragmentation of high energy mass-selected ionic hydrogen clusters. *Nuclear Instruments and Methods in Physics Research B*, **88**:86-90, 1994.

[18] S. Ouaskit, B. Farizon-Mazuy, M. Farizon, M. J. Gaillard, E. Gerlic, and M. Stern. Dissociation cross sections of ionic hydrogen clusters by collisions with helium at 60 keV/amu. *Physical Review A*, **49**:1484-1486, 1994.

[19] Yamaguchi, J. F. Gaw, and H. F. Schaefer III. Molecular clustering about a positive ion. Structures, energetics, and vibrational frequencies of the protonated hydrogen clusters H_3^+, H_5^+, H_7^+, and H_9^+. *Journal of Chemical Physics*, **78**:4074-4085, 1983.

[20] Y. Yamaguchi, J. F. Gaw, R. B. Remington, and H. F. Schaefer III. The H_5^+ potential energy hypersurface: Characterization of ten distinct energetically low-lying stationary points. *Journal of Chemical Physics*, **86**:5072-5081, 1987.

[21] H. Huber. Near-Hartree-Fock energies and structures of even- and odd-membered H_n^+ clusters ($n \leq 13$) obtained with floating basis sets. *Journal of Molecular Structure*, **121**:281-286, 1985.

[22] M. Farizon, B. Farizon-Mazuy, N. V. de Castro Faria, and H. Chermette, Ab initio calculations of hydrogen ionic clusters. *Chemical Physics Letters*, **177**:451-457, 1991.

[23] M. Farizon, H. Chermette and B. Farizon-Mazuy. Structure and energetics of hydrogen clusters. Structure of H_{11}^+ and H_{13}^+. Vibrational frequencies and infrared intensities of the H_{2n+1}^+ clusters ($n = 2 - 6$). *Journal of Chemical Physics* **96**:1325-1332, 1992.

[24] M. Farizon, N. V. de Castro Faria, B. Farizon-Mazuy, and M. J. Gaillard. Interaction of hydrogen H_n^+ clusters with thin carbon foils. *Physical Review A*, **45**:179-183, 1992.

[25] J. B. Anderson. A random walk simulation of the Schrödinger equation: H_3^+. *Journal of Chemical Physics*, **63**:1499-1503, 1975.

[26] J. B. Anderson. Quantum chemistry by random walk. H ^2P, H_3^+ D_{3h} $^1A_1$, H_2 Σ_u^+, H_4 $^1\Sigma_g^+$, Be ^1S. *Journal of Chemical Physics*, **65**:4121-4127, 1976.

[27] P.J. Reynolds, D.M. Ceperley, B.J. Alder and W.A. Lester, Jr. Fixed-node quantum Monte Carlo for molecules. *Journal of Chemical Physics*, **77**:5593-5603, 1982.

[28] D.M. Ceperley and B.J. Alder. Quantum Monte Carlo for molecules: Green's function and nodal release. *Journal of Chemical Physics*, **81**:5833-5844, 1984.

[29] D. Frye, A. Preiskorn, G. C. Lie, and E. Clementi. Gaussian functions in Hylleraas-configuration interaction calculations. V. An accurate ab initio H_3^+ potential-energy surface. *Journal of Chemical Physics*, **92**:4948-4955, 1990.

[30] J. B. Anderson. Quantum chemistry by random walk: Higher accuracy for H_3^+. *Journal of Chemical Physics*, **96**:3702-3706, 1992.

[31] M. T. Elford. The heat of dissociation of H_5^+ derived from measurements of ion mobilities. *Journal of Chemical Physics*, **79**:5951-5959, 1983.

[32] R. J. Beuhler, S. Ehrenson, and L. Friedman. Hydrogen cluster ion equilibria. *Journal of Chemical Physics*, **79**:5982-5990, 1983.

A COUPLED MONTE CARLO/EXPLICIT EULER METHOD FOR THE NUMERICAL SIMULATION OF A FOREST FIRE SPREADING MODEL

ANTONINI PUPPIN MACEDO
Agência Espacial Brasileira
Departamento de Desenvolvimento Técnico Científico
SBN - Ed. Eng. Paulo Mauricio sala 409
70040-950 Brasília, DF, Brazil
antonini@enm.unb.br

ANTONIO C. P. BRASIL JUNIOR
Universidade de Brasília
Departamento de Engenharia Mecânica
70910-900 Brasília, DF, Brazil
brasil@enm.unb.br

ABSTRACT

This paper presents an application of a general Monte Carlo integration method, coupled with an explicit euler time discretization to the numerical solution of an integro-differential forest fire propagation model. In this model, the time variation of the enthalpy of the fuel bed is given in terms of radiative heat transfer and radiative and convective heat loss terms. The radiative heat exchange is given by a local flame model. This approach results in a convolution integral between the heat generated by combustion and an empirical shape function.

1-INTRODUCTION

The physical phenomena involved in the combustion of forest biomass, during natural or accidental fires, are extremely complex and the behavior involves advanced knowledge related to fluid dynamics and heat transfer in reacting media. The current importance of the study of this kind of problem, both by mathematical modeling and numerical simulation, as well as advanced experimentation, are encouraged by the growth of international concerns with environmental questions. Points of interest are the evaluation of the amount of emissions to the atmosphere, especially of CO, CO_2 and particulates, and the prevention of environmental and material destruction caused by forest fires.

In Brazil, during the dry season, vegetation fires reach the peak of about 6,000 fires a day, as detected by the NOAA series satellites (Setzer 1994). One can list some types of great importance: the after cutting burning in the Amazon (*Queimada*); the natural fires in the Brazilian savanna like region, the *Cerrado*; pasture maintenance fires in the Amazon and sugar cane culture burns.

Among works related to the simulation and modeling of biomass burning phenomena, the Russian literature is especially rich. Works by Dorrer (1984a), (1984b) present models of the propagation of forest fires that are adopted as a basis for the present work. Works of Brabander et al. (1989) and Grishin et al.(1985) use a mixture theory approach to describe *in situ* observations of fires in homogeneous forests. Works of Konev (1984), on the propagation of flame fronts in surface fires, and Grishin et al (1984), about the effects of thermals during forest fires, present some specifical points that can be used in mathematical models for forest fires spreading.

Other works present relevant aspects of heat transfer in forest fires; e.g papers by Rothermel (1972) and Albini (1976). In the latter, the heat transfer is quantified by a Green's function integral. This model is used by Dorrer cited above. Anderson (1968) proposed a simplified model of forest fires propagation based on discrete convolution methods which were improved by Richards (1988).

The evaluation of the various parameters involved in the phenomena, such as specific heat and heat content of the fuel, characteristics of the flame, moisture content, mass decay rates and others can be find in the works of Baines (1990), de Mestre et al. (1989) and Ventura et al. (1990).

The use of Monte Carlo integration techniques for the numerical simulation of such processes is well cited in the works of Kaviani & Fatehi (1992). Therefore, the present work intents to present a method that is able to approach the problems cited above, from a numerical simulation point of view. To perform it, starting from a proper mathematical model, a straightforward Monte Carlo Integration / Explicit Euler Discretization is presented.

2-MATHEMATICAL MODEL

The propagation of a surface forest fire is considered as a combustion reaction occurring in a biomass fuel layer distributed in a 2D domain $\Omega \subset \Re^2$. If in a generic point $P(x, y) \in \Omega$ there is combustion, the mass of the fuel $m(x, y, t)$ decays as follows:

$$\frac{\partial m(x, y, t)}{\partial t} = -K \tag{1}$$

This equation gives the kinetics of the Pyrolisis reaction and the rate of reaction, K, is given by an Arrhenius-like law:

$$K = A([O_2], T) \exp(-\frac{E}{RT}) \tag{2}$$

In this equation, $[O_2]$ is the concentration of oxygen in the air layer in contact with the material in combustion, T is the gas temperature and R the gas constant. A and E are characteristic constants of the combustion of forest material (Kaviani & Fatehi (1992)). These constants can also be evaluated from the experimental exponential decay plots of Rothermel (1972) or Ventura et al. (1990).

The heat produced during the combustion process at this point is given by:

$$\Phi(x, y, t) = -K\Delta H_m \tag{3}$$

where ΔH_m is the heat content of the biomass fuel.

Considering surface forest fires, the flame front propagates through the fuel bed from neighbouring points by radiative heat tranfer. Combustion starts when the points reach the ignition temperature.

This heating process is given by the following energy balance:

$$\frac{\partial(\overline{\rho c}T(x,y,t))}{\partial t} = q_{rad+} - q_{rad-} - q_c \tag{4}$$

$T(x,y,t)$ is the temperature and q_{rad+} is the radiative heat flux, from the flame, that heats the fuel bed and q_{rad-} and q_c are the radiative and the convective loss terms.

These terms are given by:

$$q_{rad+} = \int\int \Phi(x',y',t)\xi(x-x',y-y')dxdy \tag{5}$$

$$q_{rad-} = \alpha\sigma(T^4 - T_\infty^4) \tag{6}$$

$$q_c = \alpha\bar{U}(T - T_\infty) \tag{7}$$

σ is the Stefan-Boltzman constant, $\bar{U}$ is the convective heat transfer coefficient and α is the absorvity coeficient,(Baines (1990), de Mestre et al. (1989)).

The term $\overline{\rho c}$ is the mean value of the product of the density and specific heat of the fuel bed. This term includes the evaporation process in a two equation model as given by de Mestre et. al. (1989).

An important point in this approach to a flame influence function given by equation (5), is the simplification of the radiative heat transfer, using an empirical relation for $\xi(x-x',y-y')$ as proposed by Dorrer (1984).

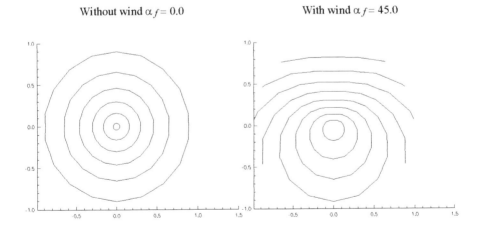

Fig. 1: Typical shape functions - $\xi(x,y)$ - contour levels.

This function is given by (in polar coordinates):

$$\xi(x - x', y - y') = \xi(r, \theta) =$$

$$a_0 \delta \left(1 - \frac{\delta}{3h_f}\right) \exp\left(\frac{r\cos^2\alpha_f}{r_0(1 + \sin^2\alpha_f \cos\theta)}\right) \qquad (8)$$

The parameters in this equation are:

a_0 : Fraction of the heat spent in the propagation of the combustion.

δ : Thickness of the fuel bed.

h_f : Flame height.

α_f : Flame inclination angle

r_0 : Effective flame effect radius

The function $\xi(x - x', y - y')$ is in polar coordinates, r and θ are the polar coordinates with origins in the point $P(x, y)$, and x' and y' are its neighbors.

Figure 1 shows some configurations and contour levels of these functions for different angles α_f, related to different wind velocities.

3-NUMERICAL METHODS

To solve the set of equations (1) to (8), a coupled Explicit Euler / Monte Carlo Integration Method was developed. This approach joins the simplicity of a finite difference scheme to the flexibility and stability of the Monte Carlo Method.

The choice of a Monte Carlo integration was made due to several particularities involving both the phenomenum and its mathematical model. Biomass fuel burning is characterized by several non-homogeneous parameters; non constant environmental variables (wind, humidity and topography of the terrain), lots of local changes in the latter, discontinuities in the fuel bed, non-homogeneous properties of the fuel and others.

Field experiments and static analysis may define important parameters such as a distribution law for the fuel bed on the surface elements. In the same way, different distribution laws may be obtained for the humidity profile of a vegetal cover (for each kind of forest, for example). Therefore, these parameters should be included in the definition of the sampling sequence for the Monte Carlo integration. In the present analysis, only homogenous sequences where used for the Monte Carlo integration, but future studies and experimental comparisons and calibrations may suggest the use of non-homogeneous sequences that would reproduce better the observed physical phenomena. Numerically, another integration method, such as Gaussian quadrature, could be used, with few differences in the final implementation of the code.

The high computational cost of this kind of simulation, especially for real field situations, suggests the use of high performance computing and optimizations of the final code wherever it is possible. So, Monte Carlo integration routines may give much better optimized codes, in particular parallel ones.

Important points observed in the implementation, as it will be seen in the following discussion, are the necessity of several calculations, and the definition of two different regions, one for the time propagation and the other for the calculation of the integrals, which increases the computional costs of the model.

Some important topics may be listed as follows:

i) The precision of the integrals does not need to be very high. Futhermore, it is not safe to use convergence criterion, such as *repeat until* certain an error bound. As the functions go through several changes during the propagation, the error bound convergence criterion may delay, unnecessarily, the calculations on some particular points on the grid. This is particularly true where the presence of strong propagation fronts and/or discontinuities on the fuel bed produce hard boundaries in the functions to be integrated.

ii) The definition, *a priori*, of a number of points to be used in each sequence for Monte Carlo calculations has been enough to ensure convergence and good precision.

iii) As the points on the Monte Carlo sequence are interpolated from the points on the time propagation grid, the use of a defined number of points for this sequence will generate a small "Monte Carlo" integration domain and few points (and small amounts of memory) are required.

iv) The approach allows many different manipulations of the grids for the inclusion of obstacles to the fire, changes in topography, etc. at relatively low computational and pre-processing costs.

v) The use of an Explicit Euler Method for time discretization with small time steps, would compensate for the loss of precision due to the generation of Monte Carlo sequences with few points. This is possible due to the very slow velocity of propagation of the phenomenum.

vi) In fact, as the propagation of the flames is very slow, it was possible to perform several studies that have lead to an optimized relationship between the time discretization grid and the domain and number of points for the Monte Carlo integration. This relationship is related to precision, convergence and stability, as well as computational costs. So, the time steps will, in general, be much smaller than the ones required for time discretization stability, but will ensure the precision of the Monte Carlo integration, as the domain of integration will be very small.

vii) Explicit Euler Methods, in general, generate codes that are easy to be implemented on parallel machines, with a domain division strategy. Therefore, the structure of the final code will be in such a way that its optimization and "parallelization" will be possible.

3.1-EXPLICIT EULER TIME DISCRETIZATION

In the model presented here, the differential terms are simple enough to permit different approaches to their integration. Before introducing any question related to the numerical or mathematical point of view, it is interesting to discuss some questions about the phenomenum itself.

Numerically, the model is essentially a non-conservative initial value problem. The generation of energy in the system must be always greater than the dissipation for propagation to occur. This suggests that the use of a non-conservative time integration scheme could be used without compromising the final conditions for stability of the results.

On the other hand, dissipation is required for the final equilibrium situation. When the propagation is stopped by the end of the available fuel, or by the encountering of an obstacle that prevents heat transfer from the propagation front to the unheated fuel, or when internal conditions of the fuel bed (like heat and moisture content) are unable to support the combustion process; the dissipation must be stable enough to ensure the proper energy decay to equilibrium.

Generally, explicit time schemes can be taken as non-conservative (and sometimes unstable) ones. The option of an explicit solver for the PDE's was done in order to permit the calculation of the propagation at small time steps with an affordable computational cost. To carry out the implementation of this time integration procedure special care was taken in order to ensure stability to preserve the characteristics of the simulated system.

A general linearized formulation may be proposed for explicit euler time discretizations:

$$\overline{\rho c} T^{t+\Delta t} = \overline{\rho c} T^t + \Delta t \left(q_{rad+}^t - q_{rad-}^t - q_c^t \right) \tag{9}$$

$$m^{t+\Delta t} = m^t - K^t \Delta t \tag{10}$$

$$\Phi^{t+\Delta t} = -K^t \Delta H_m \tag{11}$$

3.2-MONTE CARLO INTEGRATION

The convolution integral of the radiative heat flux, given by the term q_{rad+}, is the only function of the model to be integrated. The domain of integration is very easy to sample (actually it is just a circle around the burning points) and the values of the function to be integrated are known at discrete points in this domain. These features allows the use of a very simple Monte Carlo integration, little precision is required and fast integration is a goal.

To perform the integration, N random numbers ($\{x_1, ..., x_n\}$) are sampled uniformly distributed in a multidimensional volume V. Then the basic theorem of Monte Carlo integration estimates that the integral of a function f over the multidimensional volume is given by:

$$\int f dV \approx V\langle f \rangle \pm V \sqrt{\frac{\langle f^2 \rangle - \langle f \rangle^2}{N}} \tag{12}$$

Here the angle brackets denote taking the arithmetic mean over the N sample points,

$$\langle f \rangle \equiv \frac{1}{N} \sum_{i=1}^{N} f(x_i) \tag{13}$$

$$\langle f^2 \rangle \equiv \frac{1}{N} \sum_{i=1}^{N} f^2(x_i) \qquad (14)$$

The "plus or minus" term in the equation (12) is a one standard deviation error estimate for the integral, not a rigorous bound. Further, there is no guarantee that the error is distributed as a Gaussian, so the error term should be taken only as a rough indication of probable error. On the other hand, it is possible to infer that the error is inversely proportional to the square root of the number of points N by the equation (12) above.

The problem in the case considered here is that the Monte Carlo integration must be coupled with a PDE solver, and the functions to be integrated experience several changes during the propagation and are defined only in points on a discrete domain. So, it is not worth while to use a procedure that "repeats until" certain error bounds. In practice, it was found that the best way is to determine, prior to the calculations, a number of points for the random sequence that would give the desired precision.

A study of convergence was done for some typical functions (for instance, the functions presented in figure 1). Figure 2 shows plots of the value of the function against the number of points used for their integration.

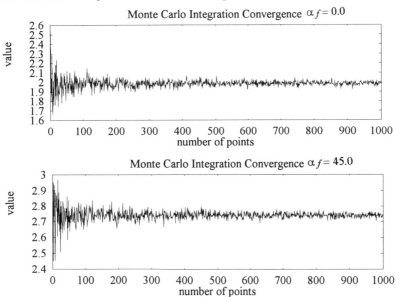

Fig. 2: Convergence study for typical shape functions - $\xi(x, y)$

In our approach, a good random generator will be the one that generates a set of numbers very fast , as well as provides homogeneous distribution. In the present case, this fact is very important as the random sequence must be generated for each point on the time discretization grid at each time step. An example is the random generation algorithm of L'Ecuyer, with Bays and Dur-ham shuffling, Press et al. (1993). This algorithm is very straightforward to implement and gives good precision and convergence. The resulting implementation is very fast as required by the coupled approach. Other generators could be cited, e.g Sobol's Pseudo-Random generator (Press et al. 1993). Using the random generator, two random sequences are generated, in polar coordinates, one for r and one for θ, which define the points on the domain of integration.

3.3-COUPLING EXPLICIT EULER / MONTE CARLO INTEGRATION

The numerical implementation of the present model must consider two points: the time discretization of the PDE's, and the calculation of the convolution between one of these equations and the radiative heat transfer shape function. As two processes are involved, they can not be separately analyzed, but one must consider both numerical processes as a whole. This means that the stability, convergence and precision of the simulations will not depend on one isolated process, but on the coupling of them.

At each time step of the propagation a random set of numbers (in polar coordinates) is set around every grid point on the space discretization grid. These points generate a random Monte Carlo *sub-grid*. At each time step new *sub-grids* are generated. It is important to remark that this procedure avoids the propagation of fronts that may have tendencies to reproduce a shape related to a sequence, if it is reproduced at each time step.

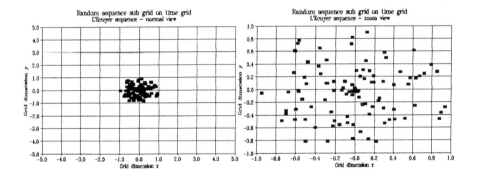

Fig. 3: Random sequence on the time discretization grid

The points in this *sub-grid* may lay on a region of several points on the main grid, thus an interpolation routine is performed for all variables involved in the calculation of the integral. As the time discretization grid used is a square grid, a generic point in the *sub grid* has only four neighbors for the interpolation. Therefore, a bilinear interpolation (Press et al. (1993) has given good precision in the calculations performed. The figure 3 shows a typical random sequence on a time discretization grid.

The precision and stability of the propagation are affected by the correct relation between the domain of integration, and the size of the time discretization grid. The domain of integration is defined by r_0 (effective flame effect radius, eq. (9)); its magnitude in relation to the size of the grid determines convergence and stability. When dx, the time discretization grid size, is much smaller than r_0, the precision is improved. On the other hand, the stability is not well ensured, as points in front of the fire front may not be equally heated. This means that spurious heated fronts may appear in front of the actual fire front and compromise the results. In fact, even complete divergence may be achieved when $r_0 >> 10dx$. After a number of numerical tests, the optimal relation was found to be; $2dx < r_0 < 5dx$.

4-RESULTS

To test the method, a sample case, from an experimental analysis (Grishin at al. 1984) was chosen. As some parameters were not detailed in this work, typical physical parameters involved in biomass burning processes were taken from Baines (1990) and de Mestre (1989). Figure 4 depicts the initial configuration of the analyzed case. The problem presented here is to determine the propagation of the fire front (from these initial condictions) and check the accuracy of the method by the temperature profile of the resulting fire front. Table 1 details the various physical parameters involved in the calculations, Table 2 details the parameters for the numerical implementation and figure 5 shows the evolution of the fire front for this case. Figure 6 shows a fire front profile comparison for the experimental case presented by Grishin e al. 1985 and the numerical results obtained with the present approach. From parameters that were easy to determine, or found in the literature, it was possible to obtain good results for surface fire spreading.

The fire front propagation shown in figure 5 presents some interesting points about the qualitative behavior of the model. The front propagates faster upwind then downwind. In front of the fire front, and behind it, a heated region with no combustion forms, as expected. As the propagation goes on, the biomass fuel is consumed and a "ring"-like front forms. Inside of this "ring", a region with no fuel, and no combustion appears. These important facts identify the good qualitative representation of the phenomena by the model and the validity of the numerical implementation. Figure 6 represents a cut view of figure 5c in its middle section, in the upwind direction. As shown in this figure, the numerical results agrees well with the experimental ones, especially in the burning zone $(T > 600K)$. In the heating zone $(T < 600K)$ some disagreement is observed, probably due to a bad representation of the value of the specific heat.

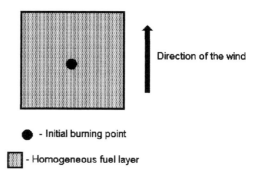

Fig. 4: Initial conditions - sample case

Table 1: Physycal parameters for the sample case

Parameter	Description	Value (units: SI)
m	biomass fuel distribution	0.5
s_f	surface / volume (of the fuel)	5700
ρ_f	dry fuel particle density	509
ρ_b	dry fuel bed density	32.1
ρ_w	water density	1000
c_f	specific heat of the dry fuel	1370
c_w	specific heat of the water	4187
l	latent heat of H_2O at 373K	2.254×10^6
L	fuel bed half length	0.46
δ	fuel bed thickness	0.08
H_f	flame height	0.7
T_a	ambient temperature	296
T_i	ignition temperature	593

Table 2: Numerical parameters for the sample case

Parameter	Value (units: SI)
Dimensions of the domain	7.5 x 7.5
Time of propagation	80
Time increment	0.5
Points in time discretization grid	15 x 15
Domain of Monte Carlo Integration (r_0)	1.0
Number of points in the random sequence	200
Relation r_0/dx	2.0
seed to initialize random generator (Press et. al. 1993)	-1

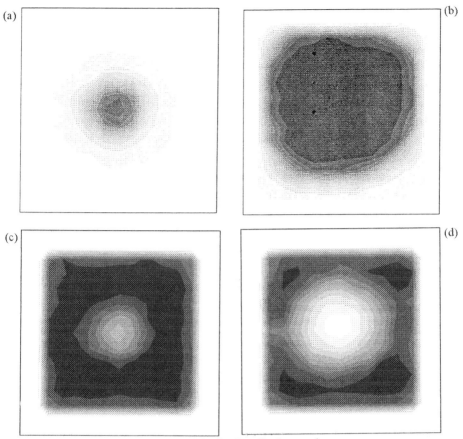

Fig. 5: Evolution of the fire front for the sample case

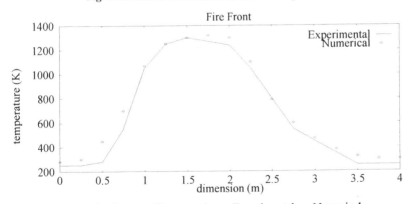

Fig. 6: Fire front profile comparisons: Experimental vs. Numerical

5-CONCLUSIONS

From the results presented here, it can be seen that the model has good qualitative and quantitative behavior. It introduces a very useful approach to the numerical solution of integro-differential models of surface vegetation fire propagation. The evolution of the propagation agrees very well with what was expected from previous works. It may be applicable to different situations such as a non-continous fuel bed, the presence of fire breaks and others.

The equations of the proposed model use physical parameters that are easy to determine from experiments or can be found in the literature. The radiative heat transfer is given by a convolution integral and gives a straightforward way to approach radiative changes in a partially-transparent medium. The calibration of the flame effect function and the convolution function coefficients must be very carefully done in order to rigorously fit the physical characteristics of the phenomenum. The use of a Green function as a flame effect function gives a powerful tool to include the effects of the wind, as its effects can be taken only as inclining the flame from the vertical. By using similar functions, the effects of the topography may be included.

The features of the functions to be integrated permit the implementation of a simple Monte Carlo Integration Method. An important point in the current implementation is the necessity of generating several random sequences (for each node of the time discretization grid) at every time step. So, fast generators are more necessary than "elaborate" ones for this kind of approach.

The coupling procedure is another key point in this approach, and the determination of optimal relations between the domain of the Monte Carlo Integration and the time discretization grid size is main factor in guaranteeing stability and convergence.The method presented here, it is possible to make modifications for the implementation on other integro-differential models. The requirements of stability and the characteristics of other models to be implemented may suggest the use of other non-linear PDE's solvers, without compromising the coupling process.

From the results presented for sample case, one finds that the model is very satisfactory. The good agreement between numerical and experimental results confirms the acceptability of the approach presented.

6-ACKNOWLEDGMENTS

The authors would like to acknowledge the CNPQ (National Council of Scientific Development of Brazil) that have gave financial support during the development of this work, as well as the referee for this paper. His/her outstanding work improved a lot our contribution.

7-REFERENCES

F. A. Albini; Estimating Wildfire Behaviour and Effects, USDA Forestry Service, Gen. Tech. Rep. INT-30, 1976.

H. E. Anderson, Sundance Fires, an Analysis of Fire Phenomena, USDA Forestry, Paper INT-56, 1968.

P. G. Baines, Physical Mechanisms for the Propagation of Surface Fires, *Math. Comput. Modelling*, 13, 12, 83-94, 1990.

O. P. Brabander, O. A. Vdovina, A. M. Grishin e A. D. Gruzin, Investigation of the Conditions of Transition of a Surface Forest Fire to A Crown Fire, *Comb. Expl. and Shock Waves*, 24, 4, 435-440, 1989.

G. A. Dorrer, A Model for Propagaton of Curvilnear Forest Fire Fronts, *Comb. Expl. and Shock Waves*, 20, 1, 8-15, 1984a.

G. A. Dorrer, A Model of the Spreading of a Forest Fire, *Heat Tranfer - Soviet. Res.*, 16, 6, 39-52, 1984b.

A. M. Grishin, N. A. Alekseyev, O. P. Brabander e V. F Zal'Mezh, Propagation of Forest Fire-Induced Thermals in the Ground Layer of the Atmosphere, *Heat Tranfer - Soviet Res.* , 16, 6, 28-38, 1984.

A. M. Grishin, A. D. Gruzin e V. G. Zverev; Study of the Structure and Limits of Propagation of the Front of an Upstream Forest Fire, *Comb. Expl. and Shock Waves*, 21, 1, 9-18, 1985.

M. Kaviani e M. Fatehi, Combustion in Porous Media, *Anais do IV ENCIT*, p17-p38, 1992.

E. V. Konev, Analysis of the Spreading of Natural and Intentionally Set Forest Fires, *Heat Tranfer - Soviet Res.* , 16, 6, 1-27, 1984.

N. J. de Mestre, E. A. Catchpole, D. H. Anderson & R. C. Rothermel, Uniform Propagation of a Planar Fire Front Without Wind, *Comb. Sci. and Tech.*, 65, 231-244, 1989.

W. H. Press, B. P. Flannery, S. A. Teukolsky and W. T. Vetterling, Numerical Recipes 2nd Edition , Cambridge Press, 1993.

A. Puppin Macêdo & A. C. P. Brasil Junior, Numerical Simulation of Ground Forest Fires, *Anais do CBEM/NNE 94*, 1994.

A. Puppin Macêdo & A. C. P. Brasil Junior, An Integro- Differential Model for the Heat Transfer in Forest Fires, *Anais do V ENCIT*, 1994.

G. D. Richards; Numerical Simulation of Forest Fires, *Int. J. of Num. Meth. Engineering*, 25, 625-633, 1988.

R. C. Rothermel, A Mathematical Model for fire Spread Prediction in Wildland Fuels, USDA Forest Service Res. Paper INT-115, 1972.

A. Setzer, A Microsatellite for Vegetation Fire Detection, Report of the Workshop on Participation Opportunity in the Brazilian Scientific Satellite, Agência Espacial Brasileira, Brasília, DF, Brazil 1994.

J. M. P. Ventura, E. C. Fernandes, D. F. G. Durao, Combustao de Residuos Florestais - Alguns Resultados, *Tecnica 90*, 021-029, Coimbra, Portugal 1990.

Microcanonical Monte Carlo

O. Schapiro, D.H.E. Gross and A. Ecker

Hahn-Meitner-Institut Berlin, Bereich Theoretische Physik, Glienickerstr.100,
D-14109 Berlin, Germany

Abstract

The generic properties of finite equilibrised many-body systems under the action of long-range forces are discussed. The *microcanonical* thermodynamics of such system is developed. As realistic example the multifragmentation of hot nuclei is investigated in some detail with the help of microcanonical Metropolis-Monte Carlo ($MMMC$). $MMMC$ is an alternative to molecular dynamics but it has the advantage that one only samples the final states one is interested in and to get the relevant branching ratios. Besides the nuclear fragmentation there are other important candidates for MMMC description like fragmenting of multiply charged metal clusters and astrophysical systems like planetary systems.

1 Introduction

Microcanonical statistics is the most fundamental basis of thermodynamics. Because of technical difficulties it is only barely developed. Even in the case of a finite isolated system it is much easier to work with the canonical ensemble, that means to ignore the unphysical fluctuations of the fundamental conserved quantities like mass, charge, energy, momentum, angular momentum etc, which eventually vanish in the thermodynamic limit of an infinite system. However, it is the only treatment for *finite* many-body systems under conditions which forbid the application of the thermodynamic limit. A typical example is nuclear multifragmentation. In a somewhat simplified view nuclear multifragmentation is a decay of a small system in statistical equilibrium under the action of *long-range forces* (e.g. Coulomb). Fragmentation of multiply charged molecular clusters and selfgravitating systems like the solar planetary system are possibly other important examples, however these are yet much less studied.

Here the conventional thermodynamics which uses "intensive" variables like temperature, pressure or chemical potential does not work. It is essential to formulate the equilibrium behaviour of these systems microcanonically in terms of the basic conserved "extensive" variables like the energy, volume, particle number etc. An imploding protostar or our solar system are prominent examples of such systems. At the moment astrophysicists discuss intensively the possibility of a statistical (equilibrium) fragmentation of the protosun - in a kind of gravitational phase-transition - by which the planetary system might have been formed e.g.[1, 2]. Numerical studies of the dynamic evolution of an imploding and rotating protostar show multifragmentation into two big stars and - in this case - a chain of nine small stars of equal size in between[3]. The similarity to the Rayleigh-Taylor instability of rotating water droplets is striking even though the

	canonical	microcanonical
Conserved quantities:	T, N system exchanges heat with a heat-bath	E, N system is isolated
weight w_i of the state i	$w_i \sim e^{\frac{E_i}{kT}}$ E_i = energy of the state i	$w_i = w_j = \frac{1}{n}$ n = number of states

Table 1: Difference between the microcanonical and canonical ensemble.

physics is fundamentally different due to the long-range gravity instead of the short-range surface-tension. These are a few illustrations that the physics of fragmentation is highly interesting and new microcanonical calculational techniques allowing to take into consideration the strict conservation laws are of great necessity. Also in solid state physics *microcanonical* approaches of first order phase transitions of some lattice systems became recently a promising tool for numerical studies and finite-size scaling extrapolations towards infinite systems [4, 5].

In this respect multifragmentation of hot nuclei can give us a lot of insight into this problem as it is the most advanced and successfull study of this kind. Fragmentations of multiply charged metallic clusters offer another exciting study of similar physical scenarios because for metallic cluster one can vary the size, the charge, and the excitation energy in wide ranges quite independently. This is not possible for hot nuclei which can only be formed by collisions. The charge/mass ratio is fixed within narrow bounds. The impact parameter and the excitation energy is widely distributed and difficult to be controlled.

2 Microcanonical thermodynamics

One may describe the thermodynamics of isolated systems under the condition that the decaying system equilibrizes at freeze-out. However, ordinary thermodynamics distinguishes intensive variables like temperature, pressure and chemical potential from extensive ones like energy, entropy, volume and particle number. These definitions are useful in the thermodynamic limit for *large* systems with short-range forces. For large systems usually the canonical ensemble is used for the calculations. This causes the unphysical fluctuations in the conserved quantities (energy, momentum, angular momentum, mass, charge), but those can be ignored for sufficiently large systems. As we are dealing with finite systems which cannot be extrapolated to the thermodynamic limit of large systems because of the long-range Coulomb-force which causes the correlations on the length scale of the system-size, the strict respectation of the conservation laws of energy and angular momentum is basically important. For this purpose we need to apply the thermodynamics in the microcanonical description. The basic differences between the two, microcanonical and canonical ensembles are shown in table 1.

We define a system to be at statistical equilibrium if it populates uniformly all accessible configurations in phase-space, i.e. all quantum states, which are consistent with the fundamental conservation laws of energy, momentum, angular momentum, charge and

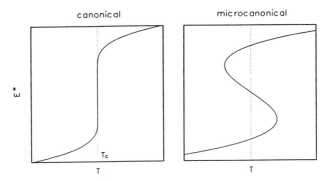

Figure 1: Signals of the phase transition in the $E^*(T)$ dependence. Left: canonical, right: microcanonical

mass. A key quantity of a statistical description of a many-body system at equilibrium is the *partition-sum, Z, the number of all accessible quantum states of the system*. Semi-classically, it is the sum of all different realisations of the system in phase-space measured in units of $(2\pi\hbar)^{d/2}$. d is the number of degrees of freedom of the system.

In the case of fragmentation of nuclei at medium energies the total number of degrees of freedom is of the order of ~ 1000 and we have a chance to calculate Z, or at least variations of it, by a clever Monte Carlo (Metropolis) routine directly [6]. E.g., the entropy of the system is given by

$$S = ln(Z) \tag{1}$$

and the temperature T by

$$\frac{1}{T} = \frac{\partial S}{\partial E} \tag{2}$$

which can be evaluated by sampling $< \partial/\partial E >$ [7]. Integrating $1/T(E)$ over E gives then S or $Z = e^S$. Other thermodynamic quantities are obtained similarly.

2.1 Microcanonical Metropolis Monte Carlo, MMMC

Of course, one cannot simply count the number Z of possible configurations, which is in the case of nuclear fragmentation typically of the order of $\sim 10^{20}$ to $\sim 10^{30}$. However, expectation-values like

$$< F >= \frac{tr(F\rho)}{tr(\rho)}, \tag{3}$$

are evaluated in a *microcanonical* Metropolis Monte Carlo ($MMMC$) [8] routine as the average of F over a sufficiently large sample of *the most important configurations* of the system at given total energy, momentum, angular momentum, mass, and charge *contributing to $< F >$* which occur in the sample with a frequency proportional to their *relative* phase-space volume. $MMMC$ allows to find this *systematically* in an acceptable CPU-time. $\rho(E)$ is the microcanonical statistical operator with $tr(\rho) = Z$. The mathematical formulation of a realistic though exact microcanonical Metropolis sampling was first given by [9].

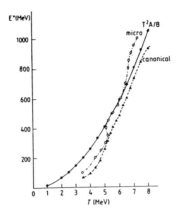

Figure 2: MMMC simulation of $E^*(T)$ for the ^{131}Xe-nucleus at constant freeze-out volume.

In the nuclear case an alternative possibility to simulate a microcanonical system is molecular dynamics (MD), or also quantum-molecular dynamics (QMD) [10]. In the region of multifragmentation (the phase transition region) many different exit channels are open simultaneously, their number is typically of the order of 10^{20}. As one does not know the important *initial* configurations which contribute to the desired *final* observable one has to compute all transitions. This is of course not possible. Moreover, in contrast to $MMMC$, MD and likewise QMD do not provide any systematic way to obtain the branching ratios (the difference of entropy) of selected fragmentation channels. This can only be achieved by a diffusion through the phase-space guided by the relative contributions to the desired exit channels (importance sampling) as in $MMMC$.

2.2 Phase-transition towards multifragmentation

In thermodynamics phase-transitions are especially interesting. The reasons are twofold: first, phase-transitions are quite dramatic changes of the thermodynamic behaviour of the system, on the other hand phase-transitions of second order are especially simple, as details of finite-range forces are unimportant compared to the long-range of the critical correlations. Many quite different systems fall into a few universality classes near to the critical point of the transition depending on the geometry only. Here we are interested in phase-transitions of *microcanonical finite* systems. In finite systems phase-transitions of first and of second order are indistinguishable when the correlation length is of the order of the size of the system. As a function of the excitation energy the entropy of the system increases quickly at the transition.

The dependence of the excitation energy E^* on temperature at a phase transition is different for microcanonical and canonical treatment. Fig.(1) shows schematically the difference. While in the canonical case the temperature is conserved, we get a constant T for rising E^* at the critical point. In the microcanonical case, where E^* is a conserved

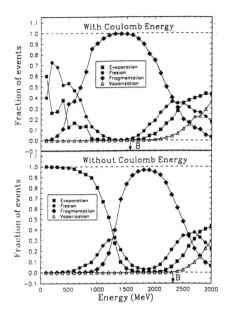

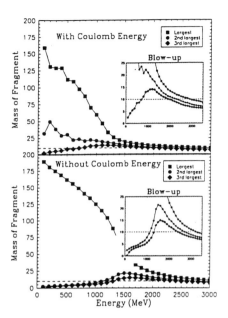

Figure 3: Relative yield of decays of ^{197}Au into different final channels distinguished by the number of fragments with $A \geq 10$, see text.

Figure 4: Mean values of the mass of the largest, second largest and third largest fragment vs. excitation energy for ^{197}Au.

quantity, the $E^*(T)$ shows an "S-shape".

The signal of phase transition in infinite systems is a singularity in heat capacity $C_v(T) = \frac{\partial E}{\partial T}$. However, the canonical calculation for *finite* systems shows only smooth maxima in $C_v(T)$ instead of singularities. The mathematical reason for this is that the partition sum Z, as well as all its derivatives, becomes a finite sum of exponentials and is therefore analytic in $\beta = \frac{1}{T}$. The canonical $E^*(T)$ dependence at transition as shown in fig.(1) occurs only in infinite systems. For finite systems the signal of the phase transition is the S-shape of $E^*(T)$ in the microcanonical description.

3   MMMC results for nuclear multifragmentation

Continuing the discussion of the phase transitions we show the $E^*(T)$ dependence for the microcanonical and canonical calculations for ^{131}Xe nucleus from ref. [8] in fig.(2). The microcanonical calculation shows a typical "S-shape" at about $E^* = 300MeV$ and a weak anomaly at about $E^* = 600MeV$, while the canonical calculation does not show any structure. The backbending of the microcanonical curve at $300MeV$ corresponds to the onset of hot fission, the anomaly at $600MeV$ to the onset of multifragmentation.

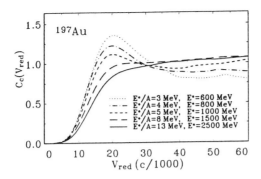

Figure 5: Correlation functions for intermediate mass fragments for the decay of ^{197}Au, integrated over all angles as function of the reduced relative velocity for $v_{red} = v_{rel}/\sqrt{Z_1 + Z_2}$ at various excitation energies. The IMF mass range is from 6 to 20.

3.1 The effect of the Coulomb force

In fig.(3) we show the relative yields of decays of $^{197}Au^*$ into final channels classified by the number of large or intermediate-mass (IMF) fragments with mass $A \geq 10$. We call these decays complete vaporisation (V) if there are no such fragments, evaporation (E), with one fragment (evap. residue) with $A \geq 10$ out of many other small ones, fission events (F), with two fragments $A \geq 10$, and finally multifragmentation (M), with three or more fragments $A \geq 10$. We see the rapid transition to multifragmentation at $E^* \approx 700MeV$.

This is quite different without Coulomb-interaction, c.f. lower part of fig(3). *The transition is shifted considerably downwards in energy from $\approx 1500MeV$ (without Coulomb interaction) to $\approx 700MeV$ by the long-range Coulomb-repulsion*, roughly by about the Coulomb self-energy. Also the production of the second or third large fragment changes considerably as can be seen in fig.(4) where the average masses of the three largest fragments are plotted for various excitation energies. Without Coulomb, only in the very narrow energy-interval of $1600MeV \leq E^* \leq 2300MeV$ the second and third largest fragments are slightly larger than just α-particles. That is the reason why we think that the repulsive long-range Coulomb-interaction triggers fragmentation into several *larger* fragments.

It is further interesting that on average over all decay channels up to about $E^* \approx 6*A$ MeV the largest fragment is at least twice as large as the second largest one. I.e. *even the hot fission is far from being symmetric*. This was also noticed in experiments [11].

3.2 Velocity-correlation functions with intermediate mass fragments (IMF)

Many-body correlations are characteristic indicators of phase-transitions. These observables go far beyond any single-particle description. They are closely related to bifurcations when new substantial parts of the phase-space become "suddenly" accessible during the time evolution of the system or when the energy is raised. Moreover, they cannot be ob-

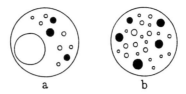

Figure 6: Typical configurations at freeze-out: (a) small E^*, (b) large E^*

tained with the normal canonical thermodynamics which gives just the mean values. Very likely, in finite systems correlation functions are more sensitive to the constraints imposed by conservation laws and long-range forces. Compared to the conventional canonical or even grandcanonical approach a microcanonical system with *sharp energy* has considerable fluctuations in its "temperature": Due to the strict constraint of a sharp total energy, the kinetic energy-distributions in the different decay channels are correlated.

One example for the many-body correlations are the IMF-IMF velocity correlations. Recently a lot of work was started to investigate those theoretically, e.g., ref.[12, 13] and experimentally, e.g. ref.[14, 15].

The two-particle correlation function is defined by

$$C(\mathbf{a}_1, \mathbf{a}_2) = \frac{\Pi(\mathbf{a}_1, \mathbf{a}_2)}{\Pi(\mathbf{a}_1)\Pi(\mathbf{a}_2)}. \tag{4}$$

$\mathbf{a}_i = (\mathbf{p}_i, m_i, Z_i, ...)$ stand for momentum, mass, charge and other variables characterising fragment i. $\Pi(\mathbf{a}_1, \mathbf{a}_2)$ is the probability of coincidence of two particles and $\Pi(\mathbf{a}_1)$ is the probability of one fragment. $C(\mathbf{a}_1, \mathbf{a}_2)$ depends on correlations existing already at the time of emission inside the source and on correlations between the final states of the two particles and the remainder of the source.

Due to the Coulomb interaction of the observed IMFs with the rest of the system the IMF-IMF correlation function can be used to get information about the rest of the source. Fig. 5 shows such correlation functions for ^{197}Au for different excitation energies. The rise of the correlation function above one at smaller relative velocities signals whether the equilibrised system is multifragmented or not. One can visualize it by the schematic pictures of the geometry at freeze-out, fig. 6. For smaller excitation energies (a) we get mostly two IMFs besides one large fragment. The relative velocity of the IMFs will be governed by their mutual repulsion. For high excitation energies we get many fragments of the size of IMFs and smaller. Most of the IMF-pairs will have many fragments inbetween. The relative velocities of these pairs are coming mostly from the Coulomb repulsion from the rest of the charge in the system. Therefore we get for excitation energies higher than the transition point flat correlation functions. Ref. [16] shows that we can reproduce the experimental correlation functions of different shape by the MMMC calculations.

References

[1] M. Kiessling. Untersuchung zur Statistischen Mechanik des Gleichgewichts Klassischer Materie mit Gravitativen Wechselwirkungen. *Dissertation, Bochum*, 1989.

[2] M. Kiessling. Statistical mechanics of classical particles with logarithmic interactions. *Comm. Pure Appl. Physics*, 46:127–56, 1993.

[3] A. Burkert and P. Bodenheimer. Multiple fragmentation in collapsing protostars. *Mon. Not. R. Astr. Soc.*, 264:798, 1993.

[4] R.W. Gerling and A. Hüller. First order phase transition studied in the dynamical ensemble. *Z. Phys. B*, 90:207–214, 1993.

[5] A. Hüller. First order phase transition in the canonical and microcanonical ensemble. *preprint, Erlangen*, 1993.

[6] N. Metropolis, A.W. Rosenbluth, M.N. Rosenbluth, A.H. Teller and E. Teller *J. Chem. Phys.* 21:1087, 1953

[7] D. H. E. Gross and X. Z. Zhang. Thermodynamics of nuclei and nuclear fragmentation. *Nucl. Phys.*, A 495:231c–244c, 1989.

[8] D. H. E. Gross. Statistical decay of very hot nuclei, the production of large clusters. *Rep. Progr. Phys.*, 53:605–658, 1990.

[9] X. Z. Zhang, D. H. E. Gross, S. Y. Xu, and Y. M. Zheng. Decay of very hot nuclei, II, microcanonical metropolis sampling of multifragmentation. *Nucl. Phys.*, A 461:668, 1987.

[10] J Aichelin, G. Peilert, A. Bohnet, A. Rosenbauer, H. Stöcker, and W. Greiner. Quantum molecular dynamics approach to heavy ion collisions. *Phys. Rev.*, C 37:2451-2468, 1988

[11] C. A. Ogilvie, J. C. Adloff, M. Begemann-Blaich, P. Boissou, J. Hubele, G. Imme, I. Iori, P. Kreutz, G. J. Kunde, S. Leray, V. Lindenstruth, Z. Liu, U. Lynen, R. J. Meijer, U. Milkau, W. F. Müller, C. Ngo, J. Pochodzalla, G. Raciti, G. Rudolf, H. Sann, A. Schüttauf, W. Seidel, L. Stuttg, W. Trautmann, and A. Tucholski. The rise and fall of multi-fragmentation. *Phys. Rev. Lett.*, 67:1214, 1991.

[12] D. H. E. Gross, G. Klotz-Engmann, and H. Oeschler. Relative velocities between fragments a key quantity in multifragmentation. *Phys. Lett.*, B 224:29–33, 1989.

[13] W. Bauer, C. K.Gelbke, and S. Pratt. Hadronic interferrometry in heavy ion collisions. *Ann. Rev. Nucl. Part. Sci.*, 42:77, 1992.

[14] Y.D. Kim, R.T. de Souza, D.R. Bowman, N. Carlin, C.K. Gelbke, W.G. Gong, W.G. Lynch, L. Phair, M.B. Tsang, and F. Zhu. Intermediate mass fragment emission in Ar+Au collisions at E/A=35 MeV. *Phys. Rev.*, C 45:338, 1992.

[15] Y. D. Kim et al. Final-state coulomb interactions for intermediate - mass fragment emission. *Phys. Rev.*, C 45:387, 1992.

[16] O.Schapiro, D.H.E. Gross *Nucl. Phys.*, A 573:143-153, 1994.

Computational Investigations of Low-Discrepancy Point Sets II

Tony T. Warnock
Los Alamos National Laboratory
Los Alamos, NM 87544

Abstract

Point sets and sequences with small L_2 discrepancy are useful in the evaluation of multiple integrals. For example, the average error in integration of all continuous functions over the unit cube (with respect to the Wiener measure) is given by the L_2 discrepancy of the point set being used.[6] The Koksma-Hlawka inequality and Zaremba's related inequality also imply the usefulness of low-discrepancy point sets.[4,7]

1 Error Bound Asymmetry

While the errors in a numerical computation do not in general depend on the coordinate system used, error estimates do. For example, in two dimensions, the best placement of one point in a square would seem to be at $(\frac{1}{2}, \frac{1}{2})$ but the minimum of the I_2 discrepancy is at (ϕ, ϕ) where $\phi = \frac{1}{2}(\sqrt{5}-1)$. If any of the other three corners are taken to be the origin, then the discrepancy is larger. As the Koksma-Hlawka inequality is a bound, the minimum of the discrepancies taken at each corner of the unit cube would be a better bound. It is too expensive to compute the discrepancy at each corner so the following compromise is used herein. The discrepancy is computed at the origin and at the opposite corner to the origin, $(1,1,\ldots,1)$. Also, the average of the discrepancies taken at each corner is computed. The minimum of these may be used in the error bound. If the average is smaller than either of the values computed for corners, then there must be some corner that has a smaller value.

Let T_0 be the discrepancy anchored at the origin, T_1 the discrepancy at the opposite point to the origin, and T_a be the average of the discrepancy at each corner. The smallest of these may be taken in the error bound. If the average is smaller than T_0 or T_1 a pseudo-minimum may be computed by spreading the excess of T_0 or T_1 to the other corners. In k dimensions the procedure is:

$$pm_0 = \frac{2^k T_a - T_0}{2^k - 1}$$
$$pm_1 = \frac{2^k T_a - T_1}{2^k - 1}$$
$$pm_2 = \frac{2^k T_a - T_0 - T_1}{2^k - 2}$$

The pseudo-minimum is the minimum of T_0, T_1, T_a, pm_0, pm_1, and pm_2. This is the value used to compare sequences.

2 Five Sequences

The sequences studied herein are designed to be extensible as to dimensionality as well as to number of points. All of them are based on using the sequence of primes as an infinite source of "independent" functions. Being extensible in dimensionality means that the sequences can be used for simulating random walks. The coordinates of each step of a given walk are independent and a calculation may be averaged over the set of walks generated. In order to get reasonable discrepancies, the result of [3] is applied heuristically. Harman's result is that the sequence of square roots of the primes (or an integer multiple thereof) is uniformly distributed (this sequence is called the Weyl sequence below). That is, for a fixed point number N, the coordinates of the Weyl sequence are uniformly distributed. A single random walk constructed from the roots of the primes will thus be a reasonable representative of the set of all random walks. In contrast, the Halton sequence has the property that for a fixed point number, the coordinates tend toward zero. Two of the sequences constructed below are designed to combine the good discrepancy (with respect to number of points) of the Halton sequence with the uniformity of the Weyl sequence for a single value of N.

Sequence 1: The Phi Sequence

The Phi sequence is the original Halton Sequence which was one of the first shown to have low discrepancy in many dimensions.[2] For a set of N points in k dimensions, the discrepancy of the Phi sequence has been shown to be of order:

$$\frac{Log(N)^k}{N}$$

This sequence (as well as others used herein) is based on the P-adic representations of the integers. The number N is expanded in base P as:

$$N = \sum_{j=0}^{\infty} a_j P^j$$

with only finite number of the a_j being non-zero. The function $\phi_P(N)$ is defined by:

$$\phi_P(N) = \sum_{j=0}^{\infty} a_j P^{(-j-1)}$$

The coordinates of the sequence in several dimensions are given by using a set of pairwise relatively prime integers for the bases P_k:

$$X_N = (\phi_{P_1}(N), \phi_{P_2}(N), ..., \phi_{P_k}(N))$$

The sequence has the disadvantage of needing many points to achieve a low discrepancy, especially in higher dimensions. For example, if two of the primes used happened to be 101 and 103, the two coordinates would fall on a line. It takes many points before these coordinates would decouple from each other. The values are always close to zero for N small relative to P_k.

Sequence 2: The Weyl Sequence

The Weyl sequence consists of multiples of the fractional parts of the square roots of the prime numbers. This sequence does not suffer as much from initial coupling as the Halton Sequence does. For large primes, the Weyl sequence does have this problem. It has the good property that the continued fraction expansion of the square root of a given prime has bounded partial quotients. This means that the one-dimensional discrepancy of each dimension will be good. It does not show that the multi-dimensional discrepancy is good. The definition is:

$$X_N = (\{N\sqrt{P_1}\}, \{N\sqrt{P_2}\}, ..., \{N\sqrt{P_k}\})$$

where the braces signify the fractional part.

The Weyl Sequence has the property that for any value of N, the values of the coordinates are uniformly distributed, even though the order of magnitude is not very good.[3] This allows the sequence to be used in random walk problems where the independence of successive steps is important; one can then average over walks.

Sequence 3: The PhiCf Sequence

The PhiCf sequence is constructed with the idea of combining the asymptotic behavior of the Halton Sequence with the initial behavior of the Weyl Sequence. This sequence is defined by starting with the base P expansion for N (as in the Halton sequence) and modifying the digits. The modification consists in replacing each a_j with $S(P)a_j$ Mod(P) where S(P) is defined to be a number such that S(P)/P is close to the fractional part of $\sqrt{P}$. As $\sqrt{P}$ is not an integer, the higher or lower integer is chosen so that the continued fraction expansion of S(P)/P has the smaller sum of partial quotients. The k'th coordinate is given by:

$$N = \sum_{j=0}^{\infty} a_j P_k^j$$
$$b_j = S(P_k)a_j \ Mod(P_k)$$
$$\varphi cf_{P_k}(N) = \sum_{j=0}^{\infty} b_j P_k^{-j-1}$$

Sequence 4: The Rho Sequence

The Rho sequence is likewise constructed with the idea of combining the asymptotic behavior of the Halton Sequence with the initial behavior of the Weyl Sequence. The idea is to subject each digit position in the expansion of N a different permutation. Each digit position has a different multiplier obtained by expanding the fractional part of $\sqrt{P}$ in base P-1, adding 1 to each digit and using these for the permutations. The expansion of $\sqrt{P}$ in base P-1 gives numbers from 0 to P-2; adding 1 gives numbers from 1 to P-1 so that no zero multipliers are created.

$$\{\sqrt{P}\} = \sum_{j=0}^{\infty} d_j (P-1)^{-j-1}$$
$$s_j = d_j + 1$$
$$N = \sum_{j=0}^{\infty} a_j P^{-j-1}$$
$$\rho_{P_j}(N) = \sum_{j=0}^{\infty} (s_j a_j) \mathrm{mod}(P) P^{-j-1}$$

Sequence 5: The Faure Sequence

Rather than using a permutation described by a multiplier, the Faure sequence uses substitution suggested by Faure[1, 7] for one-dimensional sequences. The substitution is defined for base 2 as (0,1) and is extended to other bases by the following rules:

If the base B is even, two copies of the values for B/2 are juxtaposed; these values are doubled; then the values in the second copy are increased by 1.

If the base B is odd, the values for B-1 are augmented by inserting the value B/2 in the middle and increasing each value larger than B/2 by 1.

For example:

2:	0,1
3:	0,1,2
4:	0,2,1,3
5:	0,3,2,1,4
6:	0,2,4,1,3,5
7:	0,2,5,3,1,4,6
8:	0,4,2,6,1,5,3,7.

Computational Formulas

Computing the L_2 discrepancy of a set of N points in k dimensions takes on the order of kN^2 operations. If this were done for each value of N, the total time would be proportional to kN^3 which would be prohibitive. By using a difference formula, the discrepancies for all values from 1 to N may be computed in quadratic time. It should be noted that these formulas are ill-conditioned and require 64-bit arithmetic.

The L_2 discrepancy (anchored at the origin), T_0, for a set of N points in k dimensions is given by:[4, 5]

$$N^2 T_0^2 = \sum_{m=1}^{N} \sum_{n=1}^{N} \prod_{i=1}^{k} (1 - \max(x_{m,i}, x_{n,i})) - \frac{N}{2^{k-1}} \sum_{m=1}^{N} \prod_{i=1}^{k} (1 - x_{m,i}^2) + \frac{N^2}{3^k}$$

The discrepancy at the $(1,1,\ldots,1)$ point, T_1, is given by:

$$N^2 T_1^2 = \sum_{m=1}^{N} \sum_{n=1}^{N} \prod_{i=1}^{k} \min(x_{m,i}, x_{n,i}) - \frac{N}{2^{k-1}} \sum_{m=1}^{N} \prod_{i=1}^{k} (2x_{m,i} - x_{m,i}^2) + \frac{N^2}{3^k}$$

The average discrepancy, T_a, is given by:

$$N^2 T_a^2 = \sum_{m=1}^{N} \sum_{n=1}^{N} \prod_{i=1}^{k} \frac{1 - |x_{m,i} - x_{n,i}|}{2} - \frac{N}{2^{k-1}} \sum_{m=1}^{N} \prod_{i=1}^{k} \frac{1 - 2x_{m,i} + 2x_{m,i}^2}{2} + \frac{N^2}{3^k}$$

As structure of these equations is similar, the difference formulas will be shown for T_0. Calling the double sum P(N) and the single sum Q(N) gives the formula:

$$N^2 T_0^2 = P(N) - \frac{N}{2^{k-1}} Q(N) + \frac{N^2}{3^k}$$

The difference formulas for P(N) and Q(N) are:

$$P(N) - P(N-1) = \sum_{m=1}^{N-1} \prod_{i=1}^{k} (1 - \max(x_{m,i}, x_{N,i})) + \prod_{i=1}^{k} (1 - x_{N,i})$$
$$Q(N) - Q(N-1) = \prod_{i=1}^{k} (1 - x_{N,i}^2)$$
$$P(0) = 0$$
$$Q(0) = 0$$

3 Computational Studies

The discrepancy of each of the five selected sequences was computed with a variety of bases up to 50,000 points in seven dimensions. The pseudo-minimum for each sequence is shown in 100 point increments. The graphs are of N*T(N) versus N.

4 Conclusions

The original Halton sequence decreases in efficiency as the number of dimensions gets large. The modified Halton sequences are better than the original and also better than the Weyl sequence. The PhiCf sequence is the simplest and seems to give the best results based on the numerical studies. The Faure and Rho sequences need more parameters and are not quite as good. The graphs show that 50,000 points are not enough to see asymptotic behavior. More study is needed to see why the sequences based on square roots are so good. There may also be permutations analogous to Faure's that would be even better; the ones used here were chosen for convenience. More computer studies will be useful but asymptotic results seem out of reach.

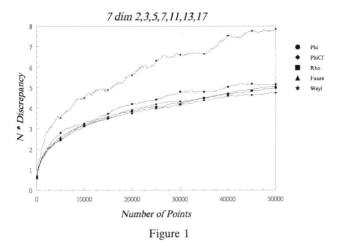

Figure 1

Figure 1 shows all five sequences using the first seven primes. The Weyl sequence is not as ood as any of the others for these bases. The others are similar with the PhiCf sequence being slightly best. As the bases are small compared to N, the sequences should be similar.

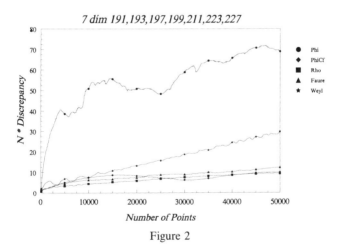

Figure 2

Figure 2 shows all five sequences using bases of 191, 193, 197, 199, 211, 227, and 229. The large value of the bases causes the Phi sequence to behave badly for small N; for these bases, an N of 50,000 is small. The Weyl sequence is better than the Phi sequence, but the other sequences are still better.

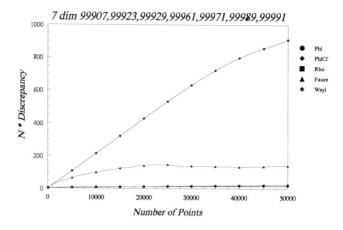

Figure 3

Figure 3 shows all five sequences using 99907, 99923, 99929, 99961, 99971, 99989, and 99991 as bases. The Phi sequence comes off badly as N is never as large as any of the bases. The Faure sequence also does a bit worse than the others. The Weyl, Rho, and PhiCf sequences have good discrepancy even though N is not as large as the base.

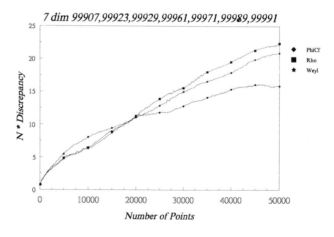

Figure 4

Figure 4 shows the PhiCf, Rho, and Weyl sequences for the above bases.

References

1   Faure, Henri: Good permutations for extreme discrepancy, *J. Num. Theory*, **42** (1992) 47-56.

2   Halton, J. H.: On the efficiency of certain quasi-random sequences of points in evaluating multi-dimensional integrals, *Numer. Math.* **2** (1960) 84-90.

3   Harman, Glyn: On the distribution of $\sqrt{p}$ modulo one, *Mathematika*, **30** (1983) 104-116.

4   Hlawka, E.: Funktionen von beschränkter Variation in der Theorie der Gleichverteilung, *Ann. Mat. Pura Appl.*(IV) **54** (1961) 325-333.

5   Warnock, T. T.: Computational investigations of low-discrepancy point sets, in S. K. Zaremba (Ed.), *Applications of Number Theory to Numerical Analysis*, Academic Press, New York (1972) 319-343.

6   Woźniakowski, H.: Average case complexity of multivariate integration, *Bull. Am. Math. Soc.* **84** (1991) 185-194.

7   Zaremba, S. K.: Some applications of multidimensional integration by parts, *Ann. Polon. Math.* **21** (1968) 85-96.

ESTIMATES FOR THE VOLUME OF POINTS OF
$(0, s)$-SEQUENCES IN BASE $b \geq s \geq 2$

Yi-Jun XIAO

Université d'Evry-Val d'Essonne, Bd des Coquibus, F-91000 Evry, France& CERMA-ENPC, La Courtine, F-93167 Noisy le Grand Cedex, France
Fax: (1) 49 14 35 77, e-mail: xy@cerma.enpc.fr.

Abstract

In this paper we suggest a notion of *volume of points* and we estimate the volume of points for $(0, s)$-sequences which are among the best sequences with low-discrepancy (cf. [1], [7], [8] and [6]).

1 Introduction

For each point $\mathbf{x}= (x_1, \ldots, x_s)$ in the s-dimensional closed unit cube $\bar{I}^s = [0, 1]^s$, let $B(\mathbf{x}) = \{(y_1, \ldots, y_s) \in I^s = [0, 1)^s | \; y_i < x_i$ for $1 \leq i \leq s\}$ and if S is a point set[1] of $\bar{I}^s$, let $B_S = \bigcup_{\mathbf{x} \in S} B(\mathbf{x})$.

Definition 1.1 *Let S be a point set of I^s, the volume of points of S is defined by the number:*

$$V(S) = \int_{I^s} 1_{B_S}(\mathbf{x})dx.$$

If D is a Jordan-measurable subset of I^s (i.e., the characteristic function of D is Riemann integrable), the upper *(resp. lower) volume of S associated with D is defined by the number:*

$$V^D(S) = \int_D 1_{B_S}(\mathbf{x})d\mathbf{x} \quad (resp. \; V_D(S) = \int_D 1_{B_{S \cap D}}(\mathbf{x})d\mathbf{x}).$$

This paper is motivated by the numerical computation of the volume of a measurable set by (quasi) Monte-Carlo methods, it means that if D is a Jordan-measurable subset of I^s and $\sigma = (\mathbf{x}_n)_{n \geq 1}$ is an I^s-valued uniformly distributed sequence, then

$$\lim_{N \to \infty} \frac{Card\{\mathbf{x}_n \in D \mid 1 \leq n \leq N\}}{N} = \int_{I^s} 1_D(\mathbf{x})dx.$$

When D is convex, an error bound for this approximation can be given via the *isotropic discrepancy* (cf. [6]). But it is also possible to approximate $\int_{I^s} 1_D(\mathbf{x})dx$ by $V_D(\sigma_N)$ which is the lower volume of σ_N associated with D with $\sigma_N = \{\mathbf{x}_n \mid 1 \leq n \leq N\}$ (this method, proposed by B. Lapeyre, is of practical use in the case where $\mathbf{x} \in D$ implies $B(\mathbf{x}) \subset D$).

---

[1] Here, a point set means a finite sequence as in the terminology of [6]

In the first approach, we study the speed of convergence towards 1 of $V(\sigma_N) = V_{I^*}(\sigma_N)$, as N tends to ∞. Estimates for a general D, together with numerical tests, will be given in a forthcoming with Lapeyre (cf. [3]).

If $\sigma = (\mathbf{x}_n)_{n \geq 1}$ is a random sequence in I^s, B.Lapeyre proves (cf. [3])

$$\mathbf{E}(1 - V(\sigma_N)) = O(\frac{(\log N)^{s-1}}{N}) \qquad \text{as } N \to \infty,$$

where $\mathbf{E}(f)$ is the *expected value* of the random variable f.

In this paper, we study the asymptotic behaviour of the volume of points for (t, s)-sequences which are among the best sequences with low-discrepancy (cf. [1], [7], [8] and [6]). We recall the definition of these sequences: (cf. [5])

Definition 1.2 *Let $b \geq 2$ be an integer. An elementary interval in base b is an interval of the form*

$$E = \prod_{i=1}^{s} [\frac{a_i}{b^{d_i}}, \frac{a_i + 1}{b^{d_i}}),$$

with integers $d_i \geq 0$ and $0 \leq a_i < b^{d_i}$ for $1 \leq i \leq s$.

Definition 1.3 *Let $b \geq 2$ and $0 \leq t \leq m$ be integers. A (t, m, s)-net in base b is a point set of b^m points in I^s such that every elementary interval in base b of volume b^{t-m} contains b^t points of this net.*

A sequence $\mathbf{x}_1, \mathbf{x}_2, \ldots,$ of points in I^s is called a (t, s)-sequence in base b if for all integers $k \geq 0$ and $m > t$ the point set consisting of the $\mathbf{x}_n$ with $kb^m < n \leq (k + 1)b^m$ is a (t, m, s)-net in base b.

(t, m, s)-nets in base b have very good geometrical properties illustrated in the following propositions (cf. [5] for proofs).

Proposition 1.4 *Let P be a (t, m, s)-net in base b, let E be an elementary interval in base b with volume b^{-u}, where $0 \leq u \leq m - t$, and let T be an affine transformation from E onto I^s. Then the points of P that belong to E are transformed by T into a $(t, m - u, s)$-net in base b.*

Proposition 1.5 *For $1 \leq r \leq s$ let $i_1, \ldots, i_r$ be r distinct elements of the set $\{1, 2, \ldots, s\}$. Let $T: I^s \to I^r$ be the map defined by*

$$T(x_1, \ldots, x_s) = (x_{i_1}, \ldots, x_{i_r}) \quad \text{for } (x_1, \ldots, x_s) \in I^s.$$

Then:

1. every (t, m, s)-net in base b is transformed by T into a (t, m, r)-net in base b;

2. every (t, s)-sequence in base b is transformed by T into a (t, r)-sequence in base b.

The purpose of this paper is to show that

$$1 - V(\sigma_N) = O(\frac{(\log N)^{s-1}}{N}) \qquad \text{as } N \to \infty.$$

We conjecture that the order of magnitude above is the best possible, i.e., there is a constant $C > 0$ such that $1 - V(\sigma_N) \geq C(\frac{(\log N)^{s-1}}{N})$ for infinitely many N. In one dimension, this is true. We think that sequences with low-discrepancy have the same asymptotic behaviour for volume of points as the random sequences; this is in contrast to discrepancy properties(cf. [4]). In §2, we give a best possible estimate for $V(\sigma_N)$ in dimension 2 and in §3 we only have an asymptotic estimate for $V(\sigma_N)$ in multidimensional case $s \geq 3$.

2 Estimates in two dimensions

Proposition 2.1 *Let m' be an integer. Let $P = [0, 1) \times [\frac{a}{b^{m'}}, \frac{a+1}{b^{m'}})$ and $P_i = [0, 1) \times [\frac{ba+i}{b^{m'+1}}, \frac{ba+i+1}{b^{m'+1}})$ $0 \leq i \leq b - 1$ be elementary intervals such that $P_i \subset P$.*

Let π be a permutation of $\{0, \cdots, b - 1\}$. We associate with π a transformation T_π from I^2 onto I^2 which is defined by

$$T_\pi(x, y) = \begin{cases} (x, y + \frac{\pi(i) - i}{b^{m'+1}}) & \text{for } (x, y) \in P_i \\ (x, y) & \text{if } (x, y) \notin P \end{cases}$$

Note that $T_\pi(P_i) = P_{\pi(i)}$. The transformation T_π will be called a row elementary interval permutation on P. Then every $(t, m, 2)$-net in base b is transformed by T_π into a $(t, m, 2)$-net in base b

Proof. We show the proposition only for $t = 0$. For $t \geq 1$, it can be proved in the same way.

Since every permutation is a product of transpositions and every transposition is a product of transpositions of consecutive integers, hence it suffices to prove the result for a transposition of two consecutive integers, say i and $i + 1$.

Let R be a $(t, m, 2)$-net in base b and $Q = T_\pi(R)$. We will show that every elementary interval I of volume b^{-m}, of the form

$$I = [\frac{u}{b^{m_1}}, \frac{u + 1}{b^{m_1}}) \times [\frac{v}{b^{m_2}}, \frac{v + 1}{b^{m_2}}) \quad \text{with } m_1 + m_2 = m,$$

contains a unique element of Q.

If $(P_i \cup P_{i+1}) \cap I = \emptyset$, it is clear that the point of R in I is not moved after the permutation, so there is a unique point of Q in I.

If $(P_i \cup P_{i+1}) \cap I \neq \emptyset$, there are two cases:

1. $m_2 > m'$

2. $m_2 \leq m'$

Note that if the intersection of two elementary intervals is not empty, then one is included in the other. So in the first case, we have

$$[\frac{v}{b^{m_2}}, \frac{v + 1}{b^{m_2}}) \subset [\frac{ba + i}{b^{m'+1}}, \frac{ba + i + 1}{b^{m'+1}}), \tag{1}$$

$$\text{or } [\frac{v}{b^{m_2}}, \frac{v + 1}{b^{m_2}}) \subset [\frac{ba + i + 1}{b^{m'+1}}, \frac{ba + i + 2}{b^{m'+1}}), \tag{2}$$

which implies that $I \subset P_i$ or $I \subset P_{i+1}$. We may assume, without loss of generality, that (1) holds. The image I' of I by the transposition is also an elementary interval of volume b^{-m} containing a unique point of R. But I is also the image of I' by the transposition so I contains a unique point of Q.

In the second case, using that if the intersection of two elementary intervals is not empty, then one is included in the other, we have

$$[\frac{v}{b^{m_2}}, \frac{v + 1}{b^{m_2}}) \supset [\frac{a}{b^{m'}}, \frac{a + 1}{b^{m'}})$$

which implies that $P_i \cap I$ is mapped on $P_{i+1} \cap I$ and conversely by the permutation and so I contains always the same number of points after the permutation. Hence, there is a unique point of Q in I. ◇

In the following, we give a lower bound for the volume of points of a $(0, m, 2)$-net in base b.

Theorem 2.2 *Let R be a $(0, m, 2)$-net in base b. Then for all $m \geq 0$ and for every elementary interval P of volume b^{-u} with $0 \leq u \leq m$*

$$V_P(R) \geq \frac{1}{b^u} - \frac{(b-1)(\frac{1}{2} + \frac{1}{b})(m-u) + 1}{b^m}. \tag{3}$$

In addition, there exists a $(0, m, 2)$-net R_{min} in base b depending on P, such that

$$V_P(R_{min}) = 1 - \frac{(b-1)(\frac{1}{2} + \frac{1}{b})(m-u) + 1}{b^m}. \tag{4}$$

Let $\sigma = (\mathbf{x}_n)_{n \geq 1}$ be an I^s-valued sequence. We note

$$EV(\sigma_N) = 1 - V(\sigma_N)$$

the *volume of points difference of σ_N*.

Before proving Theorem 2.2, we derive the following corollary.

Corollary 2.3 *Let $\sigma = (\mathbf{x}_n)_{n \geq 1}$ be a $(0, 2)$-sequence in base b, then $\forall N \geq 1$:*

$$EV(\sigma_N) \leq \frac{b[(b-1)(\frac{1}{2} + \frac{1}{b}) \log_b(N) + 1]}{N}. \tag{5}$$

Proof . $\forall N \geq 1$, there is an $m \in \mathbf{N}$ such that $b^m \leq N < b^{m+1}$. Then we have $m \leq \log_b(N)$ and

$$EV(\sigma_N) \leq EV(\sigma_{b^m}) \leq EV(R_{min}) \leq \frac{(b-1)(\frac{1}{2} + \frac{1}{b})m + 1}{b^m}.$$

The result of the Corollary follows. ◇

The idea behind the estimate in Theorem 2.2 comes from "dynamic programming". By successive transformations of a $(0, m, 2)$-net in base b, we are reduced to an extreme situation. These transformations are "row or column elementary interval permutations" introduced in Proposition 2.1.

We will show it by induction on m. The following Lemma 2.4 will be used in the case of $m = 1$ and to carry out the induction.

To make the proof easier, we may assume in the following that:

$$R \subset \{(\frac{i}{b^m}, \frac{j}{b^m}) \mid i, j = 0, \cdots, b^m - 1\}.$$

Lemma 2.4 *Let R be a $(0, m, 2)$-net in base b such that the unique point of R belonging to the elementary interval $[1 - \frac{1}{b^{m-1}}, 1) \times [\frac{b-1}{b}, 1)$ is included in $[1 - \frac{1}{b^{m-1}}, 1 - \frac{1}{b^{m-1}} + \frac{1}{b^m}) \times [\frac{b-1}{b}, 1)$. Then, there is a row elementary interval permutation on I^2 leaving $[0, 1) \times [\frac{b-1}{b}, 1)$ invariant, which transforms R into a new $(0, m, 2)$-net Q in base b having the following properties: for all $i \in \{0, \dots, b - 2\}$, there is a point of Q belonging to*

$$[1 - \frac{i+1}{b^m}, 1 - \frac{i}{b^m}) \times [\frac{i}{b}, \frac{i+1}{b})$$

In addition, if $A = [1 - \frac{1}{b^{m-1}}, 1) \times [0, \frac{b-1}{b})$, we have:

$$V_A(R) \geq V_A(Q) = \frac{(b-2)(b-1)}{2b^{m+1}}.$$

Proof. Using Proposition 2.1, these last two properties are easy to verify by permutating the elementary intervals of volume $\frac{1}{b}$ of the form

$$[0, 1) \times [\frac{i}{b}, \frac{i+1}{b}), \quad \text{with } 0 \leq i \leq b - 2.$$

$\diamond$

Remark 2.5 *Lemma 2.4 and its analogue for "columns" allow us to turn step by step to a new net Q such that $V_E(Q)$ is minimal.*

Lemma 2.6 *For all $(0, 1, 2)$-nets R in base b, we have:*

$$V(R) \geq V(R_{min}) = \frac{(b-2)(b-1)}{2b^2}. \tag{6}$$

where R_{min} is a $(0, 1, 2)$-net in base b defined by:

$$R_{min} = \{(\frac{i}{b}, \frac{b-i-1}{b}); \ i = 0, \dots, b-1\} \tag{7}$$

Proof. It suffices to move first the point $(0, \frac{i}{b})$ in $(0, \frac{b-1}{b})$, by permutating the following elementary intervals

$$[0, 1) \times [\frac{j}{b}, \frac{j+1}{b}) \quad \text{and} \quad [0, 1) \times [\frac{b-1}{b}, 1).$$

The volume of points of the new $(0, 1, 2)$-net is reduced. Then applying Lemma 2.4, the result follows.

$\diamond$

Proof of Theorem 2.2. It suffices to prove it for $P = I^2$ by Proposition 1.4.

We will show by induction on m that for all $(0, m, 2)$-nets R in base b there is a sequence of elementary interval permutations (as in proposition 2.1) which transform R into a new $(0, m, 2)$-net Q in base b such that

$$V(R) \geq V(Q) = 1 - \frac{(b-1)(\frac{1}{2} + \frac{1}{b})m + 1}{b^m}.$$

For $m = 0$, it holds. For $m = 1$, it also holds by Lemma 2.6. Suppose that the result holds for all $0 \le m' \le m$, where $m \ge 1$.

Let R be a $(0, m+1, 2)$-net in base b. We decompose I^2 into the following 6 intervals: $P_1 = [\frac{b-1}{b}, 1) \times [\frac{b-1}{b}, 1)$, $P_2 = [0, \frac{b-1}{b}) \times [0, \frac{b-1}{b})$, $P_3 = [0, \frac{b-1}{b}) \times [1 - \frac{1}{b^m}, 1)$, $P_4 = [0, \frac{b-1}{b}) \times [\frac{b-1}{b}, 1 - \frac{1}{b^m})$, $P_5 = [1 - \frac{1}{b^m}, 1) \times [0, \frac{b-1}{b})$, and $P_6 = [\frac{b-1}{b}, 1 - \frac{1}{b^m}) \times [0, \frac{b-1}{b})$. We note $E_1 = [0, \frac{b-1}{b}) \times [0, 1)$ and $E_2 = [0, 1) \times [0, \frac{b-1}{b})$.

Since there is a point for a $(0, m+1, 2)$-net in base b in the elementary intervals $[\frac{b-1}{b}, 1) \times [1 - \frac{1}{b^m}, 1)$ and $[1 - \frac{1}{b^m}, 1) \times [\frac{b-1}{b}, 1)$, we have always

$$V^{P_4}(R) = V^{P_6}(R) = \frac{b-1}{b}(\frac{1}{b} - \frac{1}{b^m}) = \frac{b-1}{b^2}(1 - \frac{1}{b^{m-1}});$$

and

$$V^{P_2}(R) = \frac{(b-1)^2}{b^2}.$$

In P_1, for each elementary interval of volume $\frac{1}{b^{m+1}}$ of the form:

$$[\frac{b-1}{b}, 1) \times [\frac{v}{b^m}, \frac{v+1}{b^m}) \quad \text{with} \quad (b-1)b^{m-1} \le v < b^m,$$

there is a unique point of R. In using a possible permutation among the elementary intervals of volume $\frac{1}{b^{m+1}}$ of the form:

$$[0, 1) \times [\frac{bv+i}{b^{m+1}}, \frac{bv+i+1}{b^{m+1}}) \quad \text{with} \quad 0 \le i \le b - 1$$

we can always place this point in the elementary interval

$$[\frac{b-1}{b}, 1) \times [\frac{bv}{b^{m+1}}, \frac{bv+1}{b^{m+1}}). \quad \text{with} \quad (b-1)b^{m-1} \le v < b^m$$

The new point set is still a $(0, m+1, 2)$-net in base b but with smaller volume of points. Similarly, we can place the point of R which is in an elementary interval of volume $\frac{1}{b^{m+1}}$ of the form:

$$[\frac{v}{b^m}, \frac{v+1}{b^m}) \times [\frac{b-1}{b}, 1) \quad \text{with} \quad (b-1)b^{m-1} \le v < b^m$$

in its sub-elementary interval

$$[\frac{bv}{b^{m+1}}, \frac{bv+1}{b^{m+1}}) \times [\frac{b-1}{b}, 1).$$

The new point set is still a $(0, m+1, 2)$-net in base b but its volume of points is smaller.

After these two transformations above, apply Lemma 2.4, and let $R^{(1)}$ denote the new $(0, m+1, 2)$-net in base b, we have

$$V_{P_3}(R) \ge V_{P_3}(R^{(1)}) = \frac{(b-2)(b-1)}{2b^{m+2}},$$

and

$$V_{P_5}(R) \ge V_{P_5}(R^{(1)}) = \frac{(b-2)(b-1)}{2b^{m+2}}.$$

For the elementary interval P_1 of volume $\frac{1}{b^2}$, if T is an affine transformation from P_1 onto I^2, then by Proposition 1.4 $T(R^{(1)} \cap P_1)$ is a $(0, m-1, 2)$-net, so by the induction hypothesis there is a sequence of elementary interval permutations (as in Proposition 2.1) which transform $T(R^{(1)} \cap P_1)$ in $R^{(2)}$, a new $(0, m-1, 2)$-net in base b such that

$$V(R^{(2)}) = 1 - \frac{(b-1)(\frac{1}{2} + \frac{1}{b})(m-1) + 1}{b^{m-1}}.$$

Thus

$$V_{P_1}(T^{-1}(R^{(2)})) = \frac{1}{b^2} - \frac{(b-1)(\frac{1}{2} + \frac{1}{b})(m-1) + 1}{b^{m+1}}.$$

To each row (resp. column) permutation on P_1, there is a corresponding row (resp. column) permutation on I^2 which changes the distribution of points in P_3 (resp. P_5). We can associate with E_1 (resp. E_2) a sequence of column (resp. row) permutations which does not change the distribution of the points in P_1 and give always

$$V_{P_3} = V_{P_5} = \frac{(b-2)(b-1)}{2b^{m+2}}.$$

In the end, writing the final net by Q, we have

$$
\begin{aligned}
V(R) \geq V(Q) &= \sum_{i=1}^{6} V^{P_i}(Q) \\
&= \frac{1}{b^2}(b-1)^2 + \frac{2(b-1)}{b^2}\left(1 - \frac{1}{b^{m-1}}\right) + \frac{(b-2)(b-1)}{b^{m+2}} \\
&\quad + \frac{1}{b^2}\left(1 - \frac{(b-1)(\frac{1}{2} + \frac{1}{b})(m-1) + 1}{b^{m-1}}\right) \\
&= 1 - \frac{(b-1)(\frac{1}{2} + \frac{1}{b})(m+1) + 1}{b^{m+1}}.
\end{aligned}
$$

◇

Remark 2.7 For a $(t, m, 2)$-net in base b and for a $(t, 2)$-sequence in base b, we can have the same estimates to within a constant.

3 Estimates in dimension $s \geq 3$

Theorem 3.1 Let $b \geq s \geq 2$ and $m \geq 1$ be integers. Then for all $(0, m, s)$-nets R in base b

$$V(R) \geq 1 - \frac{(bm)^{s-1}}{b^m}. \tag{8}$$

It is easy to derive the following corollary.

Corollary 3.2 Let $\sigma = (\mathbf{x}_n)_{n \geq 1}$ be a $(0, s)$-sequence in base b. Then for all $N \geq b$:

$$EV(\sigma_N) \leq \frac{b^s (\log_b N)^{s-1}}{N}. \tag{9}$$

Before proving Theorem 3.1, we need the following two lemmas.

Lemma 3.3 *Let R be a $(0, m, s)$-net in base b with $m \geq s$ and let $\Lambda_k = \{i_1, \ldots, i_k\}$ with $1 \leq k \leq s$ and $i_1, \ldots, i_k$ be distinct elements of the set $\{1, 2, \ldots, s\}$. We denote*

$$P_s^{b-1} = \prod_{i=1}^{s} [\frac{b-1}{b}, 1) \subset I^s \quad and \quad Q_{\Lambda_k}^{b-1} = \prod_{i \in \Lambda_k} [\frac{b-1}{b}, 1) \subset I^k.$$

We consider the following two transformations:

- T *is the affine transformation mapping P_s^{b-1} onto I^s*

- T_{Λ_k} *is the map defined by:*

$$T_{\Lambda_k}((x_1, \ldots, x_s)) = (x_{i_1}, \ldots, x_{i_k}) \quad for \, (x_1, \ldots, x_s) \in I^s.$$

Writing $E = P_s^{b-1} \bigcap R$, we have the following results:

1. *the point set $R' = T_{\Lambda_k} \circ T(E)$ is a $(0, m - s, k)$-net in base b.*

2. *for every elementary interval $P = \prod_{i=1}^{s} [\frac{a_i}{b}, \frac{a_i+1}{b})$ such that $a_i = b-1$ if and only if $i \in \Lambda_k$, then*

$$V^P(R) \geq \frac{1}{b^s} V(R'). \tag{10}$$

Proof .

1. By Proposition 1.3, $T(E)$ is a $(0, m-s, s)$-net in base b and by Proposition 1.4, $T_{\Lambda_k} \circ T(E)$ is a $(0, m - s, k)$-net in base b.

2. Since $E \subset R$, it suffices to prove that $V^P(E) = \frac{1}{b^s} V(R')$. It is clear that

$$V^P(E) = \lambda(B(E) \cap P)$$

where λ is the Lebesgue measure. But

$$B(E) \cap P = \cup_{\mathbf{x} \in E} \{\mathbf{y} \in P \mid \forall i \in \Lambda_k, \frac{b-1}{b} \leq y_i < x_i\}$$

and it is clear

$$\lambda(\cup_{\mathbf{x} \in E} \{\mathbf{y} \in P \mid \forall i \in \Lambda_k, \frac{b-1}{b} \leq y_i < x_i\}) = \frac{1}{b^s} V(R'),$$

so the result follows.

◇

Lemma 3.4 *For all $m \geq s \geq 1$*

$$\sum_{k=1}^{s-1} \binom{s}{k} \frac{m^k}{k} \leq sm^{s-1}.$$

Proof . It suffices to prove it for $m = s$. Since $\begin{pmatrix} s \\ k \end{pmatrix} \frac{s^k}{k}$ is an increasing function of k, $1 \leq k \leq s - 2$, we have for $s \geq 3$

$$\sum_{k=1}^{s-1} \begin{pmatrix} s \\ k \end{pmatrix} \frac{s^k}{k} \leq (s-2) \begin{pmatrix} s \\ s-2 \end{pmatrix} \frac{s^{(s-2)}}{s-2} + \frac{s^s}{s-1}$$

$$\leq s^{s-1}(\frac{s-1}{2} + \frac{s}{s-1}). \leq s^s$$

For $s = 2$, it is easy to verify it.

◇

Proof of Theorem 3.1. We shall prove it by induction on $s \geq 1$. For $s = 1$, it is clear that (8) is true, and for $s = 2$, it also holds thanks to Theorem 2.1. Suppose that the Theorem holds for $1 \leq k \leq s - 1$.

For $k = s \leq b$, let R_m denote a $(0, m, s)$-net in base b with $m \geq s$. We decompose I^s into b^s elementary intervals of of the form:

$$P_{a_1,\ldots,a_s} = \prod_{i=1}^{s} [\frac{a_i}{b}, \frac{a_i + 1}{b}), \text{ with } 0 \leq a_i \leq b - 1.$$

So

$$V(R_m) = \sum_{0 \leq a_1,\ldots,a_s \leq b-1} V^{P_{a_1,\ldots,a_s}}(R_m)$$

It is clear that if $0 \leq a_1, \ldots, a_s \leq b - 2$, then

$$V^{P_{a_1,\ldots,a_s}}(R_m) = \frac{1}{b^s}. \tag{11}$$

The number of all corresponding intervals is $(b - 1)^s$.

For the interval $P_{b-1,\ldots,b-1}$, of volume $\frac{1}{b^s}$, if T is an affine transformation from $P_{b-1,\ldots,b-1}$ onto I^s, then by Proposition 1.3 $R_{m-s} = T(R_m \cap P_{b-1,\ldots,b-1})$ is a $(0, m - s, s)$-net, and thus

$$V^{P_{b-1,\ldots,b-1}}(R_m) = \frac{1}{b^s}V(R_{m-s}). \tag{12}$$

For each interval $P_{a_1,\ldots,a_s} = P^{\Lambda_k}$ where $\Lambda_k = \{i_1,\ldots,i_k\} \subset \{1,\ldots,s\}, 1 \leq k \leq s - 1$ such that $a_i = b-1$ if and only if $i \in \Lambda_k$, by Lemma 3.3 and the hypothesis, there is a $(0, m-s, k)$-net R' such that

$$V^{P^{\Lambda_k}}(R) \geq \frac{1}{b^s}V(R')$$

$$\geq \frac{1}{b^s}(1 - \frac{(b(m - s))^{k-1}}{b^{m-s}}). \tag{13}$$

Since the number of all intervals of this type is $\begin{pmatrix} s \\ k \end{pmatrix} (b-1)^{s-k}$, we have by (11), (12) and (13)

$$V(R_m) \geq \frac{(b - 1)^s}{b^s} + \frac{1}{b^s}\sum_{k=1}^{s-1} \begin{pmatrix} s \\ k \end{pmatrix} (b - 1)^{s-k}(1 - \frac{(b(m - s))^{k-1}}{b^{m-s}}) + \frac{1}{b^s}V(R_{m-s})$$

$$= \frac{1}{b^s}(V(R_{m-s}) + b^s - 1 - \frac{1}{b^{m-s}}\sum_{k=1}^{s-1} \begin{pmatrix} s \\ k \end{pmatrix} (b - 1)^{s-k}(b(m - s))^{k-1}).$$

Let $m = s(n+1) + h$ with $0 \leq h \leq s - 1$ and $n \geq 0$, writing $u_n^{(h)} = V(R_{sn+h})$, we have

$$u_{n+1}^{(h)} \geq \frac{1}{b^s}(u_n^{(h)} + b^s - 1 - \frac{1}{b^{sn+h}} \sum_{k=1}^{s-1} \binom{s}{k}(b-1)^{s-k}(b(sn+h))^{k-1}).$$

But, if $(u_n)_{n \geq 0}$ and $(f_n)_{n \geq 0}$ are two real sequences satisfying

$$u_{n+1} \geq a(u_n + f_n)$$

with $a \in \mathbf{R}^+$, then

$$u_{n+1} \geq a^{n+1}u_0 + a\sum_{i=0}^{n} a^i f_{n-i}.$$

Hence

$$u_{n+1}^{(h)} \geq \frac{1}{b^{s(n+1)}}u_0^{(h)}$$

$$+ \frac{1}{b^s}\sum_{i=0}^{n}\frac{1}{b^{si}}[b^s - 1 - \frac{1}{b^{s(n-i)+h}} \sum_{k=1}^{s-1} \binom{s}{k}(b-1)^{s-k}(b(s(n-i)+h))^{k-1}]$$

$$\geq 1 - \frac{1}{b^{s(n+1)}} - \frac{1}{b^{s(n+1)+h}} \sum_{k=1}^{s-1} \binom{s}{k}(b-1)^{s-k} \sum_{i=0}^{n}(b(s(n-i)+h))^{k-1}$$

Since $b^{s-k} - (b-1)^{s-k} \geq b$ for each $1 \leq k \leq s - 1$, we have

$$\sum_{k=1}^{s-1} \binom{s}{k} b^{k-1}(b^{s-k} - (b-1)^{s-k}) \sum_{i=0}^{n}(s(n-i)+h)^{k-1} \geq b^{s-1} \geq b^h.$$

Therefore:

$$u_{n+1}^{(h)} \geq 1 - \frac{1}{b^{s(n+1)+h}} \sum_{k=1}^{s-1} \binom{s}{k} b^{k-1}b^{s-k} \sum_{i=0}^{n}(s(n-i)+h)^{k-1}.$$

Using $\sum_{i=0}^{n}(s(n-i)+h)^{k-1} \leq \frac{(s(n+1)+h)^k}{sk}$ and Lemma 3.4, we have

$$u_{n+1}^{(h)} \geq 1 - \frac{b^{s-1}}{b^{s(n+1)+h}} \sum_{k=1}^{s-1} \binom{s}{k} \frac{(s(n+1)+h)^k}{sk}$$

$$\geq 1 - \frac{b^{s-1}(s(n+1)+h)^{s-1}}{b^{s(n+1)+h}}.$$

That is

$$V(R_m) \geq 1 - \frac{(bm)^{s-1}}{b^m},$$

and it is clear that this inequality still holds for $1 \leq m \leq s - 1$. The result of the Theorem follows. $\diamond$

Remark 3.5 *For a (t, m, s)-net in base b and for a (t, s)-sequence in base b, we can have the same estimates to within a constant.*

References

[1] H.Faure, Discrépance de suites associées à un systême de numération (en dimension s) *Acta. Arithmetica. 41. (1982), 337-351.*

[2] Hua L-K and Wang Y., Application of Number Theory to Numerical Analysis, *Springer-Verlag (1981)*

[3] B. Lapeyre et Y.J. Xiao, Sur la discrépance volumique des suites, *en préparation.*

[4] H.Niederreiter, Quasi Monte Carlo methods and pseudo-random numbers. *Bull. AMS. 84,(1978), 957-1041.*

[5] H.Niederreiter, Point sets and sequences with small discrepancy, *Monatsh. Math. 104, (1987), 273-337.*

[6] H.Niederreiter, Random Number Generation and Quasi-Monte Carlo Methods, *SIAM, Philadelphia Pennsylvania, (1992).*

[7] I. M. Sobol', The distribution of points in a cube and the approximate evaluation of integrals, *Zh. Vychisl. Mat. i Mat. Fiz. 7 (1967), 784-802; USSR Comput. Math. and Math. Phys. 7 (1967), 86-112.*

[8] I. M. Sobol', "Multidimensional Quadrature Formulas and Haar Functions," *Nauka, Moscow, 1969.* [In Russian]

Lecture Notes in Statistics

For information about Volumes 1 to 19
please contact Springer-Verlag

Vol. 20: U. Herkenrath, D. Kalin, W. Vogel (Editors), Mathematical Learning Models — Theory and Algorithms: Proceedings of a Conference. xiv, 226 pages, 1983.

Vol. 21: H. Tong, Threshold Models in Non-linear Time Series Analysis. x, 323 pages, 1983.

Vol. 22: S. Johansen, Functional Relations, Random Coefficients and Nonlinear Regression with Application to Kinetic Data, viii, 126 pages, 1984.

Vol. 23: D.G. Saphire, Estimation of Victimization Prevalence Using Data from the National Crime Survey. v, 165 pages, 1984.

Vol. 24: T.S. Rao, M.M. Gabr, An Introduction to Bispectral Analysis and Bilinear Time Series Models. viii, 280 pages, 1984.

Vol. 25: E. Parzen (Editor), Time Series Analysis of Irregularly Observed Data. Proceedings, 1983. vii, 363 pages, 1984.

Vol. 26: J. Franke, W. Härdle and D. Martin (Editors), Robust and Nonlinear Time Series Analysis. Proceedings, 1983. ix, 286 pages, 1984.

Vol. 27: A. Janssen, H. Milbrodt, H. Strasser, Infinitely Divisible Statistical Experiments. vi, 163 pages, 1985.

Vol. 28: S. Amari, Differential-Geometrical Methods in Statistics. v, 290 pages, 1985.

Vol. 29: B.J.T. Morgan and P.M. North (Editors), Statistics in Ornithology. xxv, 418 pages, 1985.

Vol 30: J. Grandell, Stochastic Models of Air Pollutant Concentration. v, 110 pages, 1985.

Vol. 31: J. Pfanzagl, Asymptotic Expansions for General Statistical Models. vii, 505 pages, 1985.

Vol. 32: R. Gilchrist, B. Francis and J. Whittaker (Editors), Generalized Linear Models. Proceedings, 1985. vi, 178 pages, 1985.

Vol. 33: M. Csörgo, S. Csörgo, L. Horváth, An Asymptotic Theory for Empirical Reliability and Concentration Processes. v, 171 pages, 1986.

Vol. 34: D.E. Critchlow, Metric Methods for Analyzing Partially Ranked Data. x, 216 pages, 1985.

Vol. 35: T. Calinski and W. Klonecki (Editors), Linear Statistical Inference. Proceedings, 1984. vi, 318 pages, 1985.

Vol. 36: B. Matérn, Spatial Variation. Second Edition. 151 pages, 1986.

Vol. 37: R. Dykstra, T. Robertson and F.T. Wright (Editors), Advances in Order Restricted Statistical Inference. Proceedings, 1985. viii, 295 pages, 1986.

Vol. 38: R.W. Pearson and R.F. Boruch (Editors), Survey Research Designs: Towards a Better Understanding of Their Costs and Benefits. v, 129 pages, 1986.

Vol. 39: J.D. Malley, Optimal Unbiased Estimation of Variance Components. ix, 146 pages, 1986.

Vol. 40: H.R. Lerche, Boundary Crossing of Brownian Motion. v, 142 pages, 1986.

Vol. 41: F. Baccelli, P. Brémaud, Palm Probabilities and Stationary Queues. vii, 106 pages, 1987.

Vol. 42: S. Kullback, J.C. Keegel, J.H. Kullback, Topics in Statistical Information Theory. ix, 158 pages, 1987.

Vol. 43: B.C. Arnold, Majorization and the Lorenz Order: A Brief Introduction. vi, 122 pages, 1987.

Vol. 44: D.L. McLeish, Christopher G. Small, The Theory and Applications of Statistical Inference Functions. vi, 124 pages, 1987.

Vol. 45: J.K. Ghosh (Editor), Statistical Information and Likelihood. 384 pages, 1988.

Vol. 46: H.-G. Müller, Nonparametric Regression Analysis of Longitudinal Data. vi, 199 pages, 1988.

Vol. 47: A.J. Getson, F.C. Hsuan, {2}-Inverses and Their Statistical Application. viii, 110 pages, 1988.

Vol. 48: G.L. Bretthorst, Bayesian Spectrum Analysis and Parameter Estimation. xii, 209 pages, 1988.

Vol. 49: S.L. Lauritzen, Extremal Families and Systems of Sufficient Statistics. xv, 268 pages, 1988.

Vol. 50: O.E. Barndorff-Nielsen, Parametric Statistical Models and Likelihood. vii, 276 pages, 1988.

Vol. 51: J. Hüsler, R.-D. Reiss (Editors). Extreme Value Theory, Proceedings, 1987. x, 279 pages, 1989.

Vol. 52: P.K. Goel, T. Ramalingam, The Matching Methodology: Some Statistical Properties. viii, 152 pages, 1989.

Vol. 53: B.C. Arnold, N. Balakrishnan, Relations, Bounds and Approximations for Order Statistics. ix, 173 pages, 1989.

Vol. 54: K.R. Shah, B.K. Sinha, Theory of Optimal Designs. viii, 171 pages, 1989.

Vol. 55: L. McDonald, B. Manly, J. Lockwood, J. Logan (Editors), Estimation and Analysis of Insect Populations. Proceedings, 1988. xiv, 492 pages, 1989.

Vol. 56: J.K. Lindsey, The Analysis of Categorical Data Using GLIM. v, 168 pages, 1989.

Vol. 57: A. Decarli, B.J. Francis, R. Gilchrist, G.U.H. Seeber (Editors), Statistical Modelling. Proceedings, 1989. ix, 343 pages, 1989.

Vol. 58: O.E. Barndorff-Nielsen, P. Blæsild, P.S. Eriksen, Decomposition and Invariance of Measures, and Statistical Transformation Models. v, 147 pages, 1989.

Vol. 59: S. Gupta, R. Mukerjee, A Calculus for Factorial Arrangements. vi, 126 pages, 1989.

Vol. 60: L. Györfi, W. Härdle, P. Sarda, Ph. Vieu, Nonparametric Curve Estimation from Time Series. viii, 153 pages, 1989.

Vol. 61: J. Breckling, The Analysis of Directional Time Series: Applications to Wind Speed and Direction. viii, 238 pages, 1989.

Vol. 62: J.C. Akkerboom, Testing Problems with Linear or Angular Inequality Constraints. xii, 291 pages, 1990.